EVOLUTION
—The Whole Story—

EVOLUTION
The Whole Story

General Editor
STEVE PARKER

FIREFLY BOOKS

A FIREFLY BOOK

Published by Firefly Books Ltd. 2015

First printing

Publisher Cataloging-in-Publication Data (U.S.)

Parker, Steve.
   Evolution : the whole story / Steve Parker.
[576] pages : color photographs, illustrations, maps ; cm.
Includes index.
Summary: "Each chapter takes a major living group — from the earliest single-celled organisms to mammals and humans — and presents thematic essays discussing the evolution of particular subgroups as they appeared on earth, with reference to their anatomical legacies." — from Publisher.
ISBN-13: 978-1-77085-481-9
1. Evolution (Biology). 2. Evolution (Biology) — Popular works. I. Title.
576.8 dc23   QH367.P365 2015

Library and Archives Canada Cataloguing in Publication

Parker, Steve, 1952–, general editor
   Evolution : the whole story / Steve Parker. — First edition.
Includes index.
ISBN 978-1-77085-481-9 (bound)
   1. Evolution (Biology). 2. Evolution (Biology)—Popular works. I. Title.
QH367.P37 2015   576.8   C2015-900829-8

Published in the United States by
Firefly Books (U.S.) Inc.
P.O. Box 1338, Ellicott Station
Buffalo, New York 14205

Published in Canada by
Firefly Books Ltd.
50 Staples Avenue, Unit 1
Richmond Hill, Ontario L4B 0A7

Cover design: Erin R. Holmes / Soplari Design

Color reproduction by Portland Media Print Services Ltd.

Printed in China by Midas Printing International Ltd.

Front cover © David Carillet/Shutterstock

Back cover (from top) Fossil of Eopelobates from 50 million years ago © Joe Petersburger/National Geographic/Getty Images; An Asian gharial © AFP/Getty Images; Tyrannosaurus rex skeleton, Paul Wootton/Science Photo Library; Sword-billed Hummingbird, Yanacocha Reserve, Ecuador © Tui De Roy/Minden Pictures/Corbis

This book was designed and produced by
Quintessence Editions Ltd.
The Old Brewery
6 Blundell Street
London N7 9BH

Project Editor Sophie Blackman
Editors Sophie Blackman, Rebecca Gee,
   Frank Ritter, Dorothy Stannard
Designers Isabel Eeles, Tom Howey,
   Thomas Keenes
Production Manager Anna Pauletti
Proofreader Joanne Murray
Indexer Ruth Ellis
Editorial Directors Jane Laing, Ruth Patrick
Publisher Mark Fletcher

Note on the cladograms: these diagrams show a consensus of current scientific studies and were correct as of May 2015.

CONTENTS

FOREWORD

In 1860, a debate took place in the Oxford University Museum of Natural History. The hall where this debate took place still exists, although it's now divided by a mezzanine floor into an upper "Huxley Room," and a lower "Wilberforce Room." The names of the rooms enshrine the main protagonists of that debate.

The debate followed a lecture by an American academic on the subject of evolution, and soon burgeoned into a full-blown confrontation between science and religion. The Bishop of Oxford, Samuel Wilberforce, laid into the new theory of evolution through natural selection, expressing his doubts that species could change over time. Among the scientists in the room was Thomas Henry Huxley, who would become known as "Darwin's Bulldog." Wilberforce went on to challenge Huxley: on which side of his family did he claim descent from a monkey—on his grandmother's side or his grandfather's?

There are different accounts of precisely how Huxley responded to Wilberforce's sarcastic enquiry. Here's Huxley's own recollection of his reply:

"If then, said I, the question is put to me would I rather have a miserable ape for a grandfather or a man highly endowed by nature and possessed of great means of influence & yet who employs these faculties & that influence for the mere purpose of introducing ridicule into a grave scientific discussion, I unhesitatingly affirm my preference for the ape."

One commentator remembered perhaps what Huxley wished he'd said— muttering first "The Lord hath delivered him into my hands," before delivering this decisive blow: "I would rather be the offspring of two apes than be a man and afraid to face the truth."

Whatever was actually said, Huxley's contribution to the debate has become legendary. I think that this is partly down to one of the enduring anxieties that evolution seems to provoke. We humans have always been keen to place ourselves on a pedestal, to emphasize the gulf between us and the rest of life on the planet. The concept of humans as a special creation, made by an intelligent designer, stands in direct opposition to the idea that we are a product of unthinking natural selection: that we have evolved, like every other species on the planet. The acceptance of the fact of evolution involves the acknowledgement that humans are animals. Huxley's open recognition of his humble pedigree showed that he was willing to accept this scientific truth—even if it meant that he had to knock himself, and the rest of humanity, off a pedestal.

While humans are undoubtedly unusual animals, and unusual apes, it really is impossible to adhere to the idea of special creation if you look at the evidence around you. Our bodies and our genes are uncannily like those of other mammals, especially primates. And we now have a great fossil record of the ancestors of humans.

Five weeks before the famous debate in Oxford, Wilberforce had written a fairly damning review of Darwin's *On the Origin of Species*. He wrote about the lack of any fossil evidence to show one species gradually changing into another. To some extent, he was right. The fossil record in the mid-nineteenth century was patchy at best. Of course, there are still gaps—it is, after all, very rare for a dead organism to become fossilized. But a huge wealth of fossil evidence has come to light since Darwin's time. For example, we now have good evidence of the transition from aquatic to terrestrial existence in early amphibians; of the ancestral whales who would eventually lose their hind legs entirely; of feathered dinosaurs who were ancestral to birds; and of hominins—a whole family of apes who walked upright on two legs, including *Homo sapiens*.

These hominin fossils also show that human-ness didn't just suddenly appear. Features that we consider to be definitively, even uniquely, human, arrived in a piecemeal fashion, over vast expanses of time. And actually, it's only with hindsight that we can say that these features accumulated to a point where we can say "this is a human."

In fact, you can take any species you want and trace its evolutionary history in a similar way. And you find the same thing: a piecemeal accumulation of features until we end up with a species we're familiar with today. Tracing evolutionary histories like this also takes us back to common ancestors with other species, and indeed with whole groups of species, until we can reconstruct the huge, branching tree of life on the planet. Evolution, then, explains not just the appearance of individual species, including our own, but the entire diversity of life on Earth.

Evolution proceeds gradually, both in terms of genetic changes, and in the outward appearance of animals. But—admittedly only with the benefit of hindsight—it is possible to identify key points in evolutionary history which had huge implications for the way organisms continued to evolve. Such key moments include the appearance of the first eukaryotic cells, paving the way for the evolution of complex organisms; the evolution of animals which can survive out of water; and the appearance of flowering plants, together with the insects which evolved to pollinate them. In these pages, you can focus in on the characteristics of individual plants and animals—but you can also pull back, to look at the wider picture, to place those species in context. There's a useful timeline guide to key events popping up just when you need it.

This book sets out to present the beautiful and awe-inspiring range of biodiversity, past and present. It's a fantastic guide to life on this planet, starting with the earliest hints of living creatures preserved in rocks, then looking at plants, and at major groups of animals. In this way, we learn about not only the adaptations we see in living species, but also how those adaptations arose. And we discover how each species—including our own—is connected with others: with past ancestors and living cousins. We're all tiny twigs on the great Tree of Life.

ALICE ROBERTS
ANATOMIST AND PHYSICAL ANTHROPOLOGIST, WRITER, AND BROADCASTER
Bristol, UK

INTRODUCTION

The world today teems with at least 8 million—possibly millions more—different kinds or species of animals, plants and other living things. When Earth formed and cooled, just over 4.5 billion years ago, it was inhospitable to life. Yet within 1 billion years, life forms were appearing in the form of primitive single-celled microbes. So began evolution—which, simply put, is the changes undergone by living things through time. It took another 2 billion years or more for life to evolve beyond microbes floating in the sea. This book is an account of these first stages right through to today. As shown on the Ages of the Earth chart (see p.20), the first and greatest length of time is known as the Precambrian—"life's long, slow fuse." Chapter 1 follows the earliest developments, from single-celled forms with a relatively simple internal organization, known as prokaryotes (see p.36), into more complex cells, eukaryotes, along with examples of single-celled organisms that thrive today in the types of extreme environments that once characterized early Earth.

Around 2.3 billion years ago, those early microbial organisms triggered a build-up of oxygen in the atmosphere and oceans, known as the Great Oxygenation Event. This allowed evolution to proceed in new directions. Loose aggregates or colonies of similar cells arrived, then collectives where cells began to specialize in different functions and rely on each other. From these developed true multicelled plants (see p.56), animals and other organisms. All these events occurred within the marine realm.

Throughout this book are profiles of key species, both extinct and living, that hold a special place in our study of evolution. Some have persisted relatively unchanged for an immense amount of time, whereas others show features that are transitional between one major group and another, or are regarded as EDGE (Evolutionarily Distinct and Globally Endangered). Key information given for each species includes its formal scientific name, group, size (usually length), location, and, for living examples, its "Red List" conservation status according to the IUCN (International Union for Conservation of Nature). Appearing ahead of the profiled key species are accounts of the appearance, diversification and fate of their particular group, with a timeline of key events. The accounts are organized chronologically to follow the standard geological time scale of eras, or major time spans, subdivided into periods, then epochs.

Foremost among evidence for extinct species are fossils. The fossilized remains of early life include layered rock mounds, stromatolites (see p.42), that still form today; and the mysterious Ediacaran life-forms (see p.46). Predominantly soft-bodied and seabed-dwelling, these varied organisms stimulate scientific discussion as to their biological affinities. Efforts are made to fit them into living groups, but few fit comfortably. They may be fleeting remnants of major evolutionary pathways that faded without descendants.

Toward the end of the Precambrian supereon, 541 million years ago, and into the following Cambrian period, evolution seems to have accelerated. This may be partly because organisms with hard parts such as shells appeared, and these are much more likely to fossilize. Detailed research at the fossil-rich sites of the Burgess Shale of Canada (see p.50) and China's Chengjiang shows that marine biodiversity (the numbers and varieties of living things in the sea) had already reached modern levels. This contradicts the old idea of an "evolutionary tree" starting with just a few life forms and gradually becoming

▶ At the time of their formation, 4.5 billion years ago, Earth and the three other rocky planets of the inner solar system were inhospitable to any form of life. The protoplanets grew in size as their gravitational attraction plucked masses of rock from the space surrounding them.

▼ Photographed at the Colorado School of Mines in the United States, this fossil of calcareous algae dates from the Cambrian period, 541 million to 485 million years ago.

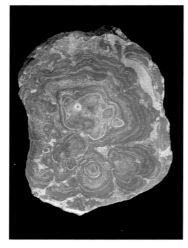

more diverse in terms of species and numbers to the present day. The tree analogy was enthusiastically championed by German biologist–scientist Ernst Haeckel (1834–1919) and has been endlessly repeated in various forms ever since, but it does not reflect reality.

As described in Chapter 2, plants are the basis of most life on Earth. By the process of photosynthesis they receive energy in the form of sunlight, change it into chemical form, incorporate it into their tissues, and become food for animals. A major transition, more than 400 million years ago, occurred when plants evolved from simple, water-dwelling algae into the first land forms, and then developed liquid-conducting vessels, as revealed by fossils such as those from Rhynie in Scotland (see p.62). Now animals that ventured onto land had both food and shelter. Plants continued to diversify as mosses, ferns and other spore-bearing groups.

Another crucial plant development was reproduction by seeds, which have a tiny embryo plant with food store, rather than by simpler spores. This led to the now-extinct seed-ferns (see p.70), the cone-bearing conifers and their cousins the cycads and ginkgoes, and by 150 million years ago (although perhaps much earlier), the flowering plants or angiosperms. This last group, with its flowers, herbs, creepers, shrubs, bushes and broad-leaved trees, now dominates plant life on Earth and comprises more than 80 percent of all known plant species.

As plants diversified into many aquatic and land habitats, they were closely followed by herbivorous animals that evolved to exploit them for food and shelter. As portrayed in Chapter 3, the first animals were invertebrates — creatures without backbones, at first in the sea. Simplest, with no muscles, nerves, heart, blood or brain, were the sponges or Porifera. They still survive today, as do the reef-building corals that assembled the first large fossil structures. Corals make up the Cnidaria, along with their relations, anemones and jellyfish. Another massive invertebrate group, with 85,000 living species, is the molluscs: extant bivalves and gastropods, extinct ammonoids, and

▼ Ferns appeared by 350 million years ago. This rock contains fossilized fronds dating from the late Carboniferous period, 323 million to 299 million years ago. Ferns were among the plants that dominated the land at this time and, when these plants died, they formed large beds of coal (a form of carbon). This led to the name of Carboniferous for this time period.

nautiloids—the living nautilus is a tiny remnant of its once greatly successful group. The echinoderms, such as starfish, urchins and crinoids, also left vast numbers of fossils and continue to thrive, with 7,000 species in today's seas.

Many different kinds of worms had also appeared by 500 million years ago. An especially curious living example is the velvet worm or onychophoran, *Peripatus* (see p.112). This has been cited as a possible model for how one type of legless worm evolved true limbs and became the first arthropod or "joint-legged" creature. The vast arthropod group includes the massively successful but now-gone trilobites (see p.118), and crustaceans such as crabs and shrimps. As animals followed plants onto land, some arthropod pioneers evolved into the first insects—by far the largest contemporary group of living things, with more than 1 million present-day recorded species. Arthropods also include millipedes, centipedes, and the arachnids—spiders, scorpions, mites and ticks. Relatives of arachnids are the horseshoe or king crabs, closely studied since they have changed so little over hundreds of millions of years, and their cousins the long-gone, fearsome sea-scorpions (see p.136).

The evolution of vertebrates, or animals with backbones, seems to have been a complex process. As explained in Chapter 4, the first vertebrates were creatures we might call "fish." But before them came worm-like creatures such as *Pikaia* (see p.176), termed chordates, with a stiffening body structure, the notochord, that has been likened to a backbone precursor. Also, the vertebrates we call "fish" do not form one complete natural group. Various kinds of fish arose at different times, probably from different ancestors. Early fish were agnathans (see p.178), that is, had no jaws, and are still represented by modern hagfish and lampreys. They also had no or few fins to control their movements. But jaws and fins were not long in appearing, culminating in the top predator of Devonian seas, 360 million years ago—the terrifying 33-foot (10-m) long member of the placoderm group, *Dunkleosteus* (see p.188).

Some branches of fish evolution led to the Chondrichthyes, with a skeleton made of cartilage. They include the elasmobranchs—sharks, rays and skates—and the holocephali, or chimaeras (ratfish). Sharks are

▲ Jewel anemones, *Corynactis viridis*, colonize water off Guillaumesse, Sark, in the British Channel Islands. Like other anemones, the jewel anemone can reproduce asexually by splitting, which leads to clusters of identically colored clones. The patches join up into extraordinary, gaudy "quilts."

▲ A school of scalloped hammerhead sharks, *Sphyrna lewini*, swim near Cocos Island, Costa Rica. The hammerhead itself evolved as a form of hydrofoil, known as a cephalofoil, to provide lift as the shark moves through water. The head also bears chemical and electrical sensors that, with the widely spaced eyes, help with detection of prey.

well-known prehistoric survivors, little changed in some 400 million years. Another direction for fish evolution was toward a skeleton of true bone and the group Osteichthyes. Within the bony fish, the major subgrouping is the actinopterygians or ray-finned fish (see p.202), which includes the great majority of enormously diverse fish species today, from tiny gobies to great marlin, swordfish and sunfish. They have stiff rod-like supports called rays to fan and fold their fins for fine control when swimming.

Far fewer than ray-fins in terms of species numbers are the bony fish called sarcopterygians or lobe-fins (see p.196). However their evolutionary importance cannot be underestimated. They include the lungfish, who developed surviving in air to a high degree, and the coelacanths. Believed long extinct, the discovery of a living coelacanth near Africa in 1938 was one of the century's greatest events in evolutionary science. Lobe-fins like the coelacanth support their fins on fleshy, muscular, stump-like lobes—from which, via a series of stages seen in fossils such as *Tiktaalik*, *Acanthostega*, *Ichthyostega* and *Pederpes*, developed the tetrapod limb (see p.206). Tetrapods are four-limbed vertebrates that include amphibians, reptiles, birds and mammals. However the idea that the living coelacanth is the ancestor of these tetrapod groups is far from correct. The actual ancestor lived further back in time, and may well not yet be known to science.

The first major tetrapod group was the amphibians. They were still tied to water, in which they laid their eggs. By 300 million years ago some of these amphibians, such as *Eryops*, had become big, powerful predators, with crocodile-like lifestyles. This has led to the Permian period (299 million to 252 million years ago) being dubbed the "Age of Amphibians." As with the Devonian (419 million to 359 million years ago) "Age of Fishes," this gives a vague idea of what was happening in one area of evolution, but many other significant changes occurred at the same time. Following the Permian and Triassic, several major groups of amphibians died out, leaving the present-day Lissamphibia, being frogs, toads, salamanders, newts and caecilians.

Before large amphibians came to dominate the land, certain of their kind evolved into the first reptiles. Chapter 5 opens with these intermediate stages and the earliest known fossil reptile, *Hylonomus*, from 312 million years ago (see p.232). It looked like a lizard, but as often in the natural world, appearances can be deceiving: *Hylonomus* belonged to a different reptile group. Reptiles of varying kinds diversified and spread, including the chelonians, or turtles and tortoises; the squamates, or lizards and snakes; and the crocodilians, being crocodiles, alligators, caimans and their extinct kin. Reptilia also includes the tuatara (see p.234), a fascinating lizard-like creature from New Zealand that has its own reptile subgroup, with closest relatives that went extinct almost 100 million years ago. Like the dawn redwood, horseshoe crab, coelacanth and others in this book selected as key species, the tuatara is sometimes called a "living fossil." But this informal, imprecise term implies that it has remained the same as its long-gone fossil ancestors, whereas in fact it has continued to evolve through time.

There were also several long-extinct reptile groups, especially in the sea, such as nothosaurs, plesiosaurs and pliosaurs, ichthyosaurs and mosasaurs, while in the skies, pterosaurs ruled. On land, one reptile grouping came to dominate life like no other creatures had done before—the dinosaurs. Most of these reptile lineages disappeared in one of the "Big Five" mass extinction events, this one being at the end of the Cretaceous period, 66 million years ago (see p.364). However, in the light of modern methods of grouping living things, dinosaurs are still with us, in the form of one of their subgroups, birds.

Birds evolved from small meat-eating saurischian dinosaurs by 150 million years ago, as depicted in Chapter 6, and include one of the most famous of all prehistoric creatures, the "early bird" *Archaeopteryx* (see p.376). The scientific details of those first stages of bird diversification are not yet clear and several groups seemed to come and go through the Cretaceous period, 145 million to 66 million years ago. In the past 50 million years bird evolution has thrown up several sets of giants, including the flightless, predatory "terror birds" such as *Phorusrhacos* (see p.410). The giant moas of New Zealand (see p.412) and the enormous elephant birds (see p.414) of Madagascar both succumbed to human persecution within the last millennium.

A very different evolutionary pathway, as described in Chapter 7, led probably from amphibian-like tetrapods to reptile-like creatures called synapsids, among them the well-known *Dimetrodon* (see p.238), and, following them, the so-called "mammal-like reptiles" or therapsids

◄ The tuatara, *Sphenodon punctatus*, here emerging from its egg, is a reptile endemic to New Zealand. Although resembling most lizards, it is part of a distinct lineage, the order Rhynchocephalia, which contains only two surviving species. Tuataras have a common ancestor with lizards and snakes and so offer insights into the evolution of those groups, and that of the earliest diapsids— the group that includes dinosaurs (and therefore also birds) and crocodiles.

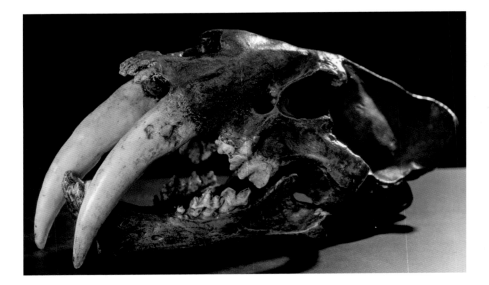

▲ This fossil skull is of a sabre-toothed cat, an extinct carnivore of the group Machairodontinae. Various sabre-tooths lived in Europe, Asia, Africa, and North and South America, from about 25 million years ago until the end of the last great ice age, 10,000 years ago. A sabre-toothed cat could open its jaw very wide, using its huge upper canine teeth to stab and slash its prey.

(see p.418). From certain of these, by 220 million years ago the first true mammals made their appearance (see p.424). Of minor importance during the time of the great dinosaurs, mammals flourished after the end-of-Cretaceous mass extinction and rapidly diversified into many groups, some of which lasted only a few tens of millions of years. Among the major mammal groups were monotremes, still represented by the egg-laying platypus and echidnas; and the therians, including the pouched mammals or marsupials (about 340 living species), and the placentals (more than 5,000 living species), where young develop inside the mother, nourished by the placenta, before birth.

One mammal group that has been relatively inconspicuous over the past 50 million years is the primates. From about 6 million years ago, one of its evolutionary offshoots led to the genus (species group) *Homo* (see p.530). In the past 200,000 years the single species *Homo sapiens*, modern humans (see p.536), has come to dominate life on Earth like no other before.

Evidence for the long, complex history of Earth's changing environments and life forms is stored in rocks, and is often referred to as the fossil record. Palaeontologists and geologists have been sampling and interpreting this record for more than 200 years to produce the current synthesis, but as a history it is far from complete. Fossil formation (see p.30) has been varied, complex and inconsistent in time and place. Furthermore, the geological processes of the rock cycle have greatly altered, diminished and fragmented the evidence for past life and its environments. Fossilization is favored in shallow seas and lakes, while creatures of deep oceans and upland terrestrial environments are poorly represented. Also fossilization is fundamentally biased toward organisms with durable hard parts, such as wood, bark, cones, roots, seeds, pollen, shells, bones, teeth, claws and horns; remains of soft bodied organisms are only preserved in special circumstances. And even these hard parts may be damaged or altered so that skeletons, for instance, are dislocated, fragmented and worn, sometimes beyond recognition. The net result is that fossilization results in a great loss of information and huge bias in the record.

Understanding these limitations is vital when making sense of the natural world and evolution, in the past and today. The biological sciences give plants, animals and other organisms an internationally recognized, two-part or binomial label incorporating its genus and species; thus, the great ice-age woolly mammoth is called *Mammuthus* (its genus) *primigenius* (its species).

The species name distinguishes the animal from other mammoths, such as the steppe mammoth, *M. trogontherii*, and the Columbian mammoth, *M. columbi*. The naming system shows how biology places like creatures with like in a system of hierarchical groups.

Methods of basic nomenclature (naming) and taxonomy (method of arrangement) have a long history. Greek philosopher Aristotle (384–322 BC) produced one of the first schemes, based on how animals reproduce, and coined the terms "species" and "genus." But it was Swedish botanist Carl Linnaeus (1707–78) who devised the foundation of modern taxonomy. His *Systema Naturae* (in editions from 1735 to 1766–68) attempted to classify all known life forms, along with some fossils. Linnaeus included more than 10,000 plant and animal species, grouping them in a low-to-high ranking of species, genus, family, order, class, phylum, and kingdom, based on similar forms.

Although there was no evolutionary implication to Linnaeus's classificatory system, it seems "pre-designed" for events that followed a century later. The notion of evolution—changes in living things through time—had been mentioned by philosophers and naturalists of ancient Greece and Rome. However, it was always overruled by the belief that plants and animals were fixed or "immutable." In his *Philosophie Zoologique* (1809), French naturalist Jean-Baptiste Lamarck (1744–1829) proposed that living beings altered from simple beginnings through increasingly sophisticated stages as a result of a "complexifying force." Lamarck argued that useful characteristics acquired during the life of an organism were strengthened and passed on to the next generation, while disused ones weakened and diminished—a notion often summarized as "use and disuse."

Fifty years later came *On the Origin of Species by Means of Natural Selection, or the Preservation of Favoured Races in the Struggle for Life* (1859) by English naturalist Charles Darwin (1809–82). In 1831–6 Darwin had traveled around the world in the survey ship HMS *Beagle* and was struck by groups of similar species in each region he visited. These included the

▼ Charles Darwin's *On The Origin of Species* (1859)—here a first-edition copy—expresses the fundamental processes that underlie the modern theory of the evolution of all living organisms.

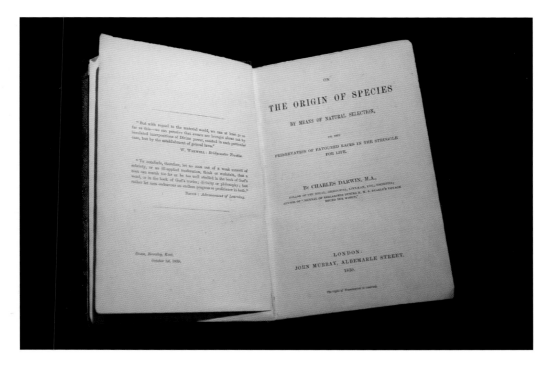

▲ The Espanola mockingbird, *Mimus macdonaldi*, is one of several closely similar species inhabiting the Galapagos Islands. Mockingbirds were among the creatures that helped Charles Darwin to form the theory that species evolve characteristics that secure them advantages in specific environments.

mockingbirds, giant tortoises, and finches of the Galapagos Islands in the Pacific. On his return, and being well aware of Lamarck's ideas on evolution, Darwin accepted the concept of transmutation over time but proposed a very different mechanism, natural selection. Briefly, the offspring of an organism are usually slightly different, and most organisms produce more offspring than can survive. If an individual has a feature, trait or adaptation that gives some advantage in facing the challenges of life—finding food and shelter, and coping with antagonistic factors such as the vagaries of the environment, competitors, predators and disease—then that "favored" individual is more likely to reach breeding age and produce its own offspring. If the advantageous adaptation is heritable (in the genes), the offspring will also have it and likewise be advantaged.

Coincidentally, another English naturalist, explorer and collector, Alfred Russel Wallace (1823–1913), had the same insight as Darwin. The two corresponded in 1856–57 and Darwin gave Wallace full credit for the contribution he made to his theories.

Darwin saw evolution as a slow and gradual process that required many generations and a great deal of time. He also recognized the role of the environment in isolating evolving populations that became new species, separate from the parent species—which may itself retain the ancestral form, change in another direction to become another new species or die out altogether. Darwin was aware of numerous objections to the theory of evolution, especially the lack of fossil evidence for transitional forms or "missing links" between major groups of organisms, and for the common ancestors from which the various branches of life diverged. Also, he did not know the mechanism for inheritance of features or traits from one generation to the next, or how new traits arose.

Darwin and most other scientists of the time were unaware that an Austrian monk, Gregor Mendel (1822–84), had been experimenting with selective breeding of pea plants. In 1866 Mendel suggested that inherited traits or characteristics are controlled by pairs of discrete "particles of inheritance," later known as genes, which came from the parents. In 1900 Dutch botanist Hugo de Vries (1848–1935) conducted similar experiments and produced similar results—only to find he was repeating Mendel's work, which thus became widely known. By this time, microscopic studies of cells had revealed the thread-like structures called chromosomes that carried Mendel's "particles of inheritance." Within a decade, US geneticist Thomas Hunt Morgan (1866–1945) was conducting experiments with fruit flies to test the link between Mendel's "discrete particles" and chromosomes. From the 1930s, the mechanism by which chromosomes exchange and alter genetic information was discovered by US geneticist Barbara McClintock (1902–92). When chromosomes pair up to form X-shaped structures, genes are shuffled around (in a process called genetic recombination) to produce new traits and greater genetic variety in the offspring.

By the 1950s, the chemical that carried genetic information was known to be DNA, or deoxyribonucleic acid. In 1953, British biophysicist and x-ray specialist Rosalind Franklin (1920–58) produced new and greatly improved images of DNA, which came to the notice of US biologist James Watson (1928–) and English scientist Francis Crick (1916–2004). They determined that the DNA molecule consists of a double helix, like a twisted ladder. The order of subunits called bases—the "rungs" of the ladder—carries the information as a chemical code. When cells divide, the double helix is "unzipped" with each half acting as a template to build a new complementary partner, thereby forming two identical copies. This is how genetic information passes from cell to cell in an organism, and also in reproduction, via sex cells—

egg and sperm—to the next generation. A change or mutation in the copying process may lead to an altered gene, and it is by this process that new traits or features arise in offspring. By the 1960s the work of Mendel and later geneticists, and Watson and Crick and other molecular biologists, had been consolidated with Darwin's proposal for evolution by natural selection. Together, Darwin's theory and this supporting evidence form the basis of the modern theory of evolution.

Molecular genetics has been particularly important in the development of the "molecular clock" (gene or evolutionary clock). In 1967, two US biochemists, Allan Wilson (1934–91) and Vincent Sarich (1934–2012), showed that the timing of evolutionary relationships between humans and other primates could be estimated using the detailed structures of their molecules, especially proteins and, later, DNA. After a species has split into two, due to the usually slow, gradual changes accumulated by nonexact copying, the structures of these molecules become different, or diverge. The longer the species have been separated, the greater the divergence. Wilson and Sarich estimated the divergence time between humans and their closest living relatives, chimpanzees, at between 4 million and 6 million years ago. This was much more recent than indicated by the fossil record and led to controversy about how fast the clock "ticked" and whether it could speed up or slow down. In today's practice, such time estimations are based on a mix of evidence from fossils, molecules like DNA, physical features and other information.

Meanwhile, German entomologist Willi Hennig (1913–76) developed a different approach to biological classification. He based grouping of organisms on synapomorphies: features that are shared by all members of the group, that are unique to the group (not found in other groups) and that have been

◄ English naturalist and geologist Charles Darwin proposed that species descended over time from common ancestors, and that distinct species evolve due to natural selection in response to environmental pressures.

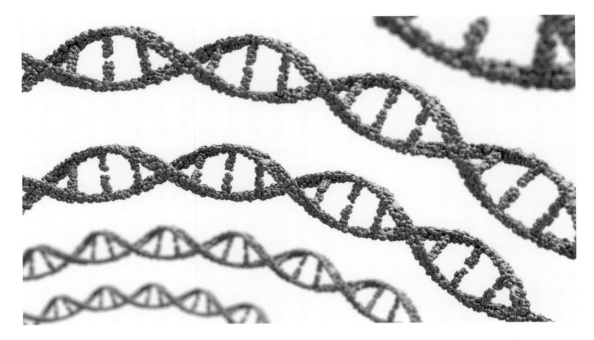

▲ This computer illustration of DNA (deoxyribonucleic acid) highlights its double-helix structure. When DNA replicates itself, the "ladder" form "unzips" down the middle. Each of the two resulting strands acts as a template for the next whole double helix; the bases of the "rungs" are matched to synthesize the new partner strands.

derived from the common ancestor of the group. Such a group is known as a clade. This system, designed specifically to reconstruct evolutionary histories, is known as phylogenetic systematics, or more commonly, cladistics. It is especially conducive to analysis by computer, which produces alternative branching diagrams—dendrograms or cladograms (for an example, see p.91). These show relationships but do not include a time element for when divergences took place. In common with most such representations, they are modified from more "pure" versions for reasons of clarity, and so, for instance, various levels or ranks of divergences may be omitted.

The flip side of the origin of species, or speciation, is extinction. It is estimated that more than 99 percent of all species that ever lived are now extinct. Evolution adapts species to current conditions, but these are usually changing—climates alter, great earth movements shift continents and ice ages alternate with tropical spells. For each species, the organisms all around—the ones it eats, competitors, predators, parasites and pathogens—are all evolving too. In this ever-continuing race to adapt, some species gain an advantage while others lose out. For example, a species may arrive on a newly formed group of islands where it has plentiful food and few enemies. Here it colonizes the islands, which each have slightly different conditions. Each group rapidly adapts to its own island, in a "burst of evolution" called adaptive radiation, and becomes a separate species. (It was during his examination of this process in the Galapagos Islands that Darwin began to form his theory about a mechanism for evolution.) Another species may then arrive and out-compete the existing ones, which go extinct, and then itself undergo adaptive radiation . . . and so on.

Modern evolutionary theory has many facets. For example, a specialized form of natural selection is sexual selection. This is where traits arise in one sex—often the male—because the other sex is attracted to the traits at breeding time as signs of evolutionary fitness and fertility. The extinct male giant elk's massive antlers and the flamboyant plumage of the living male birds of paradise are examples. Such sexually selected traits may become so extreme that they impinge heavily on general prospects of survival.

The term "convergent evolution" refers to widely different organisms developing similar features, often due to similar lifestyles or environments. For instance, bats, birds and extinct pterosaurs have wings. Superficially the wings look alike, being broad-surfaced fore limbs flapped by muscles to enable flight. But inside, the bones and other structures are very different. The bat wing is supported mainly by digits (fingers) three to five; the pterosaur wing by digit four; the bird wing has much reduced bones and relies on its feathers. Evolution caused the wings to "converge" in design to fulfil the same function.

Another evolutionary term is exaptation, sometimes called "preadaptation," in which a feature that has evolved for one purpose becomes co-opted or commandeered for another. For example, feathers may have arisen originally in certain dinosaurs for body insulation, or perhaps visual display, and then later became useful for flight in their descendants, the birds.

The scientific study of evolution is ever changing, like its subject. Genetic and molecular analyses and DNA sequencing—techniques arguably still in their infancy—are proving highly influential. The rise of the cladistic method is, for many purposes, displacing traditional Linnaean taxonomy. The "library of life" that is the fossil record grows ever greater, as does scope for reinterpretation of specimens that are already known. Evolution studies are being transformed by advancing technologies of many kinds, including new microscopes, improved fossil extraction and scrutiny, chemical isotope analysis, x-rays and scans, including CT (computed tomography).

This book also asks where this science might lead. Knowledge is progressing in leaps and bounds about evolution as a most fundamental attribute of life, and how it has shaped the history of our planet. Greater understanding might suggest—in a context of climate change, pollution, potential food shortages and disease epidemics, fewer wild habitats and many other challenges—where the natural world and ourselves are heading. It might also suggest how extinctions may be partially reversed, as in the story of the quagga (see p.558), where selective breeding has reintroduced an animal with traits considered lost for more than a hundred years.

▼ Like the extinct giant elk, the bull elk, *Cervus canadensis*, has evolved huge antlers that serve a mainly sexual purpose. To establish dominance over other males and attract females, the bulls engage in ritualized mating behaviors during the rut, including posturing, sparring with their antlers and making a series of loud vocalizations, known as bugling.

Ages of the Earth

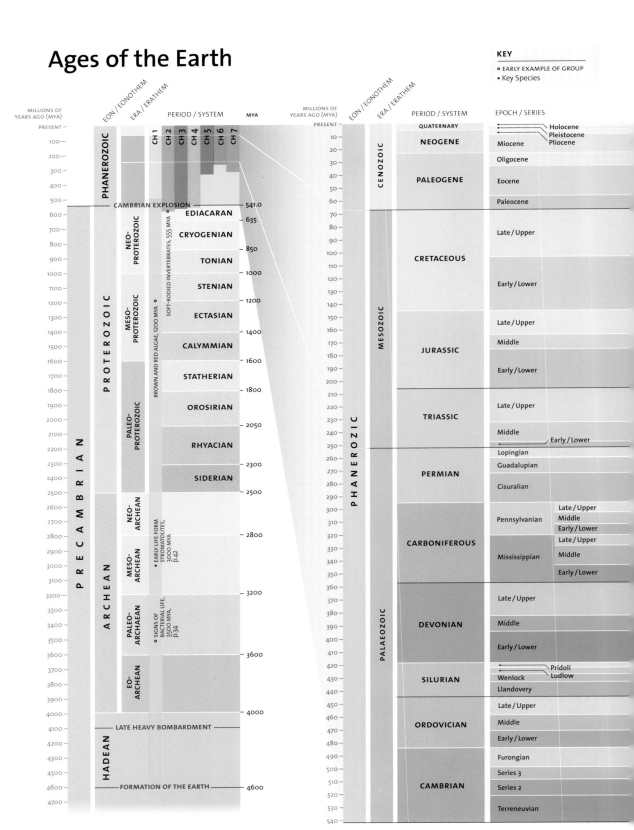

MILLIONS OF YEARS AGO (MYA)	EON / EONOTHEM	ERA / ERATHEM	PERIOD / SYSTEM								MYA
PRESENT —				CH 1	CH 2	CH 3	CH 4	CH 5	CH 6	CH 7	
100 —	PHANEROZOIC										
200 —											
300 —											
400 —											
500 —											
600 —			CAMBRIAN EXPLOSION		EDIACARAN						541.0
700 —		NEO-PROTEROZOIC		CRYOGENIAN						635
800 —											
900 —			TONIAN								850
1000 —											1000
1100 —	PROTEROZOIC		STENIAN								
1200 —		MESO-PROTEROZOIC									1200
1300 —			ECTASIAN								
1400 —											1400
1500 —			CALYMMIAN								
1600 —											1600
1700 —			STATHERIAN								
1800 —											1800
1900 —		PALEO-PROTEROZOIC	OROSIRIAN								
2000 —											2050
2100 —											
2200 —			RHYACIAN								
2300 —											2300
2400 —			SIDERIAN								
2500 —											2500

PRECAMBRIAN

SOFT-BODIED INVERTEBRATES, 555 MYA o

BROWN AND RED ALGAE, 1200 MYA o

	ARCHEAN	NEO-ARCHEAN				2800
2600 —						

■ EARLY LIFE FORM STROMATOLITES, 3000 MYA p.42

MESO-ARCHEAN

o SIGNS OF BACTERIAL LIFE, 3500 MYA, p.34

PALEO-ARCHEAN

3200

3600

EO-ARCHEAN

4000

LATE HEAVY BOMBARDMENT

HADEAN

FORMATION OF THE EARTH — 4600

MILLIONS OF YEARS AGO (MYA)	EON / EONOTHEM	ERA / ERATHEM	PERIOD / SYSTEM	EPOCH / SERIES
PRESENT —			QUATERNARY	Holocene / Pleistocene / Pliocene
10 —		CENOZOIC	NEOGENE	Miocene
20 —				
30 —				Oligocene
40 —			PALEOGENE	Eocene
50 —				
60 —				Paleocene
70 —				
80 —			CRETACEOUS	Late / Upper
90 —				
100 —				
110 —				Early / Lower
120 —				
130 —				
140 —				
150 —		MESOZOIC	JURASSIC	Late / Upper
160 —				Middle
170 —				
180 —				Early / Lower
190 —				
200 —				
210 —			TRIASSIC	Late / Upper
220 —				
230 —				
240 —				Middle
				Early / Lower
250 —	PHANEROZOIC			Lopingian
260 —			PERMIAN	Guadalupian
270 —				
280 —				Cisuralian
290 —				
300 —			Pennsylvanian	Late / Upper / Middle / Early / Lower
310 —				
320 —			CARBONIFEROUS	Late / Upper
330 —			Mississippian	Middle
340 —				
350 —				Early / Lower
360 —				
370 —		PALAEOZOIC	DEVONIAN	Late / Upper
380 —				
390 —				Middle
400 —				Early / Lower
410 —				
420 —			SILURIAN	Pridoli / Ludlow
430 —				Wenlock
440 —				Llandovery
450 —				Late / Upper
460 —			ORDOVICIAN	Middle
470 —				Early / Lower
480 —				
490 —				Furongian
500 —				Series 3
510 —			CAMBRIAN	Series 2
520 —				
530 —				Terreneuvian
540 —				

The first signs of life occurred deep in Earth's past, during the Precambrian supereon (see far left), which lasted for almost 90 per cent of the planet's existence. In this chart, the remainder of time, from the Cambrian period to the current Quaternary period (see far left, top) is expanded below and keyed into the book's chapter-by-chapter coverage. Geological time periods are related to specific series of rock formation, and these are correlated with major events, first appearances of life forms, and key species in their evolution.

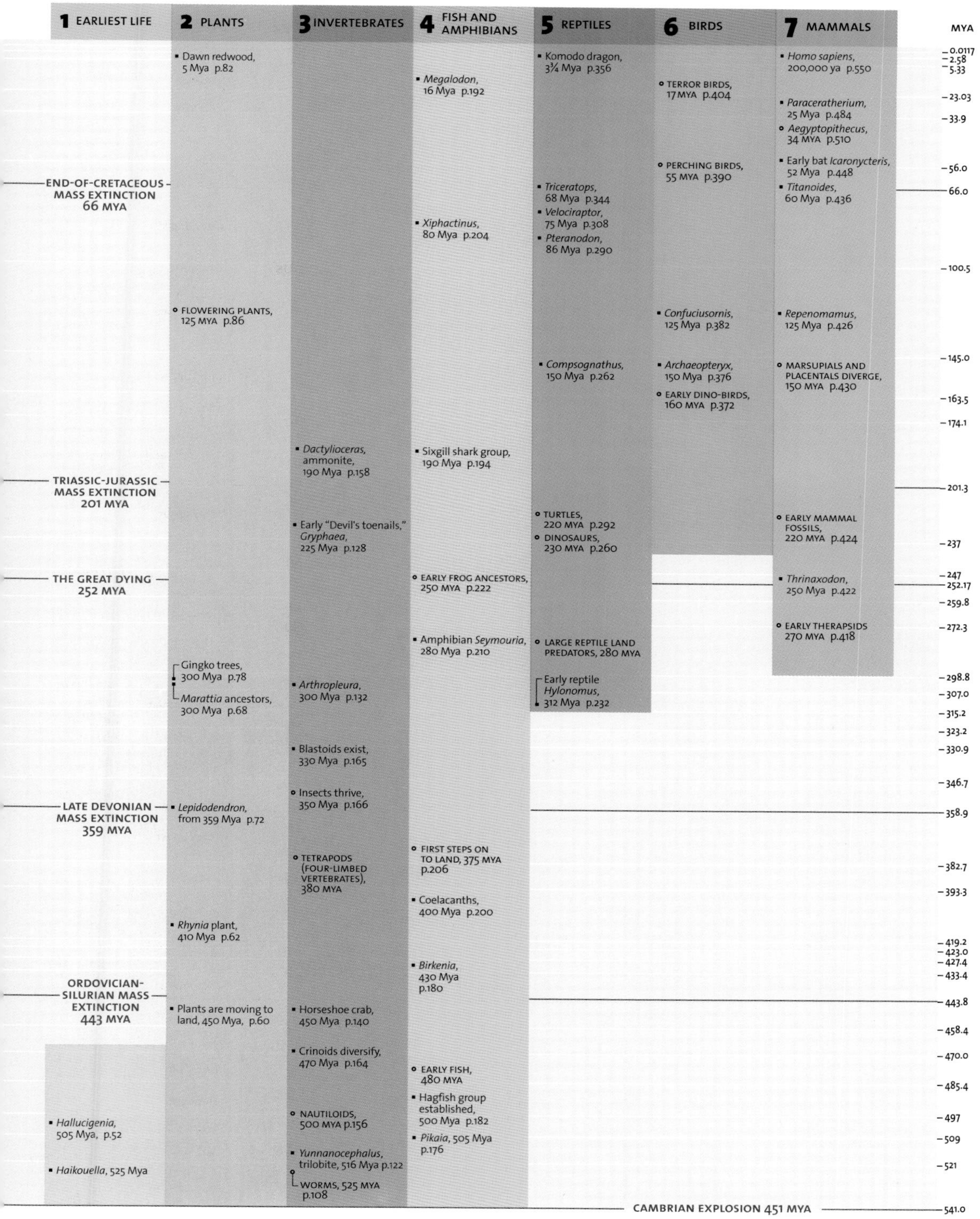

Mass extinction events (left margin)

- END-OF-CRETACEOUS MASS EXTINCTION, 66 MYA
- TRIASSIC-JURASSIC MASS EXTINCTION, 201 MYA
- THE GREAT DYING, 252 MYA
- LATE DEVONIAN MASS EXTINCTION, 359 MYA
- ORDOVICIAN-SILURIAN MASS EXTINCTION, 443 MYA

1 EARLIEST LIFE

- Hallucigenia, 505 Mya, p.52
- Haikouella, 525 Mya

2 PLANTS

- Dawn redwood, 5 Mya p.82
- FLOWERING PLANTS, 125 MYA p.86
- Gingko trees, 300 Mya p.78
- Marattia ancestors, 300 Mya p.68
- Lepidodendron, from 359 Mya p.72
- Rhynia plant, 410 Mya p.62
- Plants are moving to land, 450 Mya, p.60

3 INVERTEBRATES

- Dactylioceras, ammonite, 190 Mya p.158
- Early "Devil's toenails," Gryphaea, 225 Mya p.128
- Arthropleura, 300 Mya p.132
- Blastoids exist, 330 Mya p.165
- Insects thrive, 350 Mya p.166
- TETRAPODS (FOUR-LIMBED VERTEBRATES), 380 Mya
- Horseshoe crab, 450 Mya p.140
- Crinoids diversify, 470 Mya p.164
- NAUTILOIDS, 500 Mya p.156
- Yunnanocephalus, trilobite, 516 Mya p.122
- WORMS, 525 MYA p.108

4 FISH AND AMPHIBIANS

- Megalodon, 16 Mya p.192
- Xiphactinus, 80 Mya p.204
- Sixgill shark group, 190 Mya p.194
- EARLY FROG ANCESTORS, 250 Mya p.222
- Amphibian Seymouria, 280 Mya p.210
- FIRST STEPS ON TO LAND, 375 MYA p.206
- Coelacanths, 400 Mya p.200
- Birkenia, 430 Mya p.180
- EARLY FISH, 480 MYA
- Hagfish group established, 500 Mya p.182
- Pikaia, 505 Mya p.176

5 REPTILES

- Komodo dragon, 3¾ Mya p.356
- Triceratops, 68 Mya p.344
- Velociraptor, 75 Mya p.308
- Pteranodon, 86 Mya p.290
- Compsognathus, 150 Mya p.262
- TURTLES, 220 MYA p.292
- DINOSAURS, 230 MYA p.260
- LARGE REPTILE LAND PREDATORS, 280 MYA
- Early reptile Hylonomus, 312 Mya p.232

6 BIRDS

- TERROR BIRDS, 17 MYA p.404
- PERCHING BIRDS, 55 MYA p.390
- Confuciusornis, 125 MYA p.382
- Archaeopteryx, 150 Mya p.376
- EARLY DINO-BIRDS, 160 Mya p.372

7 MAMMALS

- Homo sapiens, 200,000 ya p.550
- Paraceratherium, 25 MYA p.484
- Aegyptopithecus, 34 MYA p.510
- Early bat Icaronycteris, 52 Mya p.448
- Titanoides, 60 Mya p.436
- Repenomamus, 125 Mya p.426
- MARSUPIALS AND PLACENTALS DIVERGE, 150 MYA p.430
- EARLY MAMMAL FOSSILS, 220 MYA p.424
- Thrinaxodon, 250 Mya p.422
- EARLY THERAPSIDS, 270 MYA p.418

MYA scale (right margin)

0.0117 — 2.58 — 5.33 — 23.03 — 33.9 — 56.0 — 66.0 — 100.5 — 145.0 — 163.5 — 174.1 — 201.3 — 237 — 247 — 252.17 — 259.8 — 272.3 — 298.8 — 307.0 — 315.2 — 323.2 — 330.9 — 346.7 — 358.9 — 382.7 — 393.3 — 419.2 — 423.0 — 427.4 — 433.4 — 443.8 — 458.4 — 470.0 — 485.4 — 497 — 509 — 521 — 541.0

CAMBRIAN EXPLOSION 451 MYA

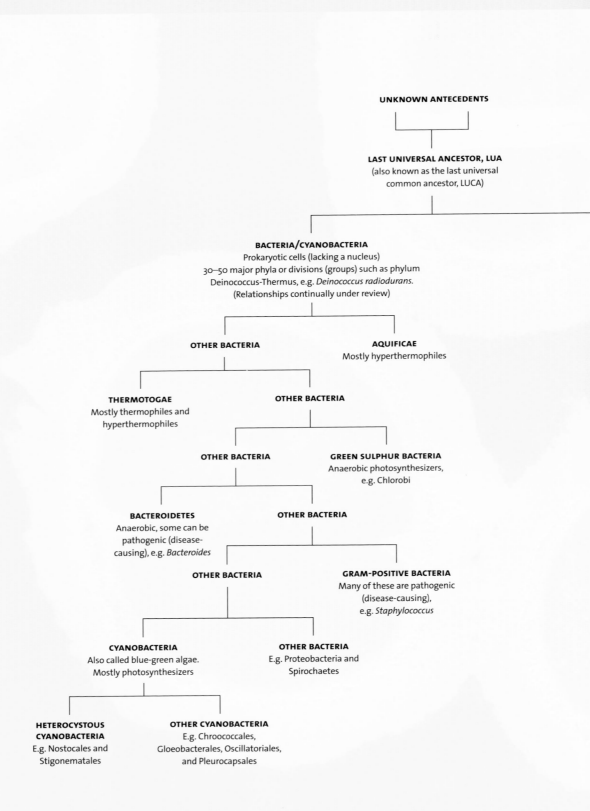

1 | EARLIEST LIFE

It is accepted that all life evolved from a kind of semi-microbial form, often referred to as the Last Universal Ancestor (LUA). Those initial stages of diversification involved simple prokaryotic cells—still represented by the immense domain of Bacteria, living almost everywhere, and Archaea, many with extremophile lifestyles. Some of these gained membrane-enclosed parts within to become eukaryotes, and then banded into communities as the first multicelled organisms that would lead to plants and animals.

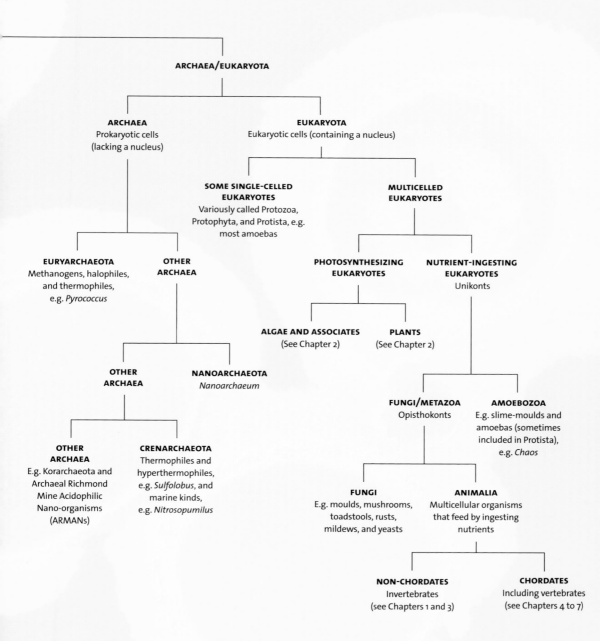

ARCHAEA/EUKARYOTA

ARCHAEA
Prokaryotic cells
(lacking a nucleus)

EUKARYOTA
Eukaryotic cells (containing a nucleus)

SOME SINGLE-CELLED EUKARYOTES
Variously called Protozoa, Protophyta, and Protista, e.g. most amoebas

MULTICELLED EUKARYOTES

EURYARCHAEOTA
Methanogens, halophiles, and thermophiles, e.g. *Pyrococcus*

OTHER ARCHAEA

PHOTOSYNTHESIZING EUKARYOTES

NUTRIENT-INGESTING EUKARYOTES
Unikonts

ALGAE AND ASSOCIATES
(See Chapter 2)

PLANTS
(See Chapter 2)

OTHER ARCHAEA

NANOARCHAEOTA
Nanoarchaeum

FUNGI/METAZOA
Opisthokonts

AMOEBOZOA
E.g. slime-moulds and amoebas (sometimes included in Protista), e.g. *Chaos*

OTHER ARCHAEA
E.g. Korarchaeota and Archaeal Richmond Mine Acidophilic Nano-organisms (ARMANs)

CRENARCHAEOTA
Thermophiles and hyperthermophiles, e.g. *Sulfolobus*, and marine kinds, e.g. *Nitrosopumilus*

FUNGI
E.g. moulds, mushrooms, toadstools, rusts, mildews, and yeasts

ANIMALIA
Multicellular organisms that feed by ingesting nutrients

NON-CHORDATES
Invertebrates
(see Chapters 1 and 3)

CHORDATES
Including vertebrates
(see Chapters 4 to 7)

CHARTING PREHISTORIC TIME

1 Jurassic sedimentary layers of strata (the total for the period forming the Jurassic system) are exposed in a Norwegian fjord.

2 Nicolas Steno compared shark tooth fossils with the teeth of living sharks.

3 Erosive forces yielding results such as this Jurassic limestone, Torcal de Antequerea reserve, Spain, have been at work through billions of years.

Most civilizations have versions of what happened during the time before recorded history—prehistoric time—an age often populated by gods, spirits, and mythical beings. As scientific reasoning and questioning spread throughout Europe, from the 15th century, people queried these traditional accounts. The age-of-the-Earth debate prompted discussions among scholars and scientists, and gradually accumulating knowledge led toward a realization that Earth was far older than most people believed. With this understanding came the need to provide a timescale. Today, the most prevalent and accepted scientific version of prehistoric time is known as the International Chronostratigraphic Chart, which is part of an ongoing project organized by the International Commission on Stratigraphy, itself a major subcommittee of the International Union of Geological Sciences. The chart is a work in progress and it is updated regularly. It aims to incorporate the latest discoveries and apply them to many areas of knowledge around the world, especially geology, paleontology, and evolutionary science. Its largest time span is the eon (during which an eonothem of rocks is laid down), subdivided into eras (erathems), periods (systems; see image 1), epochs (series), and ages (stages).

For centuries, fossils of animals and plants were often regarded as natural, physical products of rocks. They "grew" without the need to involve life forms, in

KEY EVENTS

2,750 years ago	AD 50–100	325	1650	1669	1779
From the reigns of mythical kings, ancient Babylonian scholars estimate various dates for the beginning of the world, some as far back as 400,000 years.	Han Chinese scholars predict that the cosmos is destroyed and recreated in a cycle lasting approximately 23½ million years.	Ancient Roman historian Eusebius dates the creation of the Earth, using the times that legendary kings ruled, to about 30,000 years ago.	From the genealogies in the Bible's *Genesis*, Archbishop Ussher calculates that the Earth was created in October 4004 BCE.	Nicolas Steno formulates important geological principles, including the law of superposition and the principle of original horizontality.	Georges-Louis Leclerc (1707–88) makes a sphere to mimic the Earth, measures how fast it cools, and calculates that Earth is 75,000 years old.

much the same way as nodules (lumps) of flint, precious gems, branching veins of minerals, and nuggets of gold and silver. The word "fossil" is derived from Latin for "obtained by digging" and originally meant any object taken from the ground, with no reference to being the preserved remains of organisms. Other interpretations were that fossils fell from the sky, the moon, or the stars, or that they were placed in the rocks to show the power and grandeur of the gods, or to trick nonbelievers into exposing their views.

Danish scientist Nicolas Steno (1638–86) questioned how a fossil could be embedded within another lump of rock: the problem of "solids within solids." He saw the resemblance of fossils to the shapes of items from living things today, such as sharks' teeth (see image 2), and wondered if this was how such objects originated. Around the same time, other questioning minds were thinking along similar lines. In 1692, English naturalist John Ray (1627–1705) wrote: "These [fossils] were originally the shells and bones of living fishes and other animals bred in the sea." English free-thinking scientist-naturalist Robert Hooke (1635–1703) had a similar opinion: that fossilized objects such as ammonite shells and wood were of biological origin and had been petrified by mineral-rich waters. Steno expanded his study of fossils to include the rocks they occurred in, and how these formed and were laid down. He was a founder of the sub-branch of geology termed stratigraphy: the layering or stratification of rocks, especially sedimentary and volcanic rocks. In the 1660s, Steno devised three aspects of stratigraphy that, like many great advances in science, seem obvious today but were innovative at the time. First, the law of superposition says that in a series of layers or strata, the oldest are deepest, overlain by younger and younger strata, forming a sequence in time. Second, the principle of original horizontality states that when strata formed, they settled under gravity so they were flat and horizontal; any change in angle, tilt, or folding since is the result of earth movements. Third, the principle of lateral continuity describes how rock layers extended in every direction when they formed; if similar layers are now separated by eroded valleys, split by earthquakes, or divided by volcanic force, it can be assumed that they were once continuous.

In the 1780s, Scottish geologist-scientist James Hutton (1726–97) developed the principle of uniformitarianism. This proposed that processes and events seen today in nature, such as erosion by wind (see image 3), ice, and waves, and sediment layers being consolidated into rocks, have always occurred, and that this shaped Earth through the ages: "The present is key to the past." In the 1830s, British geologist Charles Lyell (1797–1875; see p.233) introduced uniformitarianism to a wider audience in his work *Principles of Geology: Being an Attempt to Explain the Former Changes of the Earth's Surface, by Reference to Causes Now in Operation*. This proposal demanded a much greater age for Earth than estimates of the time, which ranged from a few thousand to a few million years. Hugely influential French naturalist Georges Cuvier (1769–1832) had meanwhile opened

1860s	1900	1907	1920s	1974	2006
William Thomson (1824–1907) calculates the age of Earth to be between 20 million and 400 million years; he later refines this to 40 million years.	Most scientists accept the physicists' estimates that the Earth is around 100 million years old; some geologists urge a much greater age.	Bertram Boltwood (1870–1927) publishes studies on minerals dated by radioactive decay, such as uranium to lead.	More radiometric dating techniques show the age of the Earth is measured as a few billion years, rather than millions or trillions.	The International Commission on Stratigraphy begins its long-term project to produce a global geological timescale.	The Tertiary period is replaced by the Paleogene and Neogene, necessitating adjustments to the Quaternary period.

the way to scientific acceptance of extinction, which had been denied by most followers of the church. In 1813, his *Essay on the Theory of the Earth* proposed that fossils were indeed the remains of long-gone creatures, but that the causes of their disappearance were catastrophes of biblical proportions, especially floods; there was no need for any kind of evolution. Gradually, the concepts developed of geological time being extremely long, of fossils being preserved remains of living things through time, and of this immense time span of prehistory needing some form of calendar or timetable.

Furthermore, around this time, European geologists were noting and naming distinctive rock layers and their characteristic fossils. Their work was linked to great construction projects of the industrial age, such as canal building, mining and quarrying for coal and minerals, excavating wells for water, and from the 1830s, the fast-expanding new railroads. In 1822, English geologist-clergyman William Conybeare (1787–1857) and English mineralogist William Phillips (1775–1828) published *Outlines of the Geology of England and Wales,* in which they introduced the term "Carboniferous," referring to the thick layers of coal-bearing rocks (from the Latin *carbo,* meaning "coal"). Carboniferous was the first of the names now used for time spans called periods, which are the main working units of the geological timescale. In the same year Belgian geologist Jean d'Omalius d'Halloy (1783–1875; see image 5) labeled distinctive strata from an area of Paris as "Cretaceous" (from the Latin *creta,* meaning "chalk"). The term "Jurassic" was coined a few years later in 1829 by French geologist-chemist Alexandre Brogniart (1770–1847) for extensive limestones of the Jura Mountains in France and Switzerland.

In the early 1830s, English geologist Adam Sedgwick (1785–1873)—who in 1831 was field geology tutor to the young Charles Darwin (1809–82)—and Scottish geologist Roderick Murchison (1792–1871) worked in Wales. Murchison recorded rocks with fossils that consisted of few fish but many kinds of trilobites and brachiopods. He chose the name "Silurian" from the Silures, a Celtic tribe that occupied the area in the time of the ancient Romans. Sedgwick recognized a separate system of rocks and fossils in the middle of Wales laid down before the Silurian, and named it "Cambrian" (*Cambria,* meaning "Wales"). He and Murchison formalized these two names in a scientific paper in 1835. Meanwhile, the term "Triassic" had been introduced in 1834 by German geologist Friedrich Von Alberti (1795–1878) for three distinctive layers occurring together through northwest Europe (*trias,* meaning "three"): red beds of sandstone, with chalk above, capped by black shales. A few years later, in 1839, Sedgwick and Murchison proposed the name "Devonian" for a system of rock types and fossils that had been studied in the county of Devon, southwest England, and the following year Murchison defined and named the "Permian" using rock strata from Perm, in Russia. However, it was not until 1879 that English geologist Charles Lapworth (1842–1920) devised the name "Ordovician," from the Celtic tribe Ordovices. This settled problems with rock layers in North Wales, which some experts were including in the preceding Cambrian while others were assigning to the following Silurian.

The name "Tertiary" dates back to 1759 when Italian geologist Giovanni Arduino (1714–95), studying the rocks of Tuscany, suggested that prehistoric time should be divided into four great spans: Primitive or Primary, Secondary, Tertiary, and Volcanic or Quaternary. In 1828, Lyell (see image 4) included the term "Tertiary" in his new chronology and subdivided it into Pliocene, Miocene, and Eocene—terms now given the rank of epochs—based on the changing or evolving fossils. During these pioneering times, there were many overlaps and alternatives for the various names, and it took several decades before the basic terminology was agreed. For example, in the United States, distinctive rock systems led to the use of "Mississippian" and "Pennsylvanian" for what Europeans regarded as the lower or early, and upper or late, subdivisions of the Carboniferous period.

The International Chronostratigraphic Chart is itself an evolving entity. From the 1980s doubts arose that the Tertiary could be defined in the same way, or as accurately, as other periods. In the 2000s, groups of geologists and scientists proposed that the name "Tertiary" be discontinued and replaced by two new names: Paleogene and Neogene. This caused problems with the Quaternary, a name that dates back to Arduino's proposal in 1759. The Quaternary was demoted to an informal status; however, with increasing rock and fossil data from around the world, it was redefined in 2009 as the period following the Neogene, beginning around 2.58 million years ago rather than the former 1.8 million years ago and extending to the present.

Expanding Steno's basic principles of stratigraphy, the geological periods and their order in time were broadly agreed by about 1900 to 1920. This allowed more secure geological mapping (see image 6): comparing and correlating distinctive sets of rocks and their fossils, especially using index fossils (see p.159). However, what were the actual or absolute dates, in millions of years ago, for these periods? Methods included estimating how long a particular depth of sediment took to settle and consolidate into rock. More accurate methods began in about 1907 with the advent of radiometric dating (also known as radiodating or radioactive dating; see p.25). Recent advances in this technique have allowed further refinements for prehistoric time. In 2004, the first new period for more than 120 years was added: the Ediacaran. Lasting from about 635 million to 541 million years ago, it is named from the fossils of the Ediacara Hills, Australia (see p.46). More recently, the Precambrian super-eon has come more into focus, with a series of periods named in the Proterozoic eon (see p.20). **SP**

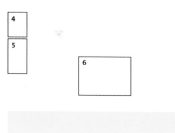

4 Charles Lyell championed uniformitarianism, a concept that has been applied widely, even to the evolution of the universe.

5 Jean d'Omalius d'Halloy mapped Cretaceous rocks in France. He was also noted by Charles Darwin as writing on "descent with modification," an early term pertaining to evolution.

6 Geological maps have progressed considerably since the early 19th century, but they still use shading, symbols, or color-coding to denote rocks of various ages.

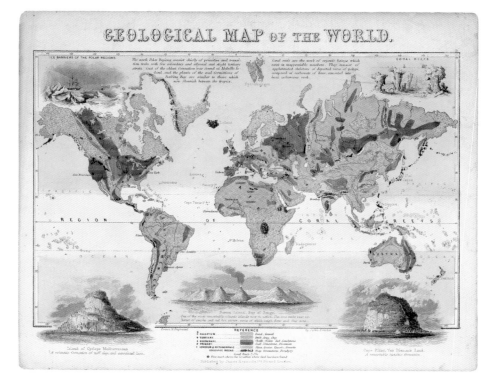

How Fossils Form
ARCHEAN EON—HOLOCENE EPOCH

SPECIES *Paraceraurus exsul*
GROUP Phacopida
SIZE 6¼ inches (16 cm)
LOCATION Russia

Fossils such as this intricate trilobite shell are not only fascinating and beautiful, but also provide inestimable evidence for how life on Earth has changed or evolved over time. Although a fossil is the remains of a living thing that has been preserved, usually by incorporation into rock, the once-living matter, such as animal bone tissue or tree bark material, is usually replaced by rock-forming minerals. This means that the original bone or bark substances have gone, but their shapes remain. Fossilization takes a long time, generally millions of years, and the conventional younger limit is 10,000 years.

Most fossils are of tough, resistant parts: plant roots, bark, cones, seeds, leaf veins and tiny pollen grains, as well as animal shells, teeth, bones, horns, claws, and spines. These parts tend to last longer against scavengers feeding on dead remains, and against worms, fungi, and microbes causing decay, thereby increasing chances of preservation. Furthermore, most fossils form in watery environments: seas, lakes, rivers, and swamps. The dead parts need to be covered by particles or sediments, such as sand, silt, mud, or clay, both to protect them from further decay and disintegration, and to begin the long fossilization process. This is why aquatic shelled animals, such as trilobites and ammonites, feature so regularly in the fossil record. As further sediments build up on top, pressures and temperatures increase, and gradually the particles and minerals in the fossil and its surroundings (matrix) are consolidated and cemented into solid rock. Consequently, fossils are found in sedimentary rocks, such as sandstone, limestone, mudstone, and shale. Igneous rocks are those that are heated so much they become molten, as with lava from volcanoes; metamorphic rocks are also greatly altered by heat and pressure, although not enough to become molten. In both cases, such conditions nearly always destroy any preexisting fossils. **SP**

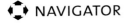

✪ NAVIGATOR

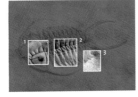

⊚ FOCAL POINTS

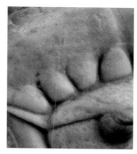

1 REPLACEMENT
Natural minerals in the once-living parts may persist, grow or be replaced by molecules of the same mineral. Growing in stages, a trilobite shed or molted its hard shell, then increased in size before the new shell beneath hardened. Many trilobite fossils are cast-off shells rather than the entire animal.

2 PERMINERALIZATION
The joints between the skeletal plates can be seen here. This is because groundwater seeped through buried parts carrying minerals that filled the air or watery spaces within. This is known as permineralization. When it happens slowly and evenly, it can conserve the shapes of microscopic parts, such as cells and their contents.

3 MATRIX
Nonfossil parts—the minerals and rock around the actual fossil—are known as the matrix. Since it may contain the same minerals as the fossil, the two can look very similar in color and texture, and so exposing the fossil itself is an extremely delicate task. This *Paraceraurus exsul* trilobite is dated to the Ordovician period.

FOSSIL TYPES

Although all fossils are evidence of once-living matter, there are several different types of fossil. Body fossils, known often simply as "fossils," are those of actual body parts. Index fossils are of common, widespread organisms that change rapidly through time, acting as indicators of evolutionary time. Trace fossils are signs or remnants left by the organism, such as footprints and coprolites (fossilized excrement), as well as bark imprints and root tunnels. Mold fossils are the empty spaces where the once-living remains have been destroyed, whereas cast fossils are mold fossils that fill with another set of minerals. Carbon film fossils are when the living matter is heated and pressurized so much that most of it is destroyed. Only its carbon atoms remain, like a dark silhouette (above).

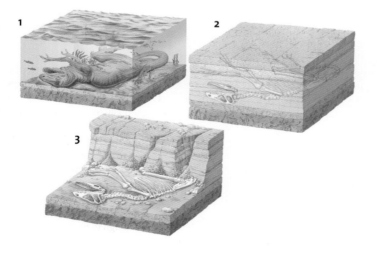

◄ Minimal scavenging or decay after death (1), followed by a swift covering with sediment, is a recipe for an organism to be fossilized in prime condition. Gradually, more sediments settle on top, burying the remains deeper (2). Rising temperature and pressure, and the slow percolation of mineral-containing groundwater, continue the process. Eventually, great earth movements and erosive forces may expose the fossils at the surface (3).

DATING ROCKS AND FOSSILS

1 The potassium–argon dating process, conducted using an electronic furnace, is one means of dating rocks by analysing their naturally occurring radioactive isotopes, created by cosmic rays.

2 Henri Becquerel was awarded the Nobel Prize in Physics in 1903 for his discovery of spontaneous radiation, shared with Pierre and Marie Curie.

3 During the process of radioactive decay, the nucleus of a uranium atom splits into smaller isotopes.

For centuries, traditions told of how digging down into the ground was like going back in time: the deeper you go, the older the rocks. Danish geologist and anatomist Nicolas Steno (1638–86) established this in a scientific way with his law of superposition: in a series of layers or strata, the oldest are the deepest and the youngest are uppermost (see p.24–5). Geologists and paleontologists began to understand how the presence of certain rock types containing specific fossils—especially common, widespread, and fast-changing ones, known as index fossils (see p.159)—could be used to determine the relative or comparative ages of the rocks and other fossils they contain.

One of the first to correlate such information across a large area was English geologist William Smith (1769–1839). As a surveyor for mining and canal construction projects, he noted similar rocks and especially their fossils across widespread regions. In 1815 he produced a geological map of England, Wales, and parts of Scotland—the first such map to cover a nation. Before Smith, in 1809, Scottish-American geologist William Maclure (1763–1840) had published a similar map for parts of east and southeast North America.

Soon after Smith, more maps followed, especially in France, Germany, and around Europe. Using such information, relative or comparative dating became

KEY EVENTS

1809	1815	1896	1904–05	1907	1927
William Maclure publishes *Observations on the Geology of the United States, Explanatory of a Geological Map.*	William Smith creates a geological map of England, Wales, and parts of Scotland, with color-coded areas for rocks from various prehistoric times.	Henri Becquerel reports on the effects of "spontaneous penetrating radiation" (radioactivity) given off by uranium-containing chemicals.	Ernest Rutherford suggests radioactive decay can be used to estimate the age of Earth, which is presumed to be 40 million years or less.	Bertram Boltwood publishes his estimates for the age of Earth at up to 2 billion years, based on uranium–lead radiometric dating.	English geologist Arthur Holmes (1890–1965) revises the latest in his series of uranium–lead dates to indicate that Earth is 3 billion years old.

established during the 19th century. However, there was still no way of knowing when a fossil formed in terms of thousands or millions of years ago: its absolute age or date. The arrival of absolute dating came from a flurry of scientific activity at the end of the 19th century. In 1896, French physicist Henri Becquerel (1852–1908; see image 2) discovered radioactivity, which led to research on the nature of matter and atoms, and why some atoms are inherently unstable. They give off radioactivity in the form of rays and particles, and as they do, they naturally change or decay from one chemical element (pure substance) to another. For example, the heavy, dense, metallic element uranium spontaneously emits radioactivity and decays, step by step, through a series of several other elements to a stable end result: lead. In a sample of a uranium-containing mineral, such as zircon, half of the form of uranium called U238 decays into the form of lead known as Pb206 over a time of 4.47 billion years. Half of another form, U235, decays into lead Pb207 over 704 million years. These times are known as the half-lives of the uranium–lead decay series (see image 3). They represent a "ticking clock" from which age can be established. By measuring the kinds and proportions of uranium, lead, and other elements in a mineral sample, the mineral's original formation time can be calculated.

Around 1905, New Zealand-British physicist Ernest Rutherford (1871–1937), of "splitting the atom" fame, proposed that radioactive decay might be used to determine the age of Earth. In 1907 one of his students, American chemist Bertram Boltwood (1870–1927), published his uranium–lead studies, which suggested that Earth was between 400 million and 2 billion years old. This was a far greater age than the tens of millions of years popularly believed at that time.

Uranium–lead radioactive or radiometric dating (see image 1) was the first absolute method to be developed and it is still widely used. Its modern techniques work best for ages between 4.5 billion and 1 million years, often to accuracies of within 1 percent. Much shorter timescales are possible with radiocarbon dating, in which one form of carbon, C14, decays into another, C12. Carbon is an essential substance in living matter—for example, being taken in by plants, in the form of carbon dioxide from air, for photosynthesis. So radiocarbon dating is especially suitable for once-living things, from wood and bones to fur, antlers, and shells.

Depending on the minerals in a fossil or its rock, radiometric dating may not be possible. In fact, with fossils and the sedimentary rocks in which they form, it is rarely possible. However, dating the strata or layers just below and/or above indicates the fossil's age range—especially in igneous rocks, which cool from molten lava and "trap" the minerals, thereby giving their start date. A combination of relative or comparative dating with radiometric methods for absolute dating, and also molecular clock techniques (see p.17), is central to our modern understanding of what lived when, how fast organisms change through time, and how evolution proceeds. **SP**

RADIOACTIVE AGE DATING OF ROCKS

- ● URANIUM (U) unstable, parent atoms
- ● LEAD (Pb) stable, daughter atoms

Radioactive Decay → Parent Atoms / Daughter Atoms

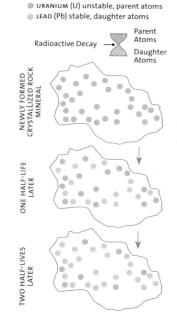

NEWLY FORMED CRYSTALLIZED ROCK MINERAL

ONE HALF-LIFE LATER

TWO HALF-LIVES LATER

1938	1948	1949	1953	1992–94	2003
A group of German chemists develop rubidium–strontium dating; the form Rb87 of rubidium has an immense half-life of 48¾ billion years.	The potassium–argon decay series is discovered, rapidly leading to the development of potassium–argon radiometric dating.	At the University of Chicago, Willard Libby (1908–80) and his team develop the method of radiocarbon dating, bringing huge progress in archeology.	Uranium–lead dating of Canyon Diablo meteorite fragments in Arizona give Earth's age at 4.5 billion years, close to the modern value.	University of California scientists argon–argon date Mojokerto child Homo erectus at 1¼ million years, far older than previously thought.	Argon–argon dates for Mojokerto child are revised to 1½ million years, highlighting the need for careful sample recording and selection.

The Grand Canyon

PALEOPROTEROZOIC ERA—EARLY TRIASSIC PERIOD

LOCATION Arizona, United States

AGE 1.85 billion to 250 million years

GEOLOGY Limestone and sandstone

One of the world's premier natural wonders, the Grand Canyon in Arizona is 275 miles (440 km) long, more than 15 miles (25 km) wide in places, and up to 6,000 feet (1,830 m) deep. Its eroded near-vertical surfaces record rock formations from more than one third of Earth's entire history. Many of these rocks are sedimentary in nature and contain myriad fossils, thereby tracking evolution of plants and animals across huge spans of time. Even after so long, many of the canyon's rock layers are still almost horizontal and undamaged due to a lack of great earth movements in the area.

Grand Canyon rocks and fossils have been dated by both relative and absolute means. The oldest go back more than 1.8 billion years to the Paleoproterozoic era. Until the 21st century the conventional view was that the Grand Canyon was worn away or eroded, chiefly between 6 million and 1 million years ago, to expose three main sets of rocks. Deepest and therefore earliest, at the bottom of the canyon, are the Vishnu Basement Rocks, dating from around 1.85 billion to 1.25 billion years ago; on top at certain sites are Grand Canyon Supergroup rocks, from 1.25 billion to as young as 650 million years ago; and above these, formed most recently, are Layered Paleozoic Rocks. In 2008, a radiometric dating study, using the method known as thermochronology, indicated that the canyon was eroded much earlier, perhaps more than 50 million years ago. Thermochronology looks at the mineral apatite and how its crystals record radioactive decay of uranium and thorium once exposed by erosion. A report in 2012 using the same technique pushed back the canyon's erosion to possibly 70 million years ago, meaning that great dinosaurs could have viewed its early stages. Another study in 2014 proposed a more complex history with most of the Grand Canyon's erosion taking place in the past few million years, having widened and deepened several preexisting parts that date back 70 million years. **SP**

FOCAL POINTS

1 AGENT OF EROSION

Most of the river cutting that formed the present canyon occurred between about 6 million and 1 million years ago. This happened as the land around was uplifted by earth movements. Also, the climate at that time was much wetter and fed large amounts of water into the ancient Colorado River and surrounding river systems. The desertlike conditions that persist today have prevented plant growth from obscuring the rock layers.

3 LAYERED PALEOZOIC ROCKS

Many of these are sedimentary, laid down in ancient seas, and they bear all kinds of fossils. For example, the Tapeats Sandstone from 520 million years ago contains fossils of trilobites and brachiopods (lampshells). The Redwall Limestone, 340 million years old and up to 800 feet (245 m) deep, preserves fossils of bigger, more evolved trilobites and also more brachiopods, as well as sponges, corals, nautiloids, and crinoids.

2 VISHNU BASEMENT ROCKS

Lowest and therefore earliest, the Vishnu Basement Rocks are absolute-dated to between 1.84 billion and 1.25 billion years old. Rather then being sedimentary in origin, they are mainly igneous and metamorphic so they contain no significant fossils. They consist of various rock and mineral types, including granite, pegmatite, gneiss, and diorite, and because of their great age they show the most deformity, such as folds and displacements.

4 RECENT FORMATIONS

The Toroweap Formation, 275 million to 272 million years old, has sandstone and limestone bearing fossils that include marine corals and mollusks, as well as terrestrial plants. This indicates that an ancient sea spread and then shrank through time. Above this is the Kaibab Limestone Formation, dating to around 270 million years. It is rich in fossils ranging from corals, trilobites, shrimp, and large nautiloids to conodonts and sharks' teeth.

GAPS IN TIME

The Grand Canyon's rocks, like those in numerous other places, have unconformities or gaps in their geological strata. These are "missing" layers or strata, where much younger rocks overlie much older ones, with no in-between layers. For example, absolute dating reveals that in some areas of the Grand Canyon, the Tapeats Sandstone, which is around 520 million years old, sits directly on Vishnu Basement Rocks that are more than 1.25 billion years older (left). At other sites, the Tapeats Sandstone is in contact with Grand Canyon Supergroup rocks where the time difference is less than 250 million years. These unconformities have resulted from the "missing" rocks being eroded away, then younger layers forming directly on older ones. The resulting breaks or gaps in the fossil record can be misleading—for example, suggesting that a great extinction has taken place followed by the sudden appearance of many new life forms.

EARLY EARTH: ORIGINS OF LIFE

1 Artistic impressions of early Earth include hot rocks and water, volcanoes, storms, meteorites, and the young moon.

2 Evidence of the Late Heavy Bombardment includes the moon's craters and similar structures on other planets and moons.

3 Quartz mineral veins grow and even multiply in rock, but lack other features of life.

Cosmology, astrophysics, planetary sciences, the relatively new field of paleobiology, and many other branches of science come together to chart the earliest phases of planet Earth. However, the details of how life began, and how it initially proceeded—evolved—from its presumably small and simple origins, are much less clear.

Earth was formed around the same time as the rest of the solar system: other planets, their moons, comets, asteroids, meteoroids, and in the center, the star we call the sun. Various radiometric dating methods (see p.31) and calculations put Earth's origin at around 4.5 bmillion years ago. It coalesced from swirling dust, gases, and similar matter, growing or accreting (clumping) under their increasing collective gravity. The main formation stage took perhaps 10 million to 15 million years, and this earliest Earth was unrecognizable from the planet today. It was as small as half of its present size and it spun around much faster, in a period of only a few hours rather than today's 24. Its surface churned with volcanic activity (see image 1), heat, and fierce jets of gases. It is theorized that soon after its formation, 4.53 billion years ago, a Mars-sized planetary body, 10 percent of Earth's mass, struck a glancing blow. Referred to as Theia, some of its matter contributed to Earth, while the rest was shattered into debris that began to orbit the Earth; this gradually coalesced under its own

KEY EVENTS

4.54 Bya	4.53 Bya	4.44 Bya	4.10 Bya	4.10–3.80 Bya	4 Bya
Earth forms as part of the solar system. It is very hot with molten rock, volcanic activity, and bombardment by other bodies orbiting the sun.	A Mars-sized body, Theia, hits Earth and disintegrates. Earth's gravity traps the debris to form the moon. Day–night time is 3 hours.	The Earth cools enough to form its first thin outer rocky crust as heavier matter such as iron and nickel sinks to form the core.	The oldest true rocks on Earth's surface date to this time. Some minerals, such as zircon crystals, are thought to predate this at 4.4 billion years ago.	The Late Heavy Bombardment sees asteroids smashing into Earth. It is wracked by craters and its crust is destroyed, but it gains mass.	The first chronological/ geological eon, the Hadean, which began when Earth first formed, gives way to the second, the Archean eon.

gravity to form the moon. In turn, the moon's gravity began to exert a dragging effect on Earth's rotation, thereby lengthening the day–night period (an effect that continues today). At this time, the sun was also in its infancy and only about two thirds of its present brightness. There was no hope of survival for any life as we know it.

Gradually Earth's outer layer (crust) cooled and became more solid. However, around 4.1 billion years ago the Late Heavy Bombardment began. This involved frequent impacts from swarms of asteroids, meteorites, comets, and similar bodies that smashed into Earth; it continued for 300 million years. Evidence of such impacts on the Earth is long gone, but the moon's craters (see image 2) represent some of the scars, and analysis of lunar rocks termed "impact melts"—returned by US Apollo spacecraft missions—provide the time frame. The Late Heavy Bombardment added mass to the Earth and also wracked its new crust. As the impacts faded, the crust reformed and the geological features that are familiar today slowly came into being, with the jigsawlike tectonic plates forming the ocean floors and carrying the continents.

Did life and its very earliest stages of evolution happen before the Late Heavy Bombardment, only to be wiped out by that series of cataclysmic pummelings and then begin again? Or did life forms somehow manage to survive the bombardment and flourish thereafter? Alternatively, did life first arise after the Late Heavy Bombardment, from 3.8 billion years ago? At present, the evidence is inconclusive. However, such uncertainties bring into focus a related question: what is life?

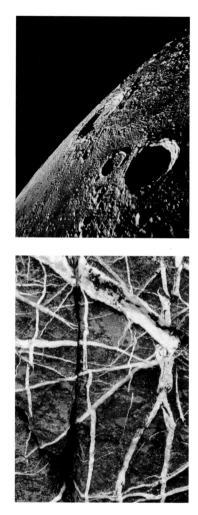

Living things—life forms or organisms—exhibit certain key features. One involves chemical changes, with atoms and molecules being shuffled around, combined, broken apart, and recombined in different configurations. To drive these changes, living things require energy; usually this comes from the sun's light energy through the process of photosynthesis in plants, and from the breakdown of energy-rich food by digestion in animals. Such chemical processing, or metabolism, produces byproducts or waste for removal, which is another of life's features. Furthermore, most organisms grow in size, by incorporating more nonliving materials into their living structures. Organisms perform all of these activities according to chemically coded instructions in their genetic material, usually DNA (deoxyribonucleic acid). Vitally, living things reproduce, or self-replicate, making more of their kind by copying and passing on their genetic material to their offspring. Over time, life forms evolve by changes or mutations in their genetic material. This suite of activities differentiates living organisms from nonliving processes that have some of the features, but not all. For example, natural mineral crystals (see image 3) "grow" by arranging more atoms and molecules onto themselves and they can also "reproduce" by seeding the growth of new, separate crystals.

3.70 Bya	3.50 Bya	3.48 Bya	2.30 Bya	2–1.5 Bya	720–600 Mya
Possible signs of life exist in rocks as two forms of carbon, C13 and C12, whose relative amounts indicated "processing" by some kind of living activity.	The oldest fossils of single-celled organisms date from this time.	The crust has re-formed and signs of life occur, as evidenced by possible fossils of microbial activity from Western Australia.	The Great Oxygenation Event sees oxygen beginning to accumulate in the atmosphere, making life conditions more favorable (see p.39).	Multicellular life forms evolve from single living cells joined together.	"Snowball (Slushball) Earth" sees the longest, most extensive cold periods the planet will ever experience. Day–night time is 20 hours.

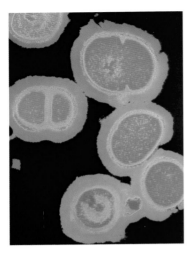

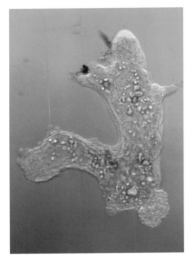

The features of life are easy to identify in plants, animals, fungi, and other familiar organisms. However, at the microscopic and molecular scales, where life began, matters are more complicated. A single cell such as a bacterium (see image 4) and the oozing blob that is an amoeba (see image 5) both display all of life's key features. Yet inside, these two are very different. The amoeba has within it separate structures called organelles, including genetic material in a nucleus or control center, and also energy-processing centers called mitochondria and digesting units known as lysosomes. Each of these structures is wrapped like a parcel in a sheetlike membrane, and the entire amoeba has a similar wrapping—the cell membrane—as its "skin." The bacterium also has a cell membrane, but inside, and it lacks membrane-wrapped organelles. Its molecules of life, including its genetic material, seem to float freely as if in a watery "soup." Furthermore, the bacterium's genetic material is organized more simply, usually as a loop or ring, rather than in separate packages known as chromosomes, as found in the amoeba.

The distinctions of lacking or possessing membrane-bound organelles, the configuration of the genetic material and the general organization of the cell interior constitute one of the most basic differences in the whole kingdom of life. It signifies an early, simple stage of life's evolution compared to a later, more sophisticated stage. Bacterial cells are known as prokaryotic, whereas amoebas and almost all other organisms—single- or multicelled, plant or animal—are called eukaryotic. The evolutionary processes by which organelles formed inside eukaryotic cells may have included some prokaryotic cells incorporating other prokaryotic cells inside them. This is termed endosymbiosis.

In the late 1970s the distinction between prokaryotes and eukaryotes, and their evolutionary histories, was amended by growing evidence for two even more basic kinds of prokaryotes. These are the life domains (fundamental groups) Bacteria and Archaea (see image 6)—the third domain is Eukaryota. Bacteria and Archaea are distinguished by differences in a substance called rRNA (ribosomal ribonucleic acid), which, in a cell, assembles subunits into proteins. Along with nucleic acids DNA and rRNA, proteins are a major class of molecules in living things. They fulfill a variety of functions, such as forming structural frameworks and (as enzymes) controlling chemical reactions. Early studies of Archaea (ancient things) showed that many were extremophiles (see p.40): living in extreme conditions, such as heat, cold, or salinity. However, many other kinds have since been discovered.

It is probable that bacteria-like organisms arose as an early stage of life on Earth. The development of simple life from nonliving material—termed abiogenesis—represents the previous stage. Conditions on the young Earth were very different to those of today. With early volcanic activity and gas emissions, the atmosphere probably contained more hydrogen than at present, less oxygen, more water vapor, and much more carbon dioxide, hydrogen sulfide, carbon monoxide, and methane—these last four are toxic to most present-day life forms. Water condensed from gas emissions and, perhaps brought by space bodies such as comets, formed the early and hot, even boiling oceans; atmospheric gases dissolved in these to make the water more acidic. In such a "primordial soup," akin to a gigantic chemical laboratory for random reactions, many kinds of atoms and molecules floated, joined, fell apart, and rejoined, countless times. Their activities were stimulated by energy, such as lightning from huge and frequent storms, by the geothermal heat from within the planet, and also by the sun's fierce light, ultraviolet, and other rays, because the early atmosphere did not shield the surface from these as it does today.

There are many theories as to how the first scraps of living matter assembled themselves and harnessed energy sources to incorporate raw materials and self-replicate. These are as much the provinces of organic chemistry, biochemistry, and molecular biology as of evolutionary biology. The favored early contender for "first life form" is RNA, which is similar to DNA in that it is a repeating chain of simple subunits (monomers). The order of chemicals called bases in these monomers works as an information-carrying system to build more RNA of the same pattern. This scenario is known as the "RNA world hypothesis"; gradually, proteins, membranes, and other complexities were added, as DNA took over from RNA. Somewhere in the process—or processes, these changes having happened many times in different ways—was the Last Universal Ancestor, or LUA (Last Universal Common Ancestor, LUCA). This was the most recent living thing to which all organisms on Earth can trace back their ancestry.

A very different suggestion is that life, or its chief building blocks, was delivered to Earth by space bodies such as comets, asteroids, and meteorites. Studies of extremophile bacteria show that certain kinds could perhaps have survived the harshest conditions of deep space. This idea moves the question of the origin of life to elsewhere in the cosmos, and also implies that life could be seeded onto many other space bodies with suitable conditions, as it was on Earth.

In the late 1970s, the manned submersible *Alvin*, exploring near the Galapagos Islands in the Pacific Ocean, discovered deep-sea hydrothermal vents. Many more have since been located. At these sites, super-heated water, sometimes exceeding 752°F (400°C), laden with dissolved minerals, gushes through seabed cracks or vents from deep in the Earth. As it meets the cold ocean, at about 35°F (2°C), the dissolved minerals become insoluble particles and cloud the water, thus giving the name "black smokers" (see image 7). Organisms here, called chemosynthetic bacteria, feed on the energy in the minerals and form the basis of food chains, supporting worms, shellfish, octopus, crabs, shrimp, fish, and other life. The vents represent living communities that, unlike nearly all others on Earth, are independent of the sun for their primary energy. It has been suggested that deep-sea chemosynthetic bacteria could represent how early bacterial cells first evolved on ancient Earth, because near the ocean surface—where sunlight was bright enough for photosynthesis—the sun's accompanying strong ultraviolet radiation would have had a destructive effect. **SP**

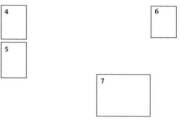

4 Prokaryote: typical bacteria—here *Staphylococcus*, some dividing in half—have a skinlike outer membrane but no membranes inside.

5 Eukaryote: a giant amoeba *Chaos* contains varied internal structures (organelles) each wrapped in a membrane.

6 Archean: the extremophile *Pyrococcus* is a thermophile, perishing at less than 158°F (70°C).

7 More than 10,000 feet (3,050 m) deep in the Atlantic Ocean, the Saracen's Head vent emits clouds of sulfur-rich minerals. Early life on Earth may have colonized such features.

The First Fossils
ARCHEAN EON

LOCATION Pilbara, Western Australia
AGE 3.48 billion years
GROUPS Bacteria, Archaea
BED DEPTH Up to 3¼ feet (1 m)

As evidence of the earliest life, the fossil record is patchy and dictated by chance. Several major factors reduced the likelihood of the earliest organisms being preserved. They were small, soft, and lived long ago, and from growing, self-replicating fragments of RNA they developed into single-celled prokaryotes resembling today's bacteria and cyanobacteria (see p.56). About 2,500 prokaryotes in a row would span 1 inch (25 mm), and more than 20,000 individuals would fit inside this printed "o."

Even so, possible fossils from 3.7 billion years ago offer a few tantalizing glimpses of the earliest life forms. A "signature" in the schist rocks of southwest Greenland could signify the earliest life. It is a substance called biogenic graphite, which is a form of the chemical element carbon. Biogenic graphite has a particular profile of the atoms of carbon called C13 and C12, which differentiates it from graphite made abiotically (by natural mineral processes that do not involve life). The Greenland schists, from the Isua Greenstone Belt, bear graphite with proportions of C13 and C12 that suggest past biological activity. The organism that left these biogenic graphite markers is unknown, but the consensus is that it was a tiny, simple, cyanobacterial-like life form.

Sandstone in Western Australia, formed 200 million years later (3.48 billion years ago) provides further evidence for life, in the form of microbially induced sedimentary structures (MISSs). Indications are that the mats of MISS-forming micro-organisms functioned like, or were, cyanobacteria, using sunlight energy trapped by photosynthesis as fuel, and emitting wastes including gas hydrogen sulfide (which smells of rotten eggs). The "mats" would have looked like slimy masses of green, purple, and brown fibers. Fossils found in South Africa, about 2.9 billion years old, contain evidence of similar structures, but they also suggest that, by 2–1.5 billion years ago, the simple single-celled prokaryotes were becoming more complex eukaryote cells, before joining together and cooperating as the earliest multicelled organisms. **SP**

◉ FOCAL POINTS

DRESSER FORMATION FOSSILS
The Dresser fossils from Pilbara, Western Australia, are estimated to be 3.48 billion years old. Thousands of single-celled marine organisms lived as microbial mats, similar to stromatolites. Surface features called polygonal oscillation cracks provide evidence that MISS growth patterns were affected by floods, droughts, and other changes in the environmental conditions.

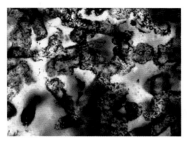

FOSSILIZED CELLS
Nestling between sand grains that were perhaps once the bottom of a shallow lagoon, the Pilbara sandstone bears possible fossils of tube-shaped bacteria-like organisms. The tubes measure around ½,₅₀₀ inch (10 μm) in diameter. At the time there would have been little oxygen in the atmosphere. The organisms would have lived communally in shallow pools.

MICROBIAL MATS
The manner of growth of the Dresser organisms has been likened to microbial mats or biofilms: thin layers of living cells mixed with inorganic particles. Such formations exist today on tidal flats and lagoons, shallow marine locations, and at playas and sabkas, which are shallow, ephemeral, mineral-rich lakes that go through cycles of flooding and drying.

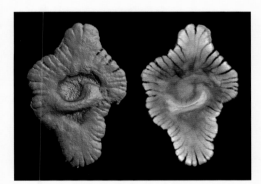

EARTH'S EVOLVING ATMOSPHERE

The gaseous make-up of early Earth's atmosphere would be deadly to today's complex life. However, over millions of years, ancient cyanobacteria produced oxygen via photosynthesis and gradually changed its composition. Around 2.3 billion years ago the Great Oxygenation Event occurred—oxygen showed a marked increase in concentration, providing favorable conditions for more varied kinds of life. Recent discoveries from just after this time are "wrinkled cookie" fossils from black shale in Gabon, West Africa, 2.1 billion years old (above). Named *Grypania*, and up to 4 inches (10 cm) long, they could be large associations of single-celled organisms. Alternatively, detailed microscanning studies imply they could be larger multicelled life forms, each with many cooperating cells of different kinds doing their own tasks for the benefit of the entire organism.

▲ Living Archaea, such as these methane-producing *Methanosarcina*, are probably little changed from archaeans of 2,000 million years ago. *Methanosarcina* would have suited early Earth's methane-rich atmosphere; it is harmed by the levels of oxygen in today's atmosphere. It can be found away from fresh air, in animal guts, sewage, landfills, and deep-sea vents.

Extremophiles
PROTEROZOIC EON—PRESENT

Most living things survive within a range of conditions: temperature, moisture, concentration of chemical substances (oxygen and carbon dioxide), nutrient availability, and acidity or alkalinity (pH). Temperature, for example, ranges from 34°F to 122°F (1°C–50°C); just above freezing point to the hottest tropics. However, certain organisms thrive outside of these ranges, in extreme conditions. These are extremophiles, and studying the ways that they have adapted their chemical processes to handle their excessive environments can reveal how early organisms survived on Earth. Thermophiles flourish in high temperatures, conventionally from 140°F to 248°F (60°C–120°C), and live in hot springs, geysers, and deep-sea geothermal vents. Cyrophiles (also known as psychrophiles) prefer temperatures as low as -4°F (-20°C) in polar regions; xerophiles inhabit the driest places; halophiles need high salt concentrations such as salt lakes; and oligophiles (oligotrophs) survive with minimal nutrients. Extremophiles are mostly single-celled microbes and many belong to the Archaea group. Examples are the Archaeal Richmond Mine Acidophilic Nano-organisms (ARMANs), which were identified from acidic water draining from Richmond Iron Mountain Mine in California. The water's pH was 1.5 or less, comparable to vehicle battery acid. Conditions such as this possibly existed in the mineral-choked waters of early Earth. The ARMAN organisms are among the smallest detected, hundreds of times smaller than typical bacteria. **SP**

◈ NAVIGATOR

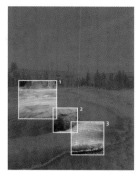

GENERA *Sulfolobus, Pyrococcus* and others
GROUPS Archaea, Bacteria, Algae
SIZE ½50 inch (0.1 mm)
LOCATION Yellowstone National Park, Wyoming

◉ FOCAL POINTS

1 HOT SPRINGS

Water from springs and geysers can exceed boiling point, 212°F (100°C), having been heated by rocks beneath. This could be due to magma being near ground level or groundwater trickling through cracks at great depth, where rocks are naturally hot, before being forced to the surface by steam pressure.

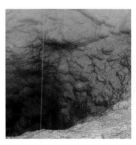

2 THERMOPHILIC ALGAE

Algae are a diverse group of plantlike organisms that grow using sunlight for energy. Thermophilic types include *Mougeotia*, which grows as green hairlike filaments at moderate temperatures. Along with cyanobacteria, such as *Oscillatoria*, they add green, blue, and purple zones to hot springs.

3 THERMOPHILIC ARCHAEA

Heat-loving Archaea grow as colorful mats, scums, and slimes, often coating rocks at pool edges. The particular kinds depend on the pH and nutrient content of the water, as well as temperature. One group is *Sulfolobus*, which tolerates high acidity and temperatures up to 194°F (90°C).

POLYEXTREMOPHILE

The bacterium *Deinococcus radiodurans* (below) is known as "the world's toughest life form." It is a polyextremophile: able to withstand radiation, excessive cold, a lack of water, very acidic situations and a vacuum, or lack of air. The metabolism of *D. radiodurans* and how it repairs its DNA when damaged by radiation exposure are revealing. Intense cold, a lack of air and radiation exposure are all encountered in space, and suggest that life could have arrived on Earth from elsewhere (see p.37). Recently discovered extremophiles are bacteria living on methane and sulfide chemicals in hot rocks 5,000 feet (1,500 m) below the deep seabed, suggesting organisms might survive on other planets or moons.

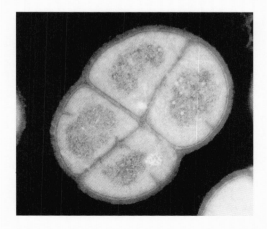

Stromatolites
ARCHEAN EON—PRESENT

👁 FOCAL POINTS

1 MINERAL ACCUMULATIONS
Cyanobacterial layers trap and accumulate minute sediment particles, which are always floating in sea (or lake) water, onto their sticky surfaces. As the cyanobacteria multiply they generate new surface layers, while those below gradually consolidate into rocklike substances, mainly limestones or dolostones. Thrombolites are similar to stromatolites but, rather than banding, have a more random "clotted" appearance.

2 MICROBIAL MATS
Stromatolitic cyanobacteria often grow in sheetlike layers called microbial mats, in combination with other microscopic life forms such as bacteria. Mingled in are other encrusting growths, such as plantlike algae, simple colonial animals known as sea mats (bryozoans or moss animals) along with coral animals or polyps and the tiny young stages of fixed-down sea creatures such as barnacles and mussels.

3 SHAPES
Each stromatolite has a unique shape molded by factors such as temperature, tides, water depths, currents, nutrient levels, particulate amounts, and sunlight intensity. Shapes include domed or flat-topped mounds, often resembling a series of stepping stones, as well as branches and columns. The often wrinkled bandings occurred from repeated laying-down of thin layers of calcium carbonate minerals.

GROUPS Bacteria, Archaea, and Algae
SIZE Mound diameters up to 6 feet (1.8 m)
LOCATION Shark Bay, Western Australia
AGE Oldest mounds 3,500 years

Some of the first organic (life-related) structures on Earth were probably similar to their equivalents today: "living rocks" or stromatolites, sometimes called stromatoliths. They are a combination of simple, mainly microscopic life forms and rocky minerals that accumulate over centuries and millennia to form layers with characteristic banded patterns. They occur in various shapes, especially rounded mounds, disks, or domes. Stromatolites chiefly develop by the action of cyanobacteria, also known as blue-green algae, which are suspected of being among Earth's early life forms (see p.56) and which form microbial mats on rocky surfaces in shallow water. These accumulate mineral particles as more layers of cyanobacteria grow on top—somewhat reminiscent of coral reef formation but on a much tinier scale.

Fossil stromatolites date back to more than 3 billion years ago, to the Archean eon. However, it is sometimes difficult to decipher whether ancient stromatolites were formed biologically (biotically; with the participation of living organisms by natural mineral processes, thereby involving life) or inorganically (abiotically; developing from minerals and similar substances alone, with no need for life's involvement). It seems that stromatolites were widespread to around 1.5 billion to 800 million years ago, and they provide useful information on seawater chemical make-up, temperatures, mineral abundance, and other conditions through time. They then began to fade from the fossil record. The main explanations for this are a possible change in the chemistry of the seas, competition with newly appearing plantlike algae and true plants, and the evolution of early animals that were able to feed on or "graze" this useful source of nutrition. In modern times, stromatolites are even more scarce. They occur mainly in shallow waters with very high salt concentrations, termed hypersaline, where few modern animals can survive. **SP**

BANDED IRON FORMATIONS

Banded iron (ironstone) formations (BIFs; right) are a major type of sedimentary rock, formed in distinctive bands, and may be rocky leftovers from the early Earth produced by cyanobacteria. Typically, a BIF has reddish layers of silicate mineral, such as chert or shale, alternating with silver or black layers of iron-rich materials, usually an oxide of iron such as magnetite or hematite. These fine layers range from 1 millimeter to several centimeters thick. BIFs date to when Earth's atmosphere had no or little oxygen. Eroded iron-rich rocks led to dissolved iron accumulating in the seas. Here it combined with oxygen produced by cyanobacteria, forming the solid iron oxide minerals— in effect, "rust." As the cyanobacteria multiplied, oxygen rose to toxic levels and killed them; silicate minerals were deposited instead. Meanwhile, dissolved iron built up once more, as did cyanobacteria, and so the cycle continued. BIFs occurred mainly from 2.5 billion to 1.8 billion years ago.

EARLY INVERTEBRATES

1 A scene in a warm, shallow Ediacaran sea. Many organisms are difficult to place in any subsequently evolved group.

2 Some Ediacaran fossils resemble living soft corals, such as the sea pen *Pteroeides*.

3 *Cyclomedusa*, up to 8 inches (20 cm) across, may have been a cnidarian.

The fossil record for most of Precambrian time, before 541 million years ago, is sparse and not always easy to interpret. It is probable that many life forms were soft and unlikely to fossilize, while the relentless rock cycle since then would have largely destroyed the remains that did occur. Trace signs of what could be animal burrows discovered since the 1950s have been dated to about 700 million years ago, the late Precambrian. Evidence such as the Chuar Group fossils from the Grand Canyon in Arizona suggests that some soft-bodied animals existed as long ago as 750 million years. In fact, the evolution of multicelled animals may presage even these finds, because some trace fossils have been dated to around 1 billion years ago, deep into the Precambrian. Before this, life on Earth probably consisted of organisms with non-nucleated cells (prokaryotes), such as bacteria (see p.36).

Indisputable fossil evidence shows that by 560 million years ago, well before the start of the Cambrian period, an array of invertebrate animals had evolved. The picture now emerging of early invertebrate life remained tantalizingly obscure until, in 1946, an amazing discovery in the Ediacara Hills, in the Flinders Ranges of South Australia (see p.46), threw open the curtain to this vital stage in evolution. The finds revealed a fascinating array of organisms,

KEY EVENTS

3.5 Bya	1 Bya	750 Mya	700 Mya	635–541 Mya	600 Mya
Life probably comprises organisms known as prokaryotes, such as bacteria. They are mostly single-celled.	Possible early multicelled animals leave evidence such as feeding marks and movement, known as trace fossils.	Soft-bodied animals are thought to exist by this time, as evidenced by the Chuar Group fossils from the Grand Canyon in Arizona.	Trace fossils that could be animal burrows indicate that early multicelled animals may have lived at this time.	The Ediacaran period spans this time. The name was officially recognized in 2004.	Date of the start of the Vendian period, a name suggested in the 1950s and still used informally, that overlaps the Ediacaran.

only some of which could easily be identified as belonging to living groups. These assemblages came to be known as the Ediacaran biotas (see image 1), after the site of their first discovery. These earliest known metazoans (multicelled animals) allowed paleontologists to fill what had been a glaring gap in the history of evolution. Since these remarkable finds, many other assemblages of Precambrian fossils have been discovered in North America, Russia, Europe, and Africa. Their ages range from 575 million to 542 million years old. The site of the first discovery gave its name to a geological time span, the Ediacaran period, which lasted from 635 million to 541 million years ago and marked the end of the Precambrian.

The soft-bodied animal fossils from the Australian Ediacaran rocks—hard body parts like shells had apparently not yet evolved—represent an impressive range of types. Some show a distinct resemblance to living groups of invertebrates. Others are so unusual that the same specimen has been regarded as an alga (simple plant), lichen, giant protozoan (single-celled life form), a natural rock formation with no relation to living things, or an unknowable extinct line of evolutionary experiment that went nowhere. The recognizable fossils, and those from similar-aged sites elsewhere, include forms that can be classified as cnidarians—jellyfish and sea pens (see image 2), the latter a form of soft coral—and various worms. Some of the wormlike animals could be flatworms (platyhelminths), while others may be segmented worms or annelids (see p.108). The genus *Cyclomedusa* (see image 3) is a possible cnidarian, as are the sea pen–like genera such as *Pteridinium* and *Rangea*. The wormlike fossils include *Dickinsonia* and *Spriggina*, which looks like a segmented worm, although some have suggested it could be an early arthropod (invertebrate with jointed legs), while *Tribrachidium* shows similarities with echinoderms (see p.162). Other fossils from the Ediacara Hills have been even more difficult to assign to living groups, or to find comparisons.

It may be significant that the earlier Ediacaran finds, such as those from Mistaken Point on the Avalon Peninsula, southeast Newfoundland, Canada, contain somewhat simpler animals such as hydroids, jellyfish, and sea pens—all soft-bodied and cnidarian-like. Dating from about 565 million years ago , they may represent an earlier stage in the evolution of those primitive invertebrates.

Among the many puzzles posed by these finds is that they are so widespread and rich, whereas fossils of soft-bodied invertebrates of more recent origin are relatively scattered and scarce. One explanation is that as more complex habitats and ecosystems emerged, with an ever more diverse array of animal types, the soft-bodied early invertebrates fell prey to their hard-shelled, carnivorous contemporaries. Soft-bodied creatures tend to be less well preserved than those whose bodies contain harder tissues, and sparse communities would have been even less likely to leave fossils. **MW**

585–580 Mya	565 Mya	550 Mya	549–542 Mya	545 Mya	541 Mya
Gaskiers glaciation, a kind of ice age, occurs in the middle of the Ediacaran. Most Ediacaran fossils date from after this time.	Ediacaran finds on the Avalon Peninsula in Newfoundland, Canada, date from this time. They are among the oldest Ediacaran fossils.	Ediacaran fossils from this time discovered on the White Sea coast of northwest Russia comprise one of the richest sites for such finds.	The Ediacaran Nama Group fossils, found in Namibia, Africa, and Oman, Arabia, date from this period.	The Baykonur glaciation, named after rock formations in Kazakhstan and Kyrgyzstan, may start to adversely affect Ediacaran life.	The Cambrian begins, when fossil numbers increase rapidly and many new forms evolve.

Ediacaran Fossils
EDIACARAN PERIOD (NEOPROTEROZOIC ERA)

👁 FOCAL POINTS

CHARNIA
This frond-shaped organism would have been about 6 inches (15 cm) tall and was attached to the seabed (see p.40: the tall leaflike shapes across the background). Experts are unsure of its exact affinity, but *Charnia* may have been a sea pen, or even some kind of alga.

DICKINSONIA
Dickinsonia was oval in shape and about 1½ inches (4 cm) long. This fossil had a central groove from which riblike features radiated (see p.40: the dark, flattened shapes on the seabed, center foreground). *Dickinsonia*'s affinities are uncertain; it was possibly some kind of flattened worm.

KIMBERELLA
Fossils of this organism, 6 inches (15 cm) long, are found at the Ediacara Hills and also at the earlier White Sea's Ust Pinega fossil beds in Russia. It is one of the first known bilaterians—with two sides, left and right, rather than a circular or radial body design—and may be a very early mollusk (see p.124).

MAWSONITES
This fossil may represent a jellyfish-like organism (see p.40: the blue floating shape, lower left of center). *Mawsonites* had a series of central disks, similar to flower petals, and spokelike polygons. Others have interpreted it as a community of microbes or a burrow system.

LOCATION Ediacara Hills, South Australia
AGE 560 million to 550 million years
GROUPS Algae, Cnidaria, Annelida, and others
BED DEPTH Up to 1½ miles (2.4 km)

The Ediacaran fossils, first described from discoveries in the Ediacara Hills in the Flinders Ranges of South Australia, are now known to exist in a number of different sites across the world. All are from the Ediacaran period, the last period of the Neoproterozoic era, between about 575 million and 540 million years ago. The Ediacaran (sometimes called Vendian) communities seem to have appeared after a series of severe ice ages in the preceding period, the Cryogenian, 850 million to 635 million years ago. Ediacaran fossils represent the earliest reasonably preserved evidence of metazoans. Until their initial discovery in Australia in 1946, there was little evidence that such complex organisms had evolved so early in the history of the Earth; scientific consensus at the time maintained that life originated in the Cambrian. Australian geologist Reg Sprigg (1919–94) stumbled upon the fossils in the Ediacara Hills while he was surveying old mines for potential uranium ore extraction on behalf of the South Australian government. His eye was caught by curious-looking rock shapes and marks. Sprigg spent more than 2 years trying to convince a skeptical scientific establishment of the fossils' great age and significance. The Ediacaran animals lived mostly on, or close to, the surface of seabed sediments, in shallow seas, or deep on the ocean floor, and some left trace fossils of their burrows. A possible trigger for the evolution of these diverse creatures could have been a rise in the proportion of oxygen in the atmosphere, produced mainly by photosynthetic cyanobacteria (see p.56). Some of the fossils are clearly animals and show marked affinities with living groups. They range in size from a few to several inches, with some of the branching types reaching up to 3 feet (1 m). **MW**

MISTAKEN POINT

Jutting into the North Atlantic Ocean, an exposed shoreline cliff at Mistaken Point, at the southern tip of the Avalon Peninsula of Newfoundland, Canada, is one of the finest sites in the world for Ediacaran fossils. The organisms appear as if they have been etched into the surfaces of the rocks (right). Most of these fossils are imprints left by soft-bodied organisms, and a range of shapes can be found, including fernlike fronds, large disk-shaped forms, and some that are long and pointed. The fossils were covered by volcanic ash, an ideal substance for radiometric dating (see p.31), therefore they can be aged accurately to around 565 million years ago, making them one of the earliest known Ediacaran biotas (communities of plant and animal life). The layered structure of the ash deposits has preserved different assemblages from various points in time. Paleontologists have determined that these organisms lived on the seabed at a great depth, thereby differing from many other Ediacaran sites, which uncover creatures from shallow marine habitats.

THE CAMBRIAN EXPLOSION

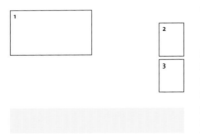

1 *Jianfengia*, a multilegged arthropod from Chengjiang, was ⅔ inch (1.7 cm) long.

2 *Parabolina* was a late Cambrian trilobite. It is sometimes used as an index fossil.

3 Around 1 inch (2.5 cm) in length, *Haikouella* was a possible chordate, a predecessor of the vertebrate group.

The most recent part of the Precambrian is known as the Proterozoic eon, which lasted from 2.5 billion years until about 541 million years ago, and then gave way to the early Cambrian period. The word "Proterozoic" means "early life," indicating that it was in this vast stretch of time that life on Earth began. Proterozoic fossils suggest that this life was almost exclusively represented by organisms with non-nucleated cells (prokaryotes; see p.36), such as bacteria, most of which were single-celled. The start of the Cambrian saw a surge in evolutionary radiation—a rapid period of change in which living things alter and diversify as they adapt to new habitats, lifestyles, and challenges, including those evolving around them. It all happened in the sea; there was no life on land yet, as far as it is known. The fossil record indicates a relatively sudden diversification among the invertebrates, as well as the presence of the first fishlike species and other early chordates (see image 3; p.174). This diversification was so rapid that the period from 30 million to 50 million years after the start of the Cambrian is often known as the "Cambrian explosion."

However, paleontologists are uncertain about the causes of this sudden burst. The wealth of Cambrian fossils is due partly to the fact that animals appeared with bodies that included harder tissues, which gave them skeletal

KEY EVENTS

2.5 Bya–541 Mya	525–520 Mya	525–520 Mya	520 Mya	520–515 Mya	518 Mya
During the Proterozoic, life evolves from simple prokaryotes through to the varied forms at the start of the Cambrian.	Communities preserved in Maotianshan Shales and other rocks at Chengjiang in Yunnan province, China, date to this time.	At Chengjiang, *Haikouella* (see p.174) and the similar *Yunnanozoon* show signs of being hemichordate or chordate animals.	The Guanshan biota, known from the Chengjiang sites, includes sponges, brachiopods, and trilobites.	Fossils are laid down at the Sirius Passet site in Greenland; it will be discovered in 1984 (the same year as Chengjiang).	*Halkiera*, a "chain-mailed slug" 3 inches (8 cm) long, is a puzzling Sirius Passet animal, possibly related to brachiopods or mollusks.

support. Calcareous (calcium-mineral) skeletons, chitinous (chitin protein) exoskeletons (hard outer casings), and hard shells all produce good fossils, and offer a thorough picture of the kind of animals that lived at that time. Also, most of the Cambrian fossils are from animals that lived close to, or on, the seabed, rather than swimmers. There were crustaceans and other arthropods (see image 1), brachiopods (lampshells), corals, echinoderms, mollusks, and sponges, mostly represented by hard body parts. To counter the "explosion" theory, increasing finds from the Ediacaran period (see p.46), before the Cambrian, have indicated that soft-bodied life forms had already started to diversify. Many of the major animal groups known today appeared. Despite this, fossils from the early Cambrian indicate a rapid change and reveal animals that have clear relationships to existing groups. This happened over a relatively short time span: 30 million years, which is less than 1 percent of time since life first appeared.

The early Cambrian explosion was recognized in the 1840s. It presented a problem to Charles Darwin (1809–82), because he realized that it could be used as evidence to contradict his theory of evolution. Darwin's vision was one of diversification that was caused by natural selection in a slow and gradual way, accumulating small changes that steadily led to larger differences, as opposed to a Cambrian "Big Bang" in which a lot of changes happened simultaneously. More recent discoveries have indeed shown that the explosion was probably part of a longer, more drawn-out process.

It has been suggested that environmental changes contributed to the plethora of Cambrian fossils. The increased levels of atmospheric oxygen could have been a key factor, because it would have caused more photosynthesis and facilitated the development of molecules such as collagen—a key structural protein found in many animal tissues. Another factor could have been volcanic activity and land erosion that increased the amount of calcium dissolved in the sea, providing essential material for the production of hard skeletons.

It is highly fortunate that the Cambrian explosion is so well preserved in the fossil record. One of the clearest assemblages of animals is provided by the Burgess Shale rocks of Canada (see p.50), which are about 510 million to 505 million years old. The fossil-rich Maotianshan Shales and nearby sites at Chengjiang, Yunnan province, China, are evidence of the Chengjiang biota (community of organisms), 10 million to 20 million years older than Burgess. The fossil site Sirius Passet, Greenland, is also from this time, and late Cambrian finds in Sweden provide more details of the creatures. Chengjiang excavations from the 1980s have revealed preserved hard and soft tissues of early sponges, corals, jellyfish, worms, arthropods, trilobites (see image 2; see p.118), brachiopods, echinoderms, and early fishlike chordates, possible forerunners of the vertebrates. All of these fossil locations also include bizarre specimens that appear to have no relationship to known groups. They were probably very early evolutionary prototypes that could not develop further. **MW**

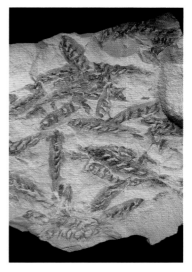

517–515 Mya	510 Mya	507 Mya	505 Mya	500 Mya	497–485 Mya
A mass extinction event at the end of the Botomian age, in the early Cambrian, kills off a large proportion of shelled animals.	Some early Burgess Shale fossils are laid down at the Burgess Pass, British Columbia, Canada.	The Wheeler Shale is established. It is a series of fossil beds similar to Burgess and is located in the House Range mountains of Utah.	The main Burgess Shale fossils are laid down, including such kinds as the early arthropod *Opabinia* and *Hallucigenia* (see p.52).	The Marjum Formation, rocks overlying the Wheeler Shale in Utah, preserves varied soft-bodied animals including trilobites.	Late Cambrian fossils, such as varied crustaceans from several sites in Sweden, including Västergötland, are laid down.

Burgess Shale Fossil Beds
MIDDLE—LATE CAMBRIAN PERIOD

LOCATION Burgess Pass, British Columbia, Canada
AGE 510 million to 505 million years
GROUPS Porifera (sponges), Annelida, Arthropoda, Mollusca, and many others
BED DEPTH 6½ feet (2 m)

About 510 million to 505 million years ago, an event took place in tropical shallow seas that resulted in the preservation of many animals in what is known as the Burgess Shale Formation, later part of the Canadian Rocky Mountains. Effectively, this froze in time an entire community of animals and other organisms that allows the reconstruction of part of an ecosystem from half a billion years ago. The event is thought to have involved an underwater avalanche of fine mud that traveled across a reef and settled at its base, taking with it many animals and burying them on the seabed. In low oxygen conditions and with no time to allow decay, the mud not only preserved the hard parts of the animals, but also filled up their bodies so that impressions were left of soft tissues, and even whole soft-bodied animals. In some cases, details of some of the internal organs are visible.

Later tectonic events in Earth's history meant that these fossil-rich rocks became part of their present site. Since then they, and other deposits of similar age elsewhere, notably in China and Greenland, have provided paleontologists with some of the richest examples of early animal fossils and insights into their evolution. The fossil-bearing layers of rock in the Canadian Burgess Shale are about 6½ feet (2 m) thick and have yielded more than 65,000 specimens, including trilobites and other arthropods, echinoderms, such as sea lilies and sea cucumbers, many kinds of worms, and even early chordates. Although some of the Burgess Shale animals are clearly related to living groups, others are very different and probably represent evolutionary lines that became extinct. Experts periodically re-examine the fossils, change their identities and classifications and merge several specimens into one new animal, such as the strange shrimp, *Anomalocaris* (see p.116). **MW**

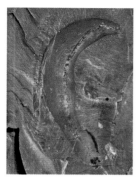

OTTOIA

Ottoia was up to 3¼ inches (8 cm) long and was worm-shaped, with a mobile front end (proboscis) armed with hooklike spines that could be everted or protruded. It was probably a priapulid worm or penis worm, similar to living types (see p.109). *Ottoia* was likely an active burrower and devoured small prey, which has been shown by fragments of animals preserved in its gut.

MARRELLA

Marrella was a primitive arthropod and one of the most abundant of the Burgess Shale animals, although it cannot confidently be assigned to any of the well-known arthropod groups, such as trilobites or crustaceans. Large spines curved over the body and provided protection; it may have filtered food using its limbs. *Marrella* was only ¾ inch (2 cm) in length.

NARAOIA

Initially, the arthropod *Naraoia*, with a large, tough, two-section body covering, was classified as a kind of crustacean, allied to crabs. Later studies looked more closely at the legs and soft parts, such as the guts beneath the hard covering, to reveal that it was a trilobite. It was unusual because it had a two-section shield arrangement, rather than the typical three-shield body (see p.118).

WAPTIA

This crustacean had a body remarkably like certain living shrimp, with a two-valved outer covering (carapace). It would have used its jointed limbs to crawl, and it probably swam efficiently in the manner of modern shrimp. Named by Walcott in 1912 after nearby Mount Wapta, *Waptia* grew to about 3 inches (8 cm) in length and left thousands of fossil specimens.

1850–79
Charles Doolittle Walcott was born in New York Mills, New York. He was interested in nature from an early age and began collecting minerals, bird eggs, and fossils. After leaving high school he became a commercial fossil collector. In 1876 he became assistant to the state paleontologist of New York; in 1879 he joined the United States Geological Survey (USGS).

1880–1908
He was active in finding new localities that yielded fossils and by 1894 he was director of the USGS and a leading expert on Cambrian deposits. In 1907 he became secretary of the Smithsonian Institution in Washington, D.C., which now houses the world's largest museum complex.

1909–27
Walcott discovered the Burgess Shale fossils in 1909. Having heard that workers on the Canadian Pacific Railway had found "stone bugs" in the Rockies, he visited a site near the Burgess Pass and excavated there for fossils, naming it the Burgess Shale after nearby Mount Burgess. Between 1910 and 1924 he returned many times and collected more than 65,000 specimens from a site now known as Walcott Quarry. These well-preserved specimens are mostly housed in the Smithsonian. After his death in 1927, his samples, notes, and photographs remained in storage until they were rediscovered by a new generation of paleontologists in the late 1960s.

BURGESS SHALE GENERA

Many different species have been found in the slate and rocks of the Burgess Shale (below), and they include worms such as *Aysheaia*, which had lobelike limbs, and *Canadia*, a polychaete annelid worm 2 inches (5 cm) long with a bristly, segmented body and tentacles on its head. Possible mollusks also inhabited the seas at this time, such as *Haplophrentis*, with its conical shell, and *Wixwaxia*, which was oval-shaped and spiny. The largest species so far discovered here is *Anomalocaris*, the possible early arthropod that spanned 3 feet (1 m) in length and hunted for prey on the ocean floor. One of the strangest is the soft-bodied *Opabinia*, which had no fewer than five eyes.

Hallucigenia
MIDDLE—LATE CAMBRIAN PERIOD

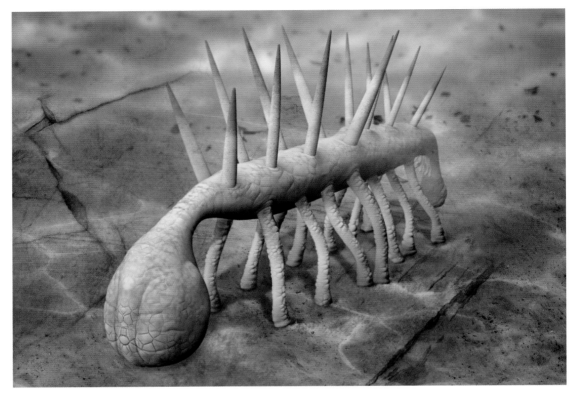

SPECIES *Hallucigenia sparsa*
GROUP Lobopodia
SIZE 1¼ inches (3 cm)
LOCATION Canada and China

The Burgess Shale fossils include some familiar creatures and others that are truly strange. A possible candidate for the oddest of all is *Hallucigenia*. It was originally described in 1911 by Charles Walcott (1850–1927), who thought it was an annelid worm so named it *Canadia sparsa*, a name that indicated its relative rarity. It was studied in more detail by English paleontologist Simon Conway Morris (b. 1951), who renamed it *Hallucigenia sparsa* in 1977. The new, quirky genus name was chosen to reflect its surreal appearance and unlikely mix of features. Morris's reconstruction of the animal was also reviewed. Some experts thought that the fossil might be part of a larger animal. As further specimens of *H. sparsa* and similar animals were uncovered—notably at Chengjiang in China, including the related *H. fortis* and the genus *Microdictyon*—it became clear that these genera bore a strong resemblance to living velvet worms, such as *Peripatus*, from the Onychophora group. It was then decided that *H. sparsa* and its cousins belong to a group that gave rise to the onychophorans, currently termed lobopods (blunt feet).

Hallucigenia was up to 1¼ inches (3 cm) long. Although the head and tail cannot be differentiated with certainty, it is thought that at the front was a long neck with two or three pairs of appendages that lacked claws at their tips. At the top of the neck was a small head with a forward-facing mouth. At the back were seven pairs of long, rigid spines, and on the underside were seven pairs of lobed legs, each with a pair of claws at its tip. Isolated fossils of back spines, and even the leg claws, have been found at various sites from the middle to late Cambrian. A recent study of the claws showed their detailed structure was similar to those of velvet worms. *Hallucigenia* and its relatives would amble along the seabed feeding on detritus, such as the decaying remains of dead creatures; they may have fed on the soft tissues of sponges. **MW**

NAVIGATOR

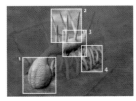

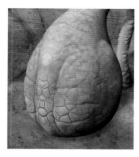

1 HEAD OR TAIL?
Distinguishing the head from the tail of *H. sparsa* was initially a problem. The original fossil had a dark, rounded area at one end and a dark, narrower mark at the other. Some scientists believe that the rounded area is merely a stain and not part of the anatomy, so it is still uncertain which end is the head and which the tail.

2 SPINES
H. sparsa fossils clearly show that its body carried two parallel rows of long, sharp spines. The seven pairs of rigid spines presumably offered some protection against predators, like the spines of echinoids or sea urchins. It is not clear if there was any movement at the joint between the spine and the body.

3 BODY
H. sparsa's tube-shaped, wormlike body, and the repeating organization of sections or units, each bearing a pair of legs (plus a pair of spines), bears resemblance to the velvet worms and other segmented animals. Internally, the body is likely to have had a straight, simple gut, although fossils of this are not clear.

4 LEGS
Later Burgess Shale fossil finds showed that *H. sparsa* had seven, or potentially eight, pairs of slender lobe-like legs, tipped by claws. These features would have helped align the creature with the living velvet worms, onychophorans. The legs appear to have lacked a hard outer covering with joints, as is found in the arthropod group.

▲ More than 100 specimens of *Hallucigenia* have been recovered from the Burgess Shale site, including some presumably young forms that were a mere ¼ inch (6 mm) in length. The creatures were soft-bodied and, while certain specimens clearly show paired series of back spines and legs, the fossils are flat and rows are sometimes buried or sheared off, giving the appearance of only a single row of spines, legs, or tentacles, as shown here.

HALLUCIGENIA MISINTERPRETED

The original Burgess fossils of *Hallucigenia* appeared to reveal an animal with two rows of spinelike processes on one surface of the body and a single row of tentacles along the center of the opposite side. In 1977, this anatomy was interpreted as an animal that walked using rigid spiny legs (right). It was thought to have fed using the softer tentacles that were in the center of its back, perhaps absorbing nutrients from water via what appeared to be tiny mouths, one at the tip of each tentacle: a truly innovative method. Interpretation was difficult at the time, as specimens were often flat. Better preserved specimens have since shown that there were two rows of tentacles and, rather than having mouths, they carried claws, so were in fact legs. The original fossils had been interpreted upside down.

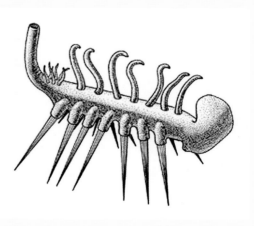

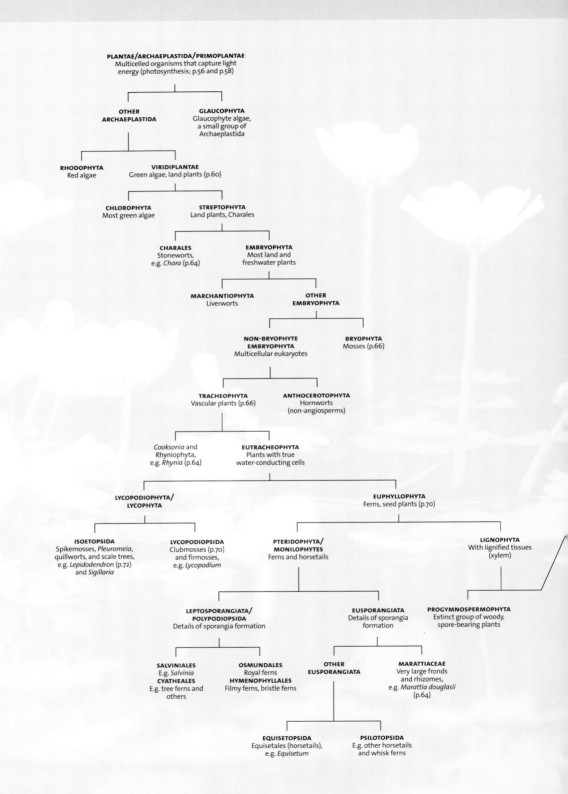

PLANTAE/ARCHAEPLASTIDA/PRIMOPLANTAE
Multicelled organisms that capture light
energy (photosynthesis; p.56 and p.58)

OTHER ARCHAEPLASTIDA

GLAUCOPHYTA
Glaucophyte algae,
a small group of
Archaeplastida

RHODOPHYTA
Red algae

VIRIDIPLANTAE
Green algae, land plants (p.60)

CHLOROPHYTA
Most green algae

STREPTOPHYTA
Land plants, Charales

CHARALES
Stoneworts,
e.g. *Chara* (p.64)

EMBRYOPHYTA
Most land and
freshwater plants

MARCHANTIOPHYTA
Liverworts

OTHER EMBRYOPHYTA

NON-BRYOPHYTE EMBRYOPHYTA
Multicellular eukaryotes

BRYOPHYTA
Mosses (p.66)

TRACHEOPHYTA
Vascular plants (p.66)

ANTHOCEROTOPHYTA
Hornworts
(non-angiosperms)

Cooksonia and
Rhyniophyta,
e.g. *Rhynia* (p.64)

EUTRACHEOPHYTA
Plants with true
water-conducting cells

LYCOPODIOPHYTA/ LYCOPHYTA

EUPHYLLOPHYTA
Ferns, seed plants (p.70)

ISOETOPSIDA
Spikemosses, *Pleuromeia*,
quillworts, and scale trees,
e.g. *Lepidodendron* (p.72)
and *Sigillaria*

LYCOPODIOPSIDA
Clubmosses (p.70)
and firmosses,
e.g. *Lycopodium*

PTERIDOPHYTA/ MONILOPHYTES
Ferns and horsetails

LIGNOPHYTA
With lignified tissues
(xylem)

LEPTOSPORANGIATA/ POLYPODIOPSIDA
Details of sporangia formation

EUSPORANGIATA
Details of sporangia
formation

PROGYMNOSPERMOPHYTA
Extinct group of woody,
spore-bearing plants

SALVINIALES
E.g. *Salvinia*
CYATHEALES
E.g. tree ferns and
others

OSMUNDALES
Royal ferns
HYMENOPHYLLALES
Filmy ferns, bristle ferns

OTHER EUSPORANGIATA

MARATTIACEAE
Very large fronds
and rhizomes,
e.g. *Marattia douglasii*
(p.64)

EQUISETOPSIDA
Equisetales (horsetails),
e.g. *Equisetum*

PSILOTOPSIDA
E.g. other horsetails
and whisk ferns

2 | PLANTS

A key advance in plants was the ability to capture the Sun's energy by photosynthesis, or "building with light." Being aquatic, early plants were supported and had little difficulty in maintaining water content or absorbing dissolved minerals and nutrients. But invasion of the land involved further evolutionary stages, such as vascularization—internal "plumbing" to distribute water and nutrients within the plant's body and provide stiffness and support in air—and, for reproduction, seeds rather than spores.

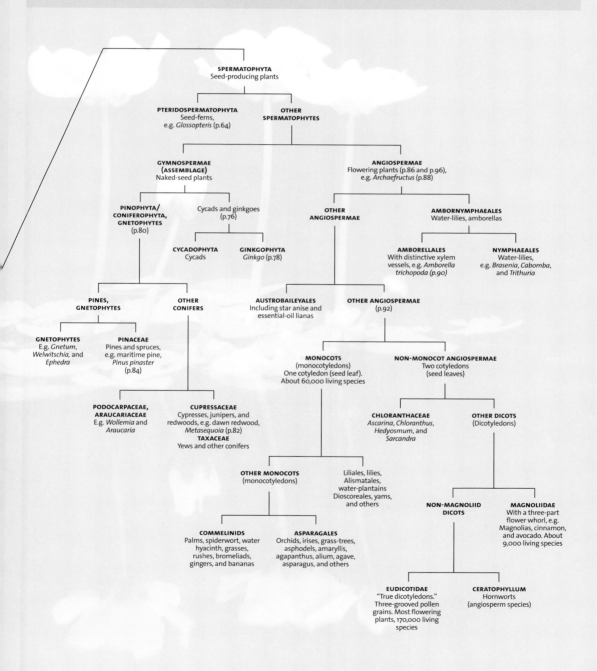

SPERMATOPHYTA
Seed-producing plants

PTERIDOSPERMATOPHYTA
Seed-ferns,
e.g. *Glossopteris* (p.64)

OTHER SPERMATOPHYTES

GYMNOSPERMAE (ASSEMBLAGE)
Naked-seed plants

ANGIOSPERMAE
Flowering plants (p.86 and p.96),
e.g. *Archaefructus* (p.88)

PINOPHYTA/ CONIFEROPHYTA, GNETOPHYTES
(p.80)

Cycads and ginkgoes
(p.76)

OTHER ANGIOSPERMAE

AMBORNYMPHAEALES
Water-lilies, amborellas

CYCADOPHYTA
Cycads

GINKGOPHYTA
Ginkgo (p.78)

AMBORELLALES
With distinctive xylem
vessels, e.g. *Amborella
trichopoda (p.90)*

NYMPHAEALES
Water-lilies,
e.g. *Brasenia, Cabomba,*
and *Trithuria*

PINES, GNETOPHYTES

OTHER CONIFERS

AUSTROBAILEYALES
Including star anise and
essential-oil lianas

OTHER ANGIOSPERMAE
(p.92)

GNETOPHYTES
E.g. *Gnetum,
Welwitschia,* and
Ephedra

PINACEAE
Pines and spruces,
e.g. maritime pine,
Pinus pinaster
(p.84)

MONOCOTS
(monocotyledons)
One cotyledon (seed leaf).
About 60,000 living species

NON-MONOCOT ANGIOSPERMAE
Two cotyledons
(seed leaves)

PODOCARPACEAE, ARAUCARIACEAE
E.g. *Wollemia* and
Araucaria

CUPRESSACEAE
Cypresses, junipers, and
redwoods, e.g. dawn redwood,
Metasequoia (p.82)
TAXACEAE
Yews and other conifers

CHLORANTHACEAE
*Ascarina, Chloranthus,
Hedyosmum,* and
Sarcandra

OTHER DICOTS
(Dicotyledons)

OTHER MONOCOTS
(monocotyledons)

Liliales, lilies,
Alismatales,
water-plantains
Dioscoreales, yams,
and others

NON-MAGNOLIID DICOTS

MAGNOLIIDAE
With a three-part
flower whorl, e.g.
Magnolias, cinnamon,
and avocado. About
9,000 living species

COMMELINIDS
Palms, spiderwort, water
hyacinth, grasses,
rushes, bromeliads,
gingers, and bananas

ASPARAGALES
Orchids, irises, grass-trees,
asphodels, amaryllis,
agapanthus, alium, agave,
asparagus, and others

EUDICOTIDAE
"True dicotyledons."
Three-grooved pollen
grains. Most flowering
plants, 170,000 living
species

CERATOPHYLLUM
Hornworts
(angiosperm species)

THE FIRST PLANTS

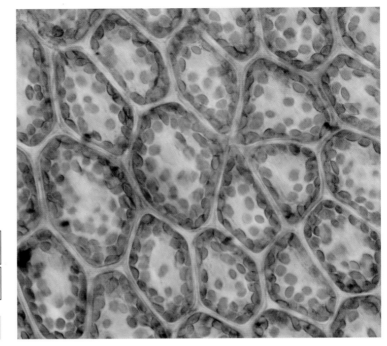

1 Boxlike living plant cells from a leaf contain small, rounded, green bodies—chloroplasts—that were perhaps once cyanobacteria and are now the driving forces behind photosynthesis.

2 *Gloeocapsa* cyanobacteria reproduce by fission, a form of asexual division in which two identical cells are produced. Cyanobacteria were the first photosynthesizers, more than 2 billion years ago.

3 Biggest of the simple plantlike algae is the giant kelp *Macrocystis*, here growing off the coast of California, United States.

The first steps toward the dominance of plants on Earth began more than 2 billion years ago and involved collaboration between very simple single-celled life forms, especially cyanobacteria (blue-green algae, see p.56 and image 2). They formed mutually beneficial relationships with one another, or with bacteria, that led to the evolution of chloroplasts (see image 1). These are the conductors of photosynthesis: the process by which plants use sunlight to convert carbon dioxide and water into sugars to provide energy.

Cyanobacteria are a group of photosynthetic organisms that were the main source of the rapidly rising levels of atmospheric oxygen around 2.3 billion years ago. Even today they contribute half the oxygen in the atmosphere. There is evidence for their existence in the form of microfossils (see p.58) and stromatolites (see p.42), rocklike mounds formed by layers of algae and trapped sediments, that stretches back almost 3 billion years. Cyanobacteria belong to the branch of life called prokaryotes (see p.36): simple single-celled organisms that, unlike more complex eukaryote cells, have no nucleus to hold their genetic material and lack other structures such as, in the case of plants, chloroplasts. Yet persuasive evidence of shared characteristics suggests that chloroplasts, which

KEY EVENTS

3 Bya	2.1 Bya	2 Bya	1.9 Bya	1.8 Bya	1.5 Bya
Colonies of bacteria microfossils (see p.54) and the first stromatolites appear. For more than a billion years, all life on Earth is bacterial.	The oldest plant-related fossils, *Grypania*, are found in rocks of this age in Gabon and Michigan.	The first eukaryotic cells, those with inner structures such as chloroplasts and mitochondria, appear through the process of endosymbiosis.	Light-eating cyanobacteria cause the concentration of oxygen in the atmosphere to increase to one part in five.	Bacterial viruses (bacteriophages) emerge around the time that prokaryotic and eukaryotic cells begin to diverge.	The eukaryote cells continue to separate into three distinct lineages that will lead to the evolution of plants, fungi, and animals.

have their own genetic make-up distinct from the other plant genes in the cell nucleus, were once free-living cyanobacteria. The photosynthetic pigments in cyanobacteria, for example, are the same as those found in the chloroplasts of algae (simple water plants) and land plants, and microstructures found inside chloroplasts are similar to those present in modern-day cyanobacteria. Furthermore, chloroplasts are about the same size as cyanobacteria and reproduce in the same way, by binary fission (dividing into two). These similarities point toward a shared ancestry between cyanobacteria and chloroplasts, a link that was explained by a theory called endosymbiosis, proposed by American biologist Lynn Margulis (1938–2011) in the 1960s.

The development of endosymbiosis is probably one of the most important events in evolution. It is a process by which one organism comes to live inside another so that the two effectively function as a single organism to their mutual benefit. It seems likely that eukaryote cells evolved through endosymbiosis when larger prokaryote cells engulfed, but did not consume or destroy, smaller cells such as cyanobacteria. The latter continued to photosynthesize and produce sugars inside the host cell. The host cell received a supply of energy while the cyanobacterium had somewhere relatively stable to live. In a similar fashion, it is believed that eukaryotic cell structures called mitochondria, which release energy for the cell, can also trace their ancestry back to free-living bacteria.

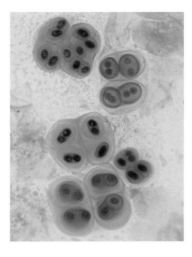

Modern simple plantlike algae consist of a range of organisms with very different structures but identical photosynthetic pigments. The explanation for this is that the different types of algae evolved when a variety of host organisms formed endosymbiotic partnerships with the same photosynthetic prokaryotic cells. A group of organisms like this, which are identified by their structural similarities rather than by one common shared ancestor, is described as polyphyletic. Living algae—the name is familiar but not scientifically useful— lack structures such as true roots and a vascular system (internal "plumbing"), or seed-producing parts such as cones or flowers. As befits their mixed origins, they are hugely diverse, ranging from single-celled diatoms (see p.59) that swarm in the trillions in lakes and seas, to larger multicelled seaweeds in the green, red, and brown groups, from sea lettuce *Ulva* to oarweeds and giant kelps (see image 3), and freshwater algae such as *Chara* (see p.64).

It was the fruitful union of endosymbiosis that began the evolutionary line that leads to the diversity of plant life today. Current thinking is that the first multicelled green plants evolved through endosymbiosis between 2.5 billion and 1 billion years ago. Sedimentary rocks from China, dating back 1.7 billion years, show fossil evidence of eukaryotes resembling present-day brown algae. Possibly the oldest plant fossil yet found is *Grypania* (see p.39), discovered in 2.1-billion-year-old rocks in the United States and Gabon, although there is some controversy as to whether it is an alga or a giant bacterium. **RS**

1.4 Bya	1.2 Bya	1.2 Bya	1 Bya	900 Mya	720–600 Mya
Stromatolites evolve into a much greater diversity of forms.	The first red and brown algae appear. They are simple plantlike organisms of one cell, or a gathering of similar cells, more complex than cyanobacteria.	Sexual reproduction begins to appear in the fossil record of single-celled eukaryotes.	The first vaucherian algae, including *Paleovaucheria*, appear. These are yellow-green and form mats both on land and in fresh water.	The first multicellular organisms appear, although some evidence suggests that they appeared much earlier, around 2.1 billion years ago.	Ice covers much of the planet—"Snowball (Slushball) Earth." Various microbes, including plant- and animal-like, evolved through this time.

Early Plant Microfossils
PALEOPROTEROZOIC ERA

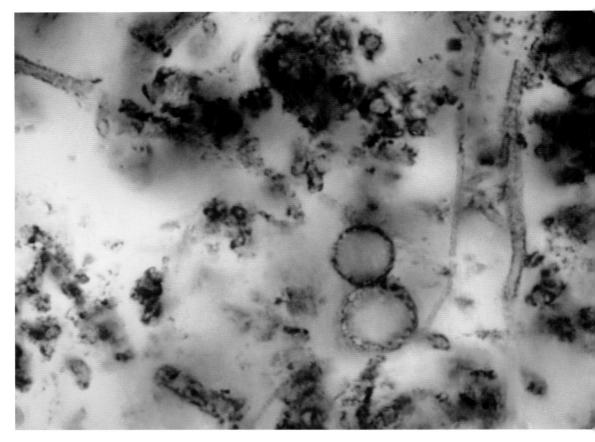

👁 FOCAL POINTS

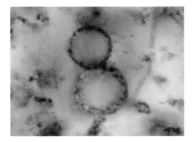

1 CYANOBACTERIA
Among the various debris in this specimen are rounded shapes that are reminiscent of the cyanobacteria living today. Some of these early cyanobacteria formed ball-shaped or spherical colonies, most of which were hollow inside. Living organisms that form colonies similar to these include the common freshwater *Volvox*, a type of green algae.

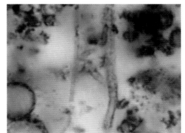

2 FILAMENTOUS ALGAE
The long, semi-transparent strands in this microfossil resemble today's filamentous cyanobacteria, simple plantlike micro-organisms that are linked end to end. The specimen is a very thin slice, or section, of a rock called Gunflint chert from near Lake Superior, Canada, and it is thought to be around 1,800 million years old.

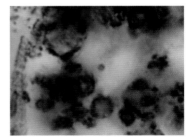

3 ALGAL COLONIES
Various single-celled algae also populated the Palaeoproterozoic habitat represented here. Some of these existed individually, while others lived in groups. The mixture of fossilized forms in such samples shows how life had already evolved and was beginning to diverge along many pathways, all of which relied on energy from the Sun.

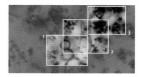
GROUPS Cyanobacteria and various algae
SIZE Mostly less than 1/250 inch (0.1 mm)
LOCATION Gunflint Chert Beds, Ontario, Canada
AGE 1.8 billion years

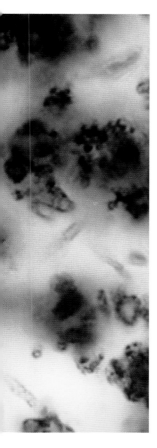

When it comes to uncovering the path of evolution, some of the most important fossils are practically invisible to the naked eye. Microfossils are traces in the rocks left by bacteria, a variety of single-celled living organisms such as diatoms and other eukaryotes, fungal cells and spores, as well as pollen and fragments of plants. They occur mainly in very fine-grained rocks in which the particles around them are so tiny that the most intricate details are preserved.

Diatoms, common marine and freshwater algae, may have existed for billions of years. Living diatoms contain several chloroplasts and carry out photosynthesis. Extraordinarily diverse, with as many as 200,000 different species (see panel), they range in size from microscopic to just large enough to see with the unaided eye. Their fossilized skeletons, known as frustules, are commonly found in rocks dating back to the Jurassic period, around 201 million to 145 million years ago. Diatoms and similar single-celled photosynthesizing organisms, such as other cellular golden and brown algae, thrived especially in warm, shallow, nutrient-rich waters, where sunlight could penetrate. They could be so abundant that hundreds of feet of rock may be composed almost entirely of their fossilized remains. There are also microfossils of parts of plants, ranging from pollen grains and the tiniest seeds to fragments of bark, cones, and roots. Pollen grains are especially numerous and resistant when fossilized. Their changing forms through time make them useful as index fossils to determine the age of the rocks and the other fossils preserved with them. **RS**

DIATOMS

Usually regarded as unicellular algae, diatoms (right) can be divided into two main groups according to their shape. Centric diatoms exhibit radial symmetry, with their parts radiating out from a central point like a wheel. Pennate diatoms show bilateral symmetry, in which the left and right sides are mirror images of one another. A diatom's cell wall is made of glasslike silica through which it absorbs nutrients and excretes waste via tiny perforations. Each species has a unique configuration of holes. Many species also have a long opening or fissure (raphe) that runs down the long length of one half of the skeleton. This exudes a mucuslike substance that allows the diatom to glide across a surface. Quite how diatoms evolved their unusual structure is uncertain, because no ancestors have been identified in the fossil record. However, their presence and evolving forms are used to determine nutrients, temperatures, and other changes in freshwater and marine environments over millions of years.

PLANTS COLONIZE THE LAND

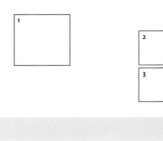

1 From the early Devonian period (400 million years ago), this *Asteroxylon* fossil shows an upright branching structure that would have been more than 14 inches (35 cm) tall.

2 *Cooksonia* was one of the earliest and most widespread land plants, with varied fossils from most continents.

3 Vascular tissues, here in the water fern or water clover, *Marsilea vestita*, are internal tubing networks that are highways for water and nutrients, and also provide stiffness and support.

When did life first move from the ocean to the land? The fossil record, which could help to pinpoint the date, is practically nonexistent in this respect. Evolutionary changes take place over millions of years, and fossilization is a rare event, especially where delicate plant tissues are involved, which was the case with *Asteroxylon* (see image 1). With little confirming evidence, it is speculated that the colonization of the land by plant life had its tentative beginnings during the Ordovician period, 485 million to 443 million years ago, but it remains possible that the process began much earlier.

Before life could take hold on the land, Earth's atmosphere had to change. Initially there were very low concentrations of oxygen in the atmosphere, perhaps less than 1 percent of the levels found today. Oxygen in the upper atmosphere forms ozone, a gas that, in terms of the development of life, performs the vitally important task of absorbing potentially harmful ultraviolet radiation from the sun. Without the ozone layer, life on land would be fully exposed to ultraviolet light, which disrupts DNA, promotes mutations, and causes cancer. It was only after aquatic photosynthesizers, such as cyanobacteria and algae, generated enough oxygen to build up the ozone layer that it became

KEY EVENTS

600 Mya	500–450 Mya	473 Mya	443–419 Mya	420 Mya	410 Mya
Levels of atmospheric oxygen increase sufficiently to form an ozone layer that shields land life from the effects of the sun's ultraviolet radiation.	The first fossil spores appear on land, although these may have blown onshore from marine algae rather than from land plants.	Sporelike structures, possibly from a type of liverwort, may be early evidence of complex plants on land.	The Silurian provides the first good evidence of the existence of plant life on land, in the form of thin coatings over rocks in humid habitats.	The first occurrence of the lycopsid plant *Baragwanathia* suggests that lycopsids were among the first land plants to evolve true leaves.	An early vascular plant makes an appearance. *Cooksonia* is small with leafless stems and has a spore-producing structure at its tip.

possible for delicate organisms to survive on the land. This was a long and slow process that took place over hundreds of millions of years. However, scientists mostly agree that plants were not the first organisms to arrive. In addition to requiring sunlight and carbon dioxide, which are readily available in the atmosphere, plants require a source of absorbable nutrients. The study of paleosols (ancient, fossilized soils) suggests that the land was inhabited by bacterial mats before true plants appeared. Cyanobacteria, algae, fungi, and possibly lichens all played a role in the formation of the first soils; today, plants source their nutrients from the soils in which they grow. Green algae of the Charophyceae group are widely regarded as being the ancestors of land plants (see p.56). Studies of the biochemical make-up and genetic material of charophytes suggest that they had more features in common with the first land plants than either would have with today's plants. The first true fossil evidence of plants on land dates from the Silurian period, 443 million to 419 million years ago. Silurian rocks also provide signs of millipedes, centipedes, and arachnids, feeding on plants and one another. This indicates an evolving terrestrial ecosystem. By colonizing the land, plants gained an evolutionary impetus that led to the diversity of plant life forms in the world today.

Although the ocean was a relatively benign and unchanging environment, the land was a hostile and challenging place. Living cells are mostly made of water, and drying out is one of the worst fates that can befall any organism. There is evidence that the arrival of land plants coincided with a time of climate change, and periods of heavy rain alternating with drought would have eliminated those forms that adapted poorly to the harsher conditions of life on land. The ebb and flow of the tides along the coastal margins would have been just as challenging. One way in which land plants adapted to reduce water loss was to develop a waxy covering, the epidermis (cuticle), that prevented evaporation. Plants also need to exchange gases with the atmosphere—they require carbon dioxide for photosynthesis, and release oxygen as a byproduct—and so plants evolved pores, called stomata, in the cuticle to allow the gases to pass in and out.

Another major adaptation was vascular tissue, which first appeared in the fossil plant record 410 million years ago, in plants such as *Cooksonia* (see image 2). Vascular tissue consists of cells that allow the transport of materials around the plant, performing a function similar to that of an animal's blood vessels. Water and nutrients absorbed by the roots are transported to the leaves, and sugar produced by the leaves is circulated. Being able to absorb water from the soil aids the plant's survival in dry conditions. Vascular tissue also provides structural support, being not only the plant's circulatory system but also its "skeleton," giving form to roots, shoots, and leaves and enabling the plant to grow upright (see image 3). Such was the advantage that plants gained with the advent of vascular tissue that they were able to spread rapidly and dominate the landscape. **RS**

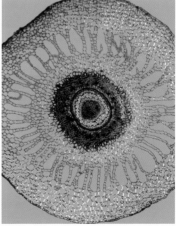

408 Mya	405 Mya	400 Mya	390 Mya	385–350 Mya	380 Mya
Aglaophyton major, a plant found in the Rhynie Chert fossil beds (see p.62), is an intermediate between nonvascular and vascular plants.	*Halleophyton*, an early vascular plant with simple leaves on its stems, lives in Yunnan province, China.	*Asteroxylon*, an early terrestrial vascular plant, has branching stems up to 14 inches (35 cm) long.	Another early and widespread vascular land plant, *Drepanophycus*, a lycopsid, has stems up to 39 inches (1 m) long.	Ferns, horsetails, and seed plants appear, some growing to more than 100 feet (30 m) in height and forming the first forests.	*Wattieza*, an early tree with woody stems (strengthened with the substance lignin) appears. It stands up to 26 feet (8 m) tall.

Rhynia and the Rhynie Chert

EARLY DEVONIAN PERIOD

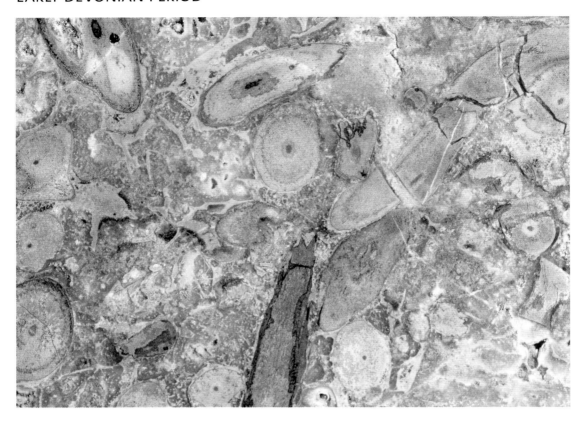

SPECIES *Rhynia gwynne-vaughanii*
GROUP Rhyniopsida
SIZE 8 to 12 inches (20–30 cm)
LOCATION Rhynie, Aberdeenshire, Scotland

◆ **NAVIGATOR**

The Rhynie Chert, near the village of Rhynie, northeast Scotland, is a remarkably well-preserved fossil community of terrestrial and aquatic plants and animals from the early Devonian, 408 million years ago. Some of the plants have been preserved in such detail that their internal structure can be studied, and the chert has become vital to understanding how plants evolved. Chert is a type of sedimentary rock comprising different varieties of quartz with finely crystalline silica minerals. The Rhynie Chert was deposited when dissolved silica from ancient hot springs formed crystals in the layers of sediment as the water cooled. Some of the silica coated and trapped plants and animals on the land, or within shallow ponds, preserving their structures in minute detail. Over millions of years the silica deposits were converted into crystalline chert rock. The kinds of organisms preserved in the chert include fungi, algae, plants such as *Rhynia gwynne-vaughanii* and *Aglaophyton*, and marine and terrestrial invertebrates, such as the early insect *Rhyniognatha* (see p.134) and a springtail, *Rhyniella*. The early vascular land plants were silicified while they were still growing, thereby preserving their pipelike systems, fragile rootlike rhizoids, stems with hairs, reproductive structures, and spores.

Other discoveries from the Rhynie Chert include the developing relationships between plants and fungi. Fossils of mycorrhizal (root-friendly) fungi have been found in vascular plants. These fungi have evolved a two-way beneficial relationship, mutualistic symbiosis, with plant roots: the plant shares its sugary high-energy substances from photosynthesis, while the fungus aids the plant's uptake of minerals and water. The opposite also occurred in the Rhynie community: an early example of parasitism in the fossil record. Cells of the alga *Paleonitella* have been found invaded by the aquatic fungus *Sorodiscus*. **RS**

👁 FOCAL POINTS

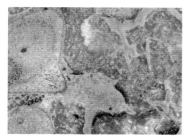

1 STEMS

One of the most abundant Rhynie plants, *Rhynia* appears to have been tolerant of a range of habitats and was able to withstand competition from other plants. Its upright stems grew to a height of around 8 to 12 inches (20–30 cm) and it spread over the surface via branching stems (rhizomes) that lay directly on the ground.

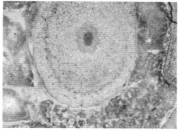

2 OUTER STEM LAYERS

Around the two parts of the core is a wide inner cortex, a dark, narrow dividing band and a lighter outer cortex, which is composed of cells packed together with air spaces in between. They are contained by the epidermis and, finally, the surface (cuticle), which has tiny pores (stomata) that allow the passage of air.

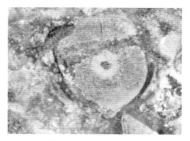

3 INNER STEM LAYERS

Vascular tissue is a plant's plumbing. The fossil of *Rhynia* has a dark central area, the xylem: long thick-walled cells specialized to carry water from the roots, or similar structures. Around this is a paler zone of phloem cells that conduct sap and similar fluids loaded with nutrients and sugars produced by photosynthesis.

◄ The chert's *Aglaophyton* was considered one of the earliest and simplest vascular plants. However, further investigation showed that it does not contain true vascular tissue, but conducting cells similar to those found in some mosses. In common with most other early plants, *Aglaophyton* had no true leaves or roots. It probably grew to about 12 inches (30 cm) in height.

DISCOVERING THE CHERT

The Rhynie Chert was discovered in 1912 by Dr. William Mackie (1856–1932), from Elgin, Scotland, during a study of the geology of Craigbeg and Ord Hill. In the area around Rhynie he spotted some unusual rocks in a drystone wall. He took samples and prepared thin sections that revealed perfectly preserved and detailed plant stems. Realizing the evolutionary significance of his findings, he made them known to paleontologists. In October 1912 an exploratory trench was dug, and in 1917 the first report on the plants of the Rhynie Chert was published by Scottish paleobotanist Robert Kidston (1852–1924). A number of discoveries were made over the years, including early insects and insectlike creatures, and in 1957, germinating spores. Continuing research includes a project to conceive the chert's ecology (left) and reconstruct a three-dimensional model of it.

Chara (Alga)
MIOCENE EPOCH—PRESENT

GENUS *Chara*
GROUP Characeae
SIZE 3 feet (1 m)
LOCATION Most continents
IUCN Not Yet Assessed

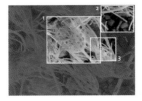

The movement of plant life from the water to the land was one of the most important events in evolution. Algae are simple plants with no true roots or leaves, and nothing approaching flowers. The charophyte algae, still found in freshwater ecosystems today, are believed to be among the closest aquatic living relatives to land plants. Unlike other green algae, charophytes reproduce sexually via gametes produced by reproductive organs—male antheridia make sperm, and female oogonia produce eggs. The resulting seedlike oospores germinate into new plants, sometimes after several years. The charophytes were much more diverse in the Ordovician than today. Among those still living are members of the *Chara* genus, plants also known as stoneworts.

Like their distant ancestors, the modern stoneworts are aquatic plants, found in swamps, rivers, streams, and estuaries. They appear to have stems and leaves but these are actually branches of the main body (thallus), like seaweed fronds. The plants are named stoneworts because they slowly become coated in encrusting limestone minerals, which give them a brittle texture. Studying the charophytes of today can give insight into what conditions may have been like in the distant past. Fossil charophytes are typically found in sedimentary chert, a quartzlike rock associated with other aquatic life forms, such as crustaceans. It is likely that the charophytes colonized relatively shallow freshwater ponds, which may have been quite alkaline. A few charophyte fossils have been found where all the stalks are swept over in one direction; this probably indicates that they were able to grow in water where there was a gentle current. **RS**

1 STALKS AND BRANCHLETS

Stoneworts are among the largest and most complex of the charophyte green algae. They have central stalks consisting of giant, multinucleate cells—cells that have more than one nucleus (control center)—with whorled branchlets, each growing at a node. Growth occurs at the tips of the central stalk and branchlets, which themselves produce new whorls in time.

2 RHIZOIDS

The plant body is anchored to the bed of its watercourse or pond by colorless rhizoids that resemble hairs. These are the equivalent of roots in later plants. In some charophyte species, structures called bulbils form at the joints of the rhizoids. These become buried in the sediment and can generate a new plant by vegetative growth, just as a plant rhizome would do.

3 SEXUAL REPRODUCTION

Biochemical studies of the charophytes have revealed characteristics they share with land plants, notably their methods of reproduction. They reproduce sexually via gametes produced by reproductive organs—male antheridia (red, above) make sperm, and female oogonia produce eggs. The resulting seedlike oospores eventually germinate and become new plants.

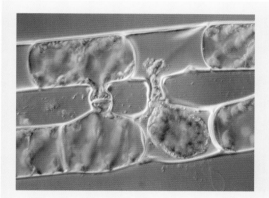

ZYGNEMATALES

Genetic analysis supports the theory closely linking *Chara* to land plants, but another genetic study, published in 2011, has thrown doubts on that thesis. Another group of algae, the Zygnematales, which include the hairlike pond inhabitant *Spirogyra*, employs a method of reproduction called conjugation in which the sex cells (sperm and egg) are of equal size and meet in a fertilization tube, where they join (above). This and other lines of evidence point to the Zygnematales as being the closest living relatives of land plants. The earlier thinking identified *Chara* as the closest relatives because they share a similar method of fertilization, called oogamy, with small sperm and a large egg; in flowering plants, the sperm is inside pollen grains.

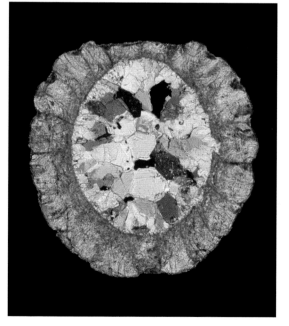

▲ Many charophytes produce resistant organic shells around their oospores. Called gyrogonites, these are frequently found in the form of fossils, formed by intracellular calcification of the spiral cells surrounding the oospore. Gyrogonites are the only link between living and fossil charophytes. It is on the evidence of these gyrogonites that charophytes may be dated to the late Silurian, about 425 million years ago. The presence of gyrogonites is also a good indicator that a site was under fresh water, rather than a saline sea, at a given time.

MOSSES AND FERNS

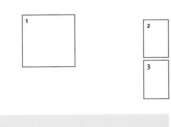

1 Ferns are regarded as among the first vascular plants and can cope with low light levels, exemplified by these lady ferns, *Athyrium filix-femina*, in a European beech forest.

2 These are spore-releasing capsules of the moss *Funaria*, the less prominent of the two generations in the moss life cycle.

3 These fossilized fern fronds date from the Jurassic period. The species is *Matonidium goeperti*, which reached about 2 feet (60 cm) in height.

Mosses, liverworts, and hornworts together make up the bryophytes, the second largest group of living plants, and their history stretches back further than any other modern-day land plants. Fossil records show that many of their features are similar to those of the long-extinct plants that made the transition from water to land. Like those early pioneers, bryophytes lack a well-developed cuticle (waxy covering) to limit water loss, and have no vascular system (internal plant plumbing) to distribute water through the plant. They are dependent on moist conditions for their survival.

One of the major challenges for ancient organisms, plant or animal, moving onto land was reproduction. Sexual reproduction in aquatic plants and animals had evolved a system whereby male and female sex cells (gametes) joined to form a single-cell zygote (fertilized egg), which divided again and again until it formed a new organism. In a watery medium, be it a pond or an ocean, it was easy for the male gametes (sperm) to swim to the waiting female gamete (the ovum, or egg cell), but it was less feasible on dry land. It was a

KEY EVENTS

473 Mya	470–450 Mya	419–359 Mya	410 Mya	400 Mya	390–320 Mya
Mosses and liverworts emerge with fossil cryptospores (sporelike structures) in rocks from Argentina, but they are difficult to assign to either group.	The first good evidence for plant life on land dates from this time, as partial matlike and strandlike fossils within land-based rocks.	The earliest members of the Equisetopsida ferns appear. The only living genus, *Equisetum*, includes the horsetails.	Plants with fernlike characteristics appear mainly in Europe, East Asia, and Australia during the Devonian period.	Early ancestors of hornworts start to display differences from ferns.	The woody, treelike cladoxylopsids, thought to be ancestors of ferns and horsetails, flourish.

problem that some groups never did solve: bryophytes and amphibians, for example, still rely on water to complete their reproductive cycles.

As a partial solution, bryophytes evolved a process of reproduction that takes place in two phases, or generations. In the first phase, the sporophyte generation, the organism reproduces by means of spores (see image 2), which are seedlike units; in the second phase, the gametophyte generation, sex cells are involved. This pattern, known as the alternation of generations, is readily observed in living mosses. At certain times of the year, stalks are seen protruding from the green mat of a moss. These stalks are the sporophytes, from which spores are produced and blown away in the wind. If they land in a suitable place for growth, the spores develop into a new moss plant, the gametophyte. Structures within the gametophyte—the equivalent of animal sex organs— produce male and female gametes. The male sperm swim through a watery film to the eggs and fertilize them, after which each egg, if it can grow, produces a new sporophyte stalk, and the process repeats. Water is needed for the sperm to reach the eggs, which is why mosses live in moist environments.

The gametophyte generation is the dominant phase in bryophyte life cycles. In vascular plants, such as ferns (see image 1), which evolved later, the opposite is true and the sporophyte or seed generation is most prominent. The development of vascular tissue not only enabled water and nutrients to circulate through the plant but also provided a supporting structure, which was important for land-based plants. Simple aquatic plants such as seaweed, surrounded and supported by water, have no need of a vascular system. Among the earliest plants to develop vascular tissues were the lycophytes (see p.70), today represented by the clubmosses, which evolved about 440 million years ago. Lycophytes came to dominate the landscape during much of the Devonian and Carboniferous periods, from 410 million to 320 million years ago.

Ferns or pteridophytes (see image 3), also among the earliest vascular plants, are well adapted to life in shady forests. In part this is due to the presence of a gene that makes the protein neochrome, which gives the fern the ability to sense and grow toward red light. This is important because, under the forest canopy, many other wavelengths of light are filtered out by the foliage. In 2014, researchers discovered that fern neochrome is shared by hornworts (plants that produce hornlike sporophytes). In evolutionary terms, ferns—such as *Marattia* (see p.68)—and hornworts diverged around 400 million years ago. If neochrome had been present in a common ancestor it would almost certainly be present in other plant families, too, but it is not. Instead, the likelihood is that the fern neochrome was acquired from the hornwort version about 180 million years ago. The most probable explanation is that the hornwort gene was somehow transferred to the ferns—a process called horizontal gene transfer— perhaps by bacteria or viruses. It allowed ferns to flourish and diversify in the competitive environment that came with the advent of flowering plants. **RS**

385–359 Mya	370 Mya	350 Mya	299–252 Mya	180 Mya	120 Mya
Fern-like *Archaeopteris* forms early forests during the late Devonian, evidenced by worldwide fossils, although initially with less woody stems.	*Archaeopteris* now has a woody trunk and looks similar to modern conifers but with fernlike foliage and spores instead of seeds.	The first true ferns appear during the Carboniferous and diversify greatly.	The first unambiguous evidence of mosses dates from this time, in rocks from the Permian period, chiefly from Asia and North America.	The hornwort neochrome gene, facilitating growth at low light levels, appears to be transferred to the ferns.	Many kinds of ferns are fading to extinction due to competition from flowering plants. The neochrome ferns survive this battle.

Marattia
CARBONIFEROUS PERIOD—PRESENT

SPECIES *Marattia douglasii*
GROUP Marattiaceae
SIZE 6 feet (1.8 m)
LOCATION Hawaii
IUCN Not Yet Assessed

◆ NAVIGATOR

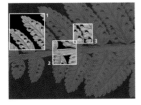

A genus of ferns found in the pantropical regions of Central and South America, Asia, Australasia, and Africa, *Marattia* diverged onto its own evolutionary path very early in prehistory. It is also one of the first examples of vascular plants in the fossil record. Indeed, 300-million-year-old fossils from the Carboniferous show that the ferns differed very little from their present-day descendants, such as *Marattia douglasii*.

Ferns have rhizomes—underground stems, that can produce the shoot and root systems of a new plant. This means that the plant can spread vegetatively, without sex, and also reproduce sexually and spread through spores. Because they are protected underground, rhizomes give the plant the ability to survive unfavorable conditions. In ferns, the rhizomes also grow above ground. In some fern species, rhizomes form more than 80 percent of the biomass (the total weight of their living matter), with the visible fronds representing most of the rest.

The main Marattiaceae family contains variable numbers of genera, depending on the classification system, including *Angiopteris, Archangiopteris, Christensenia, Danaea, Macroglossum, Marattia,* and *Protomarattia*. The number of species is also very variable, from a few dozen to a few hundred. The fact that *Marattia* ferns are so similar to their ancestors, and have no close living relatives, has led to them being described as living fossils, a controversial term coined by Charles Darwin (1809—82), to describe a species or group that stopped evolving. "Living" certainly, so not a "fossil," *Marattia* can still provide insight into how the early Carboniferous ferns would have looked. **RS**

👁 FOCAL POINTS

1 FRONDS
The leaves of ferns are often called fronds. The underside of each frond has simple unbranching veins running through. Some of *M. douglasii*'s allied species have fronds that are not much bigger than a human hand, while in *Angiopteris evecta*, the giant fern, they reach 23 feet (7 m).

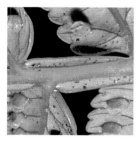

2 FROND STEMS
The stipe is the vertical stalklike part of the frond that branches up from the ground-based, rootlike rhizome. The stipe in turn branches into veins. Each vein is known as the axis or rachis (horizontal part, left). From it, in turn, branch several leaflets or pinnae, in *M. douglasii* with lobed edges.

3 SPORANGIA
Sporangia are the bodies in which the spores are produced. *M. douglasii* sporangia are thick-walled and grow in two rows on either side of the main vein. An inner layer of the sporangium called the tapetum nourishes the spores, which are released when dry conditions cause the sporangia to split.

⊥ MARATTIA RELATIONSHIPS

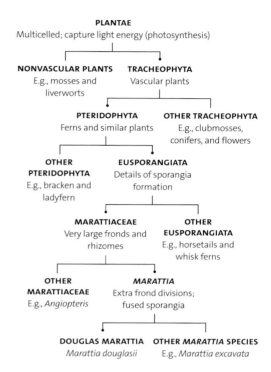

PLANTAE
Multicelled; capture light energy (photosynthesis)

NONVASCULAR PLANTS
E.g., mosses and liverworts

TRACHEOPHYTA
Vascular plants

PTERIDOPHYTA
Ferns and similar plants

OTHER TRACHEOPHYTA
E.g., clubmosses, conifers, and flowers

OTHER PTERIDOPHYTA
E.g., bracken and ladyfern

EUSPORANGIATA
Details of sporangia formation

MARATTIACEAE
Very large fronds and rhizomes

OTHER EUSPORANGIATA
E.g., horsetails and whisk ferns

OTHER MARATTIACEAE
E.g., *Angiopteris*

MARATTIA
Extra frond divisions; fused sporangia

DOUGLAS MARATTIA
Marattia douglasii

OTHER *MARATTIA* SPECIES
E.g., *Marattia excavata*

▲ Much of the grouping of ferns, or pteridophytes, is based on detailed leaf and reproductive features. These include the mode of development of sporagia—the structures on fertile leaves, sporophylls, that house developing spores for reproduction.

FOSSIL FERN CELLS

In 2014, researchers in Sweden reported a 180-million-year-old fossil fern so well preserved that cell details, including nuclei (control centers) and even individual chromosomes (gene parcels), could be observed (right). The reason for the fern's astonishing degree of preservation is that it had been buried suddenly by a volcanic lava flow. X-ray analysis and a variety of microscopic techniques even show cells at different stages of division. The fossil fern belonged to the Osmundaceae, the royal ferns, a family that still exists today. By comparing the size of the cell nuclei in the fossilized plant with its living relatives, the researchers were able to show that the royal ferns have barely changed over millions of years. Pollen and spores fossilized in the same rocks reveal large amounts from coniferous trees, including cypress, and cycads as well as ferns, indicating a well-vegetated area and probably a hot and humid climate.

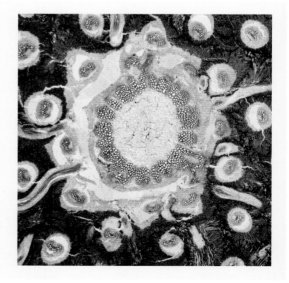

LYCOPHYTES, SCALE TREES, AND SEED FERNS

1 The living *Lycopodium* clubmosses (ground pines) can trace their group ancestry back to the Devonian.

2 The fossilized trunk of a *Lepidodendron* scale tree has scars on the bark where specialized leaves, or microphylls, once grew. Only at maturity did scale trees develop the leaf-covered branches that eventually produced cones.

3 Fronds of the seed fern *Neuropteris*, fossils of which are commonly found in the Carboniferous coal deposits of Pennsylvania.

The present-day clubmosses (see image 1) are the oldest living representatives of the spore-producing lycophytes, thought to be the first vascular plants: able to distribute water and minerals via veins. Today, they are an inconspicuous group, but 350 million years ago they dominated the landscape of the Carboniferous period for 40 million years. They formed extensive forests that could stand 115 feet (35 m) tall. The oldest lycophytes appear in the fossil record in the early Devonian period, around 410 million years ago. Many of the plants included in this group appear to have evolved from the zosterophylls, a now extinct group of plants that had an important place in the evolution of land plants. Zosterophylls lacked true leaves and roots, instead having stems covered by scalelike flaps called enations that lacked the vascular tissue (internal tubing network) of true leaves. They also had their spore-bearing parts, for reproduction, on small stalks arranged in clusters along branches and these characteristics were also seen in the lycophytes.

Although their spores are plentiful, the smaller herbaceous lycophytes have left little evidence of their mature forms in the fossil record. There are, however, some well-preserved specimens of related tall, tree-sized species. These include

KEY EVENTS

420 Mya	408–360 Mya	400 Mya	385–380 Mya	383–359 Mya	383–359 Mya
Zosterophylls have simple-branched, spine-covered stems. They were most probably terrestrial.	Evidence shows that mycorrhizal fungi live in a symbiotic relationship with plant roots.	*Eophyllophyton* appears in China. It is one of the oldest known plants with true vascular (megaphyllous) leaves.	*Wattieza*, an early tree at 26 feet (8 m) tall, has fronds rather than leaves. It is related to modern-day ferns and horsetails.	During the late Devonian, plants with true leaves and roots evolve, including lycophytes, sphenophytes, ferns, and progymnosperms.	The first true trees to form forests, *Archaeopteris* appear during the late Devonian.

the scale trees—the *Lepidodendron* group (see p.72)—the fossilized remains of which formed the basis of the coal seams that powered the Industrial Revolution millions of years later. The most significant feature of lycophytes and scale trees were their leaves, called microphylls (see image 2), which appear to have evolved independently from the leaves of other vascular plant groups. The microphyll has only a single vein of vascular tissue, whereas the leaves of other plants, called megaphylls, have multiple branching veins. A major extinction event, some 305 million to 299 million years ago, known as the CRC (Carboniferous Rainforest Collapse), involving advancing glaciers and drought conditions, brought an end to the dominance of the scale trees and other large lycophytes.

Among the most important living things on Earth are the spermatophytes (seed plants). Conifers, such as pines and firs, are seed plants, as are all of the familiar flowering plants (angiosperms), from grasses to orchids and oak trees. The differences between seeds and spores are technical. Generally seeds have a food store and a tiny embryo plant and they germinate into the sporophyte phase, while spores are smaller and simple, with only one living cell, which grows into the gametophyte (sexual) stage. The early seed-bearing plants were relatively small in comparison to the mighty scale trees that dominated the hot and humid swamps where they both grew. One of the oldest known seed plants is a seed fern, *Elkinsia*, which first appears in the fossil record toward the end of the Devonian, from 365 million years ago.

The seed ferns, also known as pteridosperms, are a loose grouping of seed-bearing plants that flourished during the Carboniferous and Permian periods, about 360 million to 250 million years ago (see image 3). The name "seed fern" is a little misleading because they were not actually ferns but early forms of the gymnosperms, the cone-bearing plant family that includes today's conifers (see p.80). Some were vinelike; others grew into tall tree-shaped forms, such as *Glossopteris* (see p.74). All had foliage very much like that of ferns, but unlike ferns, seed ferns reproduced by seeds rather than spores. The earliest seed-bearing plants lacked the specialized cones and flowers of modern-day seed plants. Their seeds were produced singly or in pairs along the branches of the plant and they were enclosed by a hard protective tissue called the integument. This feature is still found in all seeds, and it develops into the seed coat in modern plants. The seed was held within a loose, fleshy, cuplike structure called a cupule. Both integument and cupule are thought to have evolved from leaves that gradually reduced in size and adapted to the task. As the integument continued to evolve and to enclose the seed more tightly, an opening called the micropyle was left at one end, thereby allowing pollen to enter so that fertilization could take place. The pollen sacs of the seed ferns could be large and complex, and there has been speculation that they were pollinated by animals. The seeds of some species were large, too—up to 3 inches (8 cm)—and perhaps also dispersed by animals. **RS**

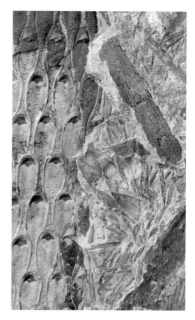

365 Mya	359 Mya	359–299 Mya	359–299 Mya	320–300 Mya	305–299 Mya
Elkinsia, a very early seed fern and one of the first known seed plants, appears in West Virginia.	The early seed plants begin to spread from the end of the Devonian.	During the Carboniferous, rich deposits of coal are formed by fossil remains of extensive swamp forests.	The high point of the lepidodendrales or scale trees, this period saw these relatives of clubmosses reach heights of more than 100 feet (30 m).	The Medullosales, seed plants related to the cycads, flourish in tropical wetlands. They grow 33 feet (10 m) tall with fronds measuring 23 feet (7 m).	The CRC signals the demise of the scale trees and large lycophytes.

Lepidodendron
CARBONIFEROUS PERIOD

Lepidodéndreas.   Sigilarias.   Estigmarias.

PAISAJES DEL MUNDO PRIMITIVO. — CUADRO TERCERO: TIPOS LEPIDODENDROS.

O ne of the most abundant plants of the Carboniferous, 359 million to 299 million years ago, was *Lepidodendron*, which means "scale tree." It was a member of the lycophytes, a group of spore-bearing vascular plants and it is a broad term for treelike fossil lycophytes that are similar in appearance but difficult to divide into individual species. *Lepidodendron* grew as an unbranched trunk covered by photosynthesizing leaf scales. Only when it matured did it produce a crown of long, thin leaves that grew directly from the stem near the growing tip. These trees could reach 130 feet (40 m) or more in height and 6½ feet (2 m) in diameter; they grew at an extraordinary rate, reaching full size in only 10 to 15 years. Since *Lepidodendron* produced no branches until toward the end of its life, individual trees could grow close together without blocking their neighbors' light, forming extremely dense stands.

Fast growth and short lifespans resulted in the formation of thick layers of dead tree remains. Over millions of years these remains were buried, and in time, transformed by pressure, heat, and chemical reactions into coal. Most of the world's coal reserves are the remains of *Lepidodendron* along with other extinct lycophytes. However, by the time of the Jurassic period, scale trees such as *Lepidodendron* had waned in the face of climate change and the upcoming gymnosperms, including cycads and conifers. **RS**

◈ NAVIGATOR

GENUS *Lepidodendron*
GROUP Lycophyta
SIZE 130 feet (40 m)
LOCATION Worldwide

◉ FOCAL POINTS

1 LEAVES
Resembling huge blades of grass, the leaves of *Lepidodendron* were around 3 feet (1 m) in length and were arranged in spirals around the stem. As the plant grew, it shed leaves from lower parts of the stem, leaving the diamond-shaped leaf bases that gave it the name "scale tree."

2 STEM
The stem of lycophytes had a central core of vascular tissue, in which a strand of water-carrying xylem tissue was surrounded by nutrient-carrying phloem tissue. The bulk of the stem was formed from layers of bark, believed to have been resistant to decay as an adaptation to its wet habitats.

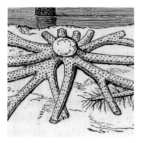

3 ROOTS
Lepidodendron had an impressive root system. This consisted of four or more radiating arms called stigmaria that could extend to 40 feet (12 m) in length but did not penetrate deeply into the ground. The stigmaria were not true roots but an intermediate between roots and stems.

COAL SWAMPS

So-called coal swamps were one of the primary terrestrial ecosystems of the Carboniferous and Permian. During the Carboniferous, large parts of what would become Europe and North America were in an equatorial position and extensive swamp forests developed in the wetland areas at the continental margins. Most of these forests were made up of lycophytes such as *Lepidodendron* (above), which grew fast, died young, and accumulated massively, laying down the basis for coal. Among their trunks lived giant insects such as griffinflies (see p.170) and the huge millipede *Arthropleura* (see p.132). Analysis has shown that up to 70 percent of coal mined in the world today is composed of lycophyte remains.

Glossopteris
PERMIAN PERIOD

GENUS *Glossopteris*
GROUP Glossopteridales
SIZE 26 feet (8 m)
LOCATION India, South America, Australia, Africa, and Antarctica

⏣ NAVIGATOR

The largest and best-known member of the Glossopteridales, an extinct group of seed ferns that first appeared during the Permian, 299 million to 252 million years ago, *Glossopteris* was a dominant part of the flora on the great southern continent of Gondwana but had become extinct by the end of the Triassic period, 201 million years ago. Fossils of more than 70 species, including *G. browniana*, have been found in India, along with other species in South America, Australia, Africa, and Antarctica, but none has been identified with certainty in the northern hemisphere.

The name *Glossopteris* (meaning "tongue") derives from the shape of the leaf, but apart from its tongue-shaped leaves, no one is certain what the various species of *Glossopteris* looked like. No large specimens have been preserved intact. They may have been large shrubs or small trees, possibly similar to a magnolia or young ginkgo, some reaching a height of 100 feet (30 m). Many fossils found with them indicate that they thrived in semi-aquatic habitats such as swamps. Thick mats of glossopterid leaves have been preserved in large numbers, suggesting that they were deciduous trees that shed their leaves in the fall. Other evidence of this includes an abscission zone—the site where leaves detach in deciduous plants today. Some species had tiny scale leaves that may be leaf buds. There are also growth rings in the trunks, which are common in trees that grow fast and then slower with changing seasons. **RS**

⊙ FOCAL POINTS

1 REPRODUCTIVE PARTS

The female and male organs were carried on the leaves, rather than as separate structures, such as flowers, which would evolve much later. In some *Glossopteris* species, the leaf edges appear to have rolled over to form an enclosing chamber around the developing seeds that were produced on the underside of the leaf.

2 OVERALL FORM

Fossils indicating the shape and form of a whole *Glossopteris* plant are yet to be found. One suggestion is that it was a tall, straight tree tapering toward a narrow point at the top, similar to many conifers in colder regions today. Another theory is that it had more of a bushy configuration with a rounded crown, as is shown here.

3 LEAVES

The tongue-shaped mature leaves of *Glossopteris* varied from around 4 to 40 inches (10–100 cm) in length. They are distinguished from the leaves of other seed plants of the time by their pronounced central vein and network of smaller veins. Other leaves of the time had secondary veins running in parallel but no midvein.

▲ The roots of the glossopterids had a distinctive structure. They are termed vertebrarian because the regularly spaced partitions along the roots give them the appearance of a backbone. The roots also have woody strands along their centers, with lobelike spaces at intervals, which is highly unusual among plants.

CONTINENTAL DRIFT

In 1912, German meteorologist and geophysicist Alfred Wegener (1880–1930) put forward his theory of continental drift. He proposed that the continents were not fixed to the surface of Earth but moved around very slowly. He had noticed the similarity of rocks on opposite sides of the Atlantic and he knew of fossil species that were distributed across the southern continents, such as the reptile *Mesosaurus* (p.241) in Africa and South America, and *Glossopteris* in Africa, South America, India, Australia, and Antarctica. The ill-fated Scott Expedition to the South Pole in 1911 to 1912, for example, obtained rocks from the Transantarctic Mountains that contained *Glossopteris* remains. It was inconceivable that these species could have evolved separately. Wegener's explanation, now accepted, was that the continents were once joined and had subsequently drifted apart (right).

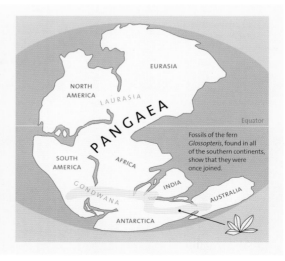

Fossils of the fern *Glossopteris*, found in all of the southern continents, show that they were once joined.

LYCOPHYTES, SCALE TREES, AND SEED FERNS 75

CYCADS, GINKGOES, AND CONIFERS

1 These fossil cycad leaves from the Cretaceous period were preserved in fine-grained claystone.

2 The king sago palm, *Cycas revoluta*, is a living decorative cycad (not a true palm, which is an angiosperm).

3 Conifers such as pines have needlelike leaves and bear their pollen and seeds in structures called woody strobili, commonly known as cones.

The cycads are an ancient plant group. Fossil cycads found in China date back to the early Permian period, 280 million to 270 million years ago. Thought to have evolved from the tree ferns, they are recognized as the sister group to all other seed plants in existence today: the ginkgoes, conifers, flowering plants (angiosperms), and a few minor groups. With their large, divided leaves, they resemble palms or tree ferns in appearance; individual plants are dioecious (either male or female), and they reproduce by means of seeds, which develop on special seed-bearing leaves on the female plant.

Present-day cycads are a small group comprising some 250 to 300 plant species found in tropical and subtropical regions around the world. Although their species number only a few today, at their height, in the Mesozoic era, especially during the Triassic and Jurassic periods, 252 million to 145 million years ago, they were diverse and made up as much as one fifth of the world's terrestrial plant life. Cycad fossils (see image 1) have been found on every continent, because the warm and humid Mesozoic climate allowed them to spread across the surface of the globe. For this reason the Jurassic is sometimes known as the Age of Cycads. Although sometimes cited as living fossils, the three surviving cycad families cannot trace their lineage back to their most ancient ancestors, although most can be recognized in fossils from the early Paleogene period, 60 million to 50 million years ago. Some of today's cycad species (see image 2)

KEY EVENTS

359–299 Mya	300 Mya	300 Mya	280 Mya	260–250 Mya	252 Mya
Conifers begin to show signs that they will take over from previously dominant vegetation, such as scale trees.	Several groups of gymnosperms have evolved by this time, some from the fernlike Progymnospermophyta.	The Cordaitales, an important group of early gymnosperms, grow in brackish coastal waters. They are related to cycads and conifers.	The first cycads appear. They come to dominate many land habitats through to the Jurassic, more than 100 million years later.	Gigantopterids, possibly ancestors of the flowering plants, flourish. They are among the most advanced land plants of the time.	The "Great Dying" mass extinction at the end of the Permian period (see p.224) opens up new evolutionary opportunities.

originated around 10 million years ago. In the past, the cycads have been classified as gymnosperms—the "naked seed" plant group, which includes conifers and ginkgoes. In gymnosperms the developing seeds are relatively unenclosed and unprotected, compared to the seeds of angiosperms, which are enclosed within a protective ovary. However, recent research has identified some cycads as being more closely related to angiosperms than to gymnosperms.

One gymnosperm, the ginkgo tree (see p.78), is a familiar sight in parks and gardens all over the world. It is a unique tree with no living relatives. The fossil record shows that it has remained virtually unchanged for the past 50 million years and bears a strong resemblance to its ancestors, which flourished in the Jurassic forests nearly 200 million years ago. Fossil discoveries from the 121-million-year-old Yixian Formation in northeast China reveal only subtle differences between the living ginkgo species and the early Jurassic members. Today, the venerable ginkgo is all but extinct in the wild. It owes its survival to its cultivation by Buddhist monks in China. The ginkgo's vascular system (internal plant plumbing) of water- and sap-carrying tubes is very similar to that of the conifers (see p.80), which suggests that the two groups have a common ancestor. Conifers, the familiar cone-bearing trees (see image 3), such as yews, pines, and cedars, are the most common and abundant of the gymnosperms and they are the tallest and largest living things ever to appear on land. Their predecessors were the Cordaitales, an extinct group of gymnosperms that had long, straplike leaves and often lived on mudflats in brackish water, probably growing in a similar way to modern mangroves. The earliest true conifers appeared around 300 million years ago and had branches and needles that resembled those of present-day *Araucaria*, a genus that includes the Chilean pine or monkey puzzle tree, although their seed cones were much more loosely assembled than the tightly packed cones of modern conifers.

Another curious group of gymnosperms is the Gnetales, encompassing the genera *Gnetum*, *Welwitschia* (see p.85), and *Ephedra*. The *Gnetum* are tropical lianas (woody vines), shrubs, or small trees with broad and glossy evergreen leaves; *Ephedra* are shrubby gymnosperms occurring on most continents, especially along seashores (common names include sea-grape), several species of which are prized in traditional medicine; and *Welwitschia*, one of the world's strangest plants, which lives wild only in and near the Namib Desert of southwest Africa, and has just two leaves growing throughout life.

The ancestors of most modern conifers originated in the Mesozoic, 252 million to 66 million years ago, when dinosaurs dominated the land. They formed mixed forests with the cycads and ginkgoes. However, what the evolutionary connection is between the various groups of gymnosperms—if any—remains uncertain. Based on examinations of evidence such as genes of living species and the structure of fossil seeds, the different gymnosperm groups may be descended from a common, as yet unknown, distant ancestor. **RS**

250 Mya	250–200 Mya	240 Mya	190 Mya	70 Mya	66 Mya
The bennettitaleans (see p.80), another order of seed plants once thought to be flowering plant ancestors, appear.	The Gnetales, which share characteristics with angiosperms, appear. *Welwitschia mirabilis* is a surviving example.	Gigantopterids, some reaching 2 feet (60 cm) in height, fade, having flourished only briefly.	Ginkgo trees reach their zenith, becoming common in many habitats, but only one species survives to the present day.	The bennettitaleans begin to fade and nearly all will be gone by the end of the Cretaceous.	The end-of-Cretaceous mass extinction (see p.364) affects many land habitats and favors the evolution of flowering plants.

Ginkgo
CRETACEOUS PERIOD—PRESENT

SPECIES *Ginkgo biloba*
GROUP Ginkgoaceae
SIZE 100 feet (30 m)
LOCATION China (in the wild)
IUCN Endangered

The present-day ginkgo tree, *Ginkgo biloba*, is the sole representative of a plant lineage that first appeared in the Permian, 299 million to 252 million years ago, and reached maximum diversity in the early Jurassic, 190 million years ago. At their height, there were at least 16 genera of the Ginkgoales and they made up a large part of the world's vegetation, alongside cycads, conifers, and ferns. Recent studies point toward cycads as the ginkgo's nearest living relatives. One of the ginkgo's microfeatures is that its sperm have flagella (tails), which gave them the ability to move. In living seed plants this is a feature found only in the ginkgo and the cycads.

Changes to the way in which the seeds are formed are among the ginkgo's few evolutionary developments. Jurassic specimens had sprays of stalks, each with a single seed; later specimens from the Cretaceous, 120 million years ago, had multiple seeds on a single stalk; while present-day trees make a few seeds on a single stalk, only one of which reaches maturity. By the early Paleocene epoch, 66 million to 60 million years ago, the Ginkgoales had been reduced to a single species, whose leaves were virtually identical to those of the present-day *Ginkgo biloba*, hence its reputation as a living fossil. The Paleocene species was common in the northern hemisphere, but as Earth's climate cooled, the ginkgo's distribution moved south and it eventually disappeared in North America and Europe. **RS**

👁 FOCAL POINTS

1 LEAVES
The leaves of ginkgo trees have changed in appearance over the course of their evolution. In the Jurassic they were divided into lobes, similar to the leaves of a modern chestnut tree. Approximately 50 million years later the lobes had joined to form the familiar fan shape of the present day.

2 POLLEN CONES
Ginkgo trees are dioecious— either male or female. Pollen grains from the cone of the male tree are carried by the wind to the female parts, which are not true cones but ovules, on a separate tree. Fertilization occurs in the fall as the seed falls from the tree and the seed coat rots away.

3 SEEDS
Ginkgo seeds are large in comparison with those of other trees and they look like yellow cherries, with a fleshy outer covering surrounding a woody nut. They fall from the tree in autumn just before the leaves. The seeds are attractive to animals such as squirrels, which aid in their dispersal.

SAVED FROM EXTINCTION

It is likely that the ginkgo may have become extinct if not for its cultivation by Buddhist monks. Although it vanished from elsewhere in its range, it survived in Buddhist monasteries in China, where monks cultivated it for medicinal purposes and the nutritious qualities of its nutlike seeds (above). The seeds were taken to Japan and Korea, and subsequently to Europe and the United States in the 18th century. The name "ginkgo" is derived from the Japanese word *ginkyo*, meaning "silver apricot," in reference to the fruit of the tree. Today nearly every arboretum or botanical garden in a temperate or subtropical climate has at least one ginkgo tree. There are some stands of ginkgo trees in China that molecular evidence suggests may be of natural origin; if so, they are the last survivors of the ginkgo tree in the wild.

THE AGE OF CONIFERS

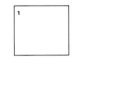

1 *Williamsonia* was one of the most widespread of the Bennettitales, with fossils known from almost every continent.

2 The living Wollemi pine, *Wollemia nobilis*, belongs to a genus that was thought to be extinct for more than 2 million years, until it was discovered in Australia in 1994. It is distantly related to the monkey puzzle tree, *Araucaria araucana*.

3 The willow podocarp, *Podocarpus salignus*, is a member of the successful, mainly southern hemisphere, family of conifers, Podocarpaceae, many of which have broad leaves.

The world was a very different place when the first conifers began to flourish. At the beginning of the Triassic period, 250 million years ago, the continents were gathered together into the vast supercontinent of Pangaea. This had a huge effect on the Triassic climate. Pangaea straddled the equator: much of it was high above sea level; there was little in the way of inland seas and lakes; and the interior of the vast landmass was too far from the surrounding ocean to feel the cooling effect of seawater or to receive much rain. Away from the coasts, therefore, the climate was generally hot and arid.

Over the following millions of years Pangaea began to break apart and form, in the south, Gondwana (present-day South America, Africa, India, Antarctica, and Australia) and, in the north, Laurasia (North America and Eurasia). As the continents drifted apart, the sea level rose and large forests of gymnosperms (see p.77) grew up along rivers and lakes. The locations of these forests are indicated by the presence of the rich coal seams that formed from their remains. The northern forests at the start of the Triassic were dominated by conifers, ginkgoes, cycads, and bennettitaleans such as *Williamsonia* (see image 1), a type of seed plant that became extinct toward the end of the Cretaceous period.

KEY EVENTS

300 Mya	200 Mya	200 Mya	200–150 Mya	150 Mya	150 Mya
The supercontinent Pangaea is formed and surrounded by the superocean Panthalassa. Climate is affected worldwide.	Modern conifers such as pines and yews appear and begin to diversify.	The Wollemi pine, actually an araucarian, exists. It is not discovered by science until 1994.	The first redwoods or sequoias (a subgroup of the cypress family) appear in the northern continents, especially North America.	Southern hemisphere podocarps begin to evolve broad, flat leaves to compete with flowering plants.	The sauropods (see p.264), great long-necked dinosaurs, evolve to eat leaves of conifers such as araucarians.

As the continents continued their slow drift, moving farther apart by the start of the Cenozoic era, 66 million years ago, species in the northern hemisphere had to cope with colder conditions, and changing seasons, rather than the unvarying conditions of the equatorial tropics. The resulting pressures spurred evolution, as the species adapted to the changing climates they faced. The majority of today's conifer species belong to lineages that diversified during the Cenozoic. New species appeared with greater frequency in the north than in the south. Firs, hemlocks, larches, pines, spruces (Pinaceae family), junipers, and cypresses (Cupressaceae) are now found across the northern hemisphere, whereas the oldest of the seven living families of conifer, Araucariaceae (see image 2) and Podocarpaceae (see image 3), are distributed primarily within the southern hemisphere, in Argentina, Australia, New Guinea, and New Zealand, where the persistence of milder, wetter habitats favored their survival.

In 2012, researchers at Yale University examined the fossil record and genetic make-up of 489 of the more than 600 living conifer species. They discovered that while the major conifer families could trace their origins back to the Mesozoic era, before 66 million years ago, most of the northern hemisphere species had appeared as recently as 5 million years ago.

Changing climate was not the conifers' only obstacle. Their position as the dominant flora was challenged by the advent of the angiosperms (flowering plants; see p.86) in the Cretaceous. Studies published by the University of Adelaide in 2011 show how the appearance of angiosperms pushed the conifers into the far north and more mountainous regions. However, one family of podocarps in the southern hemisphere, the plum pines, was able to evolve and adapt. By evolving flattened leaves, rather than the needles typical of conifers, they became one of the most successful conifers. Using fossil and genetic evidence, the researchers showed that the change from needle to flat leaf began between 150 million and 90 million years ago, and the plum pines' rate of evolution reached its peak around the time that flowering plants appeared. The broader leaves of the podocarps allowed them to compete for light with flowering plants in the emerging tropical rainforests. In effect, the podocarps' broad leaf design copied one of the flowering plants' most successful developments.

The study of conifer evolution may also help researchers to narrow down the date for the origin of modern rain forests—something that is still a cause of disagreement. The plum pines' evolution was spurred by the pressures of natural selection placed on them by the emergence of the rain forests, so determining when one event took place will provide a good pointer to the timing of the other. As to why other conifers did not make the same adaptive changes, researchers speculate that it might be because they are a very old group that has quite simply reached the limit of its potential in some areas of its continuing evolution. **RS**

145 Mya	140–120 Mya	130 Mya	100 Mya	100 Mya	10–5 Mya
Seed-producing gymnosperms begin to dominate the landscape as the seed ferns decline into extinction.	Flowering plants begin to dominate. They start to challenge conifers in many habitats.	The oldest known pine cones (strobili) date from this time, the early Cretaceous. Some conifers evolve male and female cones (see p.84).	The supercontinent of Pangaea begins to split into the continents we recognize today, thus changing the climate zones worldwide.	*Metasequoia*, the dawn redwood (see p.82), grows in western Canada, as evidenced by fossils.	Many northern hemisphere conifer tree groups undergo a burst of evolution and diversify into species still alive today.

Dawn Redwood
MIDDLE CRETACEOUS PERIOD—PRESENT

SPECIES *Metasequoia glyptostroboides*
GROUP Cupressaceae
SIZE 164 feet (50 m)
LOCATION China (in the wild)
IUCN Endangered

C ommonly known as the dawn redwood, *Metasequoia* is one of the few living plants that was first known as a fossil. It is a member of the Cupressaceae family, along with juniper and red cedar. The earliest *Metasequoia* fossils were found in western Canada and date from the middle Cretaceous, roughly 100 million years ago. In 1941, Shigeru Miki (1903–74), a Japanese paleobotanist, examined 5-million-year-old Pliocene fossils of the tree and placed them in a new genus, *Metasequoia*. Previously, the fossils had been confused with bald cypress (*Taxodium*) and redwood (*Sequoia*). In 1943, a forester discovered a deciduous conifer tree near a village in China's Sichuan province. He collected a specimen, which he later gave to an assistant at the National Central University in Chongqing. It became apparent that the specimen was something new, and in 1946 the tree's specimens were matched with *Metasequoia*, the fossils established by Miki in 1941.

Metasequoia is now grown in Europe and North America as an ornamental tree. Wild populations in its native China are, however, extremely small and the species is classified as endangered. The tree prefers a damp habitat, such as along the banks of rivers. Evidence suggests that it was once a dominant and widespread member of northern deciduous forests until a cooling and drying climate change that began around 65 million years ago drove it back to its present distribution. **RS**

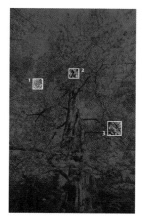

FOCAL POINTS

1 LEAF PATTERN
The relatively flexible, needlelike leaves of the dawn redwood are about 1 inch (2.5 cm) long and are arranged along their stalk in opposite pairs, with the stalks themselves similarly paired. The leaves are bright green during the growing season, then become red-brown before being shed in the fall.

2 CONES
The dawn redwood is monoecious—male and female cones are borne on different branches of the same tree. In general, trees begin producing female cones when they reach 30 to 49 feet (9–15 m) in height, and male cones when they reach heights of 59 to 89 feet (18–27 m).

3 DECIDUOUS LEAVES
Unusually for a conifer, the dawn redwood is deciduous. Temperatures were much higher during the Cretaceous than they are today, and the deciduous habit may have evolved in response to changing light levels at high latitudes rather than as a response to changing temperatures.

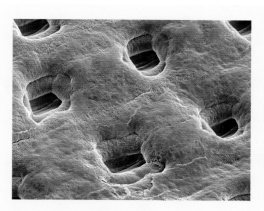

GLOBAL WARMING

Fossil remains of the dawn redwood can be studied to indicate carbon dioxide levels in the ancient atmosphere. The higher the levels of carbon dioxide, the fewer the number of stomata—tiny leaf pores (shown in spruce leaves, above)—are needed for gas exchange between leaves and atmosphere. The number of stomata in the leaves of present-day plants can be compared with those in largely unchanged fossil specimens, like the dawn redwood and ginkgo (see p.78), to gain an idea of what carbon dioxide levels were like in the past. This is important, because the role of carbon dioxide in the atmosphere as a factor influencing climate not only helps to reconstruct the past, but also can help to predict human-induced levels of carbon dioxide and climate change in the future.

Maritime Pine
NEOGENE PERIOD—PRESENT

The reproductive structure in gymnosperms is the strobilus, or "cone"—in essence, a branch that has been modified by evolution. The name gymnosperm, meaning "naked seed," comes from the fact that the seeds are not enclosed within fruits as they are in flowering plants. The conifers of pine and spruce forests, and the maritime or cluster pine, *Pinus pinaster*, have woody cones, but other conifers, such as podocarps, yews, and junipers, produce berrylike cones. The latter fruits are attractive to animals, which disperse the seeds. Another much rarer plant is the *Welwitschia* (see panel), a conifer that has salmon-colored oblong male cones and larger female cones.

The production of seeds was a huge leap forward in the evolution of plants because a seed contains a young plant that is already well developed, with an embryonic root, stem, and leaves, as well as nutritive tissue surrounded by a protective coat. After germination the young plant is nourished by food stored in the seed until it becomes self-sufficient. Seed production gave gymnosperms a competitive advantage over the ferns and early spore-bearing vascular plants (see p.70), allowing them to become the dominant form of plant life for much of the Mesozoic, 252 million to 66 million years ago. In the case of the maritime pine, the competitive advantage remains; the tree is notoriously invasive, particularly in the South African shrublands where it has already reduced native biodiversity. **RS**

✦ NAVIGATOR

SPECIES *Pinus pinaster*
GROUP Pinaceae
SIZE 100 feet (30 m)
LOCATION Europe and Africa
IUCN Least Concern

👁 FOCAL POINTS

1 CONE STRUCTURE
P. pinaster cones comprise a cluster of scales, called sporophylls, arranged around a central axis. The reproductive parts are under the scales. Pine trees are monecious, with male and female cones on the same tree. Female cones grow on the upper branches, and the male cones on the lower ones.

2 MALE CONES
Also known as microstrobili, male (pollen) cones are generally small and inconspicuous, like the immature female cones. Depending on the species, they are green or brown, in lumpy or tubelike clusters. They usually drop off the tree after releasing their pollen.

3 FEMALE CONE
Bigger than male cones, the megastrobili (female or seed cones) of *P. pinaster* have woody scales. The cones' shape and arrangement differentiate them from those of other species. When dry and "ripe," which may take 2 or more years, the scales bend and separate to release the seeds.

WELWITSCHIA

Thought to be little changed since it first appeared in the Jurassic period, *Welwitschia mirabilis* is the only surviving member in its family of Gnetales or gnetophytes. It is such an unusual plant that Friedrich Welwitsch (1806–72), the Austrian botanist who discovered it in the Namib Desert in 1859, could hardly believe what he had found and scarcely dared touch it. It has just two leathery leaves, which can grow up to 30 feet (9 m) long and 6½ feet (2 m) wide, although they are usually torn into strips by the desert winds. It also has a stem base and roots. *Welwitschia*, which can live up to an astonishing 1,500 years, is dioecious, with male and female cones (above) produced on separate plants. The evolutionary relationships of the plant probably lie with the pine group of conifers, but how it has attained such a unique form is unclear.

THE FIRST FLOWERING PLANTS

1 Seed-producing *Williamsonia* had leaflike bracts in a flowerlike arrangement.

2 Since its appearance in the Cretaceous, the rose family Rosaceae has evolved many species, with almost 3,000 today.

3 Magnolias, such as the star magnolia, *Magnolia stellata*, were one of the early groups of angiosperms to appear.

Charles Darwin (1809–82) called the origin of flowering plants (angiosperms) "an abominable mystery." Today, angiosperms are the most widespread, diverse, and numerous of the plant groups, yet it remains uncertain how they evolved. It is widely accepted that angiosperms (with enclosed or protected seeds) must have evolved from gymnosperms (naked or unprotected seeds), such as cycads and conifers, but it is not clear when. Flowering plants appear quickly in the fossil record, seemingly out of nowhere. The oldest known flowering plant fossils are pollen grains. Being small, tough, and produced in great numbers, they were more likely to fossilize than delicate leaves and flowers. Definite evidence of fossilized flower pollen dates from the early Cretaceous period, 140 million to 130 million years ago, and it has been assumed that flowering plants evolved around this time. This has been supported by *Amborella* (see p.90), one of the oldest known angiosperm lineages. However, in 2011, researchers from Switzerland and Germany reported finding pollen that closely resembled the flower pollen from the Cretaceous in samples from the Triassic period dating back 240 million years. As of yet, no fossil pollen has turned up to fill the 100-million-year gap between the Triassic and Cretaceous.

KEY EVENTS

240 Mya	150–130 Mya	140 Mya	130 Mya	125 Mya	125 Mya
Pollen grains apparently resembling those from flowers are known from fossils existing at this time, the Triassic.	The first eudicot plants appear (see p.94), with grooved pollen grains. They will diversify into families such as roses and beech.	The monocot group of flowering plants, such as grasses, orchids, and palms, diverge from the eudicots.	*Amborella* flourishes in East Asia and will persist until modern times.	Nymphales, members of an early plant group that will give rise to water lilies, water shields, and fanworts, spread.	*Archaefructus*, the oldest undisputed flowering plant fossil yet found, dates from this time.

In 2002, a team of paleobotanists from China and Florida announced the discovery of the Archaefructaceae (see p.88), a family of very ancient flowering plants with long thin stems and finely divided leaves. They had found five virtually complete fossils of the plants, including flowers, seeds and fruits, in northeast China. Believed to be between 145 million and 125 million years old, the fossils are the most complete early angiosperm fossils discovered so far, but the researchers do not believe they represent the first flowering plants.

Some lines of evidence suggest that the evolutionary adaptations leading to the emergence of flowering plants began around 250 million years ago. One possible ancestor group is the seed ferns (see p.70), some of which bore their seeds in specialized structures and had pollen-bearing parts similar to the anthers of flowering plants. They were most likely wind-pollinated, although some had large pollen grains that could have been transferred by insects and other animals. However, the reproductive organs of some cycads also had a resemblance to those of flowering plants. In some cases, for example the bennettitalean *Williamsonia* (see image 1), they were surrounded by leaflike bracts, which may have evolved into a flowerlike structure that attracted insect pollinators. Another candidate for the angiosperm ancestor is a group called the gigantopterids. These plants did not flower but had a similar leaf and stem structure to the flowering plants. They also seem to have produced similar chemical compounds to those used by angiosperms to protect their various parts from insects.

Flowers had to evolve from something and the potential can be seen in each of the aforementioned groups. Molecular and gene studies have discovered that genes involved in flower development are allied to those concerned with the development of leaves and stems, suggesting that the earliest flowers were simply modified leaves. The adaptive advantage of the first flowerlike structures probably came from their attraction to other organisms, especially insects, which helped pollinate the plant (see p.96). This is an example of a key process in evolution called exaptation: when a structure evolves to perform one function, in this case a leaf that evolved as a photosynthetic organ to capture light energy, and then becomes modified to take on another function, such as reproduction.

With insects pollinating their flowers and ensuring a healthy genetic mix, and animals carrying their seeds via their guts into new areas for colonization, the angiosperms spread and diversified. Between 125 million and 66 million years ago, many woody angiosperms evolved, including several modern groups, such as the rose (see image 2), magnolia (see image 3) and sycamore families. Herbaceous plants such as the water lilies and grasses probably originated in this period too. By around 95 million years ago the angiosperms were out-competing the conifers and other gymnosperms in many areas and were well on their way to becoming the dominant form of plant life. **RS**

115 Mya	100–90 Mya	100–65 Mya	95 Mya	70 Mya	35–40 Mya
Flowering-type plants along with many other organisms from this time are preserved in the Koonwarra Beds of Australia.	Angiosperms begin to dominate land habitats, prevailing over conifers and other gymnosperms.	The Cretaceous swamps are dominated by a mixture of conifers and eudicot angiosperms.	The Magnoliaceae family (ancient relatives of the modern magnolias), are present.	Studies of fossilized dinosaur droppings show that grasses may have evolved by this time.	Grasses—flowering plants with small and specialized flowerheads—begin to spread widely due to climate change.

Archaefructus
MIDDLE CRETACEOUS PERIOD

A|though almost certainly not the first flowering plant, *Archaefructus* is the oldest flowering plant fossil yet found. Discovered in fossil-rich rocks in northeast China, and named in 2002, *Archaefructus* caused great excitement among botanists when it was originally dated to the Jurassic period, more than 150 million years ago. Subsequently the date was revised to around 125 million years ago during the Cretaceous. The fossil showed the plant to have long thin stems and finely divided leaves and to be a member of a now-extinct genus of aquatic flowering plants. The paleobotanists who discovered *Archaefructus* thought it might be close to the physical form and ecology of the earliest flowering plants, while others, pointing to its similarity to modern water lilies, suggest that it is simply one of the main (crown) groups of angiosperms that became specialized for life in an aquatic habitat. Even if it is no longer considered to be the oldest flowering plant, *Archaefructus* is still a significant fossil find because the specimens are beautifully preserved, complete with roots, shoots, leaves, flowers, and seeds. The genus is based on the first discovered species, *A. liaoningensis*, found in the renowned Yixian Formation rocks of Liaoning, north China, where many other important fossils have been discovered, notably feathered nonavian dinosaurs. Two other species of *Archaefructus* have since been identified: *A. eoflora* and *A. sinensis*. **RS**

✦ NAVIGATOR

SPECIES *Archaefructus liaoningensis*
GROUP Archaefructaceae
SIZE 20 inches (50 cm)
LOCATION China

◉ FOCAL POINTS

1 FLOWERS
A. liaoningensis and its sister species, *A. sinensis*, had unisexual flowers, with male ones bearing pollen separate from the female flowers bearing ovules. But, *A. eoflora* had bisexual flowers, with both male and female organs on the same flower. It is likely that organs extended above the water.

2 FRUITS
A. liaoningensis had no petals. However, its seeds were enclosed in fruit (*Archaefructus* means "ancient fruit"), a characteristic that marks it out as an angiosperm. Each fruit held two to four seeds, which were probably dispersed by water currents before germinating in shallow water.

3 SHOOTS AND STEMS
The herbaceous shoots of *A. liaoningensis* were no more than 20 inches (50 cm) tall. Its stems were long, thin, and curving, suggesting that they were supported by water. The leaves were divided into narrow segments, which indicates that it was aquatic, as do the fish fossils found in the same rocks.

IDENTIFYING ANGIOSPERMS

One of the biggest unanswered questions about the early angiosperms is what type of plant they were. Were they small, delicate, and herbaceous, like *Archaefructus* and the modern water crowfoot (*Ranunculus aquatilis*, below), or did they evolve from woody gymnosperm trees and shrubs? The Paleoherb Hypothesis suggests that the earliest (basal) angiosperms were tropical herbaceous flowering plants with rapid life cycles, simple flowers, and a mix of monocot and dicot features (see p.94). The alternative Woody Magnoliid Hypothesis proposes that the first angiosperms were trees with slower life cycles, long broad leaves, and large flowers. Such features suggest that they occupied dark understory habitats, and that the evolution of the flower was not the change that drove rapid angiosperm diversification.

Amborella

LATE CRETACEOUS PERIOD—PRESENT

The genus *Amborella* is seen as a sister group to all other living flowering plants—its ancestry can be traced back further than any other angiosperm today. The single living species, *A. trichopoda*, has a low, shrubby form and grows up to 26 feet (8 m) tall, with wavy-edged leaves about 4 inches (10 cm) long and small cream-colored flowers.

A. trichopoda is the sole remaining species of an ancient angiosperm lineage, the Amborellaceae, which appeared after the first flowering plants evolved. On the angiosperm evolutionary tree, *Amborella* is on a low branch, having diverged from the other angiosperms 130 million years ago. In some ways it is closer to the gymnosperms than any other flowering plant, yet scientists do not view it as a "missing link" between the two groups. Like most gymnosperms, *Amborella* is dioecious, meaning each plant is either a female or a male. By comparing *Amborella* genes with those of gymnosperms, researchers can investigate how the genes that control the growth of a pine cone, for example, modified to control the growth of a flower. A flower's female reproductive organs are the carpels. The seeds of the plant develop inside an ovary that is held within the carpel. *Amborella* has carpels, but they are incompletely closed and sealed with secretions. Researchers believe this is significant because the carpel is thought to have evolved from a flat, leaflike structure that eventually rolled inward and created a hollow, almost completely enclosed ovary. **RS**

◆ NAVIGATOR

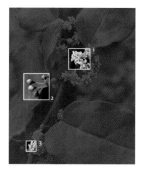

SPECIES *Amborella trichopoda*
GROUP Angiosperms
SIZE 26 feet (8 m)
LOCATION New Caledonia
IUCN Not Yet Assessed

◉ FOCAL POINTS

1 FLOWERS
Like many angiosperms, *A. trichopoda*'s flowers appear attractive to certain animal pollinators, such as insects, but they can also be pollinated by wind, like gymnosperm flowers. Each flower is either carpellate or female, or staminate being male; each plant has only one sex of flowers at a time.

2 FRUITS
The plant's carpellate or female flowers produce the type of fruit called a drupe (apricots and cherries are drupes) about t ¼ inch (6 mm) in diameter. However, an individual plant can change sex, producing only male flowers at one flowering, then solely female flowers at a later flowering.

3 CARPELS
The female reproductive organs are incompletely closed, perhaps a partway stage between the original flat, leaflike structure and the more highly evolved, hollow, enclosed ovary of later angiosperms. The carpels of the earliest angiosperms were probably not quite fused shut like those of *A. trichopoda*.

┴ AMBORELLA RELATIONSHIPS

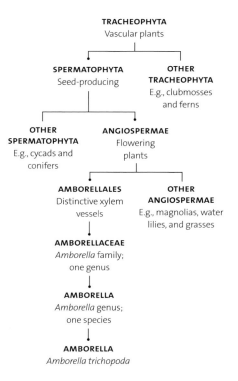

TRACHEOPHYTA
Vascular plants

SPERMATOPHYTA
Seed-producing

OTHER TRACHEOPHYTA
E.g., clubmosses and ferns

OTHER SPERMATOPHYTA
E.g., cycads and conifers

ANGIOSPERMAE
Flowering plants

AMBORELLALES
Distinctive xylem vessels

OTHER ANGIOSPERMAE
E.g., magnolias, water lilies, and grasses

AMBORELLACEAE
Amborella family; one genus

AMBORELLA
Amborella genus; one species

AMBORELLA
Amborella trichopoda

▲ The relationships shown here emphasize how *Amborella* diverged from the main evolutionary pathway that included all other flowering plants in the angiosperm group. A key feature is its unique "transitional" water-conducting vessels: xylem.

THE SPREAD OF FLOWERING PLANTS

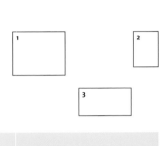

1 The world's largest bloom, the *Rafflesia* (corpse flower), has adapted to low light levels on the tropical forest floor by parasitizing vines for nourishment.

2 The seed of a sunflower, *Helianthus*, germinates, showing the two cotyledons that provide nutrients at this critical stage of development.

3 Woodland floor flowers, such as bluebells, *Hyacinthoides*, have adapted so that their main growth and flowering happens during the spring, before the leaves of trees cover them in shade.

The rapid spread of flowering plants (angiosperms) across the world, and their adaptation to endless new habitats (see image 1), a process known as adaptive radiation, is one of the most remarkable phenomena in the evolution of life. Angiosperms increased from an estimated 5 to 20 percent of plant species 105 million years ago, during the middle Cretaceous period, to more than 80 percent in some habitats by 66 million years ago, the end of the Cretaceous. The early angiosperms are believed to have been understory herbs, shrubs, and small trees, which were fast-growing (see image 2) and capable of colonizing disturbed environments such as spaces cleared in the coniferous forests by herbivorous dinosaurs, fire, or flood. They most likely sprang up in places where ferns, cycads, and conifers were not abundant, but it was not long before they were supplanting such gymnosperms.

Around 125 million to 100 million years ago the conifer forests that had dominated during the Cretaceous were in drastic decline as the angiosperms took over. By the end of the Cretaceous, successful gymnosperm groups such as the bennettitaleans (cycadeoids), some of the ginkgoes, and several types of conifer had disappeared, and the Cheirolepidiaceae family of conifers, the dominant gymnosperms in the tropics, became extinct. Ferns also declined as flowering plants moved into the understory habitat (see image 3) of the conifer

KEY EVENTS

150–130 Mya	135–66 Mya	100 Mya	80 Mya	70 Mya	66 Mya
Early eudicots exist. They will diversify into many surviving families and form the majority of today's angiosperms.	A cooling global climate sees the rise of the angiosperms and the gradual decline of the gymnosperms.	The first bees appear. As feeders on, and carriers of, angiosperm pollen, they play a crucial part in the evolution of flowering plants.	As the climate continues to cool, forests begin to resemble today's forests, with oaks, hickories, and magnolias common.	Studies of dinosaur droppings (coprolites) identify tiny mineral structures (phytoliths) from this time that are characteristic of the grasses.	The Cheirolepidiaceae, a dominant plant group for nearly 200 million years, becomes extinct.

forests that was once their stronghold. Today, the conifers have largely been driven back into the high northern and southern latitudes; the equatorial and mid latitudes are the domain of the angiosperms.

What tipped the balance in favor of the angiosperms? Reasons that might explain their sudden success have included adaptive changes such as the evolution of pollination and seed dispersal by animals. Other explanations have looked at the relationship between the angiosperms and their environment. Angiosperms have higher growth rates than gymnosperms and so require more nutrients, whereas gymnosperms do well where soil nutrient levels are low. However, the leaf litter and other remains of angiosperms decompose more easily than the needles and cones of gymnosperms. As the number of flowering plants grew, so did the nutrient levels in the soil, resulting in a feedback loop that was advantageous for angiosperms and disadvantageous for gymnosperms. Research undertaken in 2009 also points toward the evolution of an improved leaf plumbing system as a vital step forward for angiosperms. Water transport efficiency and photosynthesis are closely linked, and adaptations that result in better water distribution lead to improved yields from photosynthesis, too, giving the plant a competitive advantage. By some estimates, the evolution of a dense network of veins in angiosperm leaves, between 140 million and 100 million years ago, produced a twofold increase in leaf photosynthesis.

The rise of the angiosperms had other consequences that may be harder to unravel. It might be thought that the rise in the number of flowering plants would offer new food sources and habitats that animals—especially the evolving mammals of the Cretaceous—could exploit. However, researchers at Indiana University have shown that, contrary to expectations, mammal diversity declined during the great angiosperm radiation. Most of the mammals that

66 Mya	60 Mya	60–55 Mya	56–34 Mya	30–20 Mya	10,000 Ya
The end-of-Cretaceous extinction (see p.364) sees the loss of an estimated 75 percent of species, including nonavian dinosaurs and many plants.	Angiosperms are the dominant form of plant life in most terrestrial habitats, ousting conifers and other plants.	The ginkgoes (see p.78) are reduced to only one or a few genera, chiefly *Ginkgo*, reflecting the demise of other gymnosperms.	During the Eocene, angiosperms begin to evolve larger fruits that are more attractive to animals.	Grasslands and prairies become extensive as the climate dries and the number of grazing animals increases, promoting their evolution.	Several plants are domesticated by people in the early stages of the agricultural revolution.

survived were small insectivores. It was not until after the end of the Cretaceous with the extinction of the nonavian dinosaurs 66 million years ago that mammal herbivores began to appear in force. Fortunately, one group of mammals that did thrive in the middle Cretaceous was the therians: the ancestors of most modern mammals, including humans.

Flower fossils are exceptionally rare (see image 4), but molecular studies suggest that the earliest angiosperms are represented by a single living species, *Amborella trichopoda* (see p.90), a shrub found in cloud forests on New Caledonia, an island in the South Pacific. *Amborella*'s flowers have features that are thought to be similar to those of ancient angiosperms and it lacks the water-conducting vessels found in later angiosperm groups. The next group to appear were the magnoliids, represented by the genus *Magnolia*. Considered to be one of the earlier plant groups to evolve, the magnoliids have some primitive features. They make up about 3 percent of existing plant species and include the magnolia and tulip tree, as well as water lilies. However, they are described as polyphyletic, meaning that they have evolved from separate origins rather than from a common ancestor.

By far the largest angiosperm groups are the monocotyledons (monocots), with around 65,000 species, including grasses, sedges, palms, and orchids (see image 5), and the dicotyledons (dicots), comprising some 165,000 species, including such familiar plants as roses and geraniums, and trees such as oak and sycamore. Traditionally, the difference between the monocots and the dicots is in the number of cotyledons (seed leaves) produced by the embryo plant. Monocots have one; dicots have two. Current theory is that the dicots evolved first and the monocots evolved from them later in angiosperm history, with some genetic studies placing this divergence at 150 million to 140 million years ago, during the late Jurassic to early Cretaceous periods. However, the evolution of monocots is not well understood—it is hampered by the lack of mature plant specimens in their fossil record, which is known primarily from fossil pollen. The earliest recognizable fossils of monocots such as palms date from the end of the Cretaceous.

The distinctions between dicots and monocots are long-standing but are becoming increasingly perplexing. Some plants, called paleoherbs, have a mix of monocot and dicot characteristics that are not found together in other flowering plants. One explanation for this is that paleoherbs are representatives of an ancient group of angiosperms that emerged prior to the split into monocots and dicots, and so bore the features of both groups. The dicots are evolutionarily a polyphyletic group (lacking a single ancestor). In fact, many dicot lineages seem to be more closely related to monocots. Botanists are more inclined to recognize the eudicots, or "true dicotyledons," as a group because these can be characterized by features of their pollen. The eudicots include most trees, many flowering plants, legumes, and potatoes. In this system, the monocots, magnolias, water lilies, *Amborella trichopoda*, and some other plant groups are classified as non-eudicots.

The angiosperms as a whole dominate Earth's surface. They have gained a roothold in more habitats than any other group of plants. They are the most important food source, either directly or indirectly, for countless invertebrates, birds, and mammals, including humans, and may often provide them with shelter. Very important factors in the evolution of the angiosperms are their sometimes complex relationships with other organisms. In addition to the well-known relationship between flowers and insect pollinators, there is also the association of plant roots with mycorrhizal fungi, upon which the plants depend for nutrients, and nitrogen-fixing bacteria, with which legumes have evolved a close and beneficial relationship (see image 6).

Another important animal relationship concerns the fruits and seeds of the angiosperms. For the first 70 million to 80 million years of their existence, these were small, but during the Eocene epoch, 56 million to 34 million years ago, larger, energy-rich fruits and seeds, such as acorns, chestnuts, grapes, and the first grasses with grains, made their appearance. The advent of these fruits took place over a relatively short period of time, which also saw an increasing diversity of fruit- and seed-eating animals and birds. This was a prime example of coevolution: when one group closely influences, and in turn is influenced by, the evolution of another (see p.96). Thus, as well as the mutual benefits derived by flowering plants and pollinators, angiosperms began to produce fruits and seeds that were attractive to animals (see image 7). The creatures gained the benefit of an easily obtained, energy-rich food source, and in turn acted as distribution agents for the seeds, carrying them away from the parent plant in their excrement and into new territories for growth.

Humans, too, have formed a close relationship with the angiosperms, depending on them almost totally for their plant foods. The most economically important of the plants today, they also provide commercial products such as timber, fibers for cloth, and a variety of pharmaceuticals. In the course of this relationship humans have, in some ways, imposed on them an accelerated evolution. In our ceaseless search for higher yields, better flavors, and greater pest resistance, we have transformed some plants to such an extent that they are almost unrecognizable from their wild ancestors.

Plants also play a crucial role in regulating the world's climate. First, they carry out photosynthesis, which converts carbon dioxide into plant biomass and releases oxygen. Second, plants increase rates of rock weathering, a process that also removes carbon dioxide from the atmosphere. Many researchers now believe that long before angiosperms, the arrival of land plants in the Ordovician period, between 485 million and 443 million years ago, triggered a planet-wide ice age due to there being less carbon dioxide in the atmosphere. The modern removal of large areas of the world's vegetation—with the angiosperms its major component—by felling forests and draining wetlands seems to be increasing the levels of carbon dioxide and leading to dramatic climate change, resulting in the opposite of the ancient ice age: global warming. **RS**

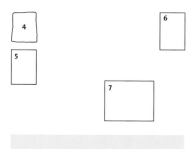

4 The delicate nature of flowers means that good fossils, such as this one found in Germany (*Porana* from the Miocene epoch), are rare.

5 Orchids, such as this monkey orchid, *Orchis simia*, are often regarded as highly evolved flowers, with complex designs and intricate relationships with insect pollinators.

6 The legume family has evolved root structures (nodules), seen here on a garden pea, *Pisum sativum*, in which helpful bacteria capture nitrogen from the atmosphere and transfer it to the plant.

7 Many animals contribute to the success of plants by spreading seeds. Here, seeds collected by leafcutter ants will germinate away from the host plant.

Flower–Insect Coevolution

CRETACEOUS PERIOD–PRESENT

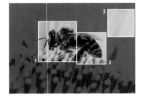

GROUPS
Angiospermae; Insecta
SPECIES About 255,000;
more than 1 million
ROLES Providers of nectar
and pollen; pollinators
LOCATION Almost all terrestrial
and many freshwater habitats

◆ NAVIGATOR

When the first plants began their colonization of the land 450 million years ago, the first plant eaters were not far behind. Within 100 million years evolution and natural selection produced an array of insects and other invertebrates with sucking, piercing, and chewing mouthparts. When seed plants began to produce pollen grains, it was not long before insects started to exploit them as a food source. Although some plants lost pollen, they gained an evolutionary boost. Relying on air currents to achieve wind pollination was haphazard, but having an insect air force to carry pollen accurately from flower to flower was advantageous. The more attractive the pollen was to insects, the more likely it would be delivered from the stamens of one flower to the waiting carpels of another, resulting in more seeds. The first bees, wasps, butterflies, and moths had evolved by the beginning of the Cenozoic era, around 66 million years ago. These insects are dependent on flowers to provide their food source as adults. This mutually beneficial coupling is an example of coevolution: when two groups of living organisms evolve together as a result of their close ecological links. Modified plant structures that were more attractive to insect pollinators were favored by natural selection, as were those insects that were quickest to exploit the pollen food source. The position of a flower's reproductive organ was important, and some flowers developed a "landing platform" suited to a particular type of bee or other insect—snapdragons, for example, open only when a bee of the correct weight lands on them. The result was the evolution of plants with distinctive flowers and scents to advertise sugar-rich nectar, and of pollinators adapted to take advantage of them. Today, more than 65 percent of angiosperms are insect-pollinated and 20 percent of insects depend on flowers for their food at some stage in their lives. **RS**

FOCAL POINTS

1 PROLIFIC POLLINATORS

Honeybees, genus *Apis*, are likely to have evolved in Southeast Asia more than 30 million years ago. These insects are almost entirely dependent on flowers to provide their food sources: pollen and nectar. During the summer months, a typical worker bee can visit up to 5,000 blooms per day, flying more than 4 miles (6 km).

2 ATTRACTION SIGNALS

Plants employ various visual signals to attract pollinators. Bees are sensitive to yellow, blue, and ultraviolet light, therefore flowers pollinated by bees are mostly yellow, such as this sunflower, *Helianthus*, or blue, with ultraviolet nectar guides to the nectaries. Flower scents are the other main attractants.

3 POLLINATION

As the bee roams the flower head, gathering nectar from the nectary glands and pollen from the stamens (male parts), pollen grains will attach to the bee's body. When it visits other flowers, some of the pollen brushes onto the carpels (female parts), and it is this transfer that begins the development of the seeds.

DARWIN'S MOTH

In early 1862, Charles Darwin (1809–82) was sent some orchids from Madagascar, including one that had a nectary of nearly 12 inches (30 cm). "Good heavens what insect can suck it," he wrote to a friend. Darwin suggested there had to be a moth that had evolved an exceptionally long tongue, or proboscis, to pollinate it. Because of the size of the flower, and his understanding of the evolution and ecology of orchids and insects, Darwin proposed the existence of an unknown insect. In 1907, more than 20 years after Darwin's death, a subspecies of the Congo moth from Madagascar (above) fulfilled Darwin's theory. It has an 8-inch (20 cm) proboscis that it carries in a coil in front of its head. In 1992, more than 130 years after Darwin first put forward his hypothesis, observations were made of the moth feeding on the orchids and transferring pollen from one flower to another.

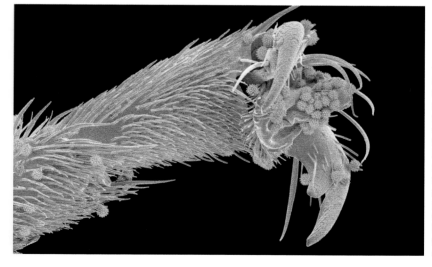

◄ A colored scanning electron micrograph (SEM) shows the leg of a honeybee, *Apis mellifera*, covered in round pollen grains. The dense coating of hairs has trapped the pollen, which is then brushed onto the stigma, the receptive area of the carpel of a flower visited by the bee. A flower that attracts specific pollinators has an advantage over other angiosperms in that its pollen is more likely to be carried to another plant of its own species.

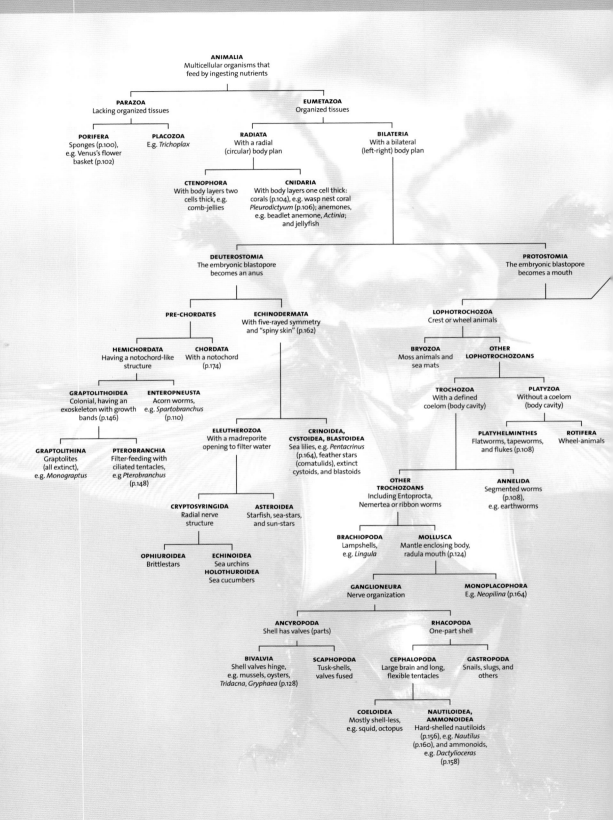

ANIMALIA
Multicellular organisms that
feed by ingesting nutrients

PARAZOA
Lacking organized tissues

EUMETAZOA
Organized tissues

PORIFERA
Sponges (p.100),
e.g. Venus's flower
basket (p.102)

PLACOZOA
E.g. *Trichoplax*

RADIATA
With a radial
(circular) body plan

BILATERIA
With a bilateral
(left-right) body plan

CTENOPHORA
With body layers two
cells thick, e.g.
comb-jellies

CNIDARIA
With body layers one cell thick:
corals (p.104), e.g. wasp nest coral
Pleurodictyum (p.106); anemones,
e.g. beadlet anemone, *Actinia*;
and jellyfish

DEUTEROSTOMIA
The embryonic blastopore
becomes an anus

PROTOSTOMIA
The embryonic blastopore
becomes a mouth

PRE-CHORDATES

ECHINODERMATA
With five-rayed symmetry
and "spiny skin" (p.162)

LOPHOTROCHOZOA
Crest or wheel animals

HEMICHORDATA
Having a notochord-like
structure

CHORDATA
With a notochord
(p.174)

BRYOZOA
Moss animals and
sea mats

**OTHER
LOPHOTROCHOZOANS**

GRAPTOLITHOIDEA
Colonial, having an
exoskeleton with growth
bands (p.146)

ENTEROPNEUSTA
Acorn worms,
e.g. *Spartobranchus*
(p.110)

TROCHOZOA
With a defined
coelom (body cavity)

PLATYZOA
Without a coelom
(body cavity)

GRAPTOLITHINA
Graptolites
(all extinct),
e.g. *Monograptus*

PTEROBRANCHIA
Filter-feeding with
ciliated tentacles,
e.g *Pterobranchus*
(p.148)

ELEUTHEROZOA
With a madreporite
opening to filter water

**CRINOIDEA,
CYSTOIDEA, BLASTOIDEA**
Sea lilies, e.g. *Pentacrinus*
(p.164), feather stars
(comatulids), extinct
cystoids, and blastoids

PLATYHELMINTHES
Flatworms, tapeworms,
and flukes (p.108)

ROTIFERA
Wheel-animals

**OTHER
TROCHOZOANS**
Including Entoprocta,
Nemertea or ribbon worms

ANNELIDA
Segmented worms
(p.108),
e.g. earthworms

CRYPTOSYRINGIDA
Radial nerve
structure

ASTEROIDEA
Starfish, sea-stars,
and sun-stars

BRACHIOPODA
Lampshells,
e.g. *Lingula*

MOLLUSCA
Mantle enclosing body,
radula mouth (p.124)

OPHIUROIDEA
Brittlestars

ECHINOIDEA
Sea urchins
HOLOTHUROIDEA
Sea cucumbers

GANGLIONEURA
Nerve organization

MONOPLACOPHORA
E.g. *Neopilina* (p.164)

ANCYROPODA
Shell has valves (parts)

RHACOPODA
One-part shell

BIVALVIA
Shell valves hinge,
e.g. mussels, oysters,
Tridacna, *Gryphaea* (p.128)

SCAPHOPODA
Tusk-shells,
valves fused

CEPHALOPODA
Large brain and long,
flexible tentacles

GASTROPODA
Snails, slugs, and
others

COELOIDEA
Mostly shell-less,
e.g. squid, octopus

**NAUTILOIDEA,
AMMONOIDEA**
Hard-shelled nautiloids
(p.156), e.g. *Nautilus*
(p.160), and ammonoids,
e.g. *Dactylioceras*
(p.158)

3 | INVERTEBRATES

From sponges—the simplest of creatures—evolution fashioned purpose-made tissues for tasks such as digestion, support, and sensation. Forms of reproduction and embryo development demarcate echinoderms and the lineages leading from the great majority of invertebrates to chordates, then vertebrates. The evolution of a hard outer body casing (exoskeleton) with jointed legs led to the great group known as arthropods, which encompasses crustaceans, insects, arachnids, and many extinct forms.

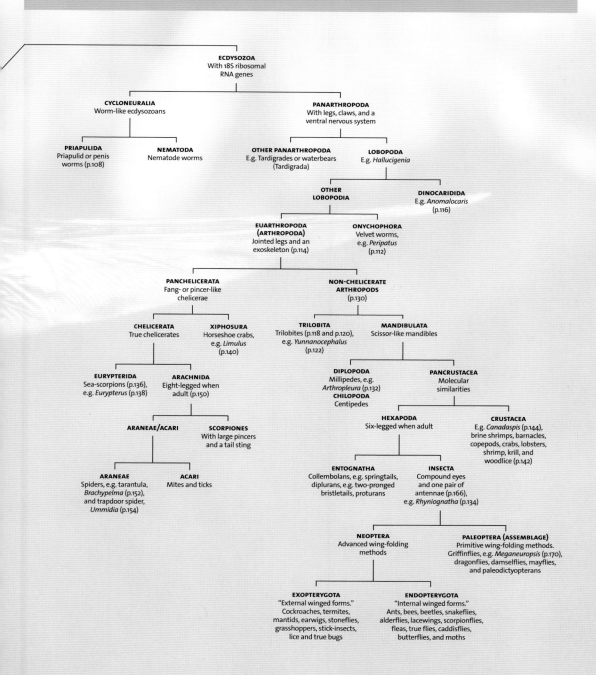

ECDYSOZOA
With 18S ribosomal RNA genes

CYCLONEURALIA
Worm-like ecdysozoans

PANARTHROPODA
With legs, claws, and a ventral nervous system

PRIAPULIDA
Priapulid or penis worms (p.108)

NEMATODA
Nematode worms

OTHER PANARTHROPODA
E.g. Tardigrades or waterbears (Tardigrada)

LOBOPODA
E.g. *Hallucigenia*

OTHER LOBOPODIA

DINOCARIDIDA
E.g. *Anomalocaris* (p.116)

EUARTHROPODA (ARTHROPODA)
Jointed legs and an exoskeleton (p.114)

ONYCHOPHORA
Velvet worms, e.g. *Peripatus* (p.112)

PANCHELICERATA
Fang- or pincer-like chelicerae

NON-CHELICERATE ARTHROPODS
(p.130)

CHELICERATA
True chelicerates

XIPHOSURA
Horseshoe crabs, e.g. *Limulus* (p.140)

TRILOBITA
Trilobites (p.118 and p.120), e.g. *Yunnanocephalus* (p.122)

MANDIBULATA
Scissor-like mandibles

EURYPTERIDA
Sea-scorpions (p.136), e.g. *Eurypterus* (p.138)

ARACHNIDA
Eight-legged when adult (p.150)

DIPLOPODA
Millipedes, e.g. *Arthropleura* (p.132)

CHILOPODA
Centipedes

PANCRUSTACEA
Molecular similarities

ARANEAE/ACARI

SCORPIONES
With large pincers and a tail sting

HEXAPODA
Six-legged when adult

CRUSTACEA
E.g. *Canadaspis* (p.144), brine shrimps, barnacles, copepods, crabs, lobsters, shrimp, krill, and woodlice (p.142)

ARANEAE
Spiders, e.g. tarantula, *Brachypelma* (p.152), and trapdoor spider, *Ummidia* (p.154)

ACARI
Mites and ticks

ENTOGNATHA
Collembolans, e.g. springtails, diplurans, e.g. two-pronged bristletails, proturans

INSECTA
Compound eyes and one pair of antennae (p.166), e.g. *Rhyniognatha* (p.134)

NEOPTERA
Advanced wing-folding methods

PALEOPTERA (ASSEMBLAGE)
Primitive wing-folding methods. Griffinflies, e.g. *Meganeuropsis* (p.170), dragonflies, damselflies, mayflies, and paleodictyopterans

EXOPTERYGOTA
"External winged forms."
Cockroaches, termites, mantids, earwigs, stoneflies, grasshoppers, stick-insects, lice and true bugs

ENDOPTERYGOTA
"Internal winged forms."
Ants, bees, beetles, snakeflies, alderflies, lacewings, scorpionflies, fleas, true flies, caddisflies, butterflies, and moths

SPONGES

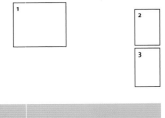

1 This fossil sponge, *Peronidella furcata*, dates from the Cretaceous period, around 136 million years ago, and was found in Berkshire, UK. The rigid skeletal structure of the sponge has helped its preservation as a fossil.

2 The leathery barrel sponge, *Geodia neptuni*, belongs to the demosponge group and can reach 3 feet (1 m) in diameter and height.

3 The spicules are an important structural element and their properties help determine the taxonomy of the sponge.

At first sight, the majority of sponges do not look like living animals. They lack a brain, heart, digestive tube, waste disposal system, and muscles for movement as well as a network of circulatory vessels and any detailed senses. However, their methods of feeding, growth patterns, detailed structure, and reproduction processes place them in the kingdom Animalia. Sponges constitute the phylum (major group) Porifera (bearer of pores). Poriferans are usually regarded as the simplest of animals. In fact, with a few other obscure types, they occupy the subkingdom Parazoa (beside the animals), while almost all other living creatures belong to the subkingdom Eumetazoa (true animals). Sponges have some of the most ancient evidence for their existence among all animals. Altered rocks that date to 640 million years ago bear "signatures" of their chemical activity, known as biomarkers. Sponge fossils date from almost 600 million years ago, during the early Ediacaran period, a time confirmed by molecular clock analysis (see p.17).

All sponges dwell in water and are sessile (fixed down). There are almost 10,000 species, most of which are marine. When they group together in large numbers, they build huge reefs similar to those of corals (see p.104). Fossil sponge reefs date back to the early Cambrian period, 540 million years ago. Some sponges have defined shapes that have influenced their common names,

KEY EVENTS

650 Mya	560 Mya	540 Mya	530 Mya	530 Mya	525 Mya
Molecular clock (see p.17) and similar studies suggest sponges appear at this time, however, fossil evidence is scant and inconclusive.	Demosponges leave fossils—including spicules, overall shapes and body cavities—in rocks in Guizo province, China.	Reefs form by large aggregations of sponges; these are the earliest known large structures on Earth built by animals.	Glass sponges thrive and leave traces of their silica spicules, especially in Australia and East Asia.	Calcareous sponges inhabit Australian seas during the early Cambrian. Their diversity will peak during the Cretaceous period.	Spongelike organisms called Archaeocyatha (ancient cups), which some experts include in the group Porifera, appear. They spread and diversify quickly.

such as barrel (see image 2), cup, horn, funnel, and vase sponges. Others have no definite shape and grow over a large surface. Sponges survive by drawing water through pores and cavities in their bodies, from which they extract dissolved oxygen and tiny nutrient particles, and into which they excrete waste. Most sponges are hermaphrodite—each is female and male together.

A typical sponge has a hollow structure with a body wall that acts like a sandwich: a layer of cells called choanocytes or collar cells (funnel-like opening cells) lines the inside; the filling is a gelatinous substance called mesohyl. The mesohyl acts as the sponge's skeleton. It is a jelly stiffened and fortified by various tiny structures, which include fibers of the protein collagen that form the tough material spongin and spicules (minute shards of calcium or silicate minerals, see image 3). These skeletal structures give the whole sponge its shape and firmness (see image 1). Another layer of cells, which may include choanocytes but is largely pinacocytes (flat cells), covers the outside. The choanocytes have long projections (flagella) that wave in unison and create water currents in the inner cavity. Water is drawn into the sponge's interior through tiny pores (ostia), which are scattered around the base or over the body. Water is then ejected through a larger opening (the osculum) near the top. As the water flows in and out, the choanocytes trap and absorb minute particles of floating nutrients.

In terms of evolution, a sponge's choanocytes are very similar to choanoflagellates, which are single-cell organisms that belong to the small grouping Filozoa and live freely as individuals, only sometimes gathering into loose colonies. This similarity indicates that sponges could have evolved from free-swimming choanoflagellates: molecular clock calculations indicate that these had probably appeared by 1 billion years ago. Some varieties of choanoflagellates formed more complex colonies and started to cooperate as one multicelled animal. This led to the four main groups of sponges—calcarean, hexactinellid, demosponges, and homoscleromorphan—with variations on the typical structure, the composition of the mesohyl skeleton and the networks of pores, internal channels, and passages for water.

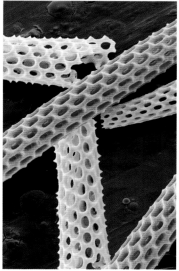

Calcarean sponges have a mesohyl strengthened with calcite spicules but have little or no spongin; they vary from tube- or vase-shaped to encrusting (they encrust hard surfaces such as rocks). Hexactinellid sponges such as Venus's flower basket (see p.102) also lack spongin but have spicules of silica (also found in sand grains and glass), which give them their informal name of glass sponges. They also vary greatly in shape. Demosponges, which have both silica spicules and spongin, form the majority of sponge species (more than 80 percent), and include common vase, barrel, trumpet, brain, and mushroom sponges, with some individuals exceeding 3 feet (1 m) in size. Homoscleromorphan sponges are the smallest group, with fewer than 100 species. These also have both silica spicules and spongin, and many are encrusting. **PB**

520 Mya	520 Mya	500 Mya	485 Mya	100 Mya	40 Mya
Sponges and many other creatures lay down fossils that will later make up the Luoping biota (see p.226) in Yunnan province, China.	Archaeocyatha thrive in many places and build reefs similar to those constructed by sponges.	With limited competition from other simple animals such as corals, sponges undergo rapid bursts of evolution and many new kinds appear.	The Archaeocyatha, which have been in rapid decline since 515 million years ago, are all extinct by the end of the Cambrian.	The glass sponges reach great diversity during the Cretaceous period. More than 300 genera exist.	Sponges, such as the demosponge *Potamophloios*, spread and colonize freshwater habitats in Canada.

Venus's Flower Basket
HOLOCENE EPOCH—PRESENT

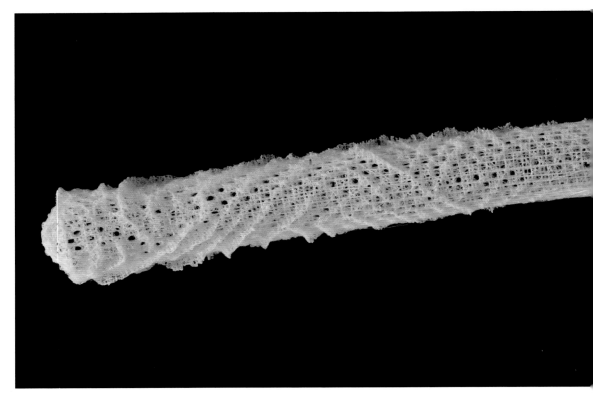

SPECIES *Euplectella aspergillum*
GROUP Hexactinellida
SIZE Up to 3 feet (90 cm)
LOCATION Western Pacific Ocean
IUCN Not Yet Assessed

Sometimes referred to as one of the most delicate and beautiful of all living things, the Venus's flower basket, *Euplectella aspergillum*, is a sponge in the hexactinellid group. Members of the group are also known as glass sponges and together they make up one of the most ancient sponge groups. Most specimens are around 1 foot (30 cm) tall, but occasional giants have been found that exceed 3 feet (1 m). This sponge species has a limited distribution in the West Pacific Ocean, near the Philippines. Other species in the genus *Euplectella* are known from as far north as the oceans surrounding Japan, and as far south as the waters surrounding Australia.

Venus's flower basket is usually found attached to the rocky seabed by specialized fibers at depths lower than 1,650 feet (500 m)—and sometimes as deep as 3,300 feet (1,000 m). Like all sponges, it creates a water current through itself by waving long projections (flagella) from inner lining cells (choanocytes), which trap and absorb the tiny particles of nutrients that are suspended in the water. The skeleton of Venus's flower basket is made up of minute star-shaped spicules of silica, a mineral also found in glass. The spicules are arranged in branching networks that lend great strength and resistance to such a fragile-looking structure. This strong skeleton resists currents and the attentions of would-be predators, such as deep-sea fish and crabs. However, certain kinds of shrimp from the family Spongicolidae (known as glass sponge shrimp or wedding shrimp) live in close partnership—mutualistic symbiosis—with Venus's flower basket. A shrimp enters the central cavity when young and feeds on nutrients discarded by its host. At the same time it keeps the host sponge clean. A male and female shrimp will often inhabit the same Venus's flower basket and then grow too large to leave through the main opening, thereby partnering for life. **PB**

◆ NAVIGATOR

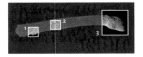

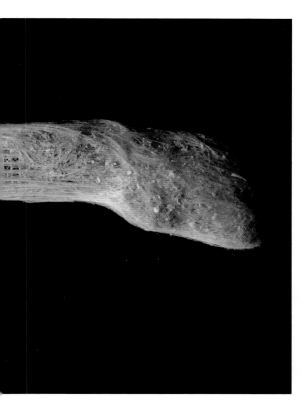

◉ FOCAL POINTS

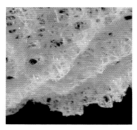

1 SPICULES
The siliceous, glassy spicules vary from milky to transparent. They are six-pointed (hence the group name hexactinellid), each formed from three rods at right angles. These structures merge to create a series of latticeworks at various angles to one another, reminiscent of a woven basket.

2 SOFT TISSUES
The glassy, latticed, skeletal component mesohyl is coated inside and out by layers of cells called a trabecular net. Cells fuse to form "super cells." The net contains elongated chambers that open into the central cavity. Choanocytes line the chambers and generate currents.

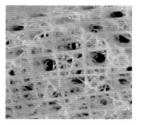

3 ATTACHMENT
Venus's flower basket is fixed to a firm area of seabed by spicule structures. These are long and fiberlike, resembling a tuft of hair. Their lengths can equal the height of the main sponge. They grow over the bottom and among small cracks and fissures to provide strong anchorage.

CARNIVOROUS SPONGE

A group of carnivorous sponges, family Cladorhizidae, has evolved a method for the capture of large items such as shrimp, prawns, and copepods. The harp sponge, *Chondrocladia lyra* (below), discovered in 2012, has horizontal stems (stolons) radiating from its fixed center, each carrying a row of branches. These are coated with curved, double-ended hooked spines that snare victims. The more a captive struggles, the more firmly it is held. A sheetlike membrane then grows over the prey and it is slowly digested.

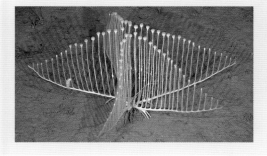

▲ The Davidson Seamount in California, provides a base of bare lava rock and a stream of passing nutrients for a vast number of corals and sponge species, including Venus's flower basket. This specimen lies 8,438 feet (2,572 m) deep.

CORALS

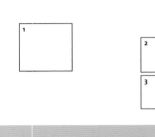

1 With the living coral polyps gone, their cuplike corallites (exoskeletons) are exposed. This is a member of the Faviidae group, with radial walls (septa) in each cup.

2 The animals that produce coral reefs are coral polyps. The feeding tentacles of these yellow cup coral, *Tubastraea*, are extended, which usually happens at night.

3 Anemones, such as the beadlet anemone, *Actinia equina*, are close cousins of corals. They are so soft they rarely leave fossils but aspects of their evolution can be inferred from corals.

Within the fossil record, corals are the main representatives of the phylum (major group) Cnidaria. The ancestors of today's corals appeared about 450 million years ago, but the origins of the Cnidaria as a whole are less clear. Almost entirely marine, the group includes jellyfish, sea anemones, and similar soft-bodied creatures. The generalized cnidarian body is a flexible sac enveloping a stomachlike, water-filled cavity with a single opening that acts as both the mouth and anus. In most cases the body has several tentacles extending around the mouth. In the free-swimming medusa form, such as a jellyfish, these tentacles hang down; in the sedentary forms called polyps, such as a sea anemone, the tentacles point upward. Corals are polyps (see image 2) and sit on the seabed. Unlike their cnidarian cousins, they secrete a calcareous hard outer casing (exoskeleton) around the base of the body (see image 1). Called a corallite, this "cup" provides a sturdy platform for the polyp and a refuge from danger. The corallite remains long after the polyp has died, and it is millions of these hard structures that form the basis of the coral fossil record. As a result, we know a lot about the evolution of corals, whereas the lineages of other cnidarians is the subject of conjecture.

KEY EVENTS

542 Mya	450 Mya	425 Mya	386 Mya	252 Mya	228 Mya
Coralomorphs, which share the most basic coral-like features, appear in shallow seas, along with primitive shelled marine creatures.	Tabulate corals appear as crusts on rocks and also on the shells and exoskeletons of other animals.	Large colonies of tabulate corals form simple reefs. Solitary rugose corals also grow on the seabed.	The world-renowned Falls of the Ohio fossil beds form on the bottom of a shallow sea that covers eastern North America.	Rugose and tabulate corals become extinct in the end-of-Permian mass extinction known as the "Great Dying."	The first scleractinian corals arise. However, these early species are solitary, not colonial and do not form reefs.

The term "coral" generally applies to members of the Scleractinia order, known as the stony corals. Rocky outcrops on the beds of shallow, warm seas, they arise from the build-up of exoskeletons. Only the surface of these reefs is alive, covered in colonies of polyps joined by tubular tissue: the coenosarc. As polyps die, more grow on top. The stony corals share the larger group Anthozoa with sea anemones and other creatures, most of which lack the calcareous exoskeleton and are therefore characterized as "soft corals" (see image 3).

Early coralomorphs—organisms that appear to share features of later corals—existed about 542 million years ago, around the boundary between the Ediacaran and Cambrian periods. However, the first fossils to be confirmed as corals appeared, around 450 million years ago, in the late Ordovician period, a time when Earth's sea levels rose and much of its surface flooded. These early corals belonged to the Tabulata order (subgroup). They were colonial, with individual polyps separated by a tabula (wall of the calcareous exoskeleton). Tabulate coral polyps were small, about half the size of modern species—on average, 1/50 to 1/30 inch (0.5–0.8 mm) wide—yet by the mid-Silurian period, about 425 million years ago, they were found in large reefs. Common examples of tabulate corals are the rounded *Favosites* and the columnar *Haylsites* genera. By this time, another family of corals had appeared: the Rugosa order, named for their rugged corallite walls. Early rugose corals, such as *Ketphyllum*, were solitary with conical or horn-shaped exoskeletons. In the Devonian period, colonial rugose corals begun to appear; the coral garden at the Falls of the Ohio in Indiana (see p.106) is a good example of this. The rugose horn coral, *Eridophyllum*, was also colonial, with physical connections between polyps. However, similar fossils may show individual horn corals growing in clusters.

The boundary of the Permian and Triassic periods, around 252 million years ago, marks the change from the Paleozoic era to the Mesozoic era, and also the most destructive extinction event in geological history. Known as the "Great Dying" (see p.224), it wiped out more than 90 percent of marine species. The tabulate and rugose corals were among them, although a few hung on for a short while. It took another 25 million years for the only surviving stony corals, the Scleractinia, to appear in the fossil record. It is thought that they may be an offshoot of the Rugosa, or may have been a rare grouping living in the shadows of the horn corals for millions of years. At first they were solitary, but later, large structures were branched, and they were fragile compared to modern reefs.

In the mid-Jurassic period, 170 million to 160 million years ago, huge reef builders took over. These included the Fungiidae family: a group of mushroom-shaped corals. Another mass extinction at the end of the Cretaceous period 66 million years ago (see p.364), which caused the extinction of many dinosaurs, impacted coral diversity. Approximately 70 percent of all corals were wiped out. The Fungiidae were reduced to a rarity as today's dominant families—including the Faviidae (the "brain" corals) and Caryophylliidae—came to the fore. **TJ**

200 Mya	185 Mya	150 Mya	66 Mya	30 Mya	3½ Mya
The first stony coral reefs appear. They are highly branched and less massive than modern coral colonies.	The development of great Triassic reefs collapses and several new families of coral begin to emerge from the remnants.	The stony corals reach their peak of diversity with at least 200 genera worldwide.	Coral diversity drops by 70 percent in the end-of-Cretaceous extinction. Two coral families—Faviidae and Caryophylliidae—begin to dominate.	Mediterranean species become extinct and coral species begin to split into two groups based in the Indo-Pacific and Atlantic regions.	The Isthmus of Panama closes, thus joining North and South America and permanently dividing the two main coral families.

Falls of the Ohio Fossil Beds
MIDDLE DEVONIAN PERIOD

LOCATION Indiana
AGE 386 million years ago
GROUPS See right
BED DEPTH 15 feet (4.5 m)

Close to the widest point of the mighty Ohio River, between Louisville, Kentucky, and Clarksville, Indiana, a 16,000-square-foot (1,500 sq m) area of exposed riverbed contains the remains of a Devonian coral garden. The corals flourished 386 million years ago, when this part of the world was bathed in a shallow sea that covered most of what is now eastern North America. The region was also closer to the equator than it is now, and the warm, sun-filled sea teemed with ancient life forms. In addition to a wealth of rugose and tabulate corals, the fossil beds contain evidence of evolving groups of sponges, brachiopods (lampshells), bryozoans (sea mats), sea snails, and other mollusks (see p.124), as well as various echinoderms (see p.162). All of these are preserved in limestone, which in some places is so fossil-rich that the remains of creatures are packed together.

Until the building of a dam in the late 1920s, the fossil beds lay undiscovered beneath the Falls of the Ohio (see panel). World-renowned, it is the largest collection of fossils of its kind anywhere and a marvelous snapshot of evolution in action. The largest individual coral fossils are solitary rugose specimens, such as *Siphonophrentis*, some of which can be 12 inches (30 cm) long; their curved shape has earned them the local name of "tusk coral." The fossil beds are protected as a US National Monument, but the public can visit under the auspices of The Falls of the Ohio State Park. Water levels change throughout the year, and the beds are fully accessible only when the water level is below 13½ feet (4.1 m). Visitors are advised to bring a bucket in order to wash away any silt left by the river and get a better look at the fossils beneath. The addition of water helps to make them stand out because the surrounding limestone darkens when wet. **TJ**

◉ FOCAL POINTS

WASP NEST CORALS

Named from its resemblance to the combs of a bee or wasp nest, *Pleurodictyum* had polygonal cups or chambers, each with several almost straight sides, that were mostly larger than ⅕ inch (5 mm). This genus of tabulate corals appeared in the Silurian and thrived during the Devonian, but had disappeared by the end of the Carboniferous period 300 million years ago.

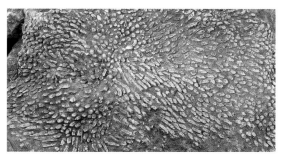

PIPE ORGAN CORALS

Several kinds of corals grew in long rod- or tubelike clusters, hence the name pipe organ coral. They included *Eridophyllum*, which formed colonies that fanned out or radiated from one small area and extended more than 16 inches (40 cm) across. Another genus, *Acinophyllum*, had projections on the corallites that may have stopped the animals from sinking into sediments.

HONEYCOMB CORALS

The most common corals on show at the Falls of the Ohio are tabulate species, which encrusted rocks and shells of other animals on the Devonian seabed. These fossils have a distinctive honeycomb appearance from the walls that divided the polyps in life. In some species of *Favosites*, the chambers were less than ⅕ inch (5 mm) across, and some were less than ¹⁄₂₅ inch (1 mm) across.

BRANCHING CORALS

Some tabulate corals grew as large branches that spread out through the bed, while others built up flattened fan- or leaf-shaped structures. *Thamnopora* (formerly classed as a species of *Favosites*) shows the classic shape resembling a tree with boughs. Its round cups or chambers were mostly about ¹⁄₁₂ inch (2 mm) across. The last tabulates perished in the end-of-Permian mass extinction.

EXPOSING THE FOSSIL BEDS

In 1803, the president Thomas Jefferson sent army officers Meriwether Lewis and William Clark to explore the land beyond the Falls of the Ohio. At that time, the falls were an unnavigable 2½-mile (4 km) rapid studded with rocks (right). Lewis and Clark's expedition spurred commercial interest in the river, and in the 1830s a canal was built to bypass the falls. A century later, in 1927, the river was tamed by a lock and dam system, and the waterline fell, thereby revealing large areas of fossil-encrusted limestone that the previously fast-flowing river had exposed through erosion. Today's Wildlife Conservation Area centers on a section of bed revealed by 20th-century engineering at the top end of the Falls, on the Indiana side.

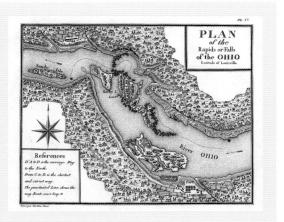

WORMS

1 One interpretation of *Cloudina* is a small, long, slim, filter-feeding animal that grows from the seabed. Its fossilized traces have been found on almost every continent.

2 The benthic (seabed-dwelling) priapulid worm, *Priapulus caudatus*, lives deep in the North Atlantic Ocean.

3 *Wiwaxia* gained some protection from predators with the two rows of spines on its back. The rest of its upper body was covered in hard plates.

Worms do not form a coherent group in the living animal kingdom nor in the fossil record. Instead, they occur on several branches of the tree of life. Today there are three main worm groups. The segmented worms (including the leeches and earthworms) make up the group Annelida. The second group is the roundworms (Nematoda): transparent creatures that are generally microscopic and are found in great numbers in water and soil. It is not uncommon to have a few living inside your digestive system (or elsewhere in the body). The third phylum is Platyhelminthes, including flatworms, flukes, and tapeworms, many of which are also internal parasites. Embryology (the study of how living organisms develop) indicates that worms occupy different stems of the evolutionary tree. Annelids and platyhelminths belong to a branch that also includes mollusks. The nematodes share a branch with animals that shed their skin as they grow, including the arthropods (joint-legged invertebrates; see p.114), such as insects, spiders, and crabs—and the velvet worms (see p.112). The latter have long bodies and multiple pairs of legs and they are frequently put forward as the missing link between primitive wormlike animals and arthropods.

Irrespective of their phylum, all worms belong to the Bilateria, an immense group that includes most animals. As the name suggests, its members have a body with bilateral symmetry—the left side is the mirror image of the right

KEY EVENTS

540 Mya	525 Mya	520 Mya	505 Mya	465 Mya	300 Mya
Cloudina, a tiny tube of calcium-based shell, resembles the tubes built by living marine annelids. It may be a link to the corals.	Found in a shale deposit in Yunnan province, China, *Cricocosmia jinningensis* is the earliest nematode fossil. Scientists believe it fed on sediments.	Priapulid worms are some of the early Cambrian's most prolific ambush hunters. They grow to around 3 inches (8 cm), some even longer.	The Burgess Shale rocks include the first polychaete worms, lobopods (ancestors of velvet worms) and acorn worms such as *Spartobranchus*.	*Palaeoscolex*, an annelid specimen from the middle Ordovician period, is a possible ancestor of the earthworms that evolved in seawater.	The bristle worm *Fossundecima* evolves a flexible feeding organ that extended from its mouth. It lived in warm seas over present-day North America.

side—plus a head at one end, where a brain (or equivalent) is located. This distinguishes such creatures from animals with a wheel-like radial symmetry, such as jellyfish and starfish (see p.162), or those with no symmetry, such as sponges. It is generally thought that the platyhelminths are the closest living representatives of the basal (original) group of the Bilateria. This means that all bilateral animals living today evolved from something like a flatworm. These worms are entirely soft-bodied and the only fossils of them are comparatively recent. The earliest is a flatworm trapped in amber from around 50 million years ago. It has been suggested that the specimen *Dickinsonia*, from the Ediacaran period, more than 550 million years ago, could be a primitive flatworm—although it could also be an annelid, a jellyfish or even a fungus. Another early fossil often referred to as a worm, or wormlike, is *Cloudina* (see image 1), which lived in oceans around 540 million years ago. Made up of a shell-like tube, it has been linked to similar fossilized structures made by living tube worms, a type of annelid that burrows into the seabed. The first confirmed tube worm fossils come from the early Jurassic period, 190 million years ago.

By the early Cambrian period, worm groups had evolved distinct lineages. The earliest nematode fossil is from 525-million-year-old shales in China. There is a distinct dark band running down the middle of the fossil, which is thought to be the gut. This would suggest the gut was filled with minerals at the time of death so these early nematodes are likely to have been sediment-feeders. One very common worm found in Cambrian shales belongs to one of the many wormlike phyla considered "minor" today, with relatively few living kinds. This is the phylum Priapulida, with just 16 surviving species. Its members (see image 2) are known as phallus or penis worms, due to their body contours. In the Cambrian they were a common predator that ambushed prey from curved burrows, striking with a spiny proboscis that poked out of the burrow.

The annelids are widely represented in the Burgess Shale rocks of Canada, which contain the remains of many animals from 505 million years ago (see p.106). These include *Canadia*, a 1¼-inch (3 cm) worm covered in bristles that resembles an extant group of annelids called polychaetes (many bristles). Another possible Cambrian polychaete is *Wiwaxia* (see image 3), which looked like a bristled slug. In addition to polychaetes, the annelids contain oligochaetes (several bristles). This group includes the earthworms and leeches but its evolution is less clear. Fossilized tubes that resemble the burrows of modern earthworms—complete with a trail of droppings—have been uncovered from 250 million years ago. If these were the work of an early earthworm, it would mean that it evolved in marine sediments. Furthermore, what appear to be hard cocoons of leech eggs have been studied from 215 million years ago. The earliest confirmed earthworm fossil is from the early Paleogene period, 65 million years ago. This was a terrestrial creature that lived in soils dominated by the roots of flowering plants, which were becoming the main land plants at that time. **TJ**

250 Mya	215 Mya	190 Mya	65 Mya	50 Mya	2,000 Ya
Fossilized burrows, found in Triassic period rocks near Sydney, Australia, are probably produced by early earthworms.	Fossils are laid down from egg cases, made out of hardened mucus, similar to that produced by leeches, in the rocks of Antarctica.	Tube worms from the annelid group lay down fossils at this time; the early Jurassic.	The first definite oligochaetes, the annelid group that includes earthworms, exist.	The earliest known flatworm fossil is laid down, entombed in amber, from the Eocene epoch.	Encysted tapeworm eggs—in tough, resistant cases known as cysts—are found preserved inside the digestive tracts of Egyptian mummies.

Spartobranchus
MIDDLE—LATE CAMBRIAN PERIOD

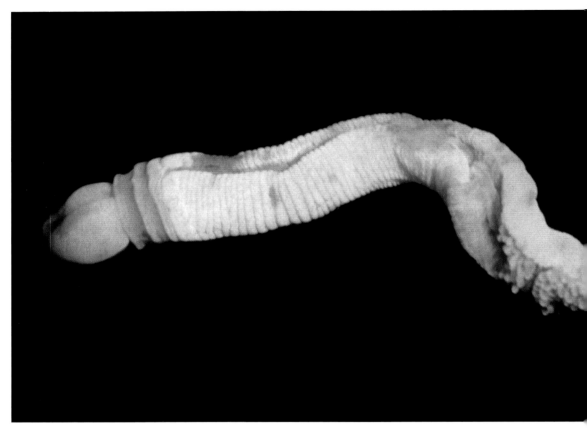

👁 FOCAL POINTS

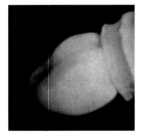

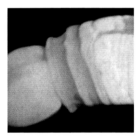

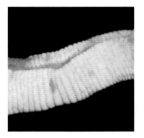

1 PROBOSCIS
The phallus-shaped, fleshy, muscular proboscis at the front end of the animal was not its head, but a digging implement. The acorn worm would have moved slowly, using rhythmic undulations of its muscles to swim. There was a cleft on the underside of the proboscis that drew food into the mouth.

2 COLLAR
The mouth was located in the collar section, a simple tube that used ciliary action—the wafting of tiny hairlike extensions on its inner surface—to draw wet sediments into the digestive system. The flow of water also supplied the gills with oxygen dissolved in the seawater.

3 TRUNK
The long, flexible trunk earned *S. tenuis* the name "worm." It accommodated a long intestine (there was no stomach), a heart and a circulatory system. The trunk ended with a swollen section not seen in living species. It has been suggested that it anchored the animal inside its tube home.

4 POSTERIOR
The "tail" of the worm contained its intestines and terminated with the anus. *S. tenuis* anchored itself to its burrow by means of multiple "legs." Fossilized burrows were branched and appear to have been lined with fibrous tube structures. Living acorn worms do not have burrows.

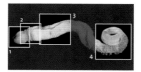
SPECIES *Spartobranchus tenuis*
(similar to living acorn worm, shown here)
GROUP Enteropneusta
SIZE 4 inches (10 cm)
LOCATION Canada

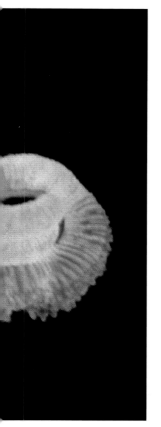

I n 2013, a species of the long-extinct acorn worm genus *Spartobranchus*, similar to the living acorn worm shown here, was discovered in the 505-million-year-old Burgess Shale rocks. University of Toronto researchers named it *S. tenuis*. Acorn worms belong to the group Hemichordata, which is on the same evolutionary line as sea anemones, starfish and vertebrates. There are around 120 living species in this group, most of which are deposit-feeders, eating sediments from the seabed to extract organic material, in the same way that earthworms eat soil. In 2013, new research into the Burgess Shale's wormlike fossils shed light on hemichordate evolution, which impacts our understanding of the ancestral form of the chordates, which eventually led to humans. This minor phylum is thought to be the group closest to Chordata, the phylum that includes the vertebrates. The discovery of *S. tenuis* pushed back the fossil record of these creatures more than 200 million years, as the earliest previously known specimen was from the late Triassic, unearthed from the Yoho National Park in Canada.

Hundreds of specimens show that *S. tenuis* lived in burrows in the Cambrian seabed. It reached 4 inches (10 cm) in length, which is small compared to modern species of acorn worm, which are limbless with long, cylindrical bodies. Unlike other worms, their bodies have three sections: a proboscis, a collar, and a trunk. *S. tenuis* had a feature that existing acorn worms lack, but is found in the other group of hemichordates: the pterobranchs (see p.148). Members of this group are smaller than acorn worms and each secretes a hard tube around its body, with the animal inside drawing in water to filter food; the tubes are often connected into colonies. *S. tenuis* fossils show that it secreted fibrous tubes around itself. This suggests that *S. tenuis* may be the missing link between the two main hemichordate groups, with a lifestyle more similar to that of modern pterobranchs than to their living acorn worm relatives. **TJ**

GROWTH PATTERNS

As hemichordates, the acorn worms are members of the Deuterostomia group, along with echinoderms—starfish (right) and urchins—and chordates, largely made up of vertebrates. Annelid, platyhelminth, and nematode worms, and all living invertebrates belong to the group Protostomia. The differences between the groups stem from their embryonic development. Protostomes grow their mouths first, followed by their anus; deuterostomes develop the anus first, and mouth second. This distinction suggests a major divergence in animal evolution, even though the genes involved in mouth and anus formation are the same.

Velvet Worm
PLIOCENE EPOCH—PRESENT

SPECIES *Peripatus novaezealandiae*
GROUP Onychophora
SIZE 3 inches (8 cm)
LOCATION New Zealand
IUCN Not Yet Assessed

Often characterized as "walking worms," velvet worms, such as *Peripatus novaezealandiae* and others, are thought to be a link between worms and arthropods. Their group first appeared in the early Cambrian, more than 500 million years ago, as specimens known as lobopods. This name means "blunt feet," due to its several pairs of unsegmented legs. In modern species, these limbs initially resemble the prolegs of a caterpillar, but closer inspection reveals a curved claw at the tip of each. There are 180 species of velvet worms alive today, forming the group Onychophora, and they live exclusively on dry land in humid, tropical areas, mostly in the southern hemisphere. Their ancestors, Cambrian lobopods, lived on the seabed, but by the end of the Silurian, 430 million years ago, they had emerged on to land. However, it is unclear whether velvet worms are direct descendants of lobopods. It seems likely that this earlier group also has links to the fully segmented arthropods, through spiny lobopods, such as *Hallucigenia* (see p.52).

Velvet worms average 2 inches (5 cm) in length with some reaching 8 inches (20 cm) in size. They are nocturnal hunters that trap their prey—ground-living insects and similar bugs, snails, and worms—by smothering them in a sticky slime fired from glands on the head. Many are viviparous, meaning that the young develop inside the female's body before birth. Female velvet worms are larger than males and have more legs. In viviparous species, such as those in the genus *Peripatus*, the genital opening of the female leads to two womb chambers (uteri) where embryos develop. **TJ**

✦ NAVIGATOR

1 HEAD AND SENSES

The head has a pair of antennae with a simple eye at the base of each antenna. These eyes are only capable of discerning light from dark. The antenna is sensitive to touch and is likely to detect chemicals as well. Some experts suggest that these have the same origins as the mouthparts of arthropods.

2 SEGMENTS

The body is made up of repeated segments. The skin forms rings that circle the body so that it is hard to discern one segment from the next, giving a sluglike appearance. Each segment has a pair of conical legs. *P. novaezealandiae* walks by extending and retracting its legs in a rhythmic cycle.

3 LEGS AND BODY

The legs and body have evolved only a thin layer of chitin cuticle that has no structural role. Instead, rigidity is achieved by hydrostatic (fluid) pressure inside the body. The only hard body part is the chitinous claw, which is retracted into the foot when at rest. When extended, the claws grip rough ground.

TARDIGRADES

Tardigrades ("slow walkers"; right) are microscopic versions of lobopods. They are thought to represent a sister group to the one that evolved into the velvet worms and arthropods. Their fossil record dates back 530 million years, and more than 1,000 species survive today. Their body's rotund appearance has earned them the name "waterbears." Few other animals have exploited so many different habitats from deep oceans to the meltwaters of mountains. When faced with hostile conditions they retract the legs and curl up into a hardy cystlike form. This resistance has served them well during 500 million years of evolution. With a maximum length of 1/50 inch (0.5 mm), they are only visible through a microscope. They have eight legs, each with between four and eight claws.

THE RISE OF ARTHROPODS

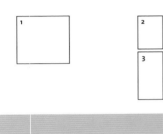

1 The emperor dragonfly, *Anax imperator*, sheds its skin. Due to their nonstretchy exoskeleton, arthropods grow by regularly shedding skin when they need to change shape (metamorphose).

2 *Paucipodia*, a lobopod from Chengjiang fossils of China, has walking limbs tipped with claws.

3 The brine shrimp, *Artemia salina*, is one of myriad small crustaceans that have evolved into almost every aquatic habitat.

Accounts of Earth's prehistory often refer to periods when mammals or dinosaurs were dominant, but if any animal group reigns the planet, it is the arthropods (joint-legged invertebrates). This huge phylum includes insects, arachnids, crustaceans, millipedes, centipedes, and a handful of more unusual creatures. They account for 80 percent of all animal species currently on Earth, while the trilobites represented just one class of extinct arthropod. Despite their huge variety of body forms and lifestyles, arthropods share some basic features. The arthropods are the main group in the Ecdysozoa, a supergroup of creatures that all shed their skin (ecdysis) as they grow (see image 1). The body is coated in a natural chemical polymer called chitin, made of numerous repeating units. Chitin provides the body with a flexible waterproof covering, but cannot enlarge, so the coating must be periodically shed to allow growth. Other animals in the ecdysozoan group include the nematodes and the velvet worms (see p.112). The common ancestor of the ecdysozoans remains unknown, but it was certainly Precambrian, before 541 million years ago.

Arthropods use chitin for more than just waterproofing. It is built up into thicker, stiffer units, creating an exoskeleton (hard outer casing) that provides

KEY EVENTS

540 Mya	520 Mya	505 Mya	480 Mya	460 Mya	443 Mya
Arthropods originate on the seabed when marine worms begin to evolve external skeletons and short legs sprout from their segments.	Lobopods, such as *Hallucigenia* and *Paucipodia*, exist, alongside *Canadaspis*, *Anomalocaris* and primitive trilobites such as *Retifacies*.	*Opabinia* and *Marrella* become common arthropods. The earliest Maxillopoda (crustaceans such as barnacles) is *Priscansermarinus*.	The earliest branchiopod crustaceans exist in marine habitats. These are represented by fairy and brine shrimp in hypersaline lakes.	The first eurypterids, predatory chelicerates also known as sea scorpions, appear.	A mass extinction at the end of the Ordovician reduces the diversity of marine life, which is mainly trilobites, by 60 percent.

protection, structural support and muscle anchorage. The structure of the exoskeleton partly gives the name arthropod, which means "jointed leg." The legs and other body parts of an arthropod are made from several articulated segments. The number of legs varies across the phylum, from four in certain butterflies, to several hundred sported by large millipedes. Research into the genetics of living species suggests that the likely Precambrian ancestors of the arthropods were creatures such as *Paucipodia* (see image 2), or "walking worms," represented today by velvet worms. *Paucipodia* and its kin are known as the lobopods, believed to have had simple bodies made up of repeating segments, each with a pair of walking limbs. Evolution has adapted this basic unit to produce an array of arthropod appendages devoted to various tasks. These include the sensory antennae, food-processing mandibles (mouthparts), external gills for breathing, and pincerlike palps with multiple uses.

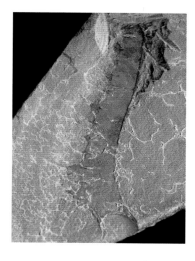

The arthropod family splits into three unequal branches—four including the trilobites. The arachnids dominate the first branch: the Chelicerata. Its members are characterized by their clawlike mouthparts. Likely ancestors are known from the early Cambrian period, 520 million years ago, but the first confirmed chelicerates were the eurypterids (sea scorpions; see p.136), which appeared in the middle Ordovician period around 460 million years ago. An early true arachnid, a ticklike creature called *Palaeotarbus*, was on land 417 million years ago. The second branch is occupied by the myriapods (many legs). As the name suggests, this group contains millipedes and centipedes. The first evidence of these are 450-million-year-old fossilized tracks made as they led the emergence of life on to land. The third arthropod branch is named Pancrustacea. It includes the insects—the biggest arthropod class—which evolved after their ancestors moved onto land. Their fossil record stretches back to *Rhyniognatha* (see p.134), 400 million years ago. The other members of the Pancrustacea are described as crustaceans but are in fact an aggregation of different groups. Crustaceans are mainly marine—crabs, barnacles, and krill. These animals have very early origins.

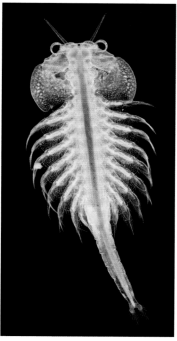

The Cambrian fossil beds from 520 million to 500 million years ago contain many specimens of *Canadaspis* (see p.144). This sea-floor animal is considered an early member of the Malacostraca, the crustacean class that includes crabs, woodlice, and shrimp (see image 3). It has also been proposed that *Canadaspis* is a basal (original) form for arthropods, meaning it could resemble the common ancestor that evolved from the lobopods. At only 2 inches (5 cm) long, *Canadaspis* crawled along the seabed on its multiple legs, using antennae to feel for prey in the sediment, protected by a shieldlike carapace. But it was prey to *Anomalocaris* (see p.116), a hunter, 3 feet (1 m) long, equipped with two sturdy tentacles fringed with spikes or hairs. Another Cambrian resident, *Marrella*, was more arthropod-like, with limb-style appendages used as gills and for movement, and long antennae; some fossils show it molting its skin. Even 500 million years ago, the arthropods were one of the most diverse animal groups on Earth. **TJ**

428 Mya	417 Mya	400 Mya	370 Mya	320 Mya	252 Mya
The millipede *Pneumodesmus* is the earliest confirmed myriapod, and lives in Scotland.	The earliest true arachnid, *Palaeotarbus*, is living in England. Centipedes are first seen at this time.	*Rhyniognatha*, the earliest insect, dates from this time. It may have wings, indicating that insects were already highly evolved.	The earliest decapod *Palaeopalaemon* exists at this time. This crustacean order includes the crabs and shrimp (see p.142).	*Tesnusocaris*, the earliest remipede, occurs. Remipedes are blind crustaceans that live in groundwater.	The last remnant trilobites, one of the greatest of the arthropods groups, die out in the end-of-Permian period extinction.

Anomalocaris
MIDDLE—LATE CAMBRIAN PERIOD

👁 FOCAL POINTS

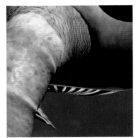

1 ARMS
The appendages (arms) were nearly 8 inches (20 cm) long. They were constructed of 12 segments of decreasing size locked together. Barbs emerged from the joins between the segments. The arms were likely prehensile (able to coil up and tightly grip objects).

2 EYES
A. briggsi had a pair of huge compound or image-forming eyes. They were positioned on stalks on either side of the head, giving the animal a wide field of vision. Each eye had 16,000 tiny separate lenses, meaning the vision of *A. briggsi* could rival any Cambrian animal.

3 SWIMMING LOBES
Without legs, *A. briggsi*'s body was analogous to that of a cuttlefish. It was fringed by a dozen fleshy lobes, which rippled to push it through the water. The lobed body formed a stable hydroplane, so a complex control system was not needed in order for it to swim straight and level.

4 MOUTH
The mouth was made up of 32 segments arranged in a ring. The inner surface was lined with serrated spikes. Reconstructions of the mouth show that it could expand and contract to grip, and even to squeeze, soft prey, but it could not be sealed shut.

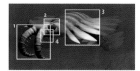
SPECIES *Anomalocaris briggsi*
GROUP Dinocaridida
SIZE 3 feet (1 m)
LOCATION China, Australia, and Canada

The alien-looking predator *Anomalocaris* ("strange shrimp") was the largest animal in the Cambrian oceans. A length of 3 feet (1 m) was typical, with some specimens double that. Fossils are common in Cambrian rocks, such as Chengjiang in China, 520 million years old; Emu Bay in Australia, 515 million years old; and the Burgess Shale of Canada (see p.50), 505 million years old. The body was protected by tough chitinous coverings that had to be molted periodically, which is one reason why *Anomalocaris* and kin are regarded as early arthropods. *Anomalocaris briggsi* is a leading member of the Dinocaridida group ("terrible shrimp"). It was an evolutionary side branch of arthropods that never progressed beyond the Devonian period so has no living representatives.

Dinocaridids are sometimes characterized as arthropods with extra-large appendages. The body of *Anomalocaris* supports this theory, as the head is dominated by two flexible arms covered in barbs that were presumably used in feeding. During the 1980s it was thought that *Anomalocaris* used these to pick up trilobites, which would have been transferred to the mouth for crushing before eating. *Anomalocaris* appears to have had a short abdomen, indicating that it ate animal foods. There is some evidence from coprolites (fossilized excrement) that trilobites were included in the animal's diet, although this is contentious because recent research has highlighted the fact that *Anomalocaris* lacked any hard, mineralized mouthparts—or any parts—that could smash into strong trilobite shells. Various theories propose how the soft-mouthed killer could crack into this shelled prey. One idea suggests that the predator would have held a trilobite in its mouth and used its arms to bend and twist its prey until it split open, similar to a human shelling a jumbo shrimp. Other experts propose that *Anomalocaris* was not the voracious hunter that others assume it to have been. Instead, it could have used its flexible appendages to probe muddy sediments for worms and jelly-bodied creatures, which would have been eaten easily. **TJ**

SPONGE-JELLYFISH FOSSILS

The story of *Anomalocaris*'s discovery is a litany of misconceptions. The first specimen was unearthed in 1892 in Canada, but only showed one of the arms. Its discoverer, Joseph Frederick Whiteaves (1835–1909), declared it was the tail of a lobster or shrimp, *Anomalocaris*. The next discovery, by Charles Doolittle Walcott (1850–1927; see p.51), was of a fossil of a round mouth, which he classified as a jellyfish. When a fossilized body was discovered, it was believed to be a sponge, and the mouth fossil found with it led people to believe that a jellyfish had been fossilized sitting on a sponge. More sponge-jellyfish fossils were unearthed, but only in the 1980s came a fossil of all three parts together (right). According to scientific conventions on naming organisms, the first name took precedence: *Anomalocaris*, which means strange shrimp.

TRILOBITES

Among the jumble of weird and wonderful early arthropods (joint-legged invertebrates) resulting from the Cambrian explosion (see p.48) are trilobites. This group of marine arthropods was a major force in the oceans for at least 170 million years. Already diverse when they appeared in the fossil record of the early Cambrian period, up to 17,000 species have been identified. This diversity was heavily impaired by the mass extinction in the late Devonian period, 359 million years ago, but a single order of trilobites clung on for another 90 million years before they became extinct in the "Great Dying" mass extinction, 252 million years ago, that marked the end of the Permian period and Paleozoic era. One reason why trilobites made their mark on the fossil record is because they had a hardened, chalky or calcified carapace—a shieldlike body covering similar to the hard shells of crabs—that was ideal for fossilization. Most trilobite specimens depict only the shed or molted hard body covering, whereas fossils containing soft body parts underneath, such as the legs and gills, are much more rare. The word "trilobite" means "three lobes" and the trilobite body comprised three lobes: a central (axial) one, and left and right pleurals. Confusingly, it also had three body sections: a semicircular head, the cephalon; a segmented midsection, the thorax; and a rounded shield at the rear end, the pygidium. Specimens fossilized mid-molt show that the old covering, or exoskeleton (external skeleton) split around the

KEY EVENTS

540 Mya	525 Mya	516–513 Mya	500 Mya	480 Mya	470 Mya
Trilobite life forms begin to appear, but their direct ancestor is unknown.	*Redlichia takooensis* is an early trilobite specimens. Even the earliest trilobites have compound eyes (see p.120).	*Yunnanocephalus* (see p.122) is one of the few trilobite specimens to show soft body parts and therefore explain trilobite anatomy.	*Ellipsocephalus*, a blind trilobite, loses its eyes as a result of adopting a deep-sea lifestyle.	Trilobite diversity increases during the Ordovician. Phacopids, which are capable of rolling themselves into tight balls, appear.	*Cheirurus*, a distinctive-looking species with long spikes on its head, appears in seas worldwide, leaving plentiful fossils.

edge of the cephalon, creating a flap. Structures on the cephalon then gripped the seabed as the trilobite flexed its body to wriggle forward out of its old skin.

With thousands of species, trilobites probably lived every conceivable lifestyle in Paleozoic oceans. They were hunters, scavengers, and grazers; some swam in open water, others patroled the seabed or burrowed into the substrate. (Parasitic trilobites have yet to be confirmed, but they probably existed.) A clue here is the hypostome, which formed a hard mouthpart in life. It surrounded the oral opening to a stomach, beneath the bulbous glabella: the part of the central lobe in the center of the head. Hypostomes of early trilobites indicate that they had generalized diets and scavenged for scraps. Later trilobites had a sturdier hypostome positioned further forward, where it braced against the head shield to deliver more of a bite. One unusual trilobite group of the Ordovician period had almost no hypostome. It is thought they absorbed nutrients from sulfur-eating bacteria that lived on the outside of the body.

The trilobite fossil record is crowded. Several species appear in the late Cambrian, already showing wide variation (see image 1). The Olenillina group from this time had smooth heads, while others had sutures, or rupture lines, which indicate where the cuticle would split during molting. The Olenillina represent early primitive trilobites that evolved during the Ediacaran period. Early Cambrian deposits in China contain soft-bodied naraoids that are thought to be a sister group to the most primitive trilobites. By the Cambrian, from 541 million years ago, there were already four major groups of trilobite; this period saw their widest diversity in terms of species numbers. However, new lineages emerged during the Ordovician, as trilobite evolution was pushed onward by the emergence of other hard-shelled organisms, such as mollusks (see p.124), echinoderms (see p.162), and graptolites (see p.146). These presented extra competition for trilobites but also new opportunities, such as reef environments.

During the Ordovician, the Phacopida order arose; these were often fossilized rolled into tight balls (see image 2). Most trilobites may have rolled defensively, but phacopids had appendages that locked the body in this position. Other novel body forms appeared, such as *Cheirurus* (see image 3), which had long spines sweeping back from the cephalon, with similar projections emerging from the thorax. Trilobite diversification meant that they came through the Ordovician extinction, 443 million years ago, relatively unscathed. Three-quarters of them survived, while 60 percent of life on Earth perished. Silurian period trilobites did not change much from their Ordovician ancestors, but many developed spiny shells as they came under pressure from new jawed fish (see p.184). The Devonian extinction, most likely precipitated by meteorite impacts that changed the chemistry of ocean water, did away with all but one trilobite group: Proetida. They were well adapted to their seabed lifestyles as they persisted for another 90 million years. However, even trilobites could not survive the end-of-Permian mass extinction, which killed more than 90 percent of all marine animal species. **TJ**

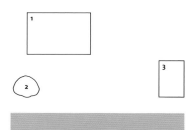

1 Some rocks of the Paleozoic era teem with preserved trilobites in mass "graveyards," such as these *Ellipsocephalus hoffi* from Europe, which could have been the victims of an underwater catastrophe, such as a mudslide.

2 Phacopida specimens are often rolled into an armadillolike ball shape.

3 Middle Ordovician *Cheirurus*, some 3 inches (8 cm) in length, sported long, swept-back horns from both the head (cephalon) and tail (pygidium).

460 Mya	443 Mya	420 Mya	359 Mya	300 Mya	252 Mya
Large aggregations of *Homotelus* are found, suggesting that trilobites formed swarms for feeding or reproduction.	The Ordovician mass extinction removes a quarter of trilobite species, including most of the Cambrian species.	*Calymene*, about 1 inch (2.5 cm) long, and *Dalmanites* are two dominant genera in the Silurian seas, each genus with numerous species.	The end-of-Devonian mass extinction events kill all but one subgroup of the trilobites.	The trilobite group Proetida, still with hundreds of species, survives mainly on seabed habitats at various depths worldwide.	The end-of-Permian mass extinction finally extinguishes the trilobites, along with many other animals and plants.

The First Compound Eyes
CAMBRIAN PERIOD

SPECIES *Trevoropyge prorotundifrons*
GROUP Phacopida
SIZE 2 inches (5 cm)
LOCATION North Africa

⬖ **NAVIGATOR**

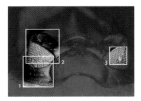

Trilobites appear in the fossil record already showing a pair of compound eyes (those with many small units) that were capable of forming an image of some detail. Along with the dinocaridids, such as the Cambrian giant *Anomalocaris*, trilobites were the first animals to evolve such eyes. This gives insight into the rapid adaptive radiation (rapid and diverse evolution; see p.92) that is often described as the Cambrian explosion. The wealth of species seen in the early and middle Cambrian is a snapshot of a process that began millions of years before. Simple eyes, little more than light-sensitive patches on the body, evolved to help animals find the most suitable habitat, from sun-drenched shallow seas to dark, sheltered crevices. Today's simple-eyed creatures are alerted by a shadow that could herald the imminent arrival of a predator. With species adopting active predatory lifestyles, trilobites needed eyes to see them coming.

This process was already well under way by the time trilobites arrived in the fossil record around 525 million years ago. Trilobite compound eyes had several thousand individual lenses, each made from transparent crystals of calcite (a form of calcium carbonate). There are three types of eye. Abathochroal eyes were the primitive form and existed in only a few Cambrian species. They were made up of about 70 lens units, each separated by a cuticle wall (sclera). Next came the holochroal eye structure, present in most Cambrian species, which indicates that they were active animals within the photic zone: the area of the sea that receives daylight. The third type of eye, schizochroal, was found only in the Phacopida trilobites from the later Ordovician, such as *Trevoropyge prorotundifrons*. It resembles the abathochroal form, but with larger and more numerous lenses, up to 700. **TJ**

👁 FOCAL POINTS

1 EYE SITES
Trilobite eyes were located on the "fixed cheeks" either side of the glabella and they were protected by a layer of transparent cuticle. Before and after molting, the trilobite would have been blinded for a short while, as the old cuticle separated and the new one hardened.

2 LENSES
A compound eye is an aggregation of image-forming units known as ommatidia. A trilobite ommatidium had two calcite lenses. A single rigid calcite crystal could not be adjusted to focus on an image, therefore two were used together. Insect eyes have a similar mosaiclike design.

3 EYE DESIGN
The dimensions of individual lenses varied from almost $\frac{1}{25}$ inch (1 mm) to less than $\frac{1}{800}$ inch (0.03 mm). Some schizochroal eye designs had as many as 700 lens units. The eye's prominence and curvature indicate whether its owner was a slow-moving grazer or a high-speed hunter.

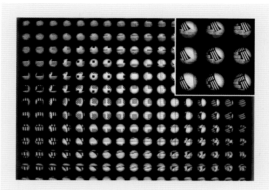

FIELD OF VISION

In most kinds of trilobite compound eye, each unit (ommatidium) detected and contributed a small point of the scene ahead, similar to a single pixel in a digital camera or on a television screen (main image, above). These tiny mosaiclike areas were combined to create one composite view. The schizochroal eye had fewer, larger lenses. Each one was fully separate from its neighbors and detected a much larger area of the scene (inset image, above). In this system, the numerous images were presumably "knitted together" in order to make sense of the whole, by comparing them for important reference points, such as lines and shapes. This is similar to the way a camera or computer can knit together several images to produce one overall image.

◄ Trilobites were among the first creatures to provide clues about the evolution of animal senses. Some Cambrian trilobites never developed eyes, while later species reduced their sense of vision as they adapted to life in deep, dark water, such as this *Paraceraurus*. There were even species with stalked eyes, perhaps to peer out of soft sediments that covered the rest of the body. The antennae were used as feelers to reach out in the dark and detect obstacles. Similarly, the tail-end cerci seen in some species would give first warning of an attack from behind. Several species have been found with pits around the head end, perhaps chemical detectors or sensors for water currents. There are also patches of thinned cuticle on the underside of some types. One theory is that these registered light and told the animal it was upside down.

Yunnanocephalus
MIDDLE CAMBRIAN PERIOD

Despite thousands of trilobite fossil specimens, it has been possible to describe the full anatomy of only 20 or so species, because so few show soft body parts. *Yunnanocephalus yunnanensis* is one example that does, with an almost three-dimensional specimen from the Chengjiang fossils of Yunnan province, China. It grew to almost 1 inch (2.5 cm) long and dates to 515 million years ago. Soft-body fossils are generally from trilobites that were buried upside down. These specimens show that trilobites had antennae and 20 pairs of limbs on the body's underside. Numerous spikes and filaments suggest that limbs had a function as touch organs and for gripping prey. Much of the ecological evidence for how trilobites lived comes from trace fossils, such as prints left by their legs in sediments and drag marks from the body. These trace fossils indicate how trilobites rested on and moved across the seabed. Some show deep furrows through the seabed that are thought to be created by deposit-feeders eating the sediment to extract organic materials, much as earthworms eat soil. Several limbs were attached to the head and most likely had a role in feeding, either clearing away sediment to reach buried food or grabbing and even chewing prey prior to swallowing. In many cases the only indication of a trilobite's diet is the glabella. Evidence suggests that species with a large glabella had a carnivorous lifestyle. **TJ**

navigator

⬡ NAVIGATOR

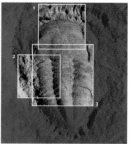

SPECIES *Yunnanocephalus yunnanensis*
GROUP Yunnanocephalidae
SIZE 1 inch (2.5 cm)
LOCATION China

👁 FOCAL POINTS

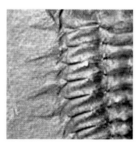

1 ANTENNAE
Trilobites had a pair of long, flexible antennae that pointed forward from under the cephalon. Some fossil specimens show trilobites also had rear-facing appendages, perhaps with a similar sensory function. The carapace of some later species had hornlike projections and spikes.

2 LIMBS
There were three or four pairs of limbs under the head, one pair for every part of the midsection and a few more on the rear end. Limbs branched in two: a longer, sturdy branch was used for walking and manipulating food; a thinner branch was used as a feathery gill or swimming organ.

3 DIGESTIVE SYSTEM
The mouth was a toothless muscular tube held in a hard plate called the hypostome. The mouth was in the bottom of the body and faced backward. Swallowed food traveled forward along the body to the stomach under the glabella, before moving to the rear end anus.

MISSZHOUIA

The Maotianshan Shales of Chengjiang in Yunnan province, China—a series of early Cambrian deposits—date from 520 million to 515 million years ago, and contain many wonderfully preserved fossils. Along with the varied trilobites are the remains of *Misszhouia* (above), a soft-bodied arthropod regarded as a link to the ancestors of trilobites. *Misszhouia* was a member of a group of bottom-dwelling arthropods known as naraoids. They had only two body sections: a rounded cephalon and a segmented pygidium. In other respects they shared many trilobite features: two antennae (but no eyes, unlike most trilobites) and 25 pairs of branched limbs with functions including feeding, breathing, and movement. Trilobite ancestors in the Ediacaran may have looked something like *Misszhouia*.

MOLLUSKS

1 Rudist mollusks once formed vast reefs, especially during the Cretaceous.

2 Gastropods, such as the snail *Helix*, went through a phase of twisting their bodies (torsion) during their evolution.

3 The giant clam, *Tridacna maxima*, belongs to the hugely successful mollusk group Bivalvia.

The mollusks are one of the most widespread and diverse groups of animals on Earth. They are found in deep ocean vents and on mountain tops, and they include slugs and snails; shellfish, such as clams, scallops, and oysters; and the big-eyed, fast-moving, predatory octopuses, squid, and cuttlefish. With 100,000 mollusk species known to science, and probably at least twice that number awaiting discovery, only the arthropods (see p.114) outdo the mollusks in species richness.

Piecing together the mollusk evolutionary tree is a challenge. However, it is fortunate for paleontologists that most mollusks live in water (mainly salty) and that nearly all have hard shells. The aquatic environment is more likely to lead to preservation and fossilization, compared to onshore terrain, and hard mollusk shells are ideal for these processes. Consequently, millions of fossils from all prehistoric periods show that mollusks have been around for a long time, perhaps for as many as 600 million years, and that they are some of the oldest multicelled animals known. Fossils found in the 1940s in the Ediacara Hills in South Australia (see p.46) date back to the Neoproterozoic era (late Precambrian) and an animal called *Kimberella*. It is the first known bilaterian: an animal with a distinctive front end, back end, and two sides. Some of its features are similar to a living group of mollusks called monoplacophorans (see p.126), once thought to be extinct.

KEY EVENTS

540 Mya	540 Mya	530 Mya	530 Mya	520 Mya	500 Mya
Monoplacophorans, one of the earliest of the mollusk groups, flourish in the early Cambrian.	The rostroconchs appear. Once thought to be early bivalves they have since been given their own mollusk class.	*Aldanella* has a coiled shell similar to those of modern snails. It lived on the seabed.	*Tannuella* is possibly the first cephalopod, or a close relative of its ancestor.	Early bivalves, such as *Fordilla*, with two valves hinged together, make their appearance.	*Odontogriphus*, a sluglike organism with a toothed tongue, resembles the shared ancestor of annelid worms and mollusks.

The earliest fossils confirmed as mollusks are very small marine mollusks with snail-like shells—the helcionelloids—which appeared at the end of the Ediacaran period, about 540 million years ago, and persisted until the early Ordovician period, 480 million years ago. By the middle Cambrian period, 500 million years ago, most of the familiar groups—including gastropods (snails and kin; see image 2), bivalves (oysters and cousins), monoplacophorans, and the now-extinct rostroconchs—are found in the fossil record. The cephalopods (octopuses and relations) are first found in the middle Cambrian; the polyplacophorans (coat-of-mail shells) in the late Cambrian; and the scaphopods (tusk shells) in the middle Ordovician, about 460 million years ago. In evolutionary terms, this diversification was rapid and extensive.

As early as the Cambrian, the first bivalves were found, but it was not until the early Ordovician that they began to diversify and become prominent in the fossil record. Their shell is made of two parts (valves) often hinged at one side in the classic design of oysters, mussels, clams (see image 3), scallops, cockles, razor shells, and others. Major extinction events at the end of the Permian (see p.224) and Cretaceous periods (see p.364) did not halt them, and today bivalves are among the prominent life forms in almost all marine habitats. However, some bivalves did become extinct at or just after the end of the Cretaceous, including *Gryphaea* (devil's toenails; see p.128) and the rudists (see image 1). The latter were a major component of reefs in shallow marine environments from the late Jurassic to the late Cretaceous; their fossils are found in the Middle East, the Mediterranean, the Caribbean, and Southeast Asia. They had two asymmetric valves with one valve attached to the sea floor. Rudists could grow to be quite large and their colonies formed "rudist reefs" that spread for hundreds of miles.

The gastropods, including spiral-shelled slugs and snails on land and conches, whelks, and "untwisted" limpets in water, are the largest group of mollusks, comprising about 80 percent of living species. Their fossil record reaches back to the Cambrian and shows evidence of periodic extinctions of major groups followed by the appearance and diversification of new groups. The cephalopods are the biggest and most intelligent of the mollusks. Their 500 million years of evolution has led to suckered tentacles to manipulate their environment, sophisticated eyes, the ability to change color, and complex social and learning skills. Squid, octopuses, and cuttlefish represent some of the 800 or so living cephalopod species, but 17,000 species have been identified in the fossil record of this hugely diverse group. Included in the extinct groups are enormous squidlike creatures—orthocone nautiloids—that swam from the Ordovician to the late Triassic. They were huge cone-shaped creatures with tentacles at one end of a straight external shell that could reach 33 feet (10 m) in length. The top predator of its day, the orthocone propelled itself through the ocean by pushing water out through a tube that ran along the cone shell. Other prominent cephalopod groups, now extinct, included other nautiloids and ammonoids. **RS**

500 Mya	490 Mya	480 Mya	460 Mya	200 Mya	100 Mya
More cephalopods appear toward the end of the Cambrian as the group gathers evolutionary momentum.	In North America, *Strepsodiscus*, about 1 inch (2.5 cm) across, is one of the earliest snail-like mollusks.	Bivalves begin to evolve and diversify, expanding to fill many new lifestyle opportunities in Ordovician seas.	The first scaphopods— with a shell shaped like an elephant tusk—are found in the middle Ordovician period.	Orthocone nautiloids, which arose in the late Cambrian, fade from the seas after a lengthy span as top predators.	Oysters five times bigger than modern species live off the coast of southern England.

Monoplacophorans
EARLY CAMBRIAN PERIOD—PRESENT

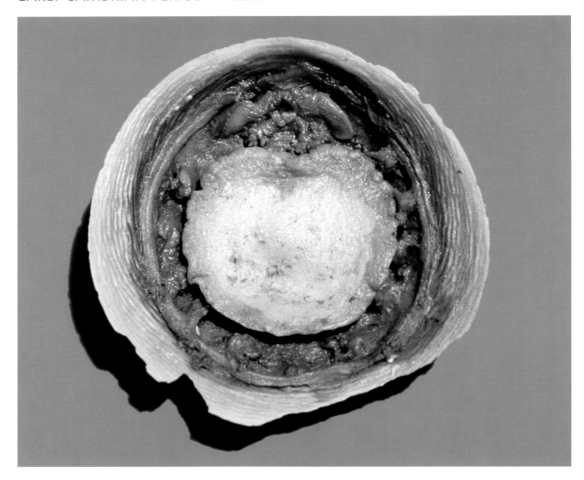

SPECIES *Neopilina galatheae*
GROUP Monoplacophora
SIZE 1½ inches (4 cm)
LOCATION East Pacific Ocean
IUCN Not Yet Assessed

⚙ NAVIGATOR

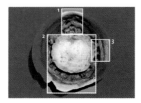

This mysterious group, resembling limpets, is often thought to be among the most primitive of mollusks. In 2007, the genetic material of a few monoplacophorans (one-plate bearer) was analyzed in order to determine where they should be placed in the mollusk family tree. The results revealed that they were closely related to the cephalopods (nautiloids, ammonoids, octopuses, and kin), which appeared to link these two unusual and distinctive mollusks as sister groups. This close evolutionary relationship indicates that they are likely to have shared an ancestor species. Paleontologists in the 1970s suggested a relationship between the monoplacophorans and the cephalopods based on the observation that early kinds of both had chambered shells. Living monoplacophorans still have shells but they no longer have chambers. Studies suggest that monoplacophorans gave rise to the polyplacophorans as well as to early cephalopods, but this is yet to be confirmed. Monoplacophorans were once thought to be long extinct, as they were known only from fossils, but living specimens were discovered during an oceanographic survey carried out by the Danish deep-sea expedition ship *Galathea* between 1950 and 1952. The expedition aimed to discover whether sea serpents existed. The serpents remained elusive but, in May 1952, 10 living specimens of monoplacophorans were discovered off the Pacific coast of Costa Rica from 11,775 feet (3,590 m) deep. They were named *Neopilina galatheae* and made headline news for evolutionists worldwide. **RS**

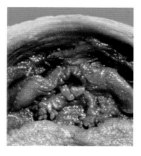

1 HEAD
The head of *Neopilina galatheae* is poorly defined with an elaborate mouth structure on the undersurface. The mouth typically has a V-shaped, thickened lip in front and small tentacles behind, the form of which varies among species. These mollusks probably crawl over the seabed feeding on anything edible in the mud and silt or that is coating the rocks.

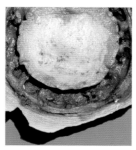

2 FOOT AND MANTLE
Below the head lies the semicircular foot by which the animal attaches itself. Five or six pairs of gills can be found between the foot and the edge of the body "cloak" (mantle). The mantle is an important feature of most mollusks and envelopes the major part of the body. It is used for protection, disguise, and movement.

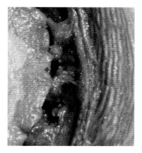

3 INTERNAL ANATOMY
There is a long, looped digestive system, two pairs of sex organs and multiple paired excretory organs functioning like kidneys. Rather than occurring once, some organs, such as the heart, are repeated throughout the body. This prompted the theory that monoplacophorans were related to segmented organisms, such as annelid worms and arthropods.

▲ Monoplacophorans have a single, caplike shell. In fossils, the opening of the shell varies in shape from almost circular to pear-shaped. Shell height can range from relatively flat to tall. In both present-day monoplacophorans, such as *N. galatheae*, and fossils, the apex of the shell is typically positioned at the front end, and in some it overhangs the leading edge of the shell. All of the 30 to 35 species occur in oceans below 650 feet (200 m) and as far down as 20,000 feet (6,000 m).

LAMPSHELLS

Lampshells or brachiopods (left) are named after a shell shape reminiscent of old oil-burning lamps. Outwardly they resemble bivalve mollusks and many live similar lifestyles, as filter-feeders on the seabed. However, detailed study of their anatomy highlights many differences between the two creatures; for example, the shell parts are on the top and bottom of the brachiopod, rather than on the left and right sides, as is the case in bivalves. Only about 335 species of brachiopods survive today, but fossil species number more than 12,000. Their origins are obscure and although some subgroups experimented with new forms, diversified and became extinct, others, such as *Lingula*, were as "conservative" in their evolution as any other animal group. Therefore, the living species are remarkably similar to their relatives from 500 million years ago.

Devil's Toenails
JURASSIC PERIOD

GENUS *Gryphaea*
GROUP Ostreidae
SIZE 4 inches (10 cm)
LOCATION Europe and North America

NAVIGATOR

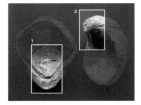

Throughout history, fossil remains have often been misinterpreted, especially those from mollusks. For example, the bullet-shaped internal shells of the extinct cephalopods called belemnoids were thought by previous generations of Europeans to be the result of lightning hitting the ground and they were named "thunderstones." Similarly, the coiled shells of ammonoids were called "snake stones" in the Middle Ages because they were believed to be snakes turned to stone. Ammonite fossils found by the early Romans were mistaken for horns, and the name "ammonite" refers to the coiled horns of the Egyptian ram god Ammon. *Gryphaea* is a genus of bivalves, related to oysters, from the Jurassic and although it is unlikely that anyone really believed that the fossil remains were devil's toenails, it is easy to see why the name arose. *Gryphaea* is a reasonably common find in Jurassic rock layers (strata). It can have a fairly flaky appearance and often takes on a black color from the clays that preserved it. With a little imagination this furrowed shell could be seen to resemble a gnarled toenail.

Oysters such as *Gryphaea* belong to the group of bivalves known as the pteriomorphs. This important, entirely marine group is made up of many of the most familiar bivalves, including the scallops (Pectinidae), oysters (Ostreidae), pearl oysters (Pteriidae), and mussels (Mytilidae); their familiarity might well correlate to their edibility. Many pteriomorphs are described as epifaunal, which means that they attach themselves to a substrate, such as the seabed, a rock, or even the hull of a ship. Other pteriomorphs burrow a short way into the seabed, pushing their way along by means of a fleshy, muscular organ known as the foot. **RS**

FOCAL POINTS

1 VALVE ASYMMETRY

As a member of the bivalve group, *Gryphaea* possessed a shell of two parts called valves. The valves are distinctly unequal in size and shape: the left or lower valve is strongly curving, whereas the right or upper valve is small and flat. *Gryphaea* lived on the seabed, with the flat right valve facing upward. This probably opened to allow water to flow in, so that oxygen and nutrients could be filtered out; most bivalves filter-feed like this.

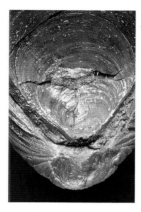

2 LEFT VALVE

The lower left valve is thick and its surface is marked with a series of concentric ridges and furrows formed as the animal grew. The curvature of the left valve appeared to become more pronounced throughout *Gryphaea*'s evolution. This led to the suggestion that the mollusk became extinct when it could no longer effectively open its upper valve. This misconception has since been debunked because it has been shown that the shell coiling actually loosened over time.

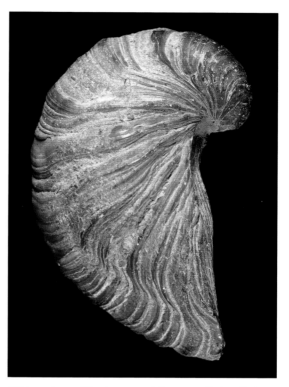

▲ This specimen of *Gryphaea arcuata* bears growth rings that usually occur with seasons, and in fossils, these indicate changes that occurred in climatic conditions.

RESILIENT OYSTER SHELLS

In 2012 researchers succeeded in sequencing the genome of the Pacific oyster, *Crassostrea gigas* (right), and consequently uncovered some clues to its evolution. The Pacific oyster has around 28,000 genes—considerably more than the 20,000 or so genes found in humans—and 8,600 of these are thought to be specific to mollusks. They show that the shell of the oyster is a surprisingly complex structure. Rather than a fairly simple matrix of calcium carbonate, its construction is an intricate process involving more than 200 kinds of protein substances, fashioned by eons of evolutionary tweaking. This gene analysis gives insight into how oysters and other mollusks have survived for so long in widely varying marine environments. For example, some genes are for a "heat shock" protein, which allows them to withstand temperature changes. If an oyster is exposed to the sun by the retreating tide, these genes kick in and allow the oyster to survive temperatures of up to 120° F (49°C). Other genes help cope with changes in salinity.

ARTHROPODS ON LAND

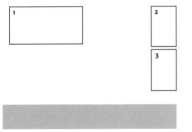

1 Dipluran bristletails are hexapods but, rather than being true insects, they are in a sister group to the Insecta.

2 Arthropleurideans evolved in the Silurian, more than 420 million years ago, and lasted for 130 million years. The underside view shows typical millipede segmented features.

3 The house centipede, *Scutigera coleoptrata*, is an insectivore, killing and eating other arthropods, such as insects and arachnids. The creature has up to 15 pairs of legs.

L ife as we know it began in water. The evolution of complex living structures requires long periods of stability, which only aquatic habitats could afford on young Earth. Perhaps surprisingly, it is thought that animals made the first forays on to land long before plants did. These pioneers were arthropods (invertebrates with an exoskeleton; see p.114), and some multilegged creatures left tracks in shoreline sands around 530 million years ago. The best guess is that they were euthycarcinoids. These arthropods had an similarity with the millipedes, but analysis of their mouthparts shows they may represent a side branch of evolution from the main lineage that gave rise to insects, dipluran bristletails (see image 1), crustaceans, millipedes, and centipedes. Why they and other arthropods were out of the water in the Cambrian period is open to conjecture. It is hard to imagine the view these creatures would have had as they hauled themselves up beyond the surf line. As there were no land plants, the dry surface would have been an eerie expanse of sand, clay, and rock—by no means a stable environment. Without plant and soil communities providing a structural integrity to the surface material, the landscape would have been harsh and ever-changing, scoured by wind, baked by sun, and torn apart by tumultuous floods.

The euthycarcinoids may have been deposit-feeders, scraping the films of bacteria and algae that grew in sheltered, tidal mudflats. It is probable that arthropods such as eurypterids (sea scorpions; see p.136), which emerged in the Ordovician period, came onto dry land as a place of refuge from waterborne predators. The arthropod exoskeleton (hard outer casing) was repurposed from

KEY EVENTS

530 Mya	460 Mya	440–420 Mya	428 Mya	417 Mya	408 Mya
Euthycarcinoids about the size of today's lobsters—20 inches (50 cm) long with 20 legs—leave tracks in dunes.	Sea scorpions evolve in the oceans, but emerge from the water periodically, perhaps to breed, lay eggs, or escape predators.	Early plants appear on land. They are nonvascular mosses and liverworts without a clear differentiation between root, stem, and leaf.	The millipede *Pneumodesmus* is one of the first confirmed land animals. It has a tube network for breathing air.	Trigonotarbids are early terrestrial predators, preying on primitive herbivores, such as the millipede *Eoarthropleura*.	Rhynie Chert reveals a terrestrial wildlife community made up of mites, crustaceans, euthycarcinoids, myriapods, and hexapods.

armor in the water into a waterproof covering on land, keeping moisture inside the body. Book gills—respiratory organs sheltered inside cavities along the underside of the body—remained damp enough to breathe the oxygen in the air.

The first fully terrestrial plants appeared as the Ordovician gave way to the Silurian period, from about 430 million years ago. There were ground-hugging mosses and liverworts that clung to their scarce water supplies in the arid terrestrial conditions. Animals soon followed; there is early evidence of a land creature, *Pneumodesmus*, a millipede dating from 428 million years ago. The piece found was ⅓ inch (1 cm) long and the whole length is estimated to be 1 inch (2.5 cm) long. The name *Pneumodesmus* means "air bands," and small holes in the specimen's outer covering, or cuticle, suggests that this creature breathed using a tracheal system: a network of tubes (trachea) that connect to the outside through pores (spiracles) and carry air into every section of the body. Today's insects and myriapods use this mode of respiration.

By the end of the Silurian, terrestrial (although still coastal) plants had become more substantial, thus providing a larger and more varied food supply for emergent arthropods. Among them were arthropleurideans (see image 2): primitive plant-eating millipedes including *Arthropleura* (see p.132). In addition to grazers, Silurian land arthropods also included predators in the form of the trigonotarbids. These were tiny arachnids that looked like mites or spiders (but they had no silk-weaving abilities) and they were mostly a few millimeters long. They preyed on other arthropods, killing them with a bite from their fangs.

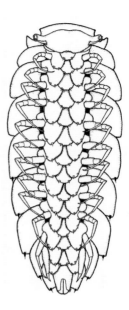

In the early Devonian period, wildlife communities had made the break with the ocean and were existing inland around bogs and ponds fed by springs, providing the moist habitat that was still required. The best example of this is the Rhynie Chert (see p.62), a layer of silica-rich rocks in eastern Scotland that dates from a 408-million-year-old peat bog around hot springs. Its fossil beds contain a wealth of terrestrial arthropods, including more trigonotarbids, and other arachnids, such as true mites and harvestmen (sometimes known as daddy-long-legs). The centipede *Crussollum*, a predator that had long legs spread widely, was similar to today's house centipede (see image 3), and *Leverhulmia* was a stumpy millipede, with around 10 legs, which ate dead plant matter.

Most significantly, Rhynie Chert was also home to hexapods (six-legged arthropods). Among them is the first confirmed insect specimen *Rhyniognatha* (see p.134), but the most common hexapod was *Rhyniella*, a springtail. Like its modern relatives, it was endowed with a jumping organ (furcula) tucked under the body. Springing it backward threw the animal into the air in an uncontrolled but spectacular leap. *Rhyniella* lived in the damp leaf litter and soil and used its spring to escape attack from centipedes and other predators. Within a few million years, the hexapods—equipped with an adequate number of legs for land, plus several other adaptable appendages—emerged from their early terrestrial communities to become the dominant invertebrates on land. **TJ**

408 Mya	395 Mya	365 Mya	340 Mya	328 Mya	50 Mya
The first confirmed insect specimen, *Rhyniognatha*, is found in a Rhynie Chert deposit in Scotland.	The first tetrapods (four-limbed vertebrate) emerge from the water to walk on land, and are related to later amphibians.	Insect flight is confirmed, although it is thought that *Rhyniognatha* and other earlier insects may have already had wings.	*Arthropleura*, the largest land arthropod ever found, is active in the forest of the Carboniferous period.	The diplurans (two-pronged bristletails), the third of the four major group of hexapods, appear in the fossil record.	Woodlice, terrestrial crustaceans belonging to the order Isopoda, are found worldwide, suggesting the terrestrial forms evolved much earlier.

Arthropleura
CARBONIFEROUS PERIOD–PERMIAN PERIOD

👁 FOCAL POINTS

1 HEAD AND MOUTHPARTS
The head of *A. armata* had simple eyes and a pair of antennae. The mouthparts are assumed to be modified limbs, as in living millipedes, although no specimens have been found. This lack of fossilized mouthparts suggests that they were not hardened or powerful like other body parts. *A. armata* was therefore unsuited to killing other animals.

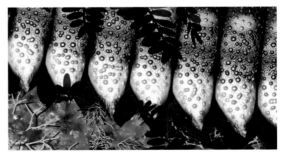

2 PLATES
A. armata specimens have a similar number of plates and body segments, yet they were not all huge in size. Smaller species reach only 1 foot (30 cm), similar to the length of large millipedes today. The plates were toughened through sclerotization, where the chitin—the tough main structural material in the exoskeleton—was modified with chemical cross links, rendering it hard and stiff.

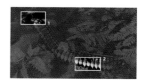

SPECIES *Arthropleura armata*
GROUP Diplopoda
SIZE 9 feet (2.8 m)
LOCATION North America, Scotland

At first glance, the vital statistics of *Arthropleura* ("jointed ribs") do not seem impressive. It was a member of the arthropod group Diplopoda, which includes today's millipedes. Some of the best evidence for *Arthropleura* comes from the tracks it left in mud. These appear as long parallel rows of small prints that swerve around obstacles, such as tree trunks and rocks. They show that *Arthropleura* stretched its body to increase its stride length when it needed to go faster. *Arthropleura*'s defining feature is that each body segment had two pairs of legs. (Members of the Chilopoda, the centipedes, have one pair of legs per segment.) It had between 25 and 30 body segments, thus making a total of 100 to 120 legs. Each segment was protected by a toughened plate that curved over the body, overlapping the one behind to create an articulated chain of armor. The plates alone meant that *Arthropleura* was well protected against attack as it patrolled the fern forests for much of the Carboniferous, into the Permian period, around 330 million to 270 million years ago. However, its size according to the time must be taken into account. *Arthropleura armata* has been estimated to have grown to 9 feet (2.8 m) long, which makes it the largest land arthropod in history, and perhaps the biggest arthropod of all, along with the giant sea scorpions. This millipede weighed 220 pounds (100 kg) and would have been able to curl up on a modern double bed with no spare room.

It has proved tempting to assume that *Arthropleura* was a voracious predator that could lift its body cobralike and bring down any forest beast it chose. However, no mouthparts have been found to confirm a carnivorous lifestyle. Analysis of the stomach has turned up fern spores and fronds, which show that *Arthropleura* was at least omnivorous, if not entirely herbivorous. Herbivores that eat tough, indigestible plant foods tend to be large: elephants are a fine example. Large bodies are more efficient and require less food per unit of body mass to survive. Nevertheless, *Arthropleura* had to eat frequently to survive. It is estimated that it consumed 1⅒ tons (1 tonne) of fronds per year. **TJ**

OXYGENATED ATMOSPHERE

Arthropleura ruled Carboniferous forests and swamps (right) and there is no evidence of another land animal that could have threatened it. The Carboniferous is named after its lasting legacy—coal—when the first woody giant plants evolved and formed Earth's earliest forests (see p.133). The massive trees boosted the oxygen content of the atmosphere much higher than today's levels, and this could be one reason why *Arthropleura* and other arthropods, such as insects, grew to such a large size. For insects, extra oxygen could diffuse, or spread naturally, through the spiracles and trachea. Today's arthropods have a natural size limit because if they grow any bigger they cannot get enough oxygen into the body. Permian climate change reduced rainfall, the coal forest communities collapsed, and *Arthropleura*, along with giant dragonflies and many other species, went with them.

Rhyniognatha
EARLY DEVONIAN PERIOD

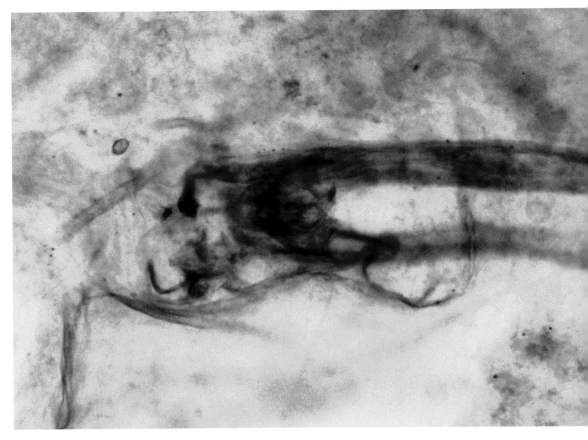

SPECIES *Rhyniognatha hirsti*
GROUP Insecta
SIZE Mouthparts and inner head structures as shown, 1/25 inch (1 mm)
LOCATION Scotland

NAVIGATOR

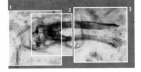

In 2004, a discovery from the rocks of Rhynie Chert in Scotland (see p.62) made headlines. Fossilized mouthparts belonging to *Rhyniognatha hirsti*—the oldest insect known to date—had been reanalyzed. The research showed that even this most ancient insect was not a primitive ancestral form, but a surprisingly specialized animal. *R. hirsti* is at least 400 million years old, placing it in the early Devonian. Prior to its discovery, the oldest insects known were from 379-million-year-old rocks located in upstate New York. These previous specimens were fragments of wingless insects, one classified as a silverfish (see panel), the other as a bristletail. Both were thought of as primitive types. Nevertheless, they are classified as members of the group Insecta because of their external mandibles (mouthparts). The name *Rhyniognatha* means "jaws of Rhynie," and the specimen shows external mouthparts, which make it an insect. However, the mouthparts are much more derived (not early, nonprimitive, or more changed) than those of the New York insects, despite being considerably older. The fragment was unearthed in 1919 and, in 1926, was thought to be the mouthparts of an insect larva. It was renamed *R. hirsti* in 1928 by English-Australian insect expert Robert "Robin" Tillyard (1881–1937). Closer analysis of its mouthparts in 2004 confirmed it was a possible winged planteater. Living insects with similar mouthparts and mode of feeding have wings. With only a mouthpart to analyze, it is impossible to know if *R. hirsti* was winged. Yet, the discovery lends credence to the theory that insects were highly diversified by the early Devonian. A study tracking changes in insect DNA puts the group as evolving 434 million years ago, in the Silurian. **TJ**

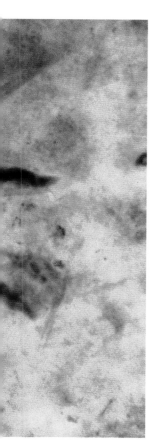

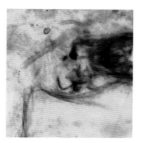

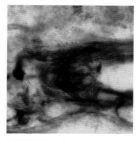

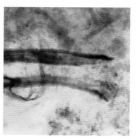

1 DIET

The mouthparts resemble those of a modern insect that feeds on plants. *Rhyniognatha* may have eaten sporangia (spore-filled organs) from the fronds of Devonian ferns. This would have been the most abundant food source at the time because flowering plants did not evolve for another 250 million years.

2 EXTERNAL MOUTHPARTS

The external mouthparts of *Rhyniognatha* can be likened to a pair of multipronged pincers, similar to human hands with curled fingers closing, here with food coming from the left. They are tiny, less than ⅟₂₅₀ inch (0.1 mm). A microscope view can only show part of them in focus at any one time.

3 HEAD

Two long, striplike parts may be cephalic apodemes. These are ingrowths of the insect exoskeleton within the head and they act as muscle anchorages. This specimen shows what was preserved, and represents the head: apodemes extend right, to the rear of the head.

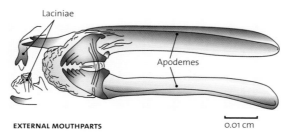

Laciniae

Apodemes

EXTERNAL MOUTHPARTS

0.01 cm

◄ This diagram interpreting *Rhyniognatha* may seem too detailed compared to the fossil specimen, but it has been composed using knowledge of other insect mouthparts and apodeme design, from both living and fossil species.

BRISTLETAILS AND SILVERFISH

Three-pronged bristletails and silverfish, often known as thysanurans, are regarded as early primitive insects with ancestors dating back 400 million years. Forming the order Archaeognatha (ancient jaws) of wingless insects, the chief feature of the bristletails is their three pronged "tail." They feed on small plants, such as mosses and lichens, which began to spread over Earth during the Silurian and have continued to facilitate their lifestyle for at least 370 million years. Silverfish (right) also sport three tails, usually shorter than those of their bristletail cousins. Relying predominantly on long antennae to feel their way around, they are either blind or have tiny eyes on their equally diminutive heads. Silverfish—named for their silvery gray and blue color—can be found mostly in soil, surviving on organic deposits, but many have also spread into human habitations to feast on leftover food, grease, and general detritus.

SEA SCORPIONS

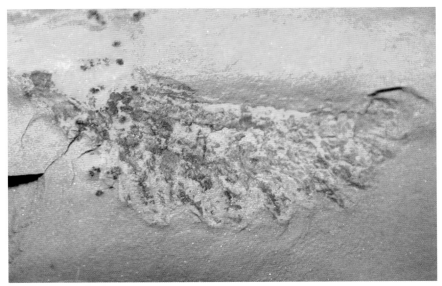

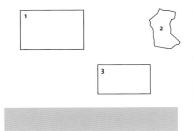

1. *Haikoucaris ercaiensis* shows large head appendages (upper left) and bristles on the rear legs (lower right). It was a possible ancestor of eurypterids.

2. *Pterygotus* fossils occur on almost every continent, dating from the Silurian to late Devonian periods, 440 million to 370 million years ago.

3. One of the largest ever arthropods, *Jaekelopterus*, from 390 million years ago, brandishes its long, clawlike chelicerae as primitive fish flee.

The eurypterids were a group of arthropods (joint-legged invertebrates; see p.114) that included the top predators of the seas in the Silurian period, 443 million to 419 million years ago. They were part of the larger group called Chelicerata, now represented by the arachnids (see p.150)—spiders, scorpions, mites, and ticks—as well as by sea spiders and horseshoe crabs (see p.140). Due to this relationship with scorpions, eurypterids are often known as sea scorpions. There are similarities—some eurypterids had large front pincers, others had a spiked tail (telson)—but they were not the scorpion's ancestors.

The sea scorpions were the earliest confirmed chelicerates. However, there are candidates for earlier forms. One possibility, and one of the oldest, is *Haikoucaris ercaiensis* (see image 1), a tiny 1-inch (2.5 cm) creature found in the shale rocks of Chengjiang in China, dating back 525 million years. *Haikoucaris* was a "great appendage" arthropod, so called because it had large arms that protruded from the head. These arms were tipped with claws, which is the key feature of the chelicerates. (The word chelicerate is derived from the Greek for "claw," and all members have a pair of clawlike appendages—the chelicerae—in front of the mouth.) *Haikoucaris*'s chelicerae were articulated arms, while a spider's are its fangs and a scorpion's are mere spikes.

By the late Cambrian period, 490 million years ago, the Chasmataspida had appeared. These little chelicerates may have been primitive forms of the eurypterids. They had a rounded prosoma (a body comprising both head and

KEY EVENTS

525 Mya	510 Mya	505 Mya	490 Mya	460 Mya	450 Mya
A tiny, 1-inch (2.5 cm) specimen called *Haikoucaris ercaiensis* dates from this time, making it the earliest chelicerate yet discovered.	*Protichnites*, a suspected primitive arthropod, leaves a fossil track on land, suggesting that it could survive out of water for short periods.	Possible ancestors of the major chelicerate groups *Sanctacaris* and *Sidneyia* exist at this time.	Chasmataspida, a group related to primitive eurypterids, appears in rocks from the late Cambrian.	The first eurypterids appear, with species such as *Megalograptus* becoming the main predators from the late Ordovician to the early Devonian.	Horseshoe crabs or xiphosurans, a sister group to the eurypterids, appear. They still survive in the oceans today.

thorax) and a tapering abdomen. The front limbs had a walking and grasping role, while the rear ones were flattened into paddles or perhaps shovels. Their structure was similar to that of the eurypterids, which appeared 30 million years later.

One of the earliest true eurypterids was *Megalograptus*, a 4-foot (1.2 m) predator of the late Ordovician period. It had a shieldlike upper covering, the carapace, over the head and thorax, with hollows for a pair of compound eyes. The long segmented abdomen tapered to a spike. The chelicerae were short, with a pair of spiked limbs behind the mouth to hold food, although prey was captured by a second pair of legs that were fringed with spikes to skewer the soft body parts. Three pairs of hind legs were devoted to movement in water, although track fossils indicate that they may have also used them to haul themselves onto land, to mate or to scavenge for carcasses.

Eurypterus (see p.138), the first sea scorpion to be described in 1825, existed 432 million years ago and retained this body form. However, not all eurypterids of this era did. The Pterygotidae, for example, had long, slender chelicerae, which ended in toothed pincers. The hindmost limbs were still paddles, but the other four pairs were walking legs with no role in prey capture. The abdominal tail (telson), was flattened into a fin. *Pterygotus* (see image 2) and its relatives were the rulers of the Silurian oceans, with many species growing to 6½ feet (2 m) long. One species, *Jaekelopterus* (see image 3), had 18-inch (45 cm) chelicerae, suggesting that it had a total length of 8 feet (2.5 m). The explosion in fish diversity in the following Devonian period put pressure on the eurypterids. By the late Carboniferous period, around 320 million years ago, they were primarily freshwater animals. In the calamitous end-of-Permian extinction, which killed more than 90 percent of aquatic species, they became extinct. **TJ**

432 Mya	410 Mya	390 Mya	380 Mya	320 Mya	252 Mya
Eurypterus, the first sea scorpion species to be identified for science, becomes a common predator during the middle Silurian.	Eurypterid diversity declines due to increased competition from large, speedy, and fast-evolving predatory fish.	*Jaekelopterus*, the largest eurypterid ever, and perhaps the largest of all arthropods, rules the seas.	The chasmataspids, an early group related to the eurypterids become extinct after more than 100 million years.	Some eurypterids move away from shallow marine habitats and seem to become primarily freshwater animals.	Eurypterids become extinct in the end-of-Permian mass extinction, known as the "Great Dying" (see p.224).

Eurypterus
LATE SILURIAN PERIOD

SPECIES *Eurypterus remipes*
GROUP Eurypteridae
SIZE 8 inches (20 cm)
LOCATION United States,
Canada, and Europe

🔆 NAVIGATOR

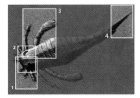

The first sea scorpion to be described by James Ellsworth De Kay (1792–1851) in 1825 was *Eurypterus*. The first fossil to be found was *Eurypterus remipes*, which is the most common of the 15 species. Its name—and that of the entire group—is derived from the Greek for "wide wing," but it was used to describe the "broad paddle" of the first specimen's swimming appendages. *Eurypterus* was seen in the fossil record between 432 million and 418 million years ago; the Silurian. At this time, it would have been living alongside other sea scorpions, but the average *Eurypterus* was a mere 8 inches (20 cm) long, although some specimens grew to more than 3 feet (1 m) in length. Its gross anatomy resembled that of the early eurypterids—short chelicerae and a pointed tail (telson) segment—but it lacked the long grappling arms of other species. As a result, it is believed to have been a generalist feeder, fighting with small prey and searching the seabed for food.

Specimens of *Eurypterus* found worldwide in the 19th century were of such high quality that paleontologists have constructed a detailed picture of this animal. For example, the wider section of the abdomen (toward the head), known as the mesosoma, contained the gills and reproductive organs. The gills were internal, covered by flat parts that are thought to have evolved from legs, and they were arranged in stacks of thin tissue, a formation known as book gills. Gas exchange occurred as water flowed through these stacks, but when the stacks of tissue retained enough moisture, the system also worked in air, as in living horseshoe crabs. The book lungs of terrestrial arachnids such as spiders are thought to have evolved from book gills. It is possible that *Eurypterus* would have come on to land to breed, which is another similarity it shares with horseshoe crabs. In the Silurian, land would have been safe and free of predators, unlike today. **TJ**

◉ FOCAL POINTS

1 FOREMOST LIMBS
E. remipes had six pairs of appendages or limbs. The front four pairs were used in feeding. The first pair are termed the chelicerae, and had three segments tipped with small pincers. (They are not visible here, as they are underneath the head in this reconstruction.) The next three pairs were short and covered with spines. This suggests that their primary role was gripping food, but they also would have helped with walking.

2 EYES
E. remipes had two curved compound eyes, meaning that each was composed of many small units (ommatidia) in a similar organization to the eyes of trilobites (see p.120). The eyes looked to the side and straight ahead. A couple of simple eyes (ocelli) were located on the top of the carapace, which it had in common with many extant chelicerates. Ocelli are capable of detecting the difference between light and dark but do not make images.

3 REARMOST LIMBS
The two pairs of hind legs were used for moving. *E. remipes* fed while standing and crawling on the seabed. The fifth pair acted as walking legs. The sixth, rearmost pair were paddles. *E. remipes* swum to escape threats and to move to new feeding grounds. Swimming was powered by tilting the paddles flat as they swung forward, then turning them side-on for the return stroke, so they pushed against the water and propelled forward.

4 TELSON
The final segment of the abdomen was the spikelike telson. *E. remipes*'s telson comprised about a quarter of the total body length. It was no doubt used as a weapon, although there is no evidence to back up suggestions that it delivered venom, similar to the sting of a scorpion. The sea scorpion may also have been able to use the long spike as an anchor, driving it into soft sediments to prevent it from being dragged by currents.

STATE FOSSIL

The first *E. remipes* fossil was unearthed in New York in 1818. The species proved common in the region, so New Yorkers made it their state fossil. This first find was initially identified as a catfish jaw. In 1825, US zoologist James Ellsworth De Kay correctly identified it as an arthropod and gave it its enduring name. However, De Kay thought it was a crustacean—a huge ancestor of the tiny fairy shrimp that live in shallow lakes, swamps, and salt pools today. Better specimens of *Eurypterus* were discovered throughout the 19th century (right). Studies by Swedish paleontologist Gerhard Holm (1853–1926) in 1898 revealed that it was in fact a member of an extinct order of chelicerates.

Horseshoe Crab
PLIOCENE EPOCH—PRESENT

SPECIES *Limulus polyphemus*
GROUP Limulidae
SIZE 2 feet (60 cm)
LOCATION Atlantic coasts of North and Central America
IUCN Near Threatened

⬡ NAVIGATOR

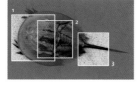

The horseshoe crab, *Limulus polyphemus*, is one of natural history's most exulted living fossils, as it originated 450 million years ago in the Ordovican and still survives today. The similarity between the four extant species and the extinct eurypterids is easy to see, and just as the sea scorpions were not scorpions, the horseshoe crabs are not crabs. Both animals are regarded as side branches of the major chelicerate lineage occupied by the eurypterids and today's arachnids. Horseshoe crabs are covered in a hard shell, with two main body sections. The head and mid-body make up the prosoma (cephalothorax), which contains the major sense organs, the mouthparts and 10 legs used in walking and feeding, with all but the last pair of legs carrying pincers. The rear of the body is made up of a segmented abdomen, tipped with a telson. Horseshoe crabs spend most of their time on the seabed, foraging for worms and shellfish. However, they can swim for short distances by turning upside down and using the carapace as a flotation device, and they can also survive in air for brief periods. To breed, they climb up to the water's edge, where they and the eggs they bury in the mud are at great risk of predation, including from licensed humans, who harvest small amounts of their unusual blood for medical research and possible applications, before returning the horseshoe crabs unharmed to the sea. **TJ**

◉ FOCAL POINTS

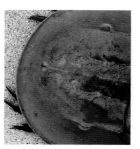

1 CARAPACE

The hard shell (carapace) must be shed periodically as the animal grows. The horseshoe-shaped section over the prosoma connects to the abdominal covering with a hinge, allowing the animal to flex its body back and forth across the middle, a tactic it deploys to shuffle down into sediment to avoid attack.

2 EYES

Limulus polyphemus has two compound eyes under the ridges that run along the side of its domed carapace, and a pair of simple eyes on the top. There are also simple eyes close to the mouth; it has been speculated that these have a role in helping the animal to remain oriented when it is swimming upside down.

3 TELSON

The telson is a rigid spike. It appears that its primary role is to turn the animal the right way up when it has flipped over on its back. This is especially important when the animals crowd together to breed in the surf. Being upside down out of the water for long periods results in the gills drying out.

◄ The mouth (center) is surrounded by the coxae (hip segments) of the first six legs. These bristle with spikes for chewing food, a system also seen in eurypterids and trilobites. Between the coxae and the tail are the gills for breathing.

SEA SPIDERS

The chelicerate group Pycnogonida, more commonly known as sea spiders (right), live on the seabed, striding about on long, spindly legs stalking prey of all kinds. Most species are tiny—no more than ½ inch (1 cm) across —although there is a giant, *Colossendeis colossea*, that reaches 3 feet (1 m) across. Most species have eight legs, but some have 10 or 12. The group is only distantly related to true spiders and other chelicerates. Today's species appear to have lost much of the complexity of those in the fossil records, which first appear at the end of the Cambrian. They have an elongated head attached to a hublike thorax, and the stomach branches into every leg. This bizarre design suggests that sea spiders are an ancient evolutionary offshoot from arthropods.

CRUSTACEANS

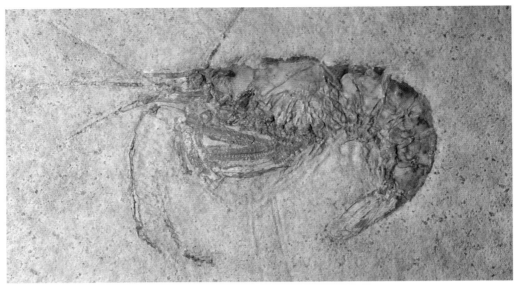

The crustaceans are an incredibly diverse group of arthropods (joint-legged invertebrates; see p.114), which have been significant members of the animal world for at least 500 million years. There are almost 70,000 living species described to date. The most familiar are the decapods—10-legged crustaceans that include crabs, shrimp, prawns, and lobsters (see image 1). However, there is a lot more to the crustaceans than seafood. The group also includes barnacles, mantis shrimp, bioluminescent copepods, and krill (see image 2). The latter exists in immense swarms that move as one through the cold oceans. Bizarrely for creatures that live in their trillions today, the fossil record does not contain a single krill specimen.

Most crustaceans are marine, although some have moved into other habitats. Brine shrimp, for example, survive in extremely saline seasonal lakes, and their eggs can lie dormant in dry salt flats for years, waiting for the water to return. Woodlice and pillbugs are terrestrial crustaceans that have existed worldwide for at least 50 million years, while land crabs led the most recent invasion of an aquatic species onto land 20 million years ago. Crustaceans form a collection of disparate classes of arthropods. Most belong to the Malacostraca and Maxillopoda classes. The aforementioned decapods, woodlice (isopods) and krill belong to the Malacostraca class; barnacles and copepods belong to the Maxillopoda. Of the minor groups, fairy, brine, and clam shrimp and water fleas make up the Branchiopoda (see image 3), while

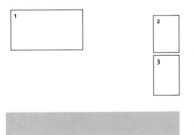

1 The prawn *Aeger tipularius* from the late Jurassic period in Germany, is remarkably similar to living species.

2 Antarctic krill, *Euphausia superba*, fulfill a huge role in the ecology of the oceans, feeding on tiny plankton and being eaten by large marine creatures.

3 Tadpole or shield shrimp, like *Triops cancriformis*, of the Branchiopoda, are another crustacean group remaining almost unchanged since the Jurassic.

KEY EVENTS

520 Mya	513 Mya	505 Mya	480 Mya	440 Mya	410 Mya
Fuxianhuia, a segmented arthropod, lives in the oceans. A proposed possible ancestor of the crustaceans, it may link to insects.	Bradoriida, a class of bivalved arthropods with similarities to today's ostracods, exists worldwide.	*Perspicaris* and *Canadaspis*, possibly early malacostracans, and *Priscansermarinus*, date from this time in Canada's Burgess Shale.	The first confirmed branchiopods and ostracods lay down fossils.	Thylacocephala, a rare class of crustaceans, exist at this time but become extinct at the end of the Cretaceous period.	*Lepidocaris*, an ancestor of modern brine shrimp, is one of the first freshwater crustaceans.

the microscopic seed shrimp (ostracods) live in sediments on the seabed and in the soils of rain forests. Both are well represented in the fossil record.

Two other crustacean groups, Remipedia and Cephalocarida, are more mysterious. The remipedes are blind and live in groundwater. They were thought to be extinct when they were first discovered in 320-million-year-old fossils in 1955. However, since 1979 some 20 living species of remipedes have been found along the coasts of most oceans and seas, especially the Indian and Atlantic oceans. The second, known as the horseshoe shrimp, which also live in sediments on the seabed, were discovered alive in 1955. They are thought to be living fossils of primitive crustaceans, even though no actual fossils of them have been found.

It is possible to trace a typical crustacean body form with three sections—head, thorax and abdomen. These are in turn divided into several segments, each of which has its own pair of appendages. They include two pairs of antennae on the head and numerous other limbs running along the underside of the body. Thoracic limbs are used for feeding and walking, abdominal ones for swimming or other purposes. Unpicking the core features of such a diverse and adaptable body plan to track the evolution of the crustaceans has not been easy. A candidate for a common ancestor with other arthropods is *Fuxianhuia*, found in rocks in China from the early Cambrian period. The head segments of this 1¼-inch (3 cm) creature differentiate it from other arthropod lineages of Chelicerata (spiders and other arachnids; see p.150) and Hexapoda (insects; see p.166). The picture becomes clearer toward the middle Cambrian with the arrival of the Bradoriida in Australian deposits about 513 million years ago. These tiny creatures had a bivalve carapace (body casing made from two parts connected by a hinge). Possible very early crustaceans, such as *Canadaspis* (see p.144) and *Perspicaris*, also had this bivalve shell, which sat over the body like a tent. Both are seen in 505-million-year-old fossils in the Burgess Shale Formation in the Canadian Rockies (see p.50).

Branchiopods (brine and fairy shrimp) and ostracods (seed shrimp) appear in the late Ordovician period, 480 million years ago, while the Devonian period, more than 70 million years later, sees the arrival of the first shrimplike decapods. However, by then pressure from new fish species was driving some crustaceans out of the oceans, with branchiopods already appearing in nonmarine habitats during the Silurian period, 443 million to 419 million years ago. The first terrestrial isopods are seen in the fossil record around 50 million years ago, but their worldwide distribution indicates that they must have made the move onto land at least 180 million years ago, when the world's land formed a single supercontinent called Pangaea. After the break-up of Pangaea, as new, shallow seas opened up around the world, the decapods diversified, reaching their peak 100 million years ago when the modern wealth of lobsters, crayfish, crabs, prawns, and shrimp appear. **TJ**

370 Mya	350 Mya	320 Mya	300 Mya	100 Mya	50 Mya
The shrimplike *Palaeopalaemon* is the earliest decapod specimen recorded.	The first copepods, a group that will eventually spread into almost all aquatic habitats worldwide, date from the early Triassic.	The earliest record of a remipede, *Tesnusocaris goldichi*, dates from this time. Remipedes are thought to be extinct until living species are later found in 1979.	Isopods, the ancestors of woodlice, appear in marine environments, later spreading to freshwater habitats.	Malacostracans proliferate, with lobsters appearing first among many decapod species.	Woodlice exist in many parts of the world, suggesting they evolved during the days of Pangaea 130 million years before.

Canadaspis
MIDDLE—LATE CAMBRIAN PERIOD

SPECIES *Canadaspis perfecta*
GROUP Canadaspididae
SIZE 2 inches (5 cm)
LOCATION North America
(other species in China)

Possibly one of the earliest crustacean species is *Canadaspis perfecta*. This is one of the most common fossils in the Phyllopod Bed, which was one of the first sites to be excavated from the 505-million-year-old Burgess Shale deposits in the Canadian Rockies. The genus name means "shield of Canada." However, despite this patriotic moniker, probable *Canadaspis* specimens have also been found in older deposits in Chengjiang, China, which date back 520 million years. The reference to a "shield" alludes to the large carapace that covered the head and thorax. The abdomen was exposed and left relatively unprotected to the rear. The animal was less than 2 inches (5 cm) long, so the exposed area would have been very small. The creature's bulky appearance led to the assumption that it could not swim, therefore shuffled over the seabed on its 10 pairs of walking legs, with the tough dome protecting against Cambrian predators. As *Canadaspis* moved forward it would have stirred up sediment, which would have then wafted beneath its body. This manufactured current supplied the gills with oxygen-bearing water, while any large edible items buried in the seabed—such as worms—would be filtered from the water by the spikes on the legs and directed to the mouth. An alternative hypothesis proposes that the leg spikes scraped algae and microbes from rocks, in a manner akin to living limpets.

During the Cambrian, there were many arthropods with carapaces, segmented bodies and numerous appendages. The structure of the head of *Canadaspis* means that it could have been the earliest crustacean, although this is disputed. The head had two pairs of antennae—a true crustacean characteristic—while a set of spines below the antennae is thought to be mandibles (mouthparts). This impedes *Canadaspis* from being included in the chelicerate (spider) side of the arthropod family tree, because chelicerates chew with appendages and their mouth does little more than suck fluid. Thus, the mouthparts of *Canadaspis* are the best evidence that it was a primitive Cambrian crustacean. **TJ**

⚙ **NAVIGATOR**

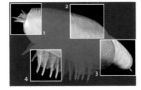

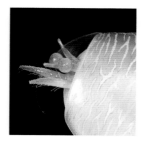

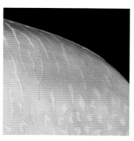

1 HEAD APPENDAGES

The head was equipped with two small eyes on stalks, both protected by short spikes. The double set of antennae may have had different functions, as in living malacostracans, where one set are feelers and the other chemical sensors. More head appendages surrounded the mouth on the underside.

2 CARAPACE

The domed carapace was bivalved—made up of two halves connected by a hinge that ran lengthwise along the top. It would have sat over the head and thorax, similar to a pitched roof. The head, with its two sets of antennae, appeared from the front opening, with the tips of the legs underneath.

3 ABDOMEN

The abdomen did not have any appendages, in contrast to living crustaceans, which use pleopods (abdominal appendages) for swimming. This suggests *C. perfecta* was an offshoot from the main line, although the abdomen conforms to the norm by ending with a spiny telson (tail).

4 LEGS

C. perfecta had 20 legs. All were biramous, meaning they split from the base near the body into a major and minor branch. The major section was spiked and used for walking and collecting food. The smaller flap-shaped extension may have been a gill or a swimming organ.

PHYLLOPOD BED

More than 4,500 *Canadaspis* fossils have been found in the Phyllopod Bed at the first and best-known dig site in Canada's Burgess Shale. The location is now known as the Walcott Quarry on Fossil Ridge, a spine of rock 7,500 feet (2,300 m) high between two mountains in British Colombia. This area was named by and for Charles Walcott (1850–1927; above), the American paleontologist who pioneered research into the abundant Cambrian fossils here between 1910 and 1917. He named one especially fossil-rich area as the Phyllopod (leaf-foot) Bed because of the many animals found there with lobed or leaflike appendages. Research into the 65,000 specimens that Walcott unearthed is still shedding new light on evolution.

▲ In addition to fossils from Canada, shown here, *Canadaspis* has appeared in the southwest United States and China. The Chinese versions have been named as a different, earlier species, *C. laevigata*. Much discussion centers on whether *Canadaspis* was a true crustacean or a surviving member of an earlier, either ancestral or parallel group.

GRAPTOLITES

1 The graptoloid, *Monograptus convolutus*, had a single unbranched stemlike form (stipe), which could be straight or curled.

2 This colony of the graptoloid *Monograptus turriculatus* displays the complex way that some remains have been pressed into tangles during fossilization.

3 The dendroid, *Dictyonema retiforme*, displays the branching pattern typical of its group.

G raptolites were a group of sea-dwelling filter-feeders, now extinct, whose identity puzzled paleontologists and evolutionary biologists for 200 years. As their name—derived from *graptos* (Greek for "writing" or "marks") and *lithos* (rock)—suggests, their fossils look like pencil marks on rocks. They are often abundant in shale-based marine rocks of the early Paleozoic era (more than 500 million years old). As a result of their rapid speciation (evolution into many kinds), graptolites, like trilobites and ammonoids, are index fossils (see p.159) that are used to date rocks and fossils around the world.

Graptolites are fossilized skeletons of colonies (rhabdosomes) of small, ribbonlike creatures called hemichordates, an early stage in the evolution of vertebrates (see p.174). The fossils are protein-based tubes that formed protective cases for the many-tentacled soft creatures called zooids within. The tubes are usually no more than 1/10 inch (2 mm) wide and up to 1 inch (2 cm) long but exceptional examples can be up to 3 feet (1 m) in length.

There were two major groups of graptolites, the dendroids (see image 3) and the graptoloids (see image 1), with a number of minor related groups such as the camaroids and tuboids. The more ancient group were dendroids, which first

KEY EVENTS

520 Mya	500 Mya	490 Mya	490–485 Mya	475 Mya	475 Mya
The first signs of graptolite fossils, dendroids, appear in rocks. They resemble scratchy marks or ancient writing.	At this point, most kinds of graptolites seem to be based on the seabed or other surfaces such as seaweeds.	By this time some graptolites are fully planktonic, needing no hard base from which to grow.	*Anisograptus* flourishes and lays down fossils in eastern North America and northern Britain.	The graptolite *Dendrograptus* occurs in rocks of the early Ordovician period in Scandinavia and eastern Europe.	Early members of the second largest group of graptolites, the graptoloids, show a branching structure.

evolved in the Cambrian period, 520 million years ago, and became extinct 170 million years later, in the early Carboniferous period. Dendroids tended to attach themselves to the seabed or to other bases, where they grew into multi-branched colonies of hundreds and sometimes thousands of individuals. Each branch was composed of three different kinds of tube-dwelling graptolite zooids: a smaller type, or bitheca, and a larger one, the autotheca, both of which had external openings; these zooids were connected by a third graptolyte type, a stolotheca. Sexuality may account for the differences between the thecal types, with the female autotheca being the largest. During the latter stages of their evolution, some dendroids became adapted to a planktonic (floating) way of life.

The graptoloid group was entirely planktonic throughout its evolution. They are also distinguished from the dendroids by their linear or straight-line colonies (see image 2) with only one type of theca having an external opening. It is assumed that both the earlier dendroids and later graptoloids had a two-stage life cycle, rather like simple creatures such as jellyfish and corals (see p.104) and some plants. The initial zooid of a colony, the siculozooid, was the result of sexual reproduction. It gave off buds, like a plant, and each of these made its own protective tube, built on to the pre-existing skeleton of the colony to form a series of overlapping tubes. These mature zooids, female and male, released eggs and sperm into the seawater. From these came free-swimming larvae; if these survived, each one grew into a siculozooid and the cycle was repeated.

Early graptoloid colonies had numerous branches, but not as many as the dendroids, so the number of individual zooids was also fewer, ranging between one or two dozen and a few hundred. The number of branches and individual zooids reduced during their evolution until single-branched colonies appeared in the Silurian period, each with only a few zooids, only to increase again later. Dendroid graptolite colonies looked superficially plantlike, whereas the graptoloid colonies had geometrical forms—unlike any living animals.

The practical use of graptolite fossils in geological mapping and the relative dating of fossils was first recognized in the late 19th century, long before their true biological nature was known. The initial assumption was that graptolites were related to hydrozoans, such as medusae or jellyfish. As soon as the detailed structure of graptolites' protein-based skeletons became known, it was evident that they were not related. Their place in evolution was a mystery until they were linked with the wormlike pterobranchs (see p.148). The tubular skeleton of the pterobranchs, secreted by the zooids in a series of overlapping protein-rich rings, shows similarities to that of the graptolites. In the middle of the 1990s these shared characteristics were sufficient for graptolites to be reclassified with the pterobranchs, as hemichordates. **DP**

450 Mya	443 Mya	440 Mya	419 Mya	380 Mya	340–330 Mya
Dicranograptus, commonly with a V- or Y-shaped form, exists in various oceans in the northern hemisphere.	A major extinction, the Hirnantian event, at the end of the Ordovician, greatly reduces the numbers and diversity of graptolites.	*Dictyonema* graptolites, a dendroid genus that may date from the Cambrian, occurs in rocks from many regions.	The graptoloid genus *Monograptus* appears. It is later used by scientists to mark the start of the Devonian period, the Lochkovian stage.	*Monograptus* fossils slowly dwindle to extinction after being relatively common for 40 million years.	The last graptolite fossils fade from rocks during the early to middle Carboniferous.

Pterobranch
PLEISTOCENE EPOCH—PRESENT

SPECIES *Rhabdopleura compacta*
GROUP Rhabdopleurida
SIZE ⅟₅₀ to ⅟₂₅ inch (0.5–1 mm)
LOCATION Most northern hemisphere oceans and seas
IUCN Not Yet Assessed

For several decades, it has been known that the extinct graptolites are probably related to the living pterobranchs, a small group of marine hemichordates, wormlike relatives of the echinoderms (starfish and urchins; see p.164), and probably the vertebrates, too. There are only 20 species of pterobranch alive today, belonging to genera such as *Rhabdopleura* and *Cephalodiscus*, all of which are fixed-down, tube-secreting colonial animals that live on the seabed. Most of these species live in deep, cold water, which has hampered their study. In the 1960s, living pterobranchs, *R. compacta*, were found in much shallower water off the coast of southwest England, from where they were collected and studied in a laboratory. Since then, colonies have also been found in warmer regions such as the Mediterranean and around Bermuda, and in water less than 33 feet (10 m) deep. Some colonies achieve a length of 1 to 2 inches (2.5–5 cm).

The hemichordate group includes pterobranchs, acorn worms, and graptolites. The general body plan in the living species of pterobranch divides into three parts—front prosome, middle mesosome, and rear metasome. The pterobranch abdomen has a structure called a stomochord, which was thought to be from the same origins as the notochord that led to the spine of vertebrates. However, these structures may be the result of convergent evolution (see p.19) rather than being developmentally related. Hemichordates are usually included with the echinoderms in the group Ambulacraria. The pterobranchs have a sporadic but long fossil record, independent of the graptolites. In 2011, a remarkable fossil pterobranch, *Galeaplumosus abilus*, was described from 525-million-year-old rocks in China. It is the oldest fossil pterobranch known and also the best preserved. A zooid can be seen with its paired arms and tentacles still attached to the stem, evidence of an astonishingly conservative evolution, with an apparently unchanged body structure and mode of life for well over 500 million years. **DP**

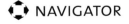

◆ NAVIGATOR

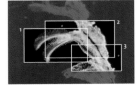

1 COLONY STRUCTURE
Most pterobranchs consist of a number of millimeter-long individuals, known as zooids, which live together in a tubular structure called the coenecium and they are interconnected by fleshy, contractile stems or stolons. The coenecium is first formed over a hard surface such as an empty shell, and from this attachment single tubes arise to house the zooids.

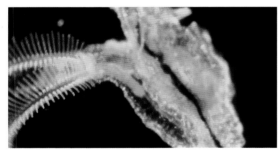

2 INDIVIDUAL ZOOIDS
Each of the three parts of the zooid has specific functions. The front, (prosome) is a muscular shield that facilitates movement and secretes the colony's protective tube (coenecium). The middle (mesosome), has one or more pairs of tentacled arms for feeding, while the rear section, the metasome, contains the reproductive and digestive organs and the stalk connecting it to the rest of the colony.

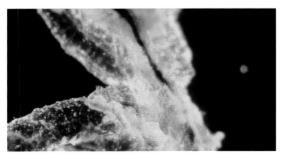

3 TUBES
From its tube opening, each filter-feeding zooid extends branched tentacle arms to feed on small organic particles drifting in the current. Cilia (tiny filaments) on the arms collect the particles from the seawater. The animal's shield-shaped fleshy prosome produces protein-rich collagen, laid down in successive rings to build the colony's protective tube, the coenecium.

⊥ PTEROBRANCH RELATIONSHIPS

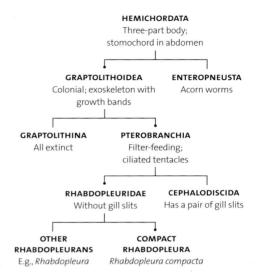

▲ The small size of pterobranchs, their scarcity and the difficulties observing them mean that relationships are still tentative. Some studies suggest that Cephalodiscida are in fact more closely related to Graptolithina than to rhabdopleurans.

FALSE ASSOCIATION

When living pterobranchs were first described in the late 19th century, they were thought to be related to the bryozoans (sea mats, above). Similarities included stuck-down colonies, tubular skeletons, and individuals with many tentacles for feeding. They were then likened to acorn worms (see p.110) because their constituent zooids reproduce sexually, releasing eggs and sperm for fertilization in the sea. Later studies of pterobranch reproduction and development revealed significant differences between pterobranchs and acorn worms, however, but the two continue to be associated within the larger grouping of the hemichordates.

ARACHNIDS

1 The orb-weaving wasp spider, *Argiope bruennichi*, shows how the legs manipulate fluid silk into stretchy strands.

2 The female *Mongolarachne jurassica* (formerly *Nephila jurassica*), from China's Daohugou fossil beds, has fore limbs 2¼ inches (5.5 cm) long.

3 A sun spider or solifugid (sun-avoider) displays its large, powerful but nonvenomous fangs (chelicerae).

Within the animal kingdom, there is one group that inspires fear in humans that is quite disproportionate to their size—the arachnids (eight-legged terrestrial arthropods). They are feared more than other arthropods (see p.114) partly because of their reputation for inflicting bites and stings. Most arachnids—including all of those big enough to be seen—are carnivorous, and many deploy venom to kill prey and defend themselves.

The most familiar members of the Arachnida are spiders, many of which spin silk (see image 1). They make up about half of the total species. The next largest subgroup contains the mites and ticks, which are tiny to microscopic in size. More obscure arachnids include the whip scorpions, whip spiders, sun spiders (see image 3), and harvestmen, which are also known as daddy-long-legs. Perhaps most formidable of all are the true scorpions, famed for the stinger they carry at the end of their flexible tail (telson) and the two menacing pincers on their pedipalps (the pair of nonwalking appendages on the front of the head, borne by almost all arachnids). The creatures all have two body sections: the front one includes the head, thorax, and legs; the rear part is the abdomen. The scorpion abdomen is further divided, with a flexible tail section, which provides clues to arachnid origins from tailed ancestors.

The arachnids are the dominant living members of the group Chelicerata, which otherwise contains only the sea spiders and horseshoe crabs (see p.140).

KEY EVENTS

505 Mya	430 Mya	420 Mya	417 Mya	408 Mya	386 Mya
Sanctacaris and *Sidneyia* are identified as possible ancestors of the major chelicerate groups.	Scorpions lay down fossils. They are believed to be aquatic forms living in shallow seas, ponds, and rivers.	*Brontoscorpio*, the largest arachnid to have lived, almost 3 feet (1 m) long, prowls the seabed and hunts for trilobites.	The first terrestrial arachnid is a tiny predator called *Palaeotarbus*, a type of trigonotarbid.	The Rhynie Chert, formed from a bog habitat around a volcanic spring, contains trigonotarbids, mites, and harvestmen.	*Attercopus* is the first arachnid to produce silk, although it does not have conventional spinnerets. It is classified as the earliest known spider.

All chelicerates have clawlike mandibles (mouthparts), in contrast to the slicing mandibles of crustaceans and insects. Chelicerates have ancestors stretching back to the early Cambrian period, beginning 540 million years ago, but the sea scorpions (eurypterids) were the first chelicerates to leave a clear trace in the fossil record, in the middle Ordovician period, around 460 million years ago (see p.136). Today's scorpions are not direct descendants of eurypterids, but they inherited a similar body plan from a common ancestor. The first true scorpions appeared in the middle Silurian period, around 430 million years ago. They were certainly aquatic and probably amphibious, able to make forays on to land. The largest known arachnid dates from 420 million years ago. It was *Brontoscorpio*, a marine scorpion almost 3 feet (1 m) long. It was by no means the biggest hunter of the time—eurypterids twice its size would have preyed upon it. Living on dry land was one way of escaping such attacks, and could have been a major driving force in the scorpions' transition to a terrestrial lifestyle.

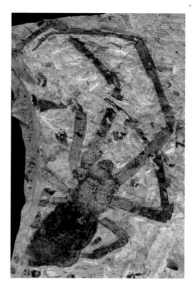

The first truly terrestrial scorpion is not confirmed in the fossil record until 370 million years ago. Long before that, arachnids from the spider branch of the evolutionary tree had moved onto land. The first terrestrial arachnids were the trigonotarbids, tiny hunters from the start of the Devonian period, around 417 million years ago. Trigonotarbids had rounded bodies, similar to those of large ticks or spiders today. They killed with a bite of the fangs and preyed on terrestrial arthropods, such as springtails and primitive millipedes. Trigonotarbids are commonly found in the Rhynie Chert fossil beds (see p.62), along with the earliest known mites, which are different from their trignotarbid neighbors because there is no clear division between their front and back body sections. The Rhynie Chert also includes harvestmen—which are also arachnids that lack a clear division between front and rear body—but harvestmen are thought to be more closely related to scorpions than to any spiderlike ancestors.

An early true spider from 386 million years ago was *Attercopus*. It had a long, pointed tail, like whip scorpions, but was able to produce silk. Around the same time, the first pseudoscorpion appeared: a tiny arachnid with scorpion-style pincers but no tail and stinger. In the Carboniferous period, the first sun spiders and whip spiders occurred, and by the end of the period the first spiders with silk-spinning spinnerets, located on the underside of the abdomen, appeared. These spiders, the Mesothelae, still have living relatives, such as liphistiids or trapdoor spiders (see p.154). By 250 million years ago, the spinnerets had moved to the tip of the abdomen, where nearly all spiders have them today. Spinning intricate spoked (orb) webs probably evolved in the Jurassic period, and the largest fossil spider known, *Mongolarachne jurassica* (see image 2), is a close match to today's orb-weaving spiders. Its fossil body is 1 inch (2.5 cm) wide, which is small when compared with tarantulas (see p.152). Spiders today dominate the arachnids, with 43,000 species in more than a hundred families. Some species, such as trapdoor spiders, are highly specialized in their web making. **TJ**

370 Mya	320 Mya	315 Mya	300 Mya	250 Mya	165 Mya
The first fully terrestrial scorpions lay down fossils. They probably prey on early insects and millipedes.	Sun spiders or camel spiders (solifugids) appear; they lack venom but become fast, strong predators with a powerful bite.	The whip spider— or tailless whip scorpion—appears in the form of *Graeophonus*, an insect hunter that mainly lives in forests.	Mesothelae is the first spider group to have spinnerets, located on the underside of the body; the trapdoor spiders are living descendants.	Spinnerets appear at the abdomen tip. One of the first spiders to have them belongs to the Hexathelidae family, today's funnel-web spiders.	*Mongolarachne jurassica* is the first known orb-web-weaving spider. With a width of 1 inch (2.5 cm), it is the largest discovered so far.

Tarantula

MIOCENE EPOCH—PRESENT

👁 FOCAL POINTS

1 URTICATING HAIRS
The rump bristles are defensive darts known as urticating hairs, each of which is tipped with barbs and has a thin, fragile base. When threatened, *B. smithi* lashes out with its rump or kicks its abdomen, so the hairs fly out.

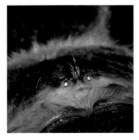

2 SETAE
B. smithi has eight eyes, but its vision is poor. Instead, like all tarantulas, it relies on touch-sensitive bristles (setae), which cover the body. These alert the spider to the slightest touch, gust of wind, vibration, sound, or even certain chemicals.

3 FANGS
The venom fangs have evolved little, pointing almost straight down. To attack, *B. smithi* rears up, holds out the front legs and stabs in venom to subdue the victim. It then pumps digestive fluids into the victim's body to make an easily consumed soup.

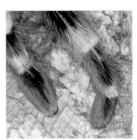

4 FEET
There are two claws on each foot, which are separated by tufts of hair. The tip of each hair divides into thousands of filaments. As the foot touches an object these brushes spread out, increasing the surface area, grip and enabling a silent tread.

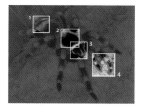

U ntil 2005, it was thought that the largest spider was a monstrously large early trapdoor type, a creature with a body the size of a soccer ball and a leg span of 20 inches (50 cm). This terrifying spider was believed to have ambushed small creatures in the undergrowth of Carboniferous forests. Perhaps disappointingly, it was then shown to be a sea scorpion, and, within that classification, not an especially big one.

In reality, living spiders are the largest known of the group. The Goliath tarantula, *Theraphosa blondi*, has a leg span of 12 inches (30 cm), large enough to conceal most of a dinner plate. This ultimate creepy-crawly is fondly known as the bird-eating spider, but in fact its diet mainly consists of large insects caught in its rain forest habitat of South America. Another large species is the popular pet, the Mexican red-knee tarantula, *Brachypelma smithi*. Like others in its group—and also like its earliest cousins—it does not spin webs; instead, it uses its silk to line a hole into the ground or a nest above ground level. The tarantula stays put there during the day but is ready to feed by dusk. It sets silken triplines around the nest to alert it to approaching prey. The 800 or so species of tarantulas and funnel webs are old-fashioned spiders. They are mygalomorphs (mouse-shaped), a reference to their rotund, hairy bodies. They have retained the primitive fangs of the first true spiders, dating from the Carboniferous. These long fangs point downward and must be raised aloft before being plunged into a target. However, the araneomorphs, the majority of spiders, possess a more "advanced" fang anatomy; their fangs open and close diagonally or sideways like pincers. **TJ**

SPECIES *Brachypelma smithi*
GROUP Theraphosidae
SIZE 6 inches (15 cm)
LOCATION Mexico
IUCN Near Threatened

WEB DESIGNS

Early spiders may have extended the silk lining in their burrows upward and outward to produce funnel web shapes, triplines to warn of passing potential victims and trapdoors to catch prey. This technique was developed further, as spiders made detached webs independently of the burrow. This then led to the evolution of dozens of different web designs, the most common of which is the orb web, with its elegant, almost perfectly geometrical, wheel-like circles. There is evidence that orb-web spinning began 150 million years ago during the late Jurassic, and experts believe that this led to other designs, such as the less tidy but more utilitarian-looking nets (above), tangle webs, domes and flasks.

▲ Similar web designs have evolved separately. For example, funnel webs such as this occur in the families Agelenidae (funnel weavers and hobo spiders), Dipluridae (funnel-web tarantulas) and Hexathelidae (including the deadly Sydney funnel-web).

Trapdoor Spider
PLEISTOCENE EPOCH—PRESENT

GENUS *Ummidia*
GROUP Ctenizidae
SIZE 2 inches (5 cm)
LOCATION North and South America, Europe, Africa, and Asia
IUCN Not Yet Assessed

◆ NAVIGATOR

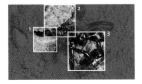

The fine, flexible, elastic, yet incredibly strong fibers we call "silk" are not a solely arachnid invention. They are used by various invertebrates, including lepidopteran larvae—butterfly and moth caterpillars—which use it to weave cocoons. In early spider evolution, the female probably used silk to wrap eggs, and the male to make a web that received sperm as he prepared to mate, as living spiders do. Since the spinning ability occurs in both males and females, it is likely to have been present at the start of the group's evolution. Next the silk was adapted to line burrows, to lay triplines, and to wrap up living prey to consume later.

The widespread genus *Ummidia* is a typical example of the trapdoor spiders of the Ctenizidae family, which are mygalomorphs, like the tarantulas. Various species prey on insects; arachnids, including other spiders; millipedes and centipedes; and even small vertebrates such as lizards. As well as a hunting lair, the burrow or tunnel functions as a safe retreat from harsh conditions and predators, and as a nursery, where the female guards her eggs. She also looks after her offspring; if prey is plentiful she may allow them to stay for several weeks until they have a better chance of independent survival. Several other spider families have independently evolved doorlike structures, such as Liphistiidae from East and Southeast Asia, brushed trapdoor spiders, Barychelidae, and wafer trapdoor spiders, Cyrtaucheniidae. **TJ**

1 BURROW

Most *Ummidia* species excavate burrows using their fangs and legs. The fangs have comb-shaped parts (rastella) to loosen soil into small lumps, which are kicked out by the back legs. When the burrow is large enough, the spider strengthens the walls with saliva, then adds the silken lining.

2 DOOR

Ummidia's well-shaped "cork" trapdoor fits neatly into the top of the burrow. Its surface is usually level with the surrounding ground. There are several other designs of trapdoor. For all types, the spider adds bits of twig, leaf, and other debris to make sure the door is well camouflaged.

3 CAPTURE

Ummidia hides in its burrow with the door closed. It may hold the door down with its fangs, on the opposite side to the hinge, legs pressed against the walls to prevent it being opened. When it detects prey the spider opens the door, grasps the victim with its fangs and front legs, then withdraws.

◄ Many trapdoor spiders look after their eggs and hatchlings in their burrows. Evolution has led to non-web wandering species, such as the spotted wolf spider, *Pardosa amentata*, taking the babies with them, piggy-back style.

PRESERVATION IN AMBER

Amber is fossilized tree resin, and some specimens date back as far as 230 million years ago. Amber is distinct from sap and other plant fluids because it lacks sugars or nutrients. It oozes out partly as a waste product, although it also serves to deter wood-boring creatures. Amber fossils are unique—in other fossils the original living tissues are replaced by rock minerals, but amber encloses the organism, sealing it off from the outside world to create a time capsule of hard and soft body parts, and a lot more as well. Amber fossils can also contain pollen grains, bacteria, hairs, seeds, fruits, flowers, small creatures—from spiders (right) to lizards— and spider silk. One Spanish sample of amber that dates from 136 million years ago contained strands of silk that could have been from an aerial orb web—this is among the earliest evidence of spiders' webs.

NAUTILOIDS AND AMMONOIDS

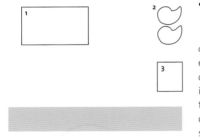

1 This specimen of a nautiloid, from the Jura mountains of Switzerland, has the straight shell characteristic of their early evolution.

2 Two views of a Jurassic fossil nautiloid, *Cenoceras pseudolineatus*, contrast the shell's outer appearance to the cut-and-polished section revealing the spiraling chambers with their central siphuncle.

3 Late Cretaceous *Nipponites* ammonoids from Japan exhibit bizarre coiling, whose functions have been much discussed.

The nautiloids, which appeared in the late Cambrian period, 500 million years ago, are among the first cephalopods (head-feet) in the fossil record. The cephalopods are the most evolved mollusks and today include octopuses, squid, and cuttlefish. They have long tentacles to catch prey, large eyes, and other acute senses, and most move by a form of jet propulsion, drawing water into the body through a wide opening and then squirting it out under pressure through a directable, nozzlelike siphon. Today, apart from the conspicuously shelled nautilus, and the reduced inner shells or pens of cuttlefish, cephalopods have lost their shells. The earliest nautiloids had straight shells (see image 1), but they underwent rapid evolution during the Ordovician period, 485 million years ago, and some developed a coiled shell shape (see image 2). Nautiloids have shells with interconnected internal chambers in which the inner surface is coated with nacre, also known as mother-of-pearl.

The coiled body form is very similar to that of many ammonoids, which appeared in the Devonian period, 419 million to 359 million years ago. The first ammonoids had straight shells, and it seems likely that they evolved from straight-shelled nautiloids. Although the ammonoids went extinct in the mass extinction at the end of the Cretaceous period (see p.364), 66 million years ago, the nautiloids survive to the present day in the shape of the nautilus (see p.160). Another similar group, the belemnoids, which also looked like squid, arose some 200 million years ago. They, too, probably evolved from nautiloids, before dying out at the end of the Cretaceous.

KEY EVENTS

500–490 Mya	480 Mya	400 Mya	390 Mya	380 Mya	359 Mya
The first nautiloids appear during the Ordovician, which is sometimes called the Age of Nautiloids.	Following the extinction of the anomalocarids at the end of the Cambrian, the nautiloids become the top ocean predators.	Nautiloids such as *Orthoceras* begin to develop partitioned shell structures such as the siphuncle.	The bactritids appear. They are a group of straight-shelled cephalopods thought to be the ancestors of the ammonoids.	The goniatites, one of the earliest ammonoid groups, make their appearance in the Devonian.	Many nautiloids disappear at this time, as part of the late Devonian mass extinction (one of the five main extinction events).

Both nautiloids and ammonoids are generally thought to have been predators. There is evidence that they consumed crustaceans and their own kind as well as fish and other aquatic prey. Their shells, straight or coiled, served as protective and supportive structures and had a hydrostatic function, enabling the animal to compensate for varying water depths and "float" in midwater with neutral buoyancy. Although nautiloids and ammonoids may look similar, they can be distinguished by certain characteristics, such as the positioning of the siphuncle. The siphuncle is an internal tube that connects the chambers of the shell. In nautiloids, it runs through the center of the shell chambers (see image 2, bottom); in most ammonoids it runs along the outer edge of the shell.

Another way to distinguish nautiloids and ammonoids is by their sutures. These are the lines of contact between the internal chamber walls of the shell and the inner shell wall. In nautiloids these lines are straight and are called simple sutures, whereas in ammonoids the sutures form undulations called lobes and saddles. As ammonoids continued to evolve, their sutures became more intricate. Those with goniatite (zigzag) suture patterns flourished during the Paleozoic era, from 380 million years ago. Ammonoids characterized by a more highly folded ceratite suture pattern replaced the goniatites and were most abundant in the Triassic period, 252 million to 201 million years ago. Most ammonoids went extinct at the end of that period, but the survivors evolved again into many diverse forms during the Cretaceous. At that time they had sutures with much more complex lobes and saddles than those of their ancestors 200 million years previously. The most complex sutures are found in the ammonites, a particularly successful type of ammonoid in the Cretaceous. Because of their rapid evolution, wide distribution and ease of recognition, nautiloids and ammonoids are important index fossils (see p.159).

The diversity of the sutures has long intrigued paleontologists. In one study, researchers discovered that simple sutures, like those of nautiloids and early ammonoids, can withstand great pressure but have poor buoyancy control: an indication that the animals lived in the deep ocean but were not fast-moving or maneuverable. In contrast, the complex sutures of Cretaceous ammonites were not so good at withstanding pressure at depth but did give effective control over buoyancy, perhaps indicating a more active lifestyle at shallower depths. Another characteristic unique to ammonoids is a wide variety of shell shapes. Although most are in the form of a flat, even spiral (or planispiral), some are wide, open-coiled, kinked, twisted (see image 3), or hooked. Experiments using computer models, and comparisons with similar living animals such as the nautilus and snails, suggest that planispiral shell shapes gave the ammonoid the ability to move through the water quickly, whereas wider and more open shell shapes slowed progress. These features suggest that planispiral kinds lived an active lifestyle in shallow waters, and that the wider-shelled ammonoids were more likely to be slow-moving bottom-dwellers. **RS**

252–201 Mya	200–145 Mya	160–100 Mya	70 Mya	66 Mya	5 Mya
The goniatites are superseded by the ceriatites as the dominant ammonoids during the Triassic.	After the extinction of the ceriatites, an ammonoid subgroup, the ammonites, begin to take their place in the Jurassic period.	During the late Jurassic and early Cretaceous, some ammonites grow larger. *Titanites* can be more than 2 feet (50 cm) in diameter.	Belemnoids begin to die out. They are the similar-looking cousins of the ammonoids that had been around for more than 120 million years.	The mass extinction event at the end of the Cretaceous causes the demise of the ammonoids after more than 300 million years of evolution.	Most of the remaining nautiloids become extinct, leaving only five or six species, the pearly or chambered nautiluses, to survive to the present day.

Dactylioceras
EARLY JURASSIC PERIOD

GENUS *Dactylioceras*
GROUP Ammonitida
SIZE 3 inches (8 cm)
LOCATION Oceans worldwide

Numerous species of the ammonoid genus *Dactylioceras* (finger horn) thrived in almost all oceans during the early Jurassic, from around 190 million years ago. They had distinctive narrow-ribbed shells that changed through time and between species. Their fossil shells can be found in large concentrations, as they were washed together by currents, often after a mass dying. This occurred, for example, after spawning, as can be seen in various living mollusks. These large concentrations, along with the various places they can be found, make *Dactylioceras* useful as index fossils (see panel).

Dactylioceras was in the group Ammonitida, one of the more advanced ammonoid groups. The shell was secreted by the mantle (fleshy body covering). As the animal grew, it moved forward in the body chamber and secreted a dividing wall, called a septum, behind it, at the rear of the mantle. Over time this created a series of chambers that made up the phragmocone. The septum was attached to the interior of the shell wall along a suture. The sutures of *Dactylioceras* and other ammonoids grew in complexity as they evolved and they are one of the main indicators for identifying species. Adding chambers in this way gives clues to the lifespan of *Dactylioceras*. The living nautilus adds a new chamber every month, on average, and a full-grown nautilus can have more than 30 chambers, so it reaches maturity in 2 to 3 years. It is also long-lived, sometimes more than 20 years. *Dactylioceras* was small; it may have matured by the age of 1 and its lifespan probably did not exceed 5 years. However, some larger ammonoid genera could have lived for more than 30 years. They were successful during their time, but the Ammonoidea are the only subclass of cephalopods to have gone extinct. Their 300 million years of existence were brought to an end in the Cretaceous mass extinction event 66 million years ago. **RS**

 **NAVIGATOR**

 FOCAL POINTS

1 RIBS

The ribs added strength to the shell of *Dactylioceras* and helped to smooth its passage through the water. Some fossils indicate two sizes of adult shell from the same species, which implies that one sex was larger than the other; the female was probably larger.

2 SHELL

The shell of *Dactylioceras* had two sections. The phragmocone was composed of a number of chambers that grew larger from the innermost to the outer part of the spiral. These were filled with a gas that gave buoyancy. The body chamber then housed and protected the ammonite.

3 PARASITES

Dactylioceras have been found with holes in the shell. Once identified as bite marks, these are now thought to have been caused by parasites. *Dactylioceras* and others have also been discovered with tube worms, small limpets, and shellfish attached to their shells.

▲ The tentacles and beak of the living ammonite protruded from the aperture of the shell. The edge of this aperture is called the peristome and the radial lines on the shell are growth marks indicating previous positions of the peristome. *Dactylioceras* was probably a predator and scavenger that stayed near the seabed; its size and shell shape suggest it was not a rapid swimmer.

INDEX FOSSILS

Index fossils—also known as marker, guide, zone, or indicator fossils—provide a basis for the geological timescale. An index fossil is a distinctive, widely distributed plant or animal characteristic of a particular, and relatively short, time period. Owing to their rapid evolution and fossil abundance, some mollusk groups are extremely useful as index fossils. Ammonoids in particular evolved so quickly that some of their fossils are used to identify zones separated by fewer than 1 million years. An example of this is *Puzosia* (right). Other examples of index fossils include corals, brachiopods, trilobites, graptolites, echinoderms such as urchins and crinoids, and the teeth of certain mammals.

Nautilus
LATE PLEISTOCENE EPOCH—PRESENT

SPECIES *Nautilus pompilius*
GROUP Nautilidae
SIZE 10 inches (25 cm)
LOCATION Pacific Ocean
IUCN Not Yet Assessed

✚ NAVIGATOR

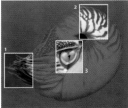

The nautiloids, members of the cephalopod group of the mollusks, first appeared in the seas of the Cambrian, 500 million years ago. Today, representatives of the group, looking little different from their ancient forebears, continue to live in the warm waters of the Pacific Ocean. The heyday of the nautiloids was during the Ordovician and Silurian, around 500 million to 400 million years ago, when they were probably the biggest and most active predators in the seas, catching and eating whatever came within reach of their tentacles. They have passed from being top predators to vulnerable prey and are now endangered by human collectors who prize their colorful shells.

The handful of nautilus species that survive in the present day all belong to a single family, the Nautilidae, which first appeared in the late Triassic, around 215 million years ago. The nautilus is the only surviving cephalopod to have retained an external shell, which provides buoyancy as well as giving protection. The pearly or chambered nautilus, *Nautilus pompilius*, the largest of the surviving species, has an external shell about 10 inches (25 cm) in diameter, which is divided internally in mature adults into about 36 separate chambers. The living nautilus adds new chambers as it grows, and so lives exclusively in the outermost chamber. The previous chambers, abandoned as it grew, are empty apart from the gases involved in keeping buoyancy in midwater. **RS**

FOCAL POINTS

1 TENTACLES

The nautilus has between 50 and 90 small contractile tentacles that it uses to explore its environment and to pluck prey from the water and the seabed, from fish to crabs, shellfish and worms. Unlike the tentacles of other cephalopods, those of the nautilus lack true suckers, but they have strong adhesive characteristics for securing slippery prey.

2 BUOYANCY "TANKS"

The empty chambers within the shell are separated by sturdy cross-walls, called septa. However, the chambers remain connected by the siphuncle, which allows the nautilus to adjust the ratio of gas and liquid inside, and hence control its buoyancy. The creature can rise and fall through the water, or hold its position at different depths.

3 POOR EYESIGHT

The eyes of the nautilus are not well developed, unlike the highly evolved eyes of other cephalopods, such as the octopus and squid. They lack a lens and are open to the environment, acting somewhat like a pinhole camera. Researchers have calculated that the visual acuity of the nautilus may be only 1⁄60 that of the octopus.

NAUTILUS RELATIONSHIPS

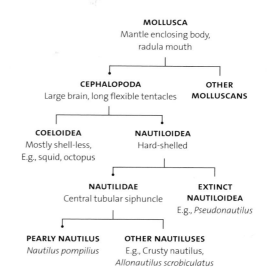

MOLLUSCA
Mantle enclosing body, radula mouth

CEPHALOPODA
Large brain, long flexible tentacles

OTHER MOLLUSCANS

COELOIDEA
Mostly shell-less, E.g., squid, octopus

NAUTILOIDEA
Hard-shelled

NAUTILIDAE
Central tubular siphuncle

EXTINCT NAUTILOIDEA
E.g., *Pseudonautilus*

PEARLY NAUTILUS
Nautilus pompilius

OTHER NAUTILUSES
E.g., Crusty nautilus, *Allonautilus scrobiculatus*

▲ Among the cephalopod mollusks, the nautilus family is usually defined by its shell geometry, the suture shape and the position and structure of the siphuncle. There are two genera of living nautiluses, *Nautilus* and *Allonautilus*.

◀ If the nautilus senses a threat, it can retreat fully into the outermost chamber of its shell, which it then seals off with a leathery hood formed from two tentacles that have become adapted for the purpose.

THE DECLINE OF THE SHELL

The nautilus, of all the surviving cephalopods, is the only one that still possesses a shell. One explanation is that other cephalopods were forced to evolve by the rise of a rival group in the oceans—bony fish. These were quick and strong enough to compete with the cephalopods for food, and to prey on the cephalopods themselves. In what may have been a response to the competition, the other cephalopods lost their shells, becoming swifter and more intelligent. Of all living invertebrates, the cephalopods have the largest brains in relation to their body and display the most complex patterns of behavior. In a remarkable use of tools, the coconut octopus, *Amphioctopus marginatus*, covers its home with a found object (right).

STARFISH AND URCHINS

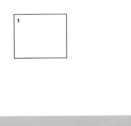

1 The echinoderm shape, as shown by the Pacific blood star, *Henricia leviuscula*, is based on a five-rayed radial pattern that is almost unique in the animal kingdom.

2 Brittle stars carpet the ocean floor, sifting edible particles. Fossils indicate they have had this lifestyle for 400 million years.

3 The protective spines of ancient urchins, similar to this living *Sphaerechinus granularis*, detached after death and formed fossils that trace their development.

Among the most iconic sea creatures, starfish (see image 1) are members of the group Echinodermata (spiny-skinned). The group contains more than 7,000 living species, all of which are saltwater-dwellers, and another 14,000 species are known from fossils, all of which are also marine. The echinoderms appeared during the Cambrian period, 500 million years ago. Their plentiful, occasionally spectacular, remains provide an extensive record of their long evolution. Certain kinds were so numerous and varied that they are now used as index fossils, to date rocks and deduce ancient conditions in shallow and deep marine environments. Echinoderms are considered to be the invertebrates most closely related to the hemichordates, such as pterobranchs (see p.148), and then to the chordates, the group containing vertebrates like ourselves.

Like the cnidarians, such as jellyfish, anemones, and coral polyps (see p.104), echinoderms are among the very few major animal groups with radial symmetry—that is, with a circular or star-shaped body plan, rather than bilateral, with a left and right side. The starfish's five "arms" show the plan is usually pentameric (based on five), although some echinoderms have a radial symmetry of seven or more. However, this radial structure is somewhat misleading: during an echinoderm's initial development, at a very early or larval stage, the animal is actually bilateral; it takes on its radial pattern as it grows.

KEY EVENTS

475 Mya	470 Mya	450 Mya	450 Mya	380 Mya	330 Mya
Possible somasteroids, a now-extinct group dating from this time, may be ancestors of both starfish and brittlestars.	The crinoids undergo rapid evolution and spread widely, adapting to many marine habitats.	The cystoid group of echinoderms flourishes; it is similar to stalked crinoids but dies out by the late Devonian period, 359 million years ago.	Fossil sea urchins date back about 450 million years ago, to the Ordovician period. Their closest living relatives are the sea cucumbers.	Species of the widespread genus *Furcaster* are among the Oegophuroidea, a now-extinct subgroup of brittlestars.	The now-extinct stalked echinoderms called blastoids thrive during the Carboniferous.

The familiar starfish, sea stars, cushion stars, and sea daisies make up the main echinoderm group known as Asteroidea or asteroids. Their close relations are the ophiuroids, which encompass the longer-armed, snaking brittlestars (see image 2) and the multibranching basket stars. Their record goes back to the Ordovician period, more than 450 million years ago. Like most echinoderms, starfish move by means of flexible, fluid-filled projections called tube-feet, which extend and contract in waves in a hand-over-hand movement. Fluid is pumped in and out of the tube-feet from a radial canal (tube) in each arm, which connects to a ring canal in the body's central disk. There is no heart or blood, only a body-wide fluid. The first portion of the abdomen, the cardiac stomach, can be turned inside out through the mouth opening to digest food, such as shellfish, after which the nutrients are taken into 10 digestive glands, two in each arm. Embedded fairly loosely in starfish skin are calcite-mineralized plates called ossicles, composed of a micromesh material called stereom and developed for protection, support, and other functions. The ossicle structure and stereom microstructure are characteristic of echinoderms. In well-preserved fossils they can be analyzed to reveal evolutionary pathways.

One of the longest surviving groups of echinoderms is the crinoids (see p.164). Resembling elaborate, feathery petaled flowers, they have left numerous remains in the fossil record, with some forming thick layers of sedimentary rocks. Also well known and still extant are the echinoids (sea hedgehogs): the spiky, needle- or thorn-covered sea urchins (see image 3), sand dollars, heart urchins, pencil urchins, sea biscuits, and others. These have a hard shell called a test made of large ossicle plates joined in interlocking patterns. The ossicles often preserve well in fossils, as do isolated defensive spines and spikes. Known from the Ordovician, urchins enjoyed great success and diversity through to the late Carboniferous period, about 320 million years ago. They then began a decline that almost ended with their disappearance during the "Great Dying" mass extinction at the end of the Permian period, 252 million years ago (see p.224). Thereafter they flourished again and adapted to new environments, with yet more patterns of spines and test plates. The fossil record of the burrowing genus *Micraster* from the late Cretaceous period is extremely detailed. Found in deep beds of chalk rocks from North America, Europe, Africa, and Antarctica, the fossils show how it evolved continuously from more than 90 million to about 60 million years ago, when it became extinct.

The closest living relatives of the urchins are holothuroideans (sea cucumbers). They fulfill the role of marine earthworms, gliding along the seabed on their tube-feet and eating mud, silt and other sediments to extract the nutrients. Unlike other echinoderms, they have re-evolved a partly bilateral body plan when adult, and in effect lie on their sides. They have leathery skin and their ossicles are small, but they show distinguishing stereom microstructures. **SP**

252 Mya	250 Mya	240 Mya	220 Mya	95 Mya	65–55 Mya
The number of sea urchins, already in decline, is further diminished in the "Great Dying," the end-of-Permian mass extinction.	After the end-of-Permian mass extinction, crinoids rebound quickly and begin another bout of evolution.	The first members of the long-lived crinoid genus *Pentacrinites* appear. The genus will survive in various guises for almost 200 million years.	*Aspiduriella* brittlestars exist in Germany. They will be the first fossils of the group to be scientifically named, as *Asterites* in 1804.	The first *Micraster* urchins appear in Cretaceous chalk rocks. They thrive and diversify over the next 30 million years.	The first true sand dollar genus, *Togocyamus*, arises during the Paleocene epoch.

Crinoids
EARLY JURASSIC PERIOD

👁 FOCAL POINTS

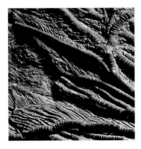

1 PINNULES
Many fossil (and all living) crinoids have pinnules, small side branches to the arms, giving them a feathery or frilled appearance. On the pinnules are tube-feet that gathered food particles, coated them with mucus and moved them to a groove on the upper side of the arm, leading to the mouth.

2 ARMS
Each movable arm has a series of linked ossicles extending to its end. Inside are extensions of the crinoid's main body systems, including nerves and the water vascular system. This worked the tube-feet, which projected upward and played a part in the process of feeding.

3 BODY
Ossicle plates support and protect the body, which had a lower cup containing the main soft organs for digestion, excretion and reproduction, and an upper section with radial plates leading to the bases of the arms. The mouth is on the upper surface in the middle, with the anus to one side.

4 STEM
The stalk has a series of disk-shaped ossicle plates known as columnals, linked by strong ligaments. Holes in the middle of the columnals align to form a canal for body fluids and nerves. In many species, a rootlike holdfast, or similar structure, at the base held the animal fast to the seabed.

SPECIES *Pentacrinus fasciculosus*
GROUP Crinoidea
SIZE 32 inches (80 cm)
LOCATION Europe

The crinoids are a major group of echinoderms and their closest relatives are the starfish. They have featherlike, mucus-covered "arms" to filter-feed and take in floating nutrient particles. The sea lilies are crinoids with a crown of arms and a stalk or stem that fastens to the seabed. Swaying in the current, they resemble miniature marine palm trees. The other crinoid group, feather stars or comatulids, either have short stalks or none at all and they are mostly free-moving. Compared to the structure of starfish, the crinoid body is upside down, with the mouth positioned on the upper side.

The sea lilies (bourgueticrinids) number fewer than 90 living species, most of which live in the deep ocean at almost 20,000 feet (6,000 m). They are rarely more than 32 inches (80 cm) long or tall. However, in the past they were hugely diverse, with some specimens stretching to 60 feet (18 m) in length, with multiple rootlike branches at the base of the stem and multibranched arms. The feather stars number around 450 living species and dwell in shallow waters. Their fossil remains are also numerous and widespread. In fact, 6,000 fossil crinoid species are known—more than 10 times the number alive today. The typical crinoid has a central cuplike body, or calyx, from which the arms extend. There are usually five arms, with each subdivided into two, giving the appearance of 10 arms. As in other echinoderms, there is a pattern of hard calcium-rich ossicles in the body, with smaller ossicles in the arms and ringlike ossicles in the stem. These plates can form up to 80 percent of the weight of an individual crinoid. They have formed innumerable fossils, both as fully articulated specimens (with all parts arranged as they would have been in life) and as disarticulated fragments. The bulk of some whole layers of sedimentary rocks, such as limestone, are formed from crinoids. **SP**

EXTINCT BLASTOIDS

The extinct Blastoidea were a group of echinoderms most closely related to the also extinct cystoids and the living crinoids, and they were similar to them in many ways. Their rounded bodies, known as thecae, were fossilized as knoblike structures (right) and they are sometimes called "sea nuts" or "sea buds." The theca sat on top of a stalk, or stem, that varied from a length of several times longer than the theca to less than half. Most blastoid species were less than 2 inches (5 cm) in diameter across the theca. Food grooves arched from the central upper mouth, around and down the five-rayed body, which was typical of echinoderms. From these food grooves, long, slim, food-gathering arms (brachioles) projected. Blastoids had appeared with other echinoderm groups by the Ordovician, around 450 million years ago. They reached large numbers and mediocre diversity during the late Carboniferous, around 300 million years ago, and then faded. The last of the group perished in the end-of-Permian mass extinction, 252 million years ago, which almost wiped out the crinoids as well.

THE RISE OF INSECTS

1 Dragonflies such as *Urogomphus*, from about 150 million years ago, have a less advanced way of holding their wings.

2 Springtails are hexapods but not true insects.

3 *Strudiella devonica* is a relatively complete fossil and may be the earliest known insect, but its classification remains controversial.

In the tree of animal life, the insects sprout out as a single branch from the other arthropod groups (invertebrates with jointed legs; see p.114). This offshoot now accounts for at least 70 percent of all animal species on Earth, and some estimates claim around 90 percent. Despite the insects being such a dominant force in the world, their precise origins are shrouded in mystery. Insects are air-breathing and chiefly terrestrial creatures. More than a million species have been described so far, but only five species of sea skater are known to survive in the open ocean. Nevertheless, many insects spend their early lives in freshwater habitats, and diverse species have adopted a fully aquatic lifestyle. This circumstantial evidence points to insects branching off from the main evolutionary line after the crustaceanlike arthropods had made the transition to land, or at least to freshwater. It was only after this that the first hexapods (six-legged arthropods) appeared, of which insects are the main type.

Although their fossil record does not stretch back that far, the assumption is that the insects and their allies arose in the middle Silurian period, from about 434 million years ago. Two factors support this. First, when hexapods appear in the fossil record in the Devonian period, from 419 million years ago, they are already exhibiting a wide range of adaptations. Second, scientists have used DNA analysis of living insects to date their evolution. These studies looked at

KEY EVENTS

435 Mya	408 Mya	379 Mya	365 Mya	330 Mya	320 Mya
Insects arise around this time, according to DNA analysis of living insects.	The first insect fossil, *Rhyniognatha hirsti*, found in the Rhynie Chert fossil bed in Scotland, dates from this time.	Apterygotes, relatives of the bristletails and silverfish, exist according to two fossil fragments identified in Gilboa, New York.	*Strudiella devonica*, from Germany, may have been an ancestor of grasshoppers.	Palaeodictyoptera, extinct six-winged insects, are among the first flying insects to lay down fossils (see p.171).	The first Neoptera appear in the form of primitive cockroaches, grasshoppers, and fast-running caloneurodeans.

how many genes various species share (the larger the number, the closer the evolutionary relationship) and at changes to DNA, which tend to occur at an approximately standard rate. It seems that today's insects shared a common ancestor from at least 434 million years ago, and it was probably a branchiopod crustacean (see p.142).

Hexapods have three body sections: a head with a pair of antennae and mouthpart appendages, a thorax with six legs, and an abdomen. They include other groups as well as insects, the largest of which is springtails (see image 2), which are wingless and have mandibles (mouthparts) inside the mouth (insects have external mouthparts). For many years the earliest hexapod known was the springtail *Rhyniella*, a common fossil in Scotland's Rhynie Chert rocks (see p.62), dating from 408 million years ago. *Rhyniella* was held up as evidence that springtails were the primitive form from which insects evolved. However, the Rhynie Chert also yielded a fragment of the head of another hexapod, *Rhyniognatha* (see p.134). A contemporary of springtail fossils, this had external mouthparts adapted to feed on ferns. This was the first evidence that insects had diversified from the scraping and scavenging springtails long before 408 million years ago, to exploit the expanding communities of terrestrial plants. Ferns were the tallest plants in the Devonian, and in order to feed on their spores and fronds insects had to be good climbers—or perhaps able flyers. No one knows whether *Rhyniognatha* had wings, and the next two insects in the fossil record do not provide any further evidence of wings.

Two fossils unearthed in Gilboa, New York, among a petrified forest of ferns dating back 379 million years, belong to a small group of insects called the Apterygota, which never grow wings. The great majority of today's insects belong to the Pterygota, the winged insects, which includes wingless members that evolved from winged ancestors. The Devonian apterygotes were similar to their modern relatives: the bristletails and silverfish. The next possible insect in the fossil record is *Strudiella devonica* (see image 3) from 365 million years ago, in the latter stages of the Devonian. Perhaps the earliest complete insect fossil, it shows a ⅓ inch (8 mm) wingless arthropod with long antennae, compound eyes for a detailed image, and mouthparts that are ideal for slicing and chewing a range of food. *Strudiella* is the sole occupant of the 50-million-year "hexapod gap" in the fossil record that extends into the early Carboniferous.

By the time the fossils reappear around 330 million years ago, the insects have diversified a great deal to exploit foods in Earth's first true forest habitats. Wing fragments provide evidence that insects had evolved flight by this time. Already the Pterygota was dividing into separate lineages. As the name suggests, the Palaeoptera are accepted as the primitive forms. They include the mayflies and dragonflies (see image 1), which are differentiated from the other winged insects, Neoptera, by the way they hold their four wings. Neopterans can fold their wings one on top of the other and then lay them over the abdomen.

310 Mya	270 Mya	220 Mya	200 Mya	135 Mya	66 Mya
The first evidence of complete metamorphosis dates from this time, in members of the now extinct Miomoptera, relatives of lacewings.	Insect evolution speeds up with the arrival of the beetles, group Coleoptera, and the true or two-winged flies, Diptera.	The first Hymenoptera appear in the form of wasps. Ants and bees will evolve later.	The first moths and butterflies, members of the Lepidoptera order of insects, lay down fossils.	Angiosperms proliferate, resulting in coevolution between plants and insects. Termites and mantids also evolve.	The end-of-Cretaceous mass extinction event extinguishes many species, but insects continue to thrive and diversify.

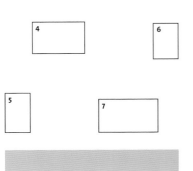

4 Living mayfly larvae may indicate how insect wings evolved from gill structures.

5 Grasshoppers, the Orthoptera group, evolved early in insect prehistory.

6 Evolution of larval forms such as the swallowtail caterpillar, *Papilio machaon*, allowed adults and their young to exploit different foods, which avoided competition between them.

7 The first pair of beetle wings have evolved into hard wing covers (elytra) as demonstrated by the burying beetle, *Nicrophorus defodiens*.

Palaeopterans lack the musculature to do this, and so hold them proud of the body—pointing upward, as in mayflies; to the side, as in dragonflies; or along the body, as damselflies do.

The palaeopterans were the first animals to achieve the power of flight. How they did this is uncertain, but one theory has arisen from observing the juvenile form of mayflies (see image 4), known as naiads or water nymphs. Living naiads have pairs of feathery gills along the abdominal segments and two pairs of wing buds on the thorax. In several species, the base of the gill is hardened to create a leathery cover that protects the flimsy tips. It may be that in the primitive preflying form of these insects, the gills covered the thorax as well as the abdomen. The animal emerged from the water to breed—like today's palaeopterans do—and the thoracic gills became modified into diaphanous aerofoils and then wings. The need to glide between food sources or away from enemies would have driven this evolutionary pressure.

The first neopterans crop up in the fossil record around 320 million years ago. They had already diversified and had similarities with today's cockroaches and grasshoppers. An extinct group of grasshopperlike insects, Caloneurodea, from this period has similarities with *Strudiella*, such as generalized mouthparts and long legs for running. The Caloneurodea also had a pair of wings that was lacking in *Strudiella*, but that may simply mean that *Strudiella* was a wingless juvenile. True grasshoppers (crickets) were seen in the Carboniferous, complete with large hind legs for leaping (see image 5). Representatives of Neoptera developed their wings on the outside of the body, as their modern descendants do today. The juvenile forms (nymphs) are wingless when they hatch, but resemble the adult form in most other ways. As they develop, they shed their skins periodically, gradually developing wing buds, but are only flight-ready in the adult form. This kind of development is called hemimetabolism or incomplete metamorphosis.

Many other insects undergo complete metamorphism (holometabolism). The egg hatches as a larva—familiar larvae include the butterfly caterpillars (see image 6), fly maggots, and beetle grubs—which, after a period of growth, transforms into the adult via an immobile phase known as the pupa. One benefit of this is that the larva has a different food source and even habitat from the adult, thereby avoiding competition between parents and offspring. While these ecological reasons are well understood, how such metamorphosis evolved is not. The first complete-metamorphosis insects in the fossil record belonged to the Miomoptera, around 310 million years ago. Now extinct, Miomopterans were rather generalized insects, with four wings and chewing mouthparts. They most closely resembled today's lacewings, which are also regarded as a primitive group.

Neopterans of both types began to diversify toward the end of the Carboniferous and through the Permian period, 350 million to 250 million years ago. The cockroaches had the biggest success, with their ancestors making up 90 percent of insects at this time. However, true bugs also appeared, with needle-shaped mouthparts for sucking sap and other liquids. The Permian also heralded the arrival of the beetles, which have forewings hardened into covers to protect the hindwings used in flight, and the first true flies, which have forewings only. The hindwings have been reduced to movement sensors that help these agile insects orientate themselves in flight.

The Triassic period, 252 million to 201 million years ago, saw the extinction of the most primitive insect groups and a severe reduction in the cockroaches, but others carry on appearing in the fossil record. The earliest wasps and stick insects evolve at this time, while the first butterflies, moths, and earwigs appear in the Jurassic period that follows. Insect evolution rocketed through the Cretaceous period, 145 million to 66 million years ago, with the arrival of nearly all of today's main groups. This was linked to the appearance of angiosperms (flowering plants) and an intense period of coevolution, as insect and plants developed in tandem to exploit one another for mutual benefit (see p.96). Termites also appeared in the Cretaceous, feeding on dead wood from fallen trees. They were the first social insects, living together in large colonies. They were followed, also in the Cretaceous, by evolved wasps and ants. Fleas emerged in the middle Cretaceous, sucking the blood of mammals and birds.

The end-of-Cretaceous mass extinction 66 million years ago (see p.364) wiped out many groups of animals but it had little impact on insects. As Earth's landmasses began to approach their current configuration, insect communities evolved independently of one another. The most successful were the beetles (coleopterans), which make up one third of insect species today (see image 7), followed by the butterflies and moths (lepidopterans); more flies (dipterans); and bees, wasps, and ants (hymenopterans). While wasps are often thought of as yellow-and-black picnic-spoilers, most species are actually tiny parasites that have evolved to attack specific groups of other arthropods, mainly insects. It seems that no insect group has seen a drop in diversity since the Cretaceous. Indeed, it is sometimes said we are living in the Age of Insects. **TJ**

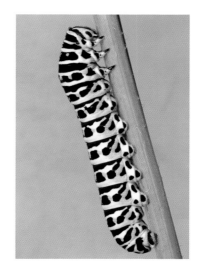

Griffinfly
EARLY PERMIAN PERIOD

SPECIES *Meganeuropsis permiana*
GROUP Meganisoptera
SIZE 17 inches (43 cm)
LOCATION North America

⬣ NAVIGATOR

Insects ruled the skies for at least 100 million years and were the only flying creatures until the arrival of the pterosaurs (see p.288) by 220 million years ago. The first airborne insects had no need to defend themselves from attack in the air, but this was not the case for very long. Griffinflies were early aerial predators, classified as the Protodonata or Meganisoptera. The former name refers to the group's similarity to the Odonata, the dragonflies of today. The most significant difference between the Meganisoptera and the Odonata is the location of the sexual organs, which are farther forward in modern species. However, members of both groups looked—and killed in—the same way. The name "Meganisoptera" hints at the enormous size of griffinflies. By the late Carboniferous, 300 million years ago, there were species with 26-inch (66 cm) wingspans. The wingspans of the early Permian species *Meganeuropsis permiana* reached 28 inches (70 cm). Why were griffinflies so big, and why are modern insects so small in comparison? First, there was no other predator to match them. Today's aerial insects are threatened by birds of prey, such as hawks and owls. The second reason is more complex; the rise in atmospheric oxygen levels during the Carboniferous meant that the first forests grew, with tree-size ferns taking in carbon dioxide from the air and locking it away in wood. For every molecule of carbon dioxide absorbed, the tree gave out one molecule of oxygen. This process resulted in a marked increase in the proportion of oxygen in the atmosphere. In turn, this enabled oxygen to permeate deeper into the bodies of insects, allowing their size to rise and leading eventually to the evolution of these primeval giants. **TJ**

FOCAL POINTS

1 LEGS

The legs of the griffinfly were equipped with short spines to hold struggling prey, important because the predator hunted on the wing and needed to hold its prey captive before chewing it to kill it. Palaeodictyoptera (see panel) were the most common prey, along with arachnids and small terrestrial vertebrates.

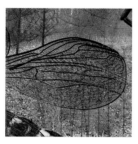

2 WINGS

Most fossil griffinfly specimens from the middle Carboniferous to the early Permian are known only from wing fragments. Unlike modern dragonflies, griffinfly wings lacked a pterostigma—the blood-filled section that adds a spot of color and is thought to act as a counterweight.

3 ABDOMEN

Griffinflies consisted of mostly wing and little body. They were 17 inches (43 cm) long, most of which was the abdomen. The male abdomen ended with a clasper to hold the female when mating. In living dragonflies, the male genitals are at the base of the abdomen, while in griffinflies they were at the tip.

▲ Fossils of griffinflies, such as this member of the Tupinae, sometimes display intricate vein networks in the wings. These tubes gave strength with flexibility to bend with aerial maneuvers and air currents.

SIX-WINGED INSECTS

The evolution of the griffinfly's great size may have been driven by an "arms race" with its chief prey: the now-extinct Palaeodictyoptera (right). These were the dominant flying insects in the Carboniferous. They had long, pointed mouthparts, perhaps to suck plant juices, but could also have been suited to animal targets. Palaeodictyoptera have been called "six-winged" insects: as well as the two pairs of wings, similar to a mayfly's, they also had a pair of flaplike winglets (paranota) attached to the pronotum (the segment of the thorax just behind the head). This may support the theory that insect wings evolved from modified gills that ran along the length of the body in aquatic ancestors. Early gliding insects could have had multiple sets of winglike appendages, which were reduced rapidly by natural selection to fewer pairs, three in the case of the Palaeodictyoptera. Like griffinflies, they grew to great sizes, with wingspans of 20 inches (50 cm).

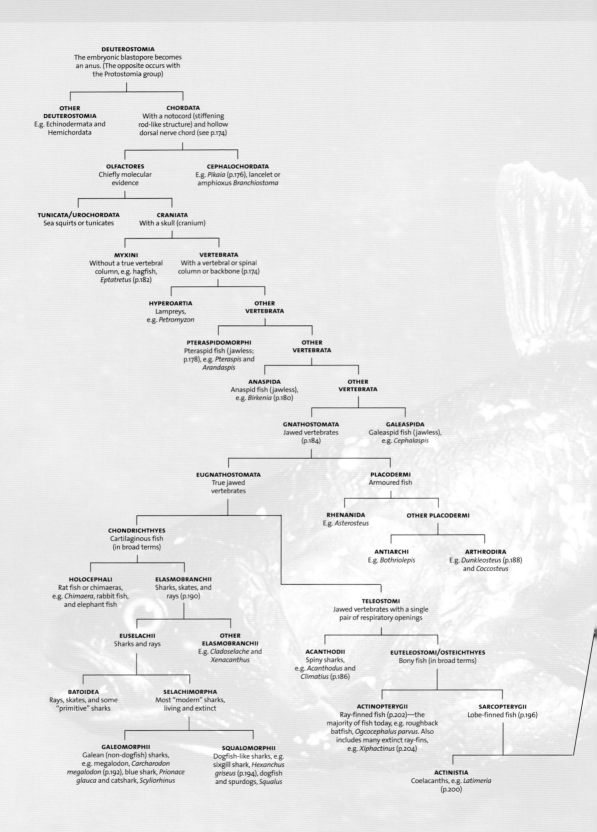

DEUTEROSTOMIA
The embryonic blastopore becomes an anus. (The opposite occurs with the Protostomia group)

OTHER DEUTEROSTOMIA
E.g. Echinodermata and Hemichordata

CHORDATA
With a notocord (stiffening rod-like structure) and hollow dorsal nerve chord (see p.174)

OLFACTORES
Chiefly molecular evidence

CEPHALOCHORDATA
E.g. *Pikaia* (p.176), lancelet or amphioxus *Branchiostoma*

TUNICATA/UROCHORDATA
Sea squirts or tunicates

CRANIATA
With a skull (cranium)

MYXINI
Without a true vertebral column, e.g. hagfish, *Eptatretus* (p.182)

VERTEBRATA
With a vertebral or spinal column or backbone (p.174)

HYPEROARTIA
Lampreys, e.g. *Petromyzon*

OTHER VERTEBRATA

PTERASPIDOMORPHI
Pteraspid fish (jawless; p.178), e.g. *Pteraspis* and *Arandaspis*

OTHER VERTEBRATA

ANASPIDA
Anaspid fish (jawless), e.g. *Birkenia* (p.180)

OTHER VERTEBRATA

GNATHOSTOMATA
Jawed vertebrates (p.184)

GALEASPIDA
Galeaspid fish (jawless), e.g. *Cephalaspis*

EUGNATHOSTOMATA
True jawed vertebrates

PLACODERMI
Armoured fish

RHENANIDA
E.g. *Asterosteus*

OTHER PLACODERMI

CHONDRICHTHYES
Cartilaginous fish (in broad terms)

ANTIARCHI
E.g. *Bothriolepis*

ARTHRODIRA
E.g. *Dunkleosteus* (p.188) and *Coccosteus*

HOLOCEPHALI
Rat fish or chimaeras, e.g. *Chimaera*, rabbit fish, and elephant fish

ELASMOBRANCHII
Sharks, skates, and rays (p.190)

TELEOSTOMI
Jawed vertebrates with a single pair of respiratory openings

EUSELACHII
Sharks and rays

OTHER ELASMOBRANCHII
E.g. *Cladoselache* and *Xenacanthus*

ACANTHODII
Spiny sharks, e.g. *Acanthodus* and *Climatius* (p.186)

EUTELEOSTOMI/OSTEICHTHYES
Bony fish (in broad terms)

BATOIDEA
Rays, skates, and some "primitive" sharks

SELACHIMORPHA
Most "modern" sharks, living and extinct

ACTINOPTERYGII
Ray-finned fish (p.202)—the majority of fish today, e.g. roughback batfish, *Ogcocephalus parvus*. Also includes many extinct ray-fins, e.g. *Xiphactinus* (p.204)

SARCOPTERYGII
Lobe-finned fish (p.196)

GALEOMORPHII
Galean (non-dogfish) sharks, e.g. megalodon, *Carcharodon megalodon* (p.192), blue shark, *Prionace glauca* and catshark, *Scyliorhinus*

SQUALOMORPHII
Dogfish-like sharks, e.g. sixgill shark, *Hexanchus griseus* (p.194), dogfish and spurdogs, *Squalus*

ACTINISTIA
Coelacanths, e.g. *Latimeria* (p.200)

4 | FISH AND AMPHIBIANS

The evolution of the backbone, or vertebral column, probably from the rod-like notochord, had hugely productive consequences, although some lineages faded into dead ends. As many kinds of fish began to flourish, some acquired jaws—a valuable adaptation for processing new sources of food. Similarly, the repurposing of fins as limbs seems to have given rise to a variety of "fishapods" in water, before tetrapods crawled out as amphibians to join plants and invertebrates on land.

RHIPIDISTIA
Lungfish and tetrapods
(p.206)

DIPNOMORPHA
Dipnoi—lungfish, e.g. *Scaumenacia*,
Fleurantia, *Neoceratodus*, *Lepidosiren*,
and *Protopterus*; Porolepiformes,
e.g. *Porolepis* and *Laccognathus*

OTHER RHIPIDISTIA

RHIZODONTIDA
E.g. *Rhizodus*

TETRAPODOMORPHA
"Tetrapod shaped"

ELPISTOSTEGALIA
Lobe-finned fish related
to tetrapods

TRISTICHOPTERIDAE
E.g. *Eusthenopteron*

TETRAPODA
Four-limbed vertebrates
(p.214)

OTHER ELPISTOSTEGALIA
Sometimes called "fishapods,"
e.g. *Panderichthys*,
Tiktaalik (p.210) and *Elpistostege*

OTHER TETRAPODA
E.g. *Ventastega*, *Acanthostega*,
and *Ichthyostega*

TETRAPODA–AMPHIBIA
Amphibia in the broad sense

WHATCHEERIIDAE
E.g. *Pederpes*
(p.212)

TETRAPODA–AMPHIBIA
Amphibia in a
narrower sense

OTHER AMPHIBIA

CRASSIGYRINIDAE
E.g. *Crassigyrinus*

BAPHETIDAE
E.g. *Eucritta*.
TEMNOSPONDYLI?
E.g. *Eryops* and *Amphibamus*
LEPOSPONDYLI?
Simple, spool-shaped vertebrae,
e.g. *Diplocaulus* (p.218)

OTHER AMPHIBIA
More derived or evolved forms

BATRACHOMORPHA
Frog-shaped amphibians

REPTILIOMORPHA
Reptile shaped amphibians

LISSAMPHIBIA
Recent and all living
amphibians (p.220)

LEPOSPONDYLI?
Simple, spool-shaped vertebrae,
e.g. *Diplocaulus* (p.218)
TEMNOSPONDYLI?
E.g. *Eryops* and *Amphibamus*

SEYMOURIAMORPHA
Reptile-like,
e.g. *Seymouria*
(p.216)

**OTHER
REPTILIOMORPHA**
E.g. *Eogyrinus*

GYMNOPHIONA/APODA
Limbless caecilians, e.g. aquatic
caecilian, *Typhlonectes*

BATRACHIA
All frogs and salamanders

AMNIOTA
Lay eggs on land or retain
fertilized egg within mother

DIADECTOMORPHA
E.g. *Diadectes* and
Limnoscelis

CAUDATA/URODELA
Salamanders and newts,
e.g. red mud salamander,
Pseudotriton rube

SALIENTIA
Early and modern frogs

SYNAPSIDA
Mammals and their
extinct relatives
(see Chapter 7)

SAUROPSIDA
"Lizard faces"—reptiles
(see Chapter 5) and birds
(see Chapter 6)

OTHER SALIENTIA
Early frogs,
e.g. *Triadobatrachus* (p.222)

ANURA
Frogs and toads

THE DEVELOPMENT OF BACKBONES

1 Resembling jellyfish, sea salps are more complex and live in all oceans. They are often strung together in colonies many feet long.

2 The lancelet, the amphioxus *Branchiostoma*, has a notochord: the precursor of a fully developed backbone.

3 *Myllokunmingia*, possibly an early chordate, may have been a transitional organism in the evolution of vertebrates.

A nimals with backbones, known as vertebrates, and those that lack them, invertebrates, are a major division of life. Vertebrates include fish, amphibians, reptiles, birds, and mammals. They comprise only 5 percent of all species, but the group contains the largest, most mobile and most intelligent animals on Earth, including humans. The earliest vertebrates were fish, which appeared around 480 million years ago. The search for their origins, and ultimately human origins, begins with precursor organisms, such as the chordates, with features that could evolve to become vertebrate characteristics. Chordates have structures that prefigure vertebrate anatomy. They have bilaterally symmetrical—or left–right, mirror-image—bodies with a single bundle of nerve fibers running in a channel along the "back." Instead of a backbone, there is a notochord: a cartilage-based stiffening rod, which runs underneath the nerve cord, supporting it. Chordates also have a row of gill openings (pharyngeal slits) along the throat behind the mouth; segmented muscles as repeated groups; and a muscular tail, which extends the body past the anal opening. All chordates show these features at some point in their lives, although many vertebrates (humans included) only have gill slits as an embryo.

The evolutionary history of chordates is immensely significant for understanding the origins of vertebrates. The animal group Chordata is ranked higher than the group Vertebrata and it contains the vertebrates along with

KEY EVENTS

555 Mya	550 Mya	541 Mya	540 Mya	534 Mya	525 Mya
Burykhia, from the Precambrian eon, is possibly an early form of sea squirt tunicate, one of the first groups of chordates.	Bilaterian animals form in water. They have a distinguishable front and back.	The first hard-shelled animals, visible to humans without a microscope, appear.	Vision develops in some early fish, both in dim light (scotopic vision) and bright light (photopic vision).	Early cephalochordate, *Cathaymyrus*, with repeating muscle blocks and a long notochord-like ridge, exists in Chengjiang, China.	Several more advanced chordate-like animals, including *Haikouella* and *Yunnanozoon* inhabit Chengjiang.

other smaller groups: tunicates and cephalochordates. Tunicates are marine filter-feeders and include the sessile (rooted) colonial sea squirts and pelagic (free-swimming) sea salps (see image 1). These baglike organisms have twin siphons through which they draw in and expel water. Although they look like invertebrates, the larval or early stage of development shows that they are chordates. With a long tail, segmented muscles, stiff notochord, nerve cord, and well-developed pharynx (throat and neck region), their "tadpole" larvae have a full set of chordate characteristics. These features are lost when the larva transforms into an adult. Some fossils and genetic studies point to the other chordate group, cephalochordates, as promising ancestors for the vertebrates. Modern cephalochordates (lancelets) include *Branchiostoma* (see image 2). These small, fishlike animals retain their notochord into adulthood. They are pointed at both ends and, unlike tunicates, the notochord continues into the head. This latter feature gave rise to the cephalochordate name, which means "head chord." Often found half-buried in sand, they swiftly move out of harm's way if disturbed, powered by blocks of muscles pulling on either side of the notochord. This same method of propulsion is also employed by fish.

Pikaia (see p.176), a 505-million-year-old resident of a Cambrian period reef community in Canada known as the Burgess Shale (see p.50), made the news in the 1980s when its paired muscle blocks and putative notochord earned it reclassification as one of the earliest chordates. *Pikaia* is important because, although it has a number of nonchordate features, its very existence shows that Cambrian marine communities included early chordates or similar creatures. In the 1990s, groundbreaking discoveries in Yunnan province, China, unveiled an unexpected slew of ancient chordates. The renowned Chengjiang biota, 525 million years old, were benthic-dwellers (creatures that live on or near the bottom of a body of water), like the Burgess Shale animals. Among the enormous diversity were eight possible chordates. *Haikouella*, a lancelet-shaped animal 1 to 1½ inches (2.5–4 cm) long, had a primitive notochord, digestive tract, tail, and pharyngeal gill arches sporting teeth. Some scientists class *Haikouella* as a wormlike hemichordate, linking invertebrates and chordates; others see it as the most primitive known chordate.

The existence of animals this complex raises the tantalizing but controversial possibility that chordates arose much earlier. Molecular clock studies (see p.17) estimate the time since two groups diverged by examining matching genetic material and assuming that changes, or mutations, happen at a predictable and steady rate. However, in the case of the early chordates, these techniques yield dates that are earlier in the Cambrian or even before. The improbability of these animals being preserved as fossils is bound to hamper the quest for the small, soft, near-defenceless invertebrate, which carefully avoided the fearsome *Anomalocaris* (see p.116) and braved the predatory trilobites to become the progenitor of vertebrates. **DG**

525 Mya	524 Mya	521 Mya	520 Mya	520 Mya	505 Mya
The pterobranch *Galaeplumosus abilus*, possibly a hemichordate link between invertebrates and vertebrates, is preserved in China.	*Myllokunmingia* (see image 3) from the Maotianshan Shales of Chengjiang is possibly an early chordate.	Early trilobites appear, which could have preyed on early chordates.	Lancelets such as the amphioxus *Branchiostoma* have probably already split away from the main line toward vertebrates.	The tunicate fossil *Shankouclava*, a type of fixed-down, leathery, baglike animal called a sea squirt, dates from this time.	*Pikaia* exists in the Burgess Shale marine community. It has all the hallmarks of being a cephalochordate.

Pikaia
MIDDLE–LATE CAMBRIAN PERIOD

SPECIES *Pikaia gracilens*
GROUP Cephalochordata
SIZE 2⅓ inches (6 cm)
LOCATION Canada

Among the many peculiar creatures of the Burgess Shale was the wormlike *Pikaia gracilens*, named in 1911 to honor nearby Mount Pika, British Columbia. Never longer than 2⅓ inches (6 cm) and clearly segmented, *Pikaia* merited little further attention and was filed away with a group of worms. However, in the late 1970s a number of the Burgess Shale fossils were re-examined and reassigned. It was discovered that the segmentation of this laterally squashed, leaflike animal was not like that of worms. Instead of regular cylindrical segments, *Pikaia* had vertical bands of muscles with a slight sinuous curve to them. It also showed what seemed to be a stiff notochord: the precursor of the backbone. With these defining characteristics, *Pikaia* was classified as a chordate and a possible ancestor of the vertebrate animals. Although this designation was provisional, most scientists acknowledge the anatomical similarities between living lancelets and *Pikaia*. Known from more than 100 specimens, *Pikaia* is not an exact match for modern chordates, which latterly suggests that it had a separate evolutionary lineage.

Pikaia had a very small head with a pair of smooth tentacles. It has been suggested that eyespots featured at their tips, but *Pikaia* might have been sightless. Behind the head on either side protruded a series of short appendages, which might have been connected to chordate gill slits for water used in respiration. These features differ from the living lancelets. *Pikaia* lived with the rest of the Burgess Shale creatures 505 million years ago, peaceably moving slowly through the water or on bottom sediments, until it was overcome by the torrents of mud that buried the entire species. **DG**

⬡ NAVIGATOR

FOCAL POINTS

1 HEAD AND APPENDAGES

The head was unique among early chordates. With its tiny mouth and lack of jaws or chewing parts, *Pikaia* would have fed on small particles carried on weak currents, or sediments. The appendages were not positioned to assist with feeding, so their function is speculative: perhaps a role in breathing, connected to proto-pharyngeal openings.

2 NONCHORDATE FEATURES

A sausage-shaped dorsal (back) organ running the length of the trunk was one of *Pikaia*'s several nonchordate features and it might have formed a storage organ. A shieldlike structure covered the head region, which has been likened to the cuticle of roundworms (nematodes). In arthropods, such a cuticle forms the extensive hardened exoskeleton.

3 MYOMERES

Pikaia's bands of muscle (myomeres) were tall and thin, with gently curved boundaries, rather than the V-shaped blocks seen in modern vertebrates. *Pikaia* had about 100 myomeres and would have swum with S-shaped curves, similar to modern chordates, the lancelets. Fish inherited this side-to-side motion from early chordates.

ACCIDENT OR DESIGN?

In his book *Wonderful Life* (1989), influential paleontologist Stephen J. Gould (1941–2002) used *Pikaia*—as one of the few Burgess Shale survivors that then led to the vertebrates—to advance a theory of "contingency" in evolution. This claimed that there is no direction in evolution, and that trends seen in the past are arbitrary, even accidental. If evolutionary history began again it would lead to a radically different world. Simon Conway Morris (b.1951, above), the paleontologist who reclassified *Pikaia*, criticized this view in his book *The Crucible of Creation* (1998). He pointed out that convergent evolution (see p.19) was a directive mechanism by which equivalent forms can arise from different stock, so replaying evolution would result each time in the same broad themes.

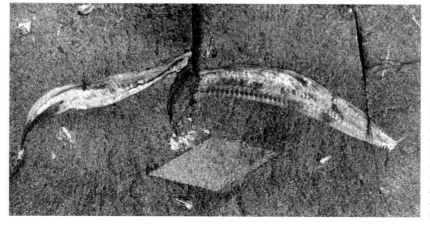

◀ Exceptionally well-preserved fossils show impressions of internal structures. A clear axial trace running along the back has been interpreted as a notochord, or a combined notochord and nerve cord. Aside from this, *Pikaia*'s place as a chordate rests mostly on its paired muscle blocks.

THE FIRST FISH

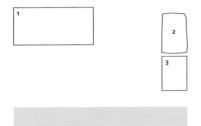

1 *Arandaspis*, around 6 inches (15 cm) long, lacked paired fins and swam with its tail, unusually flattened top to bottom (dorsoventrally).

2 At 8 inches (20 cm) in length, *Pteraspis* possessed a large front-end protective shield with swept-back horns to the sides and one on top of the head.

3 The living sea lamprey, *Petromyzon marinus*, displays horny rasping teeth in its jawless, suckerlike mouth.

The evolution of a head containing well-developed sensory organs was a major innovation in the history of life. Although the primitive chordates (see p.174) of the Cambrian seas might have carried simple organs to detect their surroundings, advances in the amount and sensitivity of sensory equipment at the front end of an animal gave it advantages when it came to finding food, negotiating territory, and avoiding predators. Chordates that have a head are known as craniates and this group led the charge into a new era, 480 million to 470 million years ago.

The protofish *Myllokunmingia* and *Haikouichthys*, found in the Chengjiang rocks of China, push back the origin of the craniates to the early Cambrian period, 530 million to 520 million years ago. These fish, with cartilage skeletal structures, a single dorsal (back) fin, and paired lateral (side) fin-folds, are often described as prevertebrates. Recently, *Metaspriggina*, another small craniate from the Burgess Shale in British Columbia, Canada, has been described as one of the earliest vertebrates. The first so-called true vertebrates were agnathans (jawless fish) and they sported a backbone that enclosed and protected the nerve cord. The widely separated fossils of *Arandaspis* from Australia (see image 1) and *Sacabambaspis* from Bolivia show that these kinds of fish enjoyed a global distribution, which reinforces the idea that their origins lie further back in time.

The Ordovician period closed 443 million years ago with the second largest of the five major extinctions in Earth's history. A sudden burst of evolution for fish followed, and they quickly found new ways of living. With great speed for

KEY EVENTS

530 Mya	480 Mya	470 Mya	453–359 Mya	448 Mya	447–443 Mya
Myllokunmingia and *Haikouichthys* in China indicate that craniates and prevertebrates exist at this time.	*Arandaspis* thrives. It is one of the earliest agnathans and a primitive vertebrate.	*Sacabambaspis*, one of the best-known arandaspids, lives on the margins of the southern supercontinent Gondwana.	Thelodonts, up to 12 inches (30 cm) long, with their distinctive spiny scales, are widespread in seas and fresh water.	Gondwana drifts over the South Pole. Global temperatures fall and a major ice age begins.	The Hirnantian end-of-Ordovician extinction event kills off 60 percent of marine genera.

evolution, more than 30 different body plans appeared, although most of these became redundant later and fell into extinction. Boom times for the jawless fish came in the following Silurian period, 443 million to 419 million years ago. Mild greenhouse conditions prevailed across the world, making it a benign place for sea life. The lack of polar ice meant that high sea levels continued throughout the Silurian, inundating continental regions. Coral reefs (see p.104) flourished in the shallow seas caused by the closing of the Iapetus Ocean in the northern hemisphere in the Silurian. Deep-bodied thelodont (nipple teeth) jawless fish scattered their spiny scales far and wide for fossilization, thereby providing evidence of their extensive distribution and evolutionary innovation.

The classic early jawless fish—typified by the arandaspids of Australia, living off the southern Gondwana landmass, and the astraspids of the northern sea basins around northern Laurentia and Baltica—were stolid creatures. Truncheon-shaped with a heavily shielded, flattened head, they had a strong tail fin but lacked paired side fins, which made them poor swimmers. This agnathan evolution produced a bizarre-looking assortment of strange head shields and nose spines, typified by the galeaspids (helmet shields). Ostracoderms (bony shields) such as *Hemicyclaspis* (semicircle plate) had a massive horseshoe-shaped frontal shield. *Pteraspis* (wing shield), a heterostracan from Belgium and England, had a tall, angled-back head horn that looked like a speedboat mast (see image 2). Despite these clumsy appendages, the jawless fish laid down much of the evolutionary groundwork for fish design. The asymmetrical heterocercal tail (with a larger upper lobe) provides lift in the water and is still seen in many sharks. More advanced ostracoderms developed paired pectoral fins and a dorsal fin to stabilize them as they swam. With paired nostrils, two prominent eyes, and a top-mounted middle eye, they had primitive chemoreception senses for smell and pressure sensors that enabled them to detect movement.

Lacking jaws, jawless fish were suspension-feeders and sucked up bits and pieces on the seabed in search of nutritious particles. These would be absorbed into the body via the gills. Anaspids (without shields), such as *Pharyngolepis* from Norway, look much more like modern fish. They did away with the head shield and were altogether lighter. Agile swimmers with circular mouths, they were more likely to be midwater filter-feeders, similar to modern basking sharks. Today's jawless fish are very different: they all lack armor, and the hagfish (see p.182) and river lampreys are scavengers, while sea lampreys are parasites that attach themselves to other fish using their teeth (see image 3). Although jawless fish were the dominant vertebrate group of the Silurian, at the start of the Devonian period and the Age of Fishes, around 420 million years ago, they started to come under predation pressure from newly evolving fish with jaws. The Devonian saw the decline and fall of the jawless fish, and very few made it through the late Devonian extinction into the Carboniferous period, which began 359 million years ago. **DG**

438–430 Mya	438–359 Mya	430–390 Mya	430–370 Mya	428–359 Mya	425 Mya
The closure of the Iapetus Ocean, as the landmasses drift, brings about major climate change.	Heterostracans—agnathan fish with large bony shields and unique "foamed bone" scales—live in marine and estuary environments.	Anaspids (without shields) such as *Birkenia elegans* (see p.180) are widespread.	Galeaspids, lacking paired fins but with numerous gill openings, live in Chinese shallow freshwater and marine environments.	Ostracoderms, the most advanced armored jawless fish, live in North America, Europe, and Russia.	The first jawed fish may be *Shimenolepis*, a member of the antiarch group of placoderm (plated skin) fish.

Birkenia
SILURIAN PERIOD—DEVONIAN PERIOD

SPECIES *Birkenia elegans*
GROUP Anaspida
SIZE 4 inches (10 cm)
LOCATION Scotland

NAVIGATOR

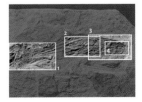

ating from the Silurian and Devonian periods, between 430 million and 390 million years ago, this extinct species of jawless fish comes from near the small town of Lesmahagow, Lanarkshire, Scotland, which at the time was covered by a shallow inland sea. The area is a world-renowned fossil site for Silurian vertebrates, and *Birkenia elegans* is one of the many well-preserved forms in its shale and sandstone rocks. It belongs to the anaspid group. Unique among fossil jawless fish, the anaspids had not evolved heavy, bony body plates. Instead, they were covered in rows of deep, overlapping scales. Fossils of anaspids are relatively rare, which makes their abundant presence—often as complete fossils—in the Silurian sediments of the Lesmahagow Inlier even more remarkable.

Birkenia was a small fish, 1 to 4 inches (2.5–10 cm) long. Its asymmetrical tail was strongly hypocercal or reverse heterocercal (with a larger lower lobe). Its slender body lacked stabilizing fins, so it is presumed to have been a clumsy swimmer, although it may have been the best-adapted inhabitant of its time. With no jaws, *Birkenia* was a filter-feeder, and its mode of feeding would have involved swimming head-down through the fine silt and other sediments on the seabed, sucking up edible particles. Anaspids—a name that means "without shields," referring to the lack of large bony plates—such as *Birkenia* were not in existence for very long; they can be regarded as a short-lived experiment in early vertebrate design. Although widespread, they were mostly small, with many species comfortably fitting into a human hand; an exception is *Rhyncholepis* from Norway, which reached a length of more than 10 inches (25 cm). By 359 million years ago, the group had been wiped out by the late Devonian extinction. **DG**

👁 FOCAL POINTS

1 HYPOCERCAL TAIL

Birkenia had evolved a hypocercal or reverse heterocercal tail—the reverse of fish such as sharks. For this reason, the fossil was displayed upside down for 25 years after discovery, until 1924, when Norwegian paleontologist Johan A. Kiær (1869–1931) turned it around the correct way.

3 HEAD AND GILLS

The head was covered with small plates. There was a single nasal opening and a large Y-shaped pineal plate containing the pineal opening—a light-sensitive structure. The eyes were simple and as in many other jawless fish, the gills were a row of holes. *Birkenia* had between six and 15 gill holes.

2 SCALES

The body was covered in rows of scales that were dorsoventrally elongated (wider than they were high). These scales overlapped, sloping downward and backward. Along the back a row of spiky scutes (waterproof scales made of keratin) served as defence, and the fifth and adjacent scutes carried a distinctive double hook.

4 MOUTH

The mouth of *Birkenia* was at the rounded end of its distinctive blunt head. This is the mouth position of a bottom-feeder, so it was long assumed that the jawless fish ploughed the seabed. However, its overall slender shape suggests that it was a mobile, active swimmer that could feed at any level in the water.

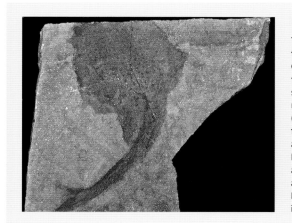

LESMAHAGOW INLIER

The fossil sites surrounding Lesmahagow in Scotland are the preeminent locations for Silurian remains. When the Geological Society of Glasgow visited the area in the 1890s, they set up Camp Siluria. During this time the society's members collected large numbers of otherwise rare specimens, such as fossil fish (left) and eurypterids (sea scorpions), many of them complete. In the 1950s, the fossil fish rock beds of Birk Knowes, Slot Burn, Dippal Burn, and Glenbuck, among others, were designated the United Kingdom's first Site of Special Scientific Interest. However, a century of avid collectors has denuded this tiny pocket and has now exhausted the Silurian remains. In 2014, a Fossil Code was launched, outlining best practice in the collection, identification, and conservation of fossils.

Hagfish
PLEISTOCENE EPOCH—PRESENT

SPECIES *Eptatretus stoutii*
GROUP Myxiniformes
SIZE 18 inches (45 cm)
LOCATION Northeastern Pacific
(Canada, Mexico, United States)
IUCN Data Deficient

NAVIGATOR

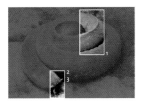

Hagfish are relics from a distant past—throwbacks in several senses. First, it is possible that they are not vertebrates because they are the only living animal to have a skull without a backbone. This makes them craniates, members of the group that contains all vertebrates, but their incomplete braincase and lack of spinal column excludes them from the Vertebrata group. They also lack jaws and paired fins and are entirely without scales. These primitive characteristics—"primitive" in that they came first, rather than were inferior or simple—are interpreted as being basal to (coming before) vertebrates. The hagfish evolutionary line split from its chordate and craniate ancestors prior to the evolution of animals with vertebrae. Modern hagfish, such as *Eptatretus stoutii*, remain largely unchanged from their 300-million-year-old ancestors.

In the craniate group, the 65 to 70 species of hagfish are the only members of the subgroup, or order, Myxiniformes. They are long and eel-like. Their smooth skin fits only loosely and has been likened to a baggy sock. The largest species, *Eptatretus goliath*, reaches more than 4 feet (1.2 m) in length, but most hagfish average around 20 inches (50 cm). Swimming, as well as looking, like eels, these scavengers inhabit cool, cold, deep-water marine environments, where they swarm over and scavenge carcasses that sink to the seabed. Analysis of hagfish stomach contents reveals that they are cosmopolitan eaters that feed on seafood including worms, shrimp, crabs, squid, octopus, fish (including sharks), whales, and even birds. They have a low metabolism and can survive for a month or more without food. However, when they find food, hagfish are energetic feeders. They are effectively "vultures" of the sea and consume dead matter from the inside, freeing up nutrients and minerals for recycling by other marine life forms. **DG**

1 NOTOCHORD

Hagfish retain their primitive notochord (precursor of the backbone) throughout their lives, unlike the other living agnathans, lampreys, in which the notochord mineralizes and is replaced by bone. It has a partial cranium but no vertebrae. Other primitive traits include a lack of scales, no lateral line system or sense organs along the sides of the body and no fins. What appears to be a tail fin is little more than a flap of skin without supporting fin rays. Hagfish kidneys are also primitive and there are multiple venous hearts.

2 EYES

The primitive eyes of the Pacific hagfish, *E. stoutii*, are positioned beneath an opaque eyepatch on either side of the head. They provide important clues to the evolutionary stages of more complex vision sensors and of the human eye. They are primitive, lensless eyespots with limited ability to resolve images, and they are also entirely without the muscles of more advanced organisms that aim the eye at a target. In several genera, the eyes are even partly covered over by the muscular body, which makes the fish almost blind.

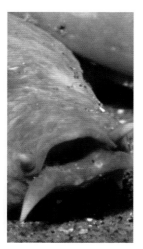

3 MOUTH AND NOSE

The Pacific hagfish has very poor eyesight—it can only detect light and dark—and consequently it uses its ability to detect chemicals in the water—aquatic smell—to locate food. A ring of eight barbules (tentacles) surrounding the slitlike mouth assists with this sense. A single nostril indicates its ancient origins—double nostrils evolved for the first time in placoderm fish about 420 million years ago. Water for breathing is inhaled through a separate opening termed the nasopharyngeal duct.

▲ Hagfish have no jaws but have horizontally grasping structures in the mouth. The "tongue" is a cartilaginous plate bearing two pairs of horny, toothlike projections, which slice up food and draw it into the gut. It bores into flesh, eating into carcasses or weak prey. Hagfish may even absorb nutrients through their smooth skin when they live in the bellies of dying or dead creatures.

SLIPKNOT TECHNIQUE

The name of the common hagfish, *Myxine glutinosa*, is very appropriate. Hagfish produce a milky secretion from their skin glands that, when combined with water, turns into a thick slime. They can convert a 5-gallon (20 L) bucket of water into slime in minutes. The gelatinous mass blocks the gills of any fish in the vicinity and, as a result of this defensive behavior, hagfish have few or no predators. Should they be caught in the jaws of a predator, they tie themselves into knots (below) and slip these along the body to squirm free. The slipknot technique rids them of the slime and restores the function of their gills. It is also a helpful technique for a scavenger, preventing unwanted food and detritus from clinging to the body.

JAWED FISH

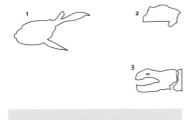

1

2

3

1 Widespread, common, and mainly freshwater, the placoderm *Bothriolepis* was relatively small: 1 foot (30 cm). It had paired pectoral fins.

2 To the right of this side view of a fossilized head of *Entelognathus* (an early, heavily armored placoderm), is the pointed snout and wide mouth; beneath is the heavily thickened lower jaw.

3 As with other arthrodires, *Coccosteus* had a jointed head-neck that enabled its jaws to open wide and swallow sizable prey.

The Devonian period, 419 million to 359 million years ago, saw a revolution in the ecology of the sea that caused the evolution of jaws in fish. The agnathans (jawless fish; see p.178) flourished both in number and diversity in the oceans of the late Silurian period, but their eventual extinction was imminent—new fish were evolving grasping jaws and sharp teeth. Jawed fish are known as gnathostomes. The group includes the acanthodians (spiny sharks) and placoderms (plated skins)—both of which are now extinct—as well as the chondrichthyans (cartilaginous fish) and the osteichthyans (bony fish), which continue to exist today. It is believed that acanthodians came first in evolutionary history, but the discovery of the placoderm fish *Entelognathus* (see image 2) in 2013 suggested that placoderms may in fact be the ancestors of jawed fish. This discovery influenced the understanding of the evolution of sharks (see p.190). Other recent fossil finds have indicated that the thelodont (nipple teeth) class of agnathans was the closest relative to the jawed fish.

Jaws evolved from the forward gill arches in the pharynx (throat and neck region). The front gill arch—a framework of bony rods that supported the gill pouch—was repurposed as a pair of jaws, while the bones behind this moved forward to stabilize the hinge. This squeezed out the gill pouch and, reduced in size, it migrated to become the small spiracle, an opening still seen in sharks and rays. Jaws may have evolved in response to increasing respiratory demands, because by opening and closing the mouth fish can more effectively force water over the gills to increase their oxygen uptake—a process known as the buccal pump mechanism. Breaking the connection

KEY EVENTS

455 Mya	440–420 Mya	420 Mya	419 Mya	419–393 Mya	394 Mya
Evidence of scales and teeth indicates that sharks may have arisen as early as the late Ordovician period.	The final closure of the Iapetus Ocean alters the global climate and ocean currents, bringing many new habitat opportunities.	Early placoderm jawed fish evolve, such as *Entelognathus* from China, which has a length of nearly 8 inches (20 cm).	The first acanthodians, such as *Acanthodes* and *Ptomacanthus*, swim in the oceans.	The acanthodian *Climatius*, measuring 3 inches (7.5 cm) long, leaves fossils across Europe and North America.	Fossils of the articulated shark *Doliodus*—the earliest known shark—have been found from this time.

between feeding and respiration resulted in a period of evolutionary experimentation and the appearance of many novel forms.

The world 400 million years ago was a place of inland waterways, lagoons, and reefs. With a warm global climate, sea levels remained high. The closure of the Iapetus Ocean had created ranges in the Caledonides mountain system of Norway, Britain, and eastern Greenland, as well as the Acadian belts of the eastern seaboard of North America. Erosion was rapid, and great rivers poured off these mountains, forming enormous delta systems and laying down great layers of Old Red Sandstone. The huge continents continued to coalesce, moving to meet one another in low latitudes. Much of North America was flooded with warm shallow seas and extensive reefs. In these plentiful and varied lagoons, jawed fish diversified into more than 200 genera. Fossil beds reveal that fish also moved to freshwater environments, but died in great numbers when lakes dried out or became anoxic (depleted of oxygen) and unable to support life. Dedicated solely to respiration, gills had evolved into the elaborate, specialized respiratory surfaces we see in modern fish.

The evolution of jaws also led to teeth. At first, these were little more than mineralized knobs, but they soon became an array of terrifying mouth weaponry, from serrated wedges to guillotine-like slicing plates. Once the first fish predators had arrived, only the jawless fish lampreys and hagfish (see p.182) survived. The placoderms were heavy-set fish with two sets of armor: one covering the head and the second over the front of the body. These thick, bony shields were most likely defence against the aggressive eurypterids (sea scorpions; see p.136). The two principal placoderm orders were the antiarchs and the arthrodires. Antiarchs, such as *Bothriolepis* (see image 1) rarely exceeded 1 foot (30 cm) in length. With crablike pectoral fins and eyes on top of the head, they were likely to have been bottom-feeders. Arthrodires were fearsome predators. With a jointed neck and sharp bony teeth, they included the *Coccosteus* (see image 3), which typically grew to 10 inches (25 cm) in length, and the huge *Dunkleosteus* (see p.188), which reached 33 feet (10 m) long and had jagged bony blades that functioned as teeth.

Although the placoderms became extinct 359 million years ago, at the end of the Devonian, the acanthodians survived until around 250 million years ago, in the late Permian period. These smallish spiny sharks, typified by *Climatius* (see p.186), had features seen in both sharks and bony fish. Typically for predators, their eyes were large and forward-placed. An active hunting lifestyle demanded good vision, speed, and maneuverability, and these environmental pressures drove fish to adopt streamlined body shapes and to develop fins. **DG**

390–380 Mya	385–360 Mya	380–360 Mya	375 Mya	360–359 Mya	252 Mya
Jawed fish undergo adaptive radiation (rapid and diverse evolution), leading to a decline in the range and numbers of jawless fish.	The massive blade-mouthed predatory placoderm *Dunkleosteus* rules the oceans.	The placoderm genus *Bothriolepis* becomes varied and widespread, with more than 80 species, chiefly in freshwater or coastal habitats.	Early tetrapod (four-limbed vertebrate) fish, such as *Tiktaalik* (see p.210), possibly begin to venture onto land.	The mass extinction at the close of the Devonian marks the end of the placoderms.	The end of the Permian sees the demise and disappearance of the acanthodians.

Climatius

EARLY DEVONIAN PERIOD

SPECIES *Climatius reticulatus*
GROUP Acanthodii
SIZE 3 inches (7.5 cm)
LOCATION Europe and North America

N amed in 1845 by Swiss biologist and geologist Louis Agassiz (1807–73), *Climatius reticulatus* was a species of small, sharklike acanthodian fish. The extinct class of acanthodians, Acanthodii, go by the name "spiny sharks," even though they were not proper sharks and were only distantly related to them (see panel). *C. reticulatus* belonged to a group with seven known species. Approximately 3 inches (7.5 cm) long, *C. reticulatus*, lived between 419 million and 393 million years ago in the early Devonian. It is well represented in the fossil record, known from complete skeletons found across Europe and North America, as well as many disarticulated spines and scutes (waterproof scales made of keratin). Like other acanthodians, *C. reticulatus* had large, forward-placed eyes, which suggests it was a visually adapted predator. Along with the placoderm fish, such as *Dunkleosteus terrelli*, these fish were among the first vertebrates equipped with jaws. This major evolutionary advance radically diversified fish in the Devonian oceans. Rather than being filter-feeding bottom-trawlers, these jawed fish became active hunters.

The emergence of fish such as *C. reticulatus* altered the ecological balance of the seas, thereby driving a wave of adaptation that saw the jawless fish lose their dominance. Strong-swimming *C. reticulatus* was typical of this new breed. A hunter of smaller fish in open waters, it had a diet that would also have included crustaceans, such as decapod and branchiopod shrimp. The paired lateral fins supported by bony spines—present in all acanthodians—were supplemented by two dorsal fins and one anal fin. These innovations, also absent in jawless fish, acted as control surfaces to keep *C. reticulatus* buoyant, finely balanced, and maneuverable in the water. Over its 20 million years of existence, the genus *Climatius* had a variable number of lateral fins as it searched for the most advantageous adaptation. **DG**

✦ NAVIGATOR

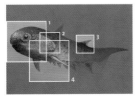

👁 FOCAL POINTS

1 HEAD

The majority of *C. reticulatus*'s adaptations were on the head. As well as its innovative jaw, which enabled active hunting, it had large, forward-placed eyes and good binocular vision. Like sharks, *C. reticulatus* continually replaced its teeth, growing new ones along the jaws as the older ones wore down.

2 OPERCULUM

The operculum was a bony flap covering the gill slits. It divided the head from the body. In some modern fish this is essential for breathing—the operculum opens as the mouth closes: this creates a pressure change in the mouth cavity that causes water to flow over the gills and out under the operculum's rear edge.

3 BONY SPINES

The sharp bony spines were the defining acanthodian feature. Arching backward, they ran along the leading edge of each fin. The fins themselves were composed of a sail of skin stretched between the spine and body. The large spines would also have discouraged bigger fish from swallowing it.

4 PECTORAL FINS

C. reticulatus and other acanthodians had the first paired pectoral fins in fish. The powerful tail fin pushed the fish into a nose-down pose (hence its name, "tilted or inclined" fish). But the hydrofoils, held out from the body and stiffened by spines, counteracted the downward push.

REWRITING FISH EVOLUTION

Although acanthodians such as *C. reticulatus* were similar in appearance to sharks, until recently they were thought to be more closely related to the bony fish rather than the cartilaginous fish. The common belief was that bony fish evolved from the spiny sharks, while the ancestors of the true sharks had much earlier roots. However, in 2009 a reappraisal of the small fish *Ptomacanthus anglicus* (above) changed this. The braincase of this early acanthodian strongly resembles that of sharks and has little in common with bony fish. Perhaps, therefore, *P. anglicus* was an early ancestor of sharklike fish or, going even further back in time, was close to the common ancestry of all jawed vertebrates. *P. anglicus* became known as "the little fish that rewrote evolution."

◀ This fine specimen of *C. reticulatus* is from the Old Red Sandstone (early or lower Devonian) formations at Tillywhandland Quarry, Angus, Scotland. It shows strong, sharp fin spines and four pairs of spines that run along each lower side of the body between the pectoral and pelvic fins.

Dunkleosteus
MIDDLE—LATE DEVONIAN PERIOD

SPECIES *Dunkleosteus terrelli*
GROUP Arthrodira
SIZE 33 feet (10 m)
LOCATION North America and Europe

The top predator of the late Devonian seas, this 4-ton (3.6-tonne) leviathan ruled the oceans between 385 million and 360 million years ago. At 20 to 33 feet (6–10 m) long, *Dunkleosteus terrelli* would easily have been the largest animal of its time and makes most modern fish look like minnows—even the great white shark is only 20 feet (6 m) in length. The beast was a placoderm (see p.184), a group of early jawed fish typified by a bulky shield covering the head and upper front body, and without any scales. With its jointed neck, *D. terrelli* was a member of the arthrodire (jointed neck) placoderms. Glittering in iridescent silver on its belly and red on its topside (known from pigments preserved in fossils from Antarctica), it was a swift hunter and preyed upon jawless fish, spiny sharks, other placoderms (including its own kind) and the large sea scorpions that terrorized the oceans. Its powerful heterocercal tail (with a larger upper lobe) would have generated terrifying bursts of speed.

There were up to 10 different species within the genus *Dunkleosteus*. It originates from Ohio but fossils are widely distributed across the United States and Europe. First described in 1873 as a species of *Dinichthys*, it was renamed in 1956 in honor of David Dunkle (1911–84), curator of vertebrate paleontology at the Cleveland Museum of Natural History, Ohio. The evolutionary history of placoderms was bright but brief. They were numerous and diverse, and are thought to have been the first fish to colonize freshwater habitats. Although not all arthrodires were as big as *D. terrelli*, their heavy armor necessitated the evolution of the first swim bladders; large livers saturated in lighter-than-water oils helped the bulky animals to maintain their buoyancy. However, despite success and diversity, the group lasted 50 million years, dying out at the end of the Devonian. **DG**

 **NAVIGATOR**

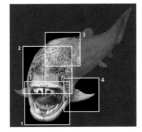

👁 FOCAL POINTS

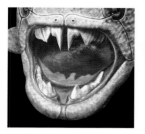

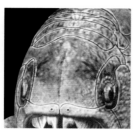

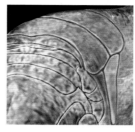

1 JAWS
The jaws were powerful, perfect for crunching through armor plating, with bite forces at the point tips of 35 tons per square inch (5 tonnes per sq cm). The dentition was made of blades formed from skull plates, which self-sharpened as the jaw opened and closed.

2 HEAD SHIELD
The head was enormous— 4¼ feet (130 cm)—and its bony armor plates 2 inches (5 cm) thick. A ring of bone protected each eye. The head shields of certain fossil placoderms bear scars that match the teeth and jaws of *D. terrelli*, showing that it attacked its own kind.

3 JOINTED NECK
A joint in the armor between the head and body was unique to *D. terrelli* and arthrodires. It allowed the head to tilt back as the jaw opened, giving a wider gape. The "nuchal gap" may have enabled them to angle their heads downward to hunt bottom-hugging agnathans.

4 PECTORAL FINS
Along with its dorsal (back) fin, pair of pelvic fins and an anal fin, *D. terrelli* had well-developed pectoral fins. The bony sockets on the sides of the shoulder shield were large and indicate mobility and agility in the water—unlike those of other acanthodians, which were rigid.

◀ The huge, thick bony head and neck armor—cephalic and thoracic shields—of *Dunkleosteus* preserved well, and many fossils are known. The head and jaw muscles and joints indicate this fish could open its jaws in milliseconds, sucking in prey.

DEMISE OF A GIANT

The late Devonian extinction 359 million years ago is one of the Big Five extinction events of the last 540 million years. This global mass event marked the boundary of the Devonian and the Carboniferous periods, and saw 80 to 85 percent of all marine species die out. Along with the trilobites, many brachiopods and corals were hit, and armored fish disappeared altogether. It was a long drawn-out event without a clear cause, but decreasing oxygen levels in the sea, global cooling, and sea-level change all contributed. *Dunkleosteus* (right) survived to the late Devonian, during which time prey evolved to become faster and more nimble, in order to escape the superpredator's jaws. The mass extinction continued this trend, making way for the osteichthyans (bony fish).

SHARKS

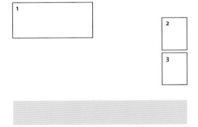

1 The living blue shark, *Prionace glauca*, is streamlined and swift with the full complement of shark fins and other features—a voracious predator.

2 The spiky "hat" and dorsal fin "anvil" of *Stethacanthus*, from 350 million years ago, typify the extraordinary evolutionary experiments in the shark group during the Carboniferous.

3 A close-up of shark skin (common dogfish, *Scyliorhinus*) reveals the tiny, pointed, toothlike scales, termed dermal denticles or placoid scales, that have been found profusely as fossils.

The origins of the class of cartilaginous fish, Chondrichthyes, which includes sharks, rays, skates, and chimaeras (ratfish), are unclear. Cartilage is the gristly tissue that stiffens human noses and ears, and also cushions joints. It preserves rarely when compared to bone, so chondrichthyans have a reduced fossil record. However, because many of these fish shed and regrow hard teeth, these are abundant in rocks. Also the fossil record of dermal denticles—the small toothlike scales that cover the bodies of sharks and other fish (see image 3)—dates back 455 million years to the late Ordovician period. An early type of placoderm (plated skin) fish found in 2013 in Yunnan province, China, shed light on the evolution of sharks. This was *Entelognathus*, a jawed fish (see p.184), but with anatomical details associated with bony fish. It means that bony fish may have retained ancient structures, while sharks evolved innovations. If typical bony fish features evolved deep in the family tree, before the lineage split from sharks, it is possible that bony skeletons came first and cartilaginous ones evolved from them, rather than the other way around as has long been assumed.

Shark evolution is a triumph of simple design. Although in the Devonian period they swam in the same seas as the placoderms and the acanthodians (spiny sharks that were not true sharks), sharks have outlasted both groups. Shark basics were laid down early and little has changed for hundreds of millions of years. The shark body plan is a sleek, streamlined shape covered in friction-reducing denticles. A prominent dorsal (back) fin maintains stability, paired pectoral fins provide control and a strong asymmetrical heterocercal tail (with a larger upper lobe) delivers propulsion (see image 1).

KEY EVENTS

455 Mya	420 Mya	409 Mya	400 Mya	370 Mya	360 Mya
The earliest known sharks occur, leaving isolated skin scales in Colorado.	*Elegestolepis* lived at this time. It is the earliest undisputed shark, known from denticle scales found in Silurian period rocks in Siberia.	*Doliodus* lived in what is now New Brunswick, Canada, leaving one of the oldest known articulated shark skeletons.	Fossil shark teeth become more common in rocks formed from marine sediments.	The *Cladoselache* genus appears. It has large, forward-placed eyes and fully attached jaws; it survives for 15 million years.	The symmoriidan sharks, such as *Stethacanthus*, appear; they last to the end of the Permian, 252 million years ago.

Although a cartilage skeleton is considerably lighter and more flexible than a bony one, sharks still needed to overcome their negative buoyancy to stop themselves sinking. They have evolved two strategies. One is a large liver filled with lighter-than-water oils. The other is the large, stiffened pectoral fins that work like wings in the water, providing lift. Disadvantages of the latter solution are that some sharks must keep swimming, otherwise they sink, and they cannot swim backward.

Shark evolution truly established itself during the Devonian. The oldest known articulated shark remains (with parts in lifelike positions), *Doliodus*, came from the early Devonian of Canada, around 409 million years ago. The first widespread cartilaginous fish was the *Cladoselache* genus from 370 million years ago. These were small and were adapted for high-speed predation, but they still looked more like fish than sharks. The late Devonian mass extinction finished off the placoderms, acanthodians, and almost all remaining agnathans (jawless fish), gifting the seas of the Carboniferous period to the bony fish and cartilaginous fish. Between 360 million and 286 million years ago, the latter responded with a flowering of diversity, and this first major adaptive radiation (rapid and diverse evolution; see p.393) saw some remarkable variations on the basic shark theme. The male *Stethacanthus* (see image 2), for example, a dogfish-sized shark living in warm, shallow seas over both present-day Scotland and Montana, carried elaborate headgear in the form of upright, spiky scales on its head and an enormous flat-topped dorsal fin. No females are found with such impressive modifications; such marked sexual dimorphism (obvious physical differences between males and females) does not occur in modern sharks.

The Carboniferous was a golden age of shark diversity, with 45 families (compared to approximately 40 today). This period also saw the emergence of the Holocephali, the other surviving subgroup of Chondrichthyes. This contains the chimaeras, rabbit fish, and elephant fish; sharks, along with rays and skates, belong to the Elasmobranchii subgroup. The mass extinction at the end of the Permian period (see p.224) decimated shark diversity. However, they persevered and undertook a second great wave of evolution that began 200 million years ago. This also produced the batoids: flattened skates and rays, whose pectoral fins enlarged into "wings." By 100 million years ago, most modern shark groups had appeared. Several species of the modern-looking genus *Squalicorax* thrived in North America, Europe, North Africa, and West Asia toward the end of the Cretaceous period, the largest reaching 17 feet (5 m) long. Next came the end-of-Cretaceous extinction event of 66 million years ago (see p.364); once again, sharks and their cousins survived. Hammerhead sharks evolved about 50 million years ago, and the filter-feeding whale, basking and megamouth sharks, as well as the manta ray, emerged between 60 and 30 million years ago. **DG**

359–299 Mya	310 Mya	201–144 Mya	155 Mya	150 Mya	16 Mya
During the Carboniferous adaptive radiation of sharks, many "experimental" forms appear.	*Helicoprion*, a shark with a bizarre spiral tooth whorl, is known from rocks in Russia, the United States, Japan, and Australia.	The "Jurassic explosion" of sharks occurs. It is the second major phase of chondrichthyan adaptive radiation and produces the sixgill shark (see p.194).	*Palaeocarcharias* is the earliest lamniform shark—the "mackerel shark" family to which many living species belong.	Early skates and rays appear, as has been shown by articulated remains of a type of guitarfish ray.	The huge, now extinct *Megalodon* (see p.192) appears. It is a great white-type shark.

Megalodon
LATE MIOCENE EPOCH—EARLY PLEISTOCENE EPOCH

SPECIES
Carcharodon megalodon
GROUP Lamnidae
SIZE 52 feet (16 m)
LOCATION All regions except
the Arctic and Antarctica

B earing teeth as big as a human hand, the giant shark megalodon (mega-tooth), 52 feet (16 m) long, menaced the seas between 16 million and 1½ million years ago, in the middle Miocene and early Pleistocene epochs. Vying in the stakes for the world's largest predator with the living sperm whale and one of its extinct cousins— *Livyatan melvillei* from 12 million years ago—as well as a few mosasaurs (see p.352) and pliosaurs (see p.282), megalodon is certainly the most fearsome. This legendary creature has a strong grip on the public imagination, given that the shark is only known from several dozen vertebrae and a few hundred teeth.

There is some discussion as to whether the prehistoric shark in fact belonged to the genus *Carcharodon* (jagged tooth), along with the great white shark, or possibly another genus, *Carcharocles* (famous shark). The older genus name of *Megalodon* is no longer favored. Megalodon is usually grouped with *Carcharodon,* because its teeth resemble those of the great white shark, but other features connect it more favorably to *Carcharocles.* Despite this, the modern great white shark is used as a model for reconstructing the body and lifestyle of megalodon. The two fearsome creatures were contemporaries, sharing the seas for several million years, but it is unlikely that these two top predators competed directly with one another. The smaller shark, the great white, may have kept to colder waters, while its oversized cousin tyrannized warmer zones. Fossil megalodon teeth have been found in Europe, Africa, North and South America, southern Asia, Indonesia, Australia, New Caledonia, and New Zealand, which indicates that the shark preferred tropical and temperate latitudes. The cooling global conditions in the Pliocene epoch signaled the onset of an ice age, which would not have been favorable for a warm-water animal such as megalodon. Just under 2 million years ago, megalodon's 14-million-year reign of terror came to an end, which allowed great white sharks the opportunity to flourish. **DG**

✪ **NAVIGATOR**

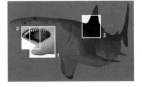

☯ FOCAL POINTS

1 JAW MECHANICS
There were four kinds of teeth in megalodon—all typically sturdy and strong—in a pair of jaws that were more than 6½ feet (2 m) wide. Complete sets of fossil teeth are extremely rare, but specimens found suggest a total of about 270 to 280 teeth ranged across five rows—one row behind the other—in a manner typical of shark jaws.

2 TEETH
The largest megalodon teeth measure 7 inches (18 cm) along one edge and can weigh nearly 1 pound (450 g). The smooth, enamelled surface was riven by vertical cracks and the edges bore about 50 regular serrations per inch. The symmetrical tooth was deep and double-rooted and, in contrast to the blade, the roots were rough to touch.

3 VERTEBRAE
The vertebrae of chondrichthyans consist of two cartilaginous tubes that encircle the notochord (precursor of the backbone) in a continuous sheath: the centra. Megalodon's vertebral centra are the size of saucers, known from partially preserved backbones from Belgium and Denmark. They range from 2 inches (5 cm) to 9 inches (23 cm) across.

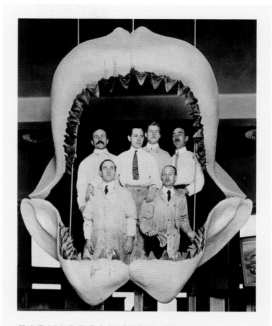

EARLY RECONSTRUCTION

Early attempts at reconstruction were made mainly from teeth; noting the similarity between those of great whites and megalodon, a scaled-up version of the former's jaws was created. Some 11½ feet (3.5 m) high and 7½ feet (2.3 m) wide, reconstructions were popular museum exhibits (above). The discovery of near-complete sets of teeth regulated the size of the jaws and altered the overall look. Work on the morphology of the jaws and skulls of extant sharks showed megalodon to be heavier set than a great white, with a wider head and larger pectoral fins.

◄ Deep gashes on the fossil bones of megalodon's victims reveal that its attack tactics differed for various prey. For medium-sized dolphins and similar fish, the focus was on crushing bones and inflicting devastating damage to vital organs. However, when hunting great whales, megalodon appears to have tried to rip up and bite off the flippers, immobilizing the prey before moving in to feed.

Sixgill Shark
MIOCENE EPOCH—PRESENT

SPECIES *Hexanchus griseus*
GROUP Hexanchidae
SIZE 18 feet (5.5 m)
LOCATION Atlantic, Pacific,
and Indian Oceans
IUCN Near Threatened

L iving in the dark and cold of the deep sea, and mostly keeping out of sight of humans, Hexanchidae is a family of secretive sharks, collectively called cow sharks. While most sharks content themselves with five pairs of gill slits, the cow sharks have six, and two species of the family have seven. There are just four living species with six gill pairs: the bluntnose sixgill, *Hexanchus griseus*, the big-eyed sixgill, *H. nakamurai*, the sixgill sawnose, *Pliotrema warreni*, and the frilled shark, *Chlamydoselachus anguineus*, which is sometimes called a living fossil and awarded its own family, Chlamydoselachidae.

These furtive animals are exclusively marine and mostly inhabit the cold depths greater than 300 feet (90 m) off the continental shelves of the Atlantic, Pacific, and Indian oceans. The bluntnose sixgill shark *H. griseus* has been recorded at depths of 6,000 feet (1,830 m) and swims vertically at night to feed on squid and crustaceans in the abundant surface waters, as shallow as 50 feet (15 m), before disappearing into the deep as the sun rises. It is a stout shark with a rounded, blunt snout, and it can grow up to 18 feet (5.5 m) long. *H. griseus* is the biggest of the hexanchoids—considered to be the last living representatives from the golden age of sharks when many different families existed. Most sharks that swim in the oceans today are relatively modern, but the sixgills have remained largely unchanged for approximately 190 million years. They offer a glimpse of what sharks looked like in the past, back in the time of the "Jurassic explosion," which was the last great burst of adaptive radiation in shark forms. In addition to the extra gill slit pair, other ancient features include the presence of an anal fin and the dorsal fin set far back near the tail. The pectoral fins each have a denticle-free "bald patch" behind them where they meet the body. **DG**

✪ NAVIGATOR

1 TEETH
H. griseus has distinctive teeth. Those in the upper jaw slant toward the back of the mouth and have smaller projections behind, giving them a thornlike look. The bottom rows are made up of flattened saw-teeth. Sixgills are opportunistic feeders and their stomach contents testify to their wide variety of prey animals.

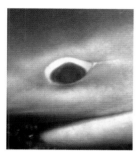

2 EYES
The light-sensitive retina contains only rod cells. These are most sensitive to short-wavelength light, which is the deep-blue part of the spectrum and has much better transmission through water compared to long-wavelength reds. This is an adaption for maximal sensitivity in the gloom of the deep ocean.

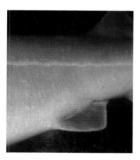

3 REPRODUCTIVE PARTS
The reproductive organs of sharks have not changed significantly for around 400 million years. The cloaca is an opening for the shark's reproductive, digestive, and excretory systems, through which eggs or young pass. Sixgills are ovoviviparous (see panel) and produce young with high viability.

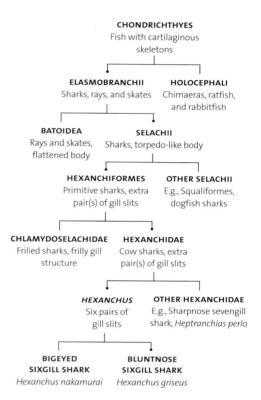

CHONDRICHTHYES
Fish with cartilaginous skeletons

ELASMOBRANCHII
Sharks, rays, and skates

HOLOCEPHALI
Chimaeras, ratfish, and rabbitfish

BATOIDEA
Rays and skates, flattened body

SELACHII
Sharks, torpedo-like body

HEXANCHIFORMES
Primitive sharks, extra pair(s) of gill slits

OTHER SELACHII
E.g., Squaliformes, dogfish sharks

CHLAMYDOSELACHIDAE
Frilled sharks, frilly gill structure

HEXANCHIDAE
Cow sharks, extra pair(s) of gill slits

HEXANCHUS
Six pairs of gill slits

OTHER HEXANCHIDAE
E.g., Sharpnose sevengill shark, *Heptranchias perlo*

BIGEYED SIXGILL SHARK
Hexanchus nakamurai

BLUNTNOSE SIXGILL SHARK
Hexanchus griseus

▲ The various cartilaginous fish groups, Chondrichthyes, are on separate evolutionary lines to the bony fish, Osteichthyes. Sixgill and sevengill sharks are regarded as more primitive, or earlier evolved, compared to most of the other sharks.

REPRODUCTION STRATEGIES

Sixgill sharks are the most productive of all shark species and regularly give birth to 100 pups at a time. Like many shark species, they are ovoviviparous, which means thin-shelled eggs are fertilized inside the female's body before hatching internally, so that she gives birth to live young (right). After hatching, but before birth, the well-formed, active pups eat unfertilized eggs, and even one another. This phenomenon is known as oviphagy and is a type of cannibalism. It is a very early form of intraspecies natural selection. Other sharks have adapted different strategies. Some are oviparous, hatching from laid eggs, and others are viviparous, developing inside the female and born fully formed.

LOBE-FINNED FISH

The Sarcopterygii (lobe-finned fish) are one of the two lineages of Osteichthyes (bony fish). The other—Actinopterygii (ray-finned fish)—is the planet's most diverse group of vertebrates (see p.202). With only a few modern representatives, the lobe-fins could be seen as remnants of a former age, clinging onto existence. However, this group of stocky fish gave rise to the first land-going vertebrates. The defining feature of the lobe-finned fish is their fins, which protrude from the body on fleshy stalks. An internal bony skeleton strengthens these appendages and they are articulated (form a joint) with a single bone. Each of the pectoral fins, the equivalent to arms, is joined to the body by the humerus (upper arm bone), which rotates in a socket in the pectoral (shoulder) girdle. Each of the pelvic fin pair, the equivalent to legs, has a femur (thigh bone) that joins to the pelvic (hip) girdle. This exact configuration is still maintained in the arms and legs of living mammals. Other characteristics of lobe-fins include teeth fully covered by enamel and a covering of "true cosmoid" scales (toughened bony scales covered in cosmine, a dentinelike tissue) that encase the fish in bony suits of armor. The tails of early lobe-fins were strongly heterocercal (with a larger upper lobe), to provide lift in the water, but modern representatives of the group have developed symmetrical tail fins.

Three groups of lobe-fins survive today. These are the coelacanths (Actinistia; see p.200), the lungfish (Dipnoi), and the tetrapods (four-limbed vertebrates) that include amphibians, reptiles, birds, and mammals. Living coelacanths caused a sensation when they were discovered off the east coast of Africa in 1938, because they were thought to have long vanished from the oceans. The six extant species of lungfish all live in fresh water and have

1 There are presently four species of African lungfish, genus *Protopterus*. The genus has remained relatively unchanged for around 100 million years.

2 *Scaumenacia curta* was alive 360 million years ago. Its long back fins formed a kind of fan, enabling it to accelerate at considerable speed.

3 *Rhizodus* was an enormous, predatory freshwater lobe-fin. Its lengthy, fanglike teeth suggest that it preyed on other fish and amphibians.

KEY EVENTS

420–395 Mya	400 Mya	400 Mya	390 Mya	386 Mya	385 Mya
Lungfish, probably the closest living relatives of early tetrapods, appear. The earliest lungfish fossils are *Diabolepis*.	The first coelacanths appear. They enjoy appreciable diversity during the Devonian and Carboniferous.	Rhipidistians, the sister group to the coelacanths, appear at around the same time.	Porolepiformes, a somewhat primitive group of rhipidistians, begin to diversify and spread into more varied habitats.	The first massive rhizodont predators, such as *Rhizodus*, appear. They eventually become extinct 86 million years later.	*Eusthenopteron* emerges, an osteolepid lobe-fin that has many basic tetrapod features.

air-breathing lungs. The Australian lungfish survive in brackish, low-oxygen waters by gulping air, while the African (see image 1) and South American lungfish bury themselves in mud to shelter through the dry season.

Bony fish evolved in the Devonian period (most probably from placoderms; see p.184) and the ray-finned and lobe-finned subgroups are thought to have diverged shortly after. The oldest bony fish known comes from Yunnan province, China: *Guiyu oneiros* (dream ghost fish) is 419 million years old and dates from the late Silurian period. By the Devonian, a period informally known as the Age of Fishes, the lobe-fins were thriving. They reached their peak diversity in the warm seas and were among the top marine predators. Soon after 400 million years ago, during the early to middle Devonian, the lobe-fins divided into two lineages that followed separate paths, evolving from a common ancestor that lived in the mouths of rivers and estuaries. The rhipidistians remained close to the shore and moved into freshwater environments, and this was the group that eventually gave rise to the lungfish, such as *Scaumenacia curta* (see image 2), which possessed both lungs and gills and the tetrapods. Meanwhile, the coelacanths descended to the security of deep marine waters.

In addition to the lungfish, which have the longest fossil record of the lobe-fins, the rhipidistian pretetrapods colonized brackish freshwater habitats and began to diversify. They included porolepiforms, long powerful fish from the middle Devonian, which may have been ambush predators, lurking in the weeds and catching passing victims with an explosive burst of speed, in the way that modern pikes hunt today. Another group, the Rhizodontiformes, were the largest of the Devonian lobe-fin predators, with the monster freshwater *Rhizodus* (see image 3) reaching 23 feet (7 m) long. Also among these groups were the Osteolepiformes—ancestors of the tetrapods that arose in the late Devonian—although they are poorly represented in the fossil record and have affinities that are difficult to pin down. Out of this modest diversity, only the lungfish and the coelacanths survived to the mass extinction event at the end of the Permian period known as the "Great Dying" (see p.224).

The evolution of the adaptations in fish that paved the way for fully land-going tetrapods is a complicated story. This branch of paleoichthyology (the study of fossil fish) reveals humanity's own deep origins, therefore scientists are eager to discover patterns of changes and to identify the key structures that made it possible. At least four separate lineages of rhipidistian lobe-fins (lungfish, rhizodonts, tristichopterids, and elpistostegalians) had tetrapod features, seemingly shared at random between them. Since these structures evolved to serve more or less the same purpose—to negotiate the challenges of living in shallow,

380 Mya	380 Mya	375 Mya	365 Mya	252 Mya	80 Mya
Tetrapodomorphs (four-limbed shapes) evolve, such as the osteolepid lobe-fin *Osteolepis*.	*Panderichthys* appears, an elpistostegalid lobe-fin fish with a large, tetrapod-like head. Such fish were closer to tetrapods than osteolepid fish.	Transitional forms of fish between lobe-fins and tetrapods are active, such as *Tiktaalik* (see p.210), with articulating wrists and "spiracle" nostrils.	The very early tetrapods appear. *Acanthostega*—one of the first limbed vertebrates—is still fully aquatic at this stage.	A massive extinction event at the end of the Permian causes all lobe-fin lineages to die out, except lungfish and coelacanths.	Fossils from this time provide the last evidence of coelacanths before their rediscovery in the 20th century.

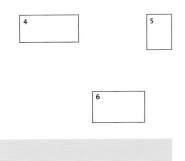

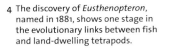

4 The discovery of *Eusthenopteron*, named in 1881, shows one stage in the evolutionary links between fish and land-dwelling tetrapods.

5 Mudskippers have adapted to survive for periods out of water. Their front fins work as "legs" but have a different skeletal pattern to tetrapods.

6 The stout pectoral and pelvic fins of the roughback batfish, *Ogcocephalus parvus*, enable it to raise its body and "walk" along solid surfaces.

possibly vegetation-choked environments that were low in oxygen—they show a high level of convergent evolution. As new fossils come to light and older fossils are revisited and reinterpreted the details of this period are constantly being revised.

One of the most celebrated of the lobe-finned protagonists of the late Devonian is *Eusthenopteron* (see image 4; see p.206). It is known from more than 2,000 fossils, many of them complete, from the Miguasha Cliffs in Canada, and usually placed in the group called tristichopterids. *Eusthenopteron* lived 385 million years ago in the late Devonian and its state of preservation is so exceptional that scientists—principally Swedish paleontologist Erik Jarvik (1907–98), who made the study of this fish his life's work—have been able to map its full skull morphology, including details such as its blood vessels, nerve placements, attachments for muscles, and shape of the brain. One of Jarvik's most important discoveries was that *Eusthenopteron* had nostrils that connected to the interior of the palate. By contrast, most fish have closed loops through which they circulate water with the aid of microhair-like cilia lining their interior surface. Primitive tetrapods, however, have orifices very similar to those in *Eusthenopteron*, called choanae. While the osteolepids maintained the typical flexible "intercranial joint" of the lobe-fins, the bones of *Eusthenopteron* covered its brain, eyes, snout, and nostrils and are a close match for those of the early tetrapods. However, the fin bones were the most eye-opening features. *Eusthenopteron* had a clear humerus, ulna, and radius (arm bones) in its pectoral fin and a femur, tibia, and fibula (leg bones) in its pelvic fin. This pattern of bones is identical to that found in modern tetrapods, including humans. These features—the choana, skull roof, and limb bone configuration—strengthen the link between lobe-fin fish and terrestrial vertebrates. *Eusthenopteron*'s importance in the history of vertebrate evolution is reflected in its nickname, the "Prince of Miguasha."

It is unclear whether, among the lobe-fins, the air-breathing lungs of lungfish and the fat-filled lungs of coelacanths evolved independently. What seems incontestable is that their common bony fish ancestor must have had a primitive gas pouch. In the ray-finned lineage, this feature developed into the gas-filled swim bladder that was an important feature of their group's overwhelming success in nearly all aquatic environments. Air-breathing lungs are necessary to life on land, but they can be regarded as an exaptation—an adaptation that evolves for another reason, then becomes co-opted for a different purpose—rather than appearing in direct response to environmental pressure. The lungs of modern lungfish, like those of humans, are toward the front of the body. This adaptation allows the fish to raise its nose above water to draw in air with minimal effort. Rather than abandoning their existing environments to seek out new habitats in a warming world and eventually leaving the water altogether—a scenario known as the "Drying Pond" hypothesis—fish probably evolved these structures to survive in shallow swampy wetlands.

There is wide variation in the design and arrangement of limblike fish fins, representing various answers to the questions posed by living in shallow water. It is likely that the problem has been solved several times. Living examples of the ray-fins include the mudskippers (see image 5) and certain kinds of batfish (see image 6). The mudskipper's pectoral fin has the usual small bones inside the body at the base of the fin, called the radials, which are found in all ray-finned fish. However, these are larger and longer and project out into the fin. At their ends they join with the fin rays (lepidotrichia)—stiff strengthening elements of the flexible fanlike fin tissue. In effect, the joints between the main skeleton and the radials work like a shoulder, and the joints between the radials and the rays equate to an elbow (or perhaps a wrist). These arrangements are entirely different to those of lobe-finned fish.

A persistent complication in the lobe-fins is the lack of fingers. Lobe-finned fish also have lepidotrichia—long, bony, fin spines at the ends of their "limbs"—while tetrapods have fingers and toes. Although most paleontologists accept the concept of adaptations to a life in the shallows as being preadaptations for life on land, the question of digits is complex. While walking underwater almost certainly predates walking on land, the origin of complex nerve wiring to coordinate digits is uncertain. However, fish with muscular fins, supported by a strong internal architecture, could thrive in brackish river deltas, coastal lagoons, and stagnant swamps choked with leaves and fallen vegetation, using their powerful fins to negotiate weed beds and fallen tree trunks, and to move along the rocky riverbed. Also, the ability to breathe air and retain a reservoir of it inside the body would have been an advantage in these habitats. Exploiting the productive shallows was probably the driving force behind the evolution of the tetrapods.

From the peak of their diversity in the Devonian and, to a lesser extent, the Carboniferous period, the lobe-fins went into a steady decline. They were hit by the end-of-Devonian mass extinction event 359 million years ago, and suffered terrible losses 252 million years ago, during the extinction event at the end of the Permian. However, as the lobe-fins declined in numbers and diversity, a new chapter was beginning for their descendants. Life on four legs had taken its first tentative steps. **DG**

Coelacanth
LATE OLIGOCENE EPOCH—PRESENT

SPECIES *Latimeria chalumnae*
GROUP Latimeriidae
SIZE 6½ feet (2 m)
LOCATION West Indian Ocean
IUCN Critically Endangered

O n December 23, 1938, Captain Hendrik Goosen (1896–1951) of the trawler *Nerine* caught a fish he had never seen before. He showed it to museum curator Marjorie Courtenay-Latimer (1907–2004) in Eastern Cape province, who declared that this fish, a coelacanth, had been assumed long extinct. The conclusion was confirmed by J. L. B. Smith (1897–1968), a university professor, who named it *Latimeria chalumnae* after Marjorie and the river near its capture. The fish from the depths of the Indian Ocean, 6½ feet (2 m) long and weighing 175 pounds (80 kg), was from a group not seen for more than 75 million years. This rare genus now contains two species: the West Indian Ocean coelacanth has been joined by the Indonesian coelacanth, *L. menadoensis*. The coelacanth, meaning "hollow spine," looks prehistoric. It is a resident of the semi-darkness, living 330 to 660 feet (100–200 m) below the surface, so its eyes are extremely sensitive to light. The heavy-set fish has thick scales and fleshy fins with internal bones. These lobe fins look ungainly, but in reality are flexible: coelacanths can swim backward with elegance, flip over and swim belly up. Other primitive features include a tail fin divided into three lobes, an oil-filled notochord—which, unlike most other vertebrates, is not replaced with a bony spine as the embryo develops—and a much-reduced opercular bone that attaches a flap of soft tissue to cover the gill. It also has a single-lobed vestigial lung that, filled with fat, serves the same function as a swim bladder. Coelacanths feed on fish, squid, and cuttlefish using spiky teeth clustered in the front of the mouth. Their fin movement is very interesting—the front fin and rear fin on opposite sides of the body move together, in the same way a lizard or a cat moves its legs. Early reports claimed that the coelacanth was a distant human ancestor. However, it was not on the direct vertebrate line of evolution but an associated lineage. **DG**

✦ NAVIGATOR

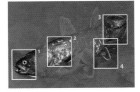

FOCAL POINTS

1 ROSTRAL ORGAN

A large sensory organ, known as the rostral organ, is in the snout. This gel-filled cavity has three pairs of canals that open to the outside of the fish—two in front of the eyes and the third on the tip of the nose. The organ is thought to form part of an electrosensory system that helps the coelacanth orient itself in space, and to locate prey in its dimly lit world.

2 COSMOID SCALES

Coelacanths have heavy scales that are similar to those found on fossils of ancient lobe-finned fish. These are known as modified cosmoid scales and they are composed of a core of dense bone surrounded by a layer of spongy bone, with an upper surface covered by the protein keratin, which forms hair, feathers, hooves, claws, and horns in other animals.

3 PAIRED LOBE FINS

The fleshy limblike fins are supported internally by a bony skeleton, which closely matches the pattern of articulating bones in the arms and legs of early tetrapods. In the water, coelacanths use their paired fins only to stabilize themselves. However, their pattern of movement is synchronized and similar to that of tetrapods.

COELACANTH RELATIONSHIPS

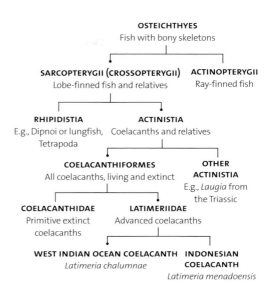

OSTEICHTHYES
Fish with bony skeletons

SARCOPTERYGII (CROSSOPTERYGII)
Lobe-finned fish and relatives

ACTINOPTERYGII
Ray-finned fish

RHIPIDISTIA
E.g., Dipnoi or lungfish, Tetrapoda

ACTINISTIA
Coelacanths and relatives

COELACANTHIFORMES
All coelacanths, living and extinct

OTHER ACTINISTIA
E.g., *Laugia* from the Triassic

COELACANTHIDAE
Primitive extinct coelacanths

LATIMERIIDAE
Advanced coelacanths

WEST INDIAN OCEAN COELACANTH
Latimeria chalumnae

INDONESIAN COELACANTH
Latimeria menadoensis

▲ The two living species of coelacanths are remnants of a formerly successful and widespread group with many dozens of extinct species. Their major group, Sarcopterygii, includes the tetrapods such as amphibians, reptiles, birds, and mammals.

4 NOTOCHORD

The notochord is a hollow, fluid-filled tube that extends the length of the body underlying the spinal cord. At depth, the fluid is under pressure, which keeps the notochord stiff, so the fish has no need for the bony spine of other vertebrates. The tail fin is divided into three lobes with an extension of the notochord in the central lobe.

THE HUNT FOR A LIVE SPECIMEN

The discovery of the coelacanth in 1938 was a scientific sensation. However, the first specimen was dead on arrival, and therefore stuffed for preservation. The hunt began for a specimen with innards intact that would provide clues to the origin of vertebrates. Despite an intensive search, the second coelacanth (right) was not found until 1952, off the Comoros Islands between Madagascar and southern Africa. The Indonesian species was discovered in 1997. It is likely to have separated from the West Indian Ocean species 15 million to 30 million years ago. More than 350 million years ago, coelacanths are likely to have split from the evolutionary lineage that led to lungfish, tetrapods, and then land vertebrates.

RAY-FINNED FISH

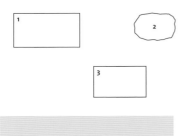

1 Gars such as the longnose gar, *Lepisosteus osseus*, are relics of the primitive but now small ray-finned fish group, Holostei.

2 *Pholidophorus* was a long-lived genus of fish that linked the holosteans and teleosts; most species were about 18 inches (45 cm) in length. Its homocercal tail is clearly visible.

3 Since its naming in 1889, many specimens of the Jurassic giant *Leedsichthys* have been found. They indicate that this wide-ranging bony fish rivaled the biggest cartilaginous fish, the shark megalodon (see p.192), in size.

Ray-finned fish, Actinopterygii, are by far the most numerous fish today. They are named after their fanlike "rayed fins"—webs of skin supported by bony spines or rays (lepidotrichia) protruding from the body. One of the earliest known is *Andreolepis*, from 420 million years ago in the Silurian period. This fish group appeared at the same time as the lobe-finned fish (Sarcopterygii). By the end of the Triassic period, 201 million years ago, ray-finned fish were dominating the world's oceans and freshwater systems, and they continue to do so today. In fact, the 30,000-plus known species make up more than half of the world's vertebrates.

Traditionally, the ray-fins include three main groups—Chondrostei, Holostei, and Teleostei—each representing one of the three basic stages or evolutionary branches of the group. Chondrosteans appeared in the early Devonian period and had peaked by the late Permian period to the early Triassic, 250 million years ago. The earliest were mostly small fish covered in thick scales, with large eyes and odd-shaped jaws (the upper jaw was fused to the cheekbone). These fish spread into most habitats and evolved into many types, from bottom feeders to reef fish. Almost all of them declined after their Permian peak and became extinct around 100 million years ago, during the Cretaceous period. The few survivors are represented by two groups: Acipenseriformes, comprising various marine and freshwater sturgeons and the plankton-feeding paddlefish of China and North America; and Polypteriformes, the African freshwater bichirs and reedfish.

The ray-fins called holosteans represent an evolutionary stage between the chondrosteans and modern teleosts (bony fish): they lacked thick body scales and their tail developed a homocercal shape (with equal-sized upper and lower

KEY EVENTS

420 Mya	260–240 Mya	250 Mya	250 Mya	240 Mya	160 Mya
Early ray-finned fish include *Andreolepis*, which was 6 to 8 inches (15–20 cm) long and lived in the Baltic region.	The sturgeon group of ray-fins, known as chondrosteans, reaches a peak in diversity during the late Permian and early Triassic.	The early chondrostean ray-fin *Perleidus* is 7 inches (17 cm) long. It is a long-lasting genus that will persist for more than 30 million years.	The Halecomorphi group of holostean ray-fins rapidly evolves and diversifies. Only one species survives today: the bowfin.	The early teleost ray-fin *Pholidophorus* develops the homocercal tail that becomes a feature of the group.	*Daitingichthys* is an early member of the teleost ray-fin herrings. The group will thrive 60 million years later in the middle Cretaceous.

lobes). These adaptations made the holosteans stronger swimmers than their chondrostean relatives. In addition, the holosteans' upper jaw was not fused to the cheekbone, thus enabling greater movement of the jaw for more efficient respiration and feeding. The holosteans reached their peak during the Jurassic and Cretaceous. The few living species of the group include the bowfin, *Amia calva*, and seven species of *Lepisosteus* (see image 1) and *Atractosteus* gars, all from North and Central America and the Caribbean region.

The majority of modern ray-fins are teleosts, which first appeared around 250 million years ago, probably from an extinct group of holosteans, Pholidophoriformes. They showed further development of the homocercal tail fin for even better swimming efficiency. *Pholidophorus* (see image 2), from around 240 million to 140 million years ago, was a fast-swimming predator with large eyes, although it retained primitive features such as a partially cartilaginous skeleton.

Teleost ray-fins rapidly became the dominant fish group, ranging from the giant *Xiphactinus* (sword ray, see p.204) of the late Cretaceous, to *Amphistium*, which lived 50 million years ago and was one of the earliest relatives of modern flatfish. Today, teleosts comprise more than 400 families and represent more than 96 percent of fish species. They live in almost every aquatic habitat, from mountain streams to tropical oceans, icy polar waters, and deep ocean trenches; new species are discovered every year. The biggest living ray-fin is a teleost: the ocean sunfish, *Mola mola*, attains 14 feet (4.3 m) in height, from the top of the dorsal (back) fin to the base of the lowest fin, and 10 feet (3 m) in length. Even larger was the middle Jurassic ray-fin *Leedsichthys* (see image 3), now extinct. At 52 feet (16 m), it was among the largest ever fish and like the whale shark, was probably a filter feeder that sifted plankton and krill from the water. **LG**

150 Mya	140 Mya	87–66 Mya	60 Mya	50 Mya	50 Mya
The elopomorph group of teleost ray-fins becomes established and eventually evolves into such varied kinds as tarpon, ladyfish, bonefish, and eels.	The teleost ray-fins known as osteoglossids (bony tongues) appear. They will eventually give rise to very large freshwater fish, such as the living arapaima.	The giant predatory teleost fish *Xiphactinus* appears during the late Cretaceous but dies out after the end-of-Cretaceous extinction event (see p.364).	Early members of the perciform (perchlike) teleosts give rise to by far the largest group of ray-fins, including two-fifths of all living fish species.	In North America, *Eosalmo driftwoodensis* is an early member of the salmon family (Salmonidae) that will also give rise to trout, chars, and graylings.	*Amphistium* is one of the earliest relatives of modern flatfish, pleuronectiforms, such as flounders, plaice, turbots, and soles.

Xiphactinus
LATE CRETACEOUS PERIOD

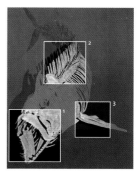
Fossils of this huge, predatory, marine fish were first found in Kansas in the 1850s, among remains of mosasaurs (see p.352), sharks (see p.190), and other creatures that existed in the Western Interior Seaway at the time. Since then, more fossil remains have been found in the United States and also in Australia, Europe, Venezuela, and Canada. Given the wide distribution of fossils, *Xiphactinus* (sword ray) is likely to have thrived throughout the world's oceans during the late Cretaceous, 87 million to 66 million years ago. It died out in the end-of-Cretaceous mass extinction. *Xiphactinus* measured up to 20 feet (6 m) in length and its upturned jaws were more than a foot (30 cm) wide in smaller specimens, but probably far wider in the largest individuals. Fossils of a possible youngster indicate a body length of only 1 foot (30 cm). It was a strong swimmer with a powerful muscular tail to propel it along at a top speed of 40 miles per hour (65 km/h). Many fossils found have had the remains of partially digested or undigested prey inside them (see panel). An *X. audax* fossil uncovered in Canada in 2010 had the flipper of a mosasaur inside its jaws. *Xiphactinus* is likely to have eaten anything it could catch, including seabirds, such as *Hesperornis*, which floated on the ocean. In turn, *Xiphactinus* was prey to Cretaceous sharks, such as *Cretoxyrhina* and *Squalicorax*, although these scavengers would likely have targeted dead or injured individuals. **LG**

SPECIES *Xiphactinus audax*
GROUP Ichthyodectidae
SIZE 20 feet (6 m)
LOCATION All oceans

👁 FOCAL POINTS

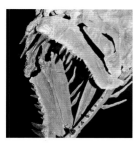

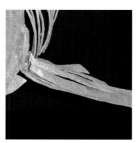

1 JAWS
The mobile jaws opened extremely wide to swallow prey whole, and were lined with teeth up to 2 inches (5 cm) long—bigger at the front. This fish may have used a bite-and-retreat tactic, weakening prey before killing. The fanglike teeth could puncture the body of larger prey to prevent escape.

2 VERTEBRAE
The backbone of *X. audax* consisted of more than 100 vertebrae. Each was shaped like a flattened oval, measuring about 1 inch (2.5 cm) wide and 1½ inches (4 cm) long. Fine rib bones extended above and below the backbone to define the long body shape of the full-grown adult.

3 PECTORAL FIN
Solid bony fin rays gave the pectoral fins a winglike appearance, while the large forked tail fin indicates that *X. audax* was a strong swimmer. Speed estimates of 40 miles per hour (65 km/h) have been made by comparing *X. audax* to modern fish, such as marlin and tuna.

A FISH IN A FISH

In 1952, US paleontologist George F. Sternberg (1883–1969), of the Sternberg fossil-hunting family, uncovered the remains of a *Xiphactinus*, 13 feet (4 m) long, in Gove County, Kansas. On closer inspection, Sternberg noticed that the fossil contained the perfectly preserved remains of another huge fish inside its stomach (above). He identified the swallowed prey as a *Gillicus arcuatus* and it was 6 feet (1.8 m) long. *Gillicus* was itself a predator and a member of the same fish family as *Xiphactinus*: Ichthyodectidae. Sternberg realized that *Xiphactinus* died shortly after eating its meal because its skeleton was so well preserved. He suggested that the *Gillicus* struggled as it was eaten and possibly ruptured one of its predator's internal organs. This fossil is now in the Sternberg Museum of Natural History in Hays, Kansas, named in honor of George Sternberg.

FROM FINS TO LIMBS

F ew images sum up ideas about evolution more than a bold prehistoric fish slowly crawling out of the shallows on weak limblike fins and struggling to gulp in oxygen from the air. In fact, this romantic vision of evolution is not too far from reality. During the Devonian period, starting some 400 million years ago, various prehistoric fish groups started the first wave of a land invasion that would ultimately lead to amphibians, dinosaurs, birds, and mammals.

The word "tetrapod" means "four feet" and refers to any vertebrate that has four legs, limbs, feet, or similar appendages, either living or extinct. It includes extinct marine reptiles, such as the ichthyosaurs (see p.244), which used their limbs as paddles to swim. It includes humans, even though they only walk on two legs. It even includes snakes, which have no limbs, since they evolved from reptiles that did. All of these examples are included in the tetrapod group because they descended from the common ancestor of amphibians, reptiles, birds, and mammals—the brave prehistoric fish described above. Experts believe that the most likely candidate came from a group of lobe-finned fish (sarcopterygians). Today, the only lobe-finned fish alive are lungfish (see image 1) and the critically endangered coelacanths (see p.200), which were thought to be extinct until a live specimen was found off the coast of South Africa in 1938.

Lobe-finned fish, such as *Eusthenopteron*, differ from their ray-finned (actinopterygian) counterparts—the most common types of fish found today—because their pectoral and pelvic fins, arranged in pairs, are supported by bones inside the body. This arrangement is necessary for fins to evolve into primitive limbs. It is likely that the main anatomical transition from fins to limbs was completed in an aquatic environment, where the limbs did not have to

KEY EVENTS

416–395 Mya	397 Mya	385 Mya	380 Mya	375 Mya	374–360 Mya
Lobe-finned fish split into the coelacanths and rhipidistians (lungfish)—the latter migrated from the oceans into shallow, freshwater habitats.	Tetrapod tracks laid down in Polish marine tidal flats indicate that the fins-to-limbs transition took place before this date.	*Eusthenopteron* occurs, one of the earliest lobe-finned fish to show tetrapod traits.	*Panderichthys* exhibits transitional features between lobe-finned fish and early tetrapods, such as distinct radial bones in the fins.	*Tiktaalik* represents the transition from fish to the earliest terrestrial tetrapods, which later develop into the amphibians.	The primitive tetrapod, *Ichthyostega*, combines a fishlike tail and gills with limbs, which it may have used to crawl along swamp bottoms in search of food.

support the entire weight of the body due to the buoyancy of the water. First, the primitive fin-limbs probably developed flexible joints, helping them to bend and push the fish forward as it crawled along in the shallows or negotiated thick waterweeds. The fin-limbs grew stronger as they developed simple knee, ankle, elbow, and wrist joints. Eventually, the primitive fish-tetrapod would have been able to crawl out of the water, where the lack of buoyancy would drive the evolution of limbs to support the weight of the animal's body. In turn, this would favor a stronger internal skeleton.

One such skeletal adaptation was the development of a rigid spinal column (backbone), with sliding and overlapping processes to link adjacent vertebrae, while providing the necessary overall flexibility. In addition to securing support for the body, the hind limbs became the actual driving force for tetrapod locomotion. This required a strong pelvic (hip) girdle to anchor these rear limbs to the backbone. The fore limbs became the steering force, and so the pectoral (shoulder) girdle—which attaches to the skull in fish—became detached in tetrapods, providing them with a flexible neck. Muscular changes accompanied these skeletal adaptations, promoting strong limb function and linking the pelvic and pectoral girdles to the spinal column.

Another requirement for the evolution of tetrapods, together with legs that work on land, is the ability to breathe air on land. Primitive air sacs existed in ray-finned and lobe-finned fish long before the first tetrapods appeared. In ray-finned fish, these air sacs were used as swim bladders to maintain buoyancy. In lobe-finned fish of the Devonian, however, the air sacs had already evolved into air-breathing lungs. Lobe-finned fish, such as living lungfish (see image 2),

1 The fossil lungfish, *Fleurantia*, from the late Devonian, had a very tall dorsal (back) fin, and marked pectoral and pelvic "prelimb" fins, which are gray in this image.

2 The Queensland or Australian lungfish, *Neoceratodus forsteri*, is the most primitive of the living lungfish. Its close relatives date back to more than 100 million years ago. Its lung is a long, thin organ that branches from the gut, so the fish effectively swallows in air and belches it out.

365 Mya	360 Mya	360–345 Mya	350 Mya	350 Mya	310–300 Mya
Early tetrapod *Acanthostega* has limbs with eight digits, as well as teeth that suggest feeding habits are partially terrestrial.	A primitive tetrapod, *Hynerpeton*, possible ancestor or cousin of amphibians, evolves from lobe-finned fish such as *Hyneria*.	The time of Romer's Gap. The lack of fossil evidence later makes it difficult to chart the tetrapod development.	*Pederpes* exists at this time. It provides important clues in the early evolution of terrestrial tetrapods.	*Crassigyrinus* represents a dead end in the evolution of terrestrial tetrapods, with well-developed hind limbs, but vestigial fore limbs.	The first true amphibians conquer the land, represented by genera such as *Eogyrinus* and *Eryops*.

breathe by gulping air into their mouths at the surface of the water. When they subsequently dive, the increasing water pressure bearing onto them forces the air in the mouth down into the lungs. The process reverses as the fish rises to the surface again, forcing stale air out of the lungs and mouth as the fish gulps in a fresh mouthful. Since this basic mechanism for respiration already existed in Devonian lobe-fins, the earliest tetrapods were already well on their way to the evolutionary stage of breathing on land.

The transition from fins to limbs was only one part of the evolutionary journey for terrestrial tetrapods. Many other anatomical changes must have taken place to help early tetrapods cope with life on the land. For example, the evolutionary change that resulted in a flexible neck increased the mobility of the head, helping them pick up food instead of grabbing items that passed by in the water. Adaptations to the skin helped conserve water and prevent dehydration, as well as providing the necessary protection against friction caused by movement across the ground. The sense organs needed to adapt to the new environment, too. The lateral line system that helps fish detect vibrations in the water would have given way to improved vision, hearing, and smell. Many of these changes are evident in the tetrapod fossil record, such as forward-facing eyes, the appearance of the nasal chamber, and small ear bones.

No one knows exactly which evolutionary forces prompted the transition from fins to limbs. The first tetrapods were probably still fully aquatic animals that lived in shallow waters. Some experts think they may have used their primitive limbs to stalk prey through the dense vegetation that lined the bottom of rivers and swamps. Others believe that environmental pressures, such as an increasingly dry climate, prompted the transition from water to land. As the heat dried up the shallow waters, primitive limbs may have helped the earliest tetrapods crawl between pools of water. Yet another hypothesis is that they were chased out of the water by bigger predatory fish. Any tetrapod that could survive out of water would have found itself in a new paradise, with an abundance of plant and insect foods and a notable absence of dangerous predators.

Although knowledge of tetrapod evolution is constantly changing, experts have identified several key species in the fossil record. *Eusthenopteron* and *Panderichthys* (see image 3) lie closer to the fish end of the evolutionary spectrum, with heavy fusiform (spindle-shaped bodies), pointed heads, and fins that included primitive limblike bones. One of the key transitional species was a tetrapodlike, lobe-finned fish called *Tiktaalik* (see p.210), which was discovered in Canada in 2004 and dates back to the late Devonian around 375 million years ago. This prehistoric creature had characteristics of both fish (gills and scales) and tetrapods (flexible neck and primitive wrist bones) and is known as a fishapod.

The transformation from fins to limbs took place over millions of years, and tetrapods evolved many times and in many different places. Fossil evidence

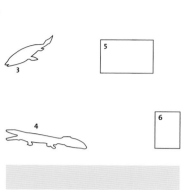

suggests that the process started in the late Devonian and continued well into the late Carboniferous period, a span of more than 50 million years. Each small step in the evolution undoubtedly helped the early fishapods move from an aquatic to a terrestrial lifestyle, but the process was not as continuous as it seemed. For example, many early tetrapods had seven or eight digits—fingers or toes—at the end of each limb. They represent dead-end lineages in the evolution of modern vertebrates, which are based on the pattern of five toes. It is therefore not yet possible to identify the direct ancestor of the modern vertebrate groups, only to study the representative fossils that have been found in order to piece together this complex evolutionary puzzle.

What makes the transition even more difficult is the gap in the fossil record during this important early stage of evolution. This period, known as Romer's Gap, spans between 360 million and 345 million years ago, from the late Devonian to early Carboniferous. Fossils dating before Romer's Gap show creatures with a mixture of fish and tetrapod features, whereas fossils after this time show a diversity of tetrapod species, with some having the characteristics of true amphibians. Recently, the gap has started to become populated with new fossil discoveries, especially *Pederpes* (see p.212), which was reclassified by Jennifer Clack (b. 1947) in 2002. Clack, an expert in tetrapod evolution, gained acclaim in her field in 1987 when, during an expedition to Greenland, she found an almost complete and well-preserved fossil of the primitive aquatic tetrapod *Acanthostega* (see image 4). This creature lived about 365 million years ago, 10 million years before *Pederpes*. It had a similar combination of fish and tetrapod characteristics, such as well-developed limb bones, but retained the lateral line sense organs running along the side of its body. It was more than 4 feet (1.2 m) long and was one of the evolutionary lines that had eight fingers on each front foot (the rear foot is not clear enough to determine the number).

At broadly the same time as *Acanthostega*, another tetrapod genus, *Ichthyostega*, lived also in Greenland, and was officially named in 1932 (see images 5 and 6). This was a much larger creature—5 feet (1.5 m) in length. In a reversal of the *Acanthostega* fossil finds, remains of the front feet of *Ichthyostega* are not known, but each rear foot had seven toes. *Ichthyostega*'s skeleton shows it was not capable of normal terrestrial, four-legged locomotion, like a crocodile or lizard. However, its front limbs and ribcage were strong enough to drag and squirm from water up onto land, perhaps to move to another pool or even to raise its body temperature in the sun. **LG**

3 *Panderichthys*, about 3 feet (1 m) long, is reconstructed from fossils from Latvia. It is one of several creatures that show parallel evolution tracing fins to limbs.

4 *Acanthostega* was among many fishapods with intermediate features that were interpreted as adaptations for feeding out of water, including changes to the limbs, teeth, and skull bones.

5 *Ichthyostega* crawls from a late Devonian swamp. Some reconstructions support the idea of a longer, slimmer creature.

6 Among the earliest fossil tetrapodlike footprints, this Devonian trackway from Valentia Island, Ireland, was possibly made by an *Ichthyostega*-type animal while it was wading in shallow water.

Tiktaalik
LATE DEVONIAN PERIOD

SPECIES *Tiktaalik roseae*
GROUP Stegocephalia
SIZE 10 feet (3 m)
LOCATION Canada

⊙ **NAVIGATOR**

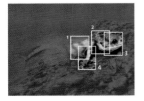

The fishapod *Tiktaalik* is a keystone fossil that may reveal the evolutionary link between fish and the first tetrapods. Although *Tiktaalik roseae* has many body features associated with fish, such as scales and gills, it also has features commonly attributed to terrestrial animals, such as limblike front fins and a sturdy ribcage. The first *Tiktaalik* fossils were uncovered in rocks on Ellesmere Island in northern Canada in 2004. Dating back more than 375 million years to the Devonian, *Tiktaalik* was a lobe-finned fish 10 feet (3 m) in length.

One of the most striking features of the Ellesmere Island remains was the structure of the fish's front fins, which had primitive armlike skeletal structures, such as a shoulder, elbow and wrist. *Tiktaalik* probably lived in shallow freshwater streams, swamps, and ponds, but may have used its specialized fins to venture onto land in the same way that present-day mudskippers do. The *Tiktaalik* fossil was also unusual because it showed that the fish lacked bony plates beside the gills, giving it a distinct neck and enabling it to move its head independently of the body. This helped *Tiktaalik* locate prey more easily, which it snapped up in well-developed jaws lined with sharp, pointed teeth. Another feature that helps position *Tiktaalik* as an intermediary between fish and tetrapods is the size of a skull bone called the hyomandibular. In fish, this bone is relatively big and connects the upper jaw with the skull. It helps fish to breathe by pumping water across their gills. In tetrapods, the hyomandibular has evolved into the stapes, one of the tiny bones in the middle ear used to aid hearing. The hyomandibular in *Tiktaalik* is intermediate in size between the two, which suggests that this ancient fish was not completely reliant on gills to breathe. **LG**

FOCAL POINTS

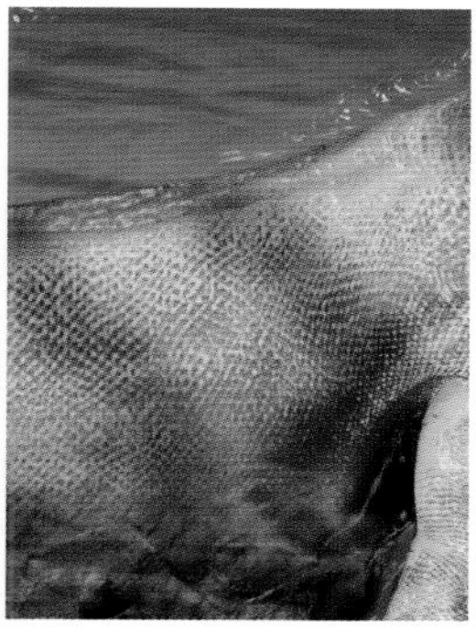

1 RIBCAGE
Another feature that suggests *Tiktaalik* breathed air using lungs is the well-developed ribcage that curved around the body and in at the base. Strong, robust ribs would help to support the weight of the fish's body out of the water. Without the water's buoyancy, the stout ribs would also protect the lungs and other internal organs from being squashed under their own weight.

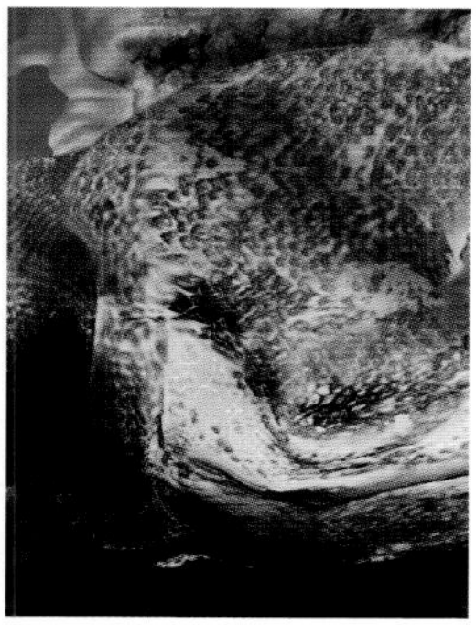

2 SPIRACLES
Spiracles (openings) on top of the skull indicate that *Tiktaalik* had primitive lungs as well as gills. *Tiktaalik* would have lived in warm, shallow waters with a poor supply of oxygen and probably used its primitive limbs to push its body out of the water and breathe in oxygen from the air. *Tiktaalik* may have even pulled its whole body out of the water to breathe on the land for short periods.

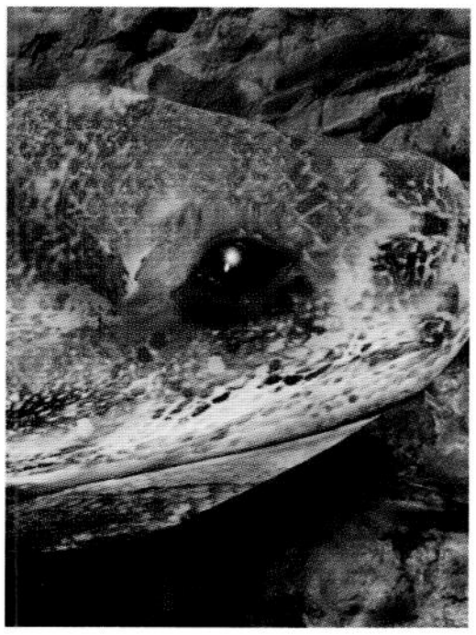

3 SKULL AND EYES
Tiktaalik had a flat skull with eye sockets on top of its head, giving this fish the appearance of a crocodile fish or alligator gar. It suggests that *Tiktaalik* spent most of its time looking upward, perhaps from its position on the bottom of the river or swamp. Another possibility is the "fish out of water" idea: it could have watched for airborne prey flying close to the water's surface.

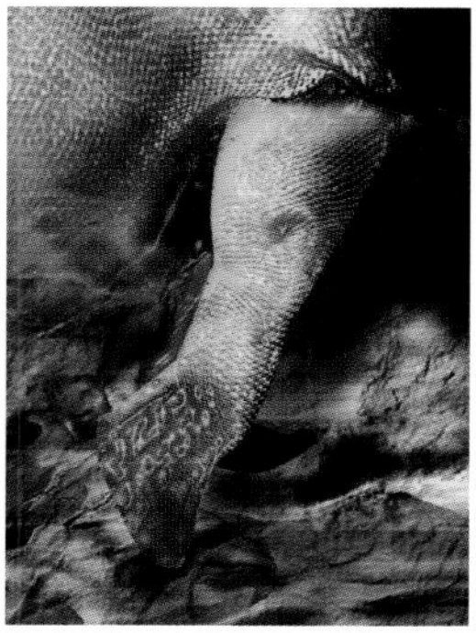

4 FRONT FINS
The front fins had simple wrist bones and a large shoulder girdle that could support the fish's body. Muscle scars on the surface of the humerus (upper arm bone) indicate where powerful fin muscles were attached. These probably helped *Tiktaalik* anchor its body on the riverbed in fast-flowing currents and perhaps, over time, helped to push the creature entirely out of the water.

TIKTAALIK RELATIONSHIPS

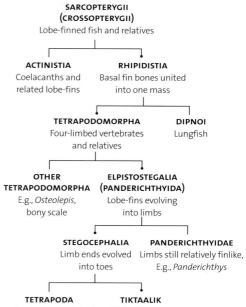

SARCOPTERYGII (CROSSOPTERYGII)
Lobe-finned fish and relatives

- **ACTINISTIA** Coelacanths and related lobe-fins
- **RHIPIDISTIA** Basal fin bones united into one mass
  - **TETRAPODOMORPHA** Four-limbed vertebrates and relatives
    - **OTHER TETRAPODOMORPHA** E.g., *Osteolepis*, bony scale
    - **ELPISTOSTEGALIA (PANDERICHTHYIDA)** Lobe-fins evolving into limbs
      - **STEGOCEPHALIA** Limb ends evolved into toes
        - **TETRAPODA** True four-limbed vertebrates
        - **TIKTAALIK** *Tiktaalik roseae*
      - **PANDERICHTHYIDAE** Limbs still relatively finlike, E.g., *Panderichthys*
  - **DIPNOI** Lungfish

▲ *Tiktaalik* is often regarded as a tetrapodomorph—the group that encompassed tetrapods (four-legged vertebrates) and their relations. But its fins had not yet evolved into "pods" (legs/feet). It was quite a long way from this innovation.

STRONG PELVIC FINS

In 2014 researchers made another important discovery about *Tiktaalik*: it may have used its rear fins to pull its body along the ground in the same way as the specially adapted front fins (see highlighted area, below). This evidence challenged the commonly held belief that the evolution of hind limbs began after the first vertebrates moved onto land. The *Tiktaalik* fossils first found on Ellesmere Island in 2004 had been incomplete and only showed the front part of the fish's body. When researchers returned to the site, they found the well-preserved pelvis and pelvic fin of the same specimen. The striking feature of the pelvis was its size, which matched the size of the bulky shoulder girdle and also featured a primitive hip socket.

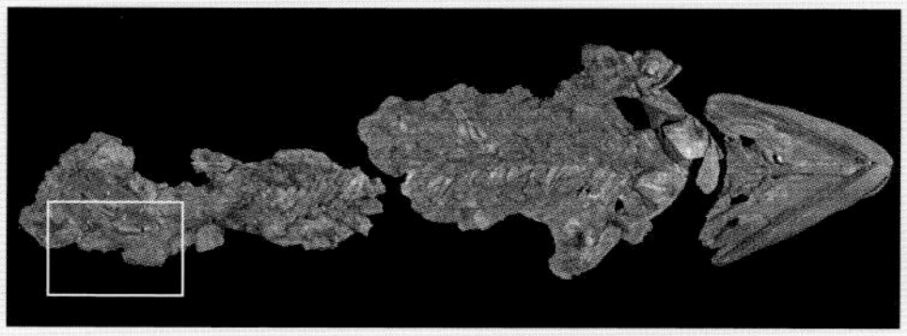

Pederpes
EARLY CARBONIFEROUS PERIOD

SPECIES *Pederpes finneyae*
GROUP Whatcheeriidae
SIZE 3 feet (1 m)
LOCATION Scotland

Norwegian paleontologist Peder Aspen (b.1942) discovered the first fossilized remains of the tetrapod *Pederpes* in 1971. The fossils were discovered in the Ballagan Limestone Formation of rocks in Dumbarton, Scotland, and date back to around 350 million years ago during the early Carboniferous period. The well-preserved skeleton was nearly complete—only the tail, parts of the skull, and various limb bones were missing. When paleontologists studied the fossil remains, they mistakenly classified the specimen as a lobe-finned fish. It took another 40 years before British paleontologist Jennifer Clack reclassified it as a primitive tetrapod. She named it *Pederpes*, meaning "Peter's foot," in honor of its Norwegian discoverer Peder, and *pes*, Latin for "foot." The name of the original, and only, species, *Pederpes finneyae*, comes from Clack's assistant, Sarah Finney, who prepared the fossil for examination. Clack's reclassification of *Pederpes* was extremely important because it bridged an empty period in the fossil record of the tetrapods: Romer's Gap (see panel).

The fossil remains show that *Pederpes* was a medium-sized tetrapod, measuring around 3 feet (1 m) long. Its head was large compared to the size of the body and triangular in shape. The structure of the bones in the hind feet clearly shows adaptations to terrestrial locomotion, with well-defined wrist and ankle joints that could bend to move the animal forward while out of the water. However, several features of the fossil specimen, such as evidence for the presence of lateral line sense organs along the side of the body and the large stapes or hyomandibular show that *Pederpes* probably spent an equal amount of time in the water as it did on the land. Despite this, most experts now agree that *Pederpes* represents the first truly terrestrial tetrapod in the fossil record, and thus is a true transitional species between the very earliest and later tetrapods. **LG**

 NAVIGATOR

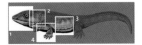

ROMER'S GAP

Pederpes is an important fossil because it bridges a gap in the tetrapod fossil record called Romer's Gap. This covered 15 million years, from 360 million to 345 million years ago (late Devonian to early Carboniferous). Specimens found before this are some of the most primitive tetrapods and share few similarities with the plethora of specimens found after. Romer's Gap was a big problem because it made it hard to chart the development of tetrapods as they moved from water to land. Romer's Gap is named after US paleontologist Alfred Romer (1894–1973; below), who organized the basic classification and evolutionary progression of vertebrates still used today.

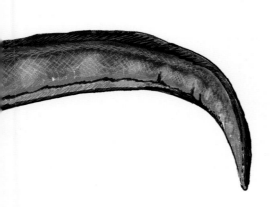

⊙ FOCAL POINTS

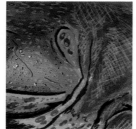

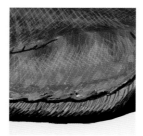

1 SKULL SHAPE
The narrow, triangular-shaped skull indicates that *Pederpes* breathed using muscular inhalations to force air into the lungs. This skull shape is most common in terrestrial animals, which means *Pederpes* may have been well adapted to life on land. Modern amphibians breathe using the buccal (mouth) cavity to pump air into the lungs (buccal pumping).

2 STAPES
The stapes bone in the *Pederpes* fossil is large compared to those of modern terrestrial tetrapods. In those animals, the bone is part of the arrangement for conducting sound through the middle ear to the tympanum (eardrum) but here it is a relatively big structure—as it is in stem tetrapods, such as *Acanthostega*—and it is not associated with a tympanum.

3 LATERAL LINE
Tubes in the bones of the fossil specimen are evidence for the presence of a lateral line system in *Pederpes*. Lateral lines are sense organs that fish use to detect vibrations in water. Their presence in the *Pederpes* fossil indicates that this creature may have spent much of its life in the water, since they are not found in modern land-dwelling tetrapods.

4 FEET
Pederpes had forward-facing feet. This is revealed by asymmetries in the fossilized bones of the hind feet and is an adaptation more commonly found in terrestrial animals. Outward-facing feet are more paddlelike and indicate an aquatic lifestyle. The hind feet had five functional digits, a common feature of all modern tetrapods.

EARLY AMPHIBIANS

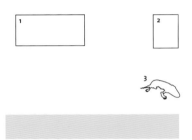

1

2

3

1 The amusingly named *Eucritta* ("true creature") was 10 inches (25 cm) long and had puzzling intermediate features between water and land dwelling.

2 *Eogyrinus* was one of several large, strong, swamp-living temnospondyl amphibians that filled the predator role that would later be taken by crocodiles.

3 At more than 6½ feet (2 m) long, bulky and powerful *Eryops* was one of the biggest land animals of the early Permian.

In the same way that fish diversified into many different groups during the so-called Age of Fishes (from the start of the Devonian period, around 420 million years ago), amphibians underwent a similar transformation in diversity from the late Carboniferous period to the end of the Permian period, 320 million to 260 million years ago. In this period, often referred to as the Age of Amphibians, many different amphibian groups came to dominate life on the supercontinent of Pangaea. The main plant life of the period, gymnosperms such as cycads, conifers, and ginkgoes, together with marine plankton, were pumping oxygen into Earth's atmosphere through the process of photosynthesis, using sunlight to convert carbon dioxide and water into oxygen and food. By the end of the Permian, atmospheric oxygen levels approaching those of today were providing an ideal environment for sustaining life on the land.

It is difficult to chart the evolutionary history of the early amphibians. Fossil evidence indicates that around 400 million years ago the first lobe-finned fish (sarcopterygians) crawled out of the late Devonian shallows using primitive limblike fins (see p.206), and it seems reasonable to suggest that those early four-limbed fish developed into early amphibians. Short, limblike fin bones appear in fossils of late Devonian transitional genera, such as *Acanthostega*, but the animals retain primary fishlike features, such as the lateral line sense organs. Fossils from later genera, such as *Eucritta* (see image 1), show more robust and well-developed limb bones. However, it is unknown whether *Eucritta* is an example of an early tetrapod (four-limbed vertebrate) that spent most of its life in water, or a representative of the first true amphibian group.

By the start of the Permian, 299 million years ago, the amphibians had become a diverse group of big, powerful animals, with lifestyles similar to

KEY EVENTS

360–345 Mya	345 Mya	330 Mya	311–300 Mya	310 Mya	310–300 Mya
Fishlike tetrapods with primitive limbs make the transition from water to land. Later fossils, such as *Pederpes* (see p.212), shed light on them.	*Eucritta* appears. It is a small, semi-aquatic tetrapod with well-developed limb bones.	The snakelike *Ophiderpeton* swims in the shallows of the Carboniferous.	Crocodile-like *Eogyrinus*, 15 feet (4 m) long, dominates the swamps of the late Carboniferous.	*Eryops* appears. This temnospondyl amphibian is 6½ feet (2 m) long and has sturdy limbs and a strong spine.	Animals that exhibit true amphibian characteristics exist at this time.

modern crocodiles. In fact, some of these ancient creatures grew to crocodile-like lengths of 16 feet (5 m) or more. Smaller prehistoric amphibians fed on fish and insects such as beetles, caddis flies, and stoneflies. The larger species, top predators of Permian wetlands, fed on smaller amphibians and early reptiles.

Like their modern descendants, female prehistoric amphibians laid their eggs in water. The eggs hatched into free-swimming larvae that used gills to breathe underwater. Slowly, each young amphibian transformed into its adult form in a process called metamorphosis. The gills gave way to lungs, enabling the adults to breathe air on land. In addition, many amphibians had moist and slimy skin that absorbed additional oxygen and passed it into the bloodstream. The fully grown adults often looked very different to the young amphibians, just as adult frogs look very different to young tadpoles. Firm evidence that the first true amphibians roamed the land comes from no earlier than the late Carboniferous, about 310 million to 300 million years ago.

Fossils have been found of *Eogyrinus* (see image 2) and *Eryops* (see image 3), two amphibians that represent a broad group that lived throughout the Carboniferous and Permian: the temnospondyls (divided vertebra) group. *Eogyrinus*, a large, slender, crocodile-like amphibian that grew up to 15 feet (4.5 m) in length, is a typical temnospondyl. Temnospondyl amphibians had big heads with massive, tooth-lined jaws, and the giant *Mastodonsaurus* had a gigantic head—6½ feet (2 m) long—that made up almost one-third of its entire length. Temnospondyls were reptilelike amphibians with short, stocky legs to support their long, muscular bodies. Genera such as *Eryops*, *Eogyrinus* and many others had dry, scaly skin—unlike the moist skin of modern amphibians. This was probably an adaptation to conserve water in the dry Permian climate.

Fossils have also been found of *Microbrachis*, which lived during the Carboniferous and represented another Permian group: the lepospondyls (husk/spoon-shaped vertebra) . The lepospondyls were much smaller creatures than the temnospondyls, often with strange body shapes. Many were semi-aquatic and had the moist, slimy skin of modern frogs. *Microbrachis* was less than 6 inches (15 cm) in length and somewhat resembled a modern salamander. It had well-developed but small limbs in relation to its body size. Another Carboniferous lepospondyl was *Ophiderpeton*, an odd snakelike creature reminiscent of today's caecilians (amphibian worm-lizards). Others were truly bizarre, such as Permian *Diplocaulus* (see p.218), 3 feet (1 m) long, whose boomerang-shaped head may have helped it to swim.

These ancient amphibian groups persisted throughout much of the Permian, but most gradually died out when the climate became increasingly harsh and dry. The changing conditions favored amniotes such as reptiles and the therapsids (mammal-like reptiles), that came to dominate the landscape of the Triassic period that followed. **LG**

305 Mya	300 Mya	280–270 Mya	275 Mya	252 Mya	250–200 Mya
Prehistoric amphibians diverge into two main groups: the large fierce temnospondyls and the sometimes bizarre-headed lepospondyls.	*Microbrachis*, a small, semi-aquatic lepospondyl, appears on the European landmass.	*Seymouria* (see p.216), a reptilelike amphibian with short, powerful limbs and scaly skin inhabits North America and Germany.	The salamander-like *Diplocaulus* appears. Its head is a distinctive boomerang shape.	The mass extinction event at the end of the Permian begins. More than 80 percent of life on Earth is doomed to extinction.	The ancestors of modern amphibians, the lissamphibians (smooth double/both-lives), appear.

Seymouria
EARLY PERMIAN PERIOD

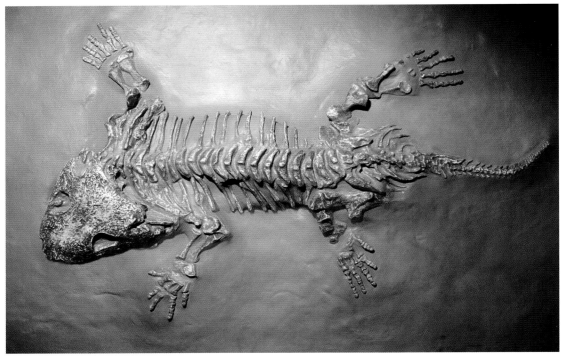

SPECIES *Seymouria baylorensis*
GROUP Seymouriidae
SIZE 2 feet (60 cm)
LOCATION North America
and Germany

An example of the reptilelike amphibians that lived in the early Permian, around 280 million to 270 million years ago, *Seymouria*, like all amphibians, would have started life as an aquatic larva or tadpolelike creature before moving onto land as an adult. No fossil larvae have been found, but that is not surprising because preservation of their soft bodies would have been an astonishingly rare occurrence. However, there are many examples of well-preserved adult *Seymouria* fossils. The first of these were found in 1906 in a town called Seymour in Baylor County, Texas—the location provided the name for the most numerous species, *Seymouria baylorensis*. Since then, many more fossils of the adult amphibian have been found scattered across the United States, in New Mexico, Oklahoma, and Utah. More recently, in 1993, fossils of two smaller specimens were uncovered in central Germany, providing evidence that its habitat was not confined to North America. The structure of the vertebrae and limb girdles in all these adult fossils indicates that *Seymouria* was well adapted to life on land. Relatively small and compact at only 2 feet (60 cm) in length, the animal had short limbs powerful enough to prop up its body and keep itself off the ground. While not very fast or agile, *Seymouria* was probably able to spend long periods away from water. On land it would undulate its backbone from side to side as it crawled or waddled across the surface in search of prey.

At the time that the newly discovered fossils were first studied, *Seymouria* appeared to be so well adapted to life on the land that it was mistaken for a reptile. Today, researchers see it as representing an important evolutionary link between amphibians and reptiles. Adult *Seymouria* may have had reptilian features that, up to now, fossils have neither confirmed nor discounted, such as a dry skin, the ability to conserve water for long periods in their bodies, and being able to excrete salt from the blood through a nasal gland. These would have been important adaptations necessary for survival in the dry, harsh climate that characterized the early Permian. **LG**

✦ NAVIGATOR

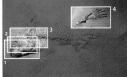

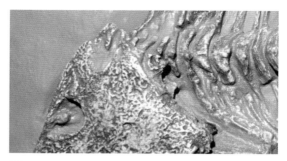

1 SNOUT, JAWS, AND DIET

S. baylorensis had two nasal openings at the front of the snout. Just below were large, robust jaws bearing many small, pointed, labyrinthodont teeth, which means they had a complex folded internal structure typical of early tetrapods. The mouth and jaws indicate that *S. baylorensis* had a varied diet that included insects, other small amphibians, and perhaps reptile eggs.

3 THICK SKULL

Fossils of *S. baylorensis* include skulls that are well preserved in three dimensions. Examination of the skulls has indicated that *S. baylorensis* had a small brain cavity and the bone was especially thick. Some paleontologists suggest that the males used their resilient heads as battering rams during mating contests. Modern mammals that fight with their heads also have thick skulls.

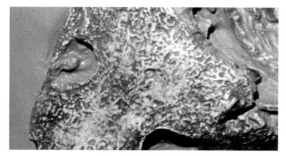

2 THIRD EYE

The roof of the skull includes a small opening for a light-receptive organ called the parietal or pineal eye. Common in many primitive vertebrates such as some fish species, frogs, lizards, and the tuatara, but absent in birds and mammals, this "eye" is linked to the pineal gland in the brain, which controls the production of hormones that regulate body temperature and the daily sleep–wake cycle.

4 SHORT LIMBS

Fossil limb bones reveal that the upper arm bone (humerus) and thigh bone (femur, above) were robust, with expanded ends to anchor the limb muscles. While the limb musculature may have been powerful enough to prop up the body, the legs themselves were relatively short compared to the length of the body. This suggests that *S. baylorensis* was unable to crawl very quickly.

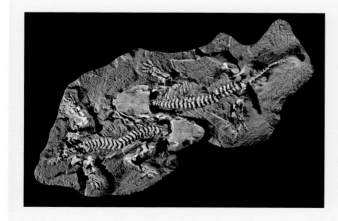

THE TAMBACH LOVERS

In 1993, US paleontologist David Berman (b. 1940) and his German colleague Thomas Martens (b. 1951) were fossil hunting in the early Permian Tambach Formation near the village of Tambach-Dietharz in central Germany. The pair found the complete fossils of two adult *Seymouria*, seemingly nestled together with one partly lying on top of the other (left). The paleontologists nicknamed the two fossils the "Tambach lovers" because it appeared that the animals died while they were mating and were subsequently fossilized. Other remains found at this site include the European *Dimetrodon teutonis*.

Diplocaulus
PERMIAN PERIOD

SPECIES *Diplocaulus salamandroides*
GROUP Nectridea
SIZE 3 feet (1 m)
LOCATION North America and Morocco

Dating from the early Permian period, *Diplocaulus* ("double stalk") was a salamander-like amphibian. The name, a reference to the skull, derived from ancient Greek. From the bony protrusions on each side of the skull it may be assumed that this amphibian had an unusual, boomerang-shaped head. It is speculated that the wide head offered protection from predators, such as *Dimetrodon* (see p.238) by being difficult to swallow. The head shape may also have uncovered prey concealed in mud. Fossils of the creature have been discovered in the Permian Red Beds of Texas, and more recently in the Argana Basin of Morocco, where *Diplocaulus* specimens were found.

Diplocaulus was one of the largest Permian amphibians, often attaining a length of 3 feet (1 m). The head of a mature *Diplocaulus* was about 1 foot (30 cm) across and was up to six times wider than it was long. The short limbs, the structure of the backbone, and the flattened body shape of *Diplocaulus* indicate that it spent almost all of its life in water, undulating its body from side to side to propel itself along the beds of rivers and swamps. However, *Diplocaulus* was not the only amphibian with a boomerang-shaped head to be found in Permian waters. *Diploceraspis* was a close relative with similar bony projections on the sides of its head. *Diploceraspis* was smaller than *Diplocaulus* but its "horns" were slightly bigger in relation to the size of its body. Both animals belonged to an order of amphibians called Nectridea, characterized in part by skull protrusions of various sizes. This group is central to a debate on the evolution of modern amphibians: frogs, salamanders, and caecilians. Most experts contend that amphibians developed from a group of the Carboniferous and Permian (360 million to 251 million years ago), known as the temnospondyls. However, a study published in 2004 suggests that, in fact, modern amphibians originate from the lepospondyls, a larger group to which the nectrideans belong. **LG**

 **NAVIGATOR**

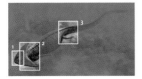

1 UPWARD-DIRECTED EYES
Eye sockets on top of the skull suggest that *D. salamandroides* probably lived on the bottom of rivers and swamps, hunting as an ambush predator. It would lie in wait in the murky depths, waiting for fish and other small amphibians to swim overhead, then leap up and use its powerful jaws to snap up the passing prey.

2 HYDROFOIL HEAD
The head of *D. salamandroides* may have acted as a hydrofoil or underwater wing, reducing drag and helping the animal to glide. Paleontologists from the University of Michigan suggest that two flaps of skin, running along the length of the body, undulated in a wavelike motion and also helped the amphibian to swim efficiently.

3 LIMBS AND TAIL
The four limbs were short and thin and the tail may have been as long as the head and body combined. Reconstructions show the animal swimming with side-to-side tail movements, but the joints between the caudal vertebrae (tailbones) suggest it could have flexed up and down, as in whales and dolphins today.

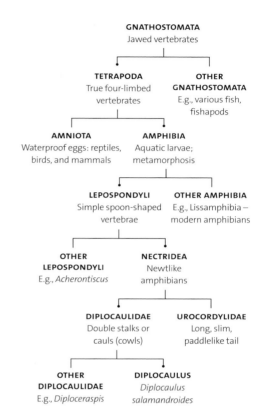

GNATHOSTOMATA
Jawed vertebrates

TETRAPODA
True four-limbed vertebrates

OTHER GNATHOSTOMATA
E.g., various fish, fishapods

AMNIOTA
Waterproof eggs: reptiles, birds, and mammals

AMPHIBIA
Aquatic larvae; metamorphosis

LEPOSPONDYLI
Simple spoon-shaped vertebrae

OTHER AMPHIBIA
E.g., Lissamphibia — modern amphibians

OTHER LEPOSPONDYLI
E.g., *Acherontiscus*

NECTRIDEA
Newtlike amphibians

DIPLOCAULIDAE
Double stalks or cauls (cowls)

UROCORDYLIDAE
Long, slim, paddlelike tail

OTHER DIPLOCAULIDAE
E.g., *Diploceraspis*

DIPLOCAULUS
Diplocaulus salamandroides

▲ *Diplocaulus* belonged to a relatively small family, Diplocaulidae, of similar boomerang-headed amphibians, which became extinct during the Permian. A larger group, Lepospondyli, comprised diverse amphibians shaped like lizards, snakes, and salamanders.

INTERNET HOAX

In 2004, photographs published on the Internet showed an unusual creature with a distinctive boomerang-shaped head (right). The caption for the photograph claimed that the live animal had been found in Malta. A few weeks later, Professor Patrick Schembri (b. 1954) of the Department of Biology, University of Malta, published an article that identified the animal as *Diplocaulus*. However, Professor Schembri expressed surprise at his own conclusion, given that the amphibian had been extinct for more than 270 million years. Schembri suggested that the photograph was in fact a hoax. His suspicions were proved correct when the photograph was later traced to a Japanese modeling enthusiast, who had made a *Diplocaulus* model in 1992 to enter in a competition and had not intended for the photograph to become an Internet sensation.

TOWARD MODERN AMPHIBIANS

1 A specimen of the temnospondyl *Amphibamus* from Ohio, displays the wide skull typical of many amphibians.

2 The aquatic caecilian, *Typhlonectes natans*, mostly respires through its skin but can also breathe at the water surface.

3 All adult amphibians, salamanders, and newts—including this red mud salamander, *Pseudotriton rube*—are carnivores and hunt small creatures.

The 7,000 or so species of living amphibians belong to a group called the lissamphibians. They include frogs and toads (anurans), salamanders and newts (caudatans or urodeles), and the rarer and lesser-known wormlike caecilians (apodans or gymnophions). The first true amphibians appeared around 310 million years ago during the late Carboniferous period. These early amphibians were represented by a group of giant crocodilelike creatures, such as *Eogyrinus* and *Mastodonsaurus*, the temnospondyls, as well as smaller, bizarre-looking lepospondyls, such as the boomerang-headed *Diplocaulus* (see p.218).

As is often the case with evolutionary biology, no one can be sure how these prehistoric amphibians developed into the common ancestor of the modern lissamphibians (smooth double/both-lives). Some experts point to *Amphibamus* (see image 1), 8 inches (20 cm) in length, which lived 300 million years ago in the swamplands of Europe and North America. Others believe that the lissamphibians branched off from a common ancestor that appeared much later, between the middle Permian and early Triassic periods. One candidate is the early Permian amphibian *Gerobatrachus*, 4¼ inches (11 cm) long. With its froglike head and salamanderlike tail, *Gerobatrachus* was nicknamed "frogmander" by the press when the fossils were first discovered in 2008 in Texas. This discovery prompted debate among experts, who suggested that frogs and salamanders developed from temnopspondyl ancestors, while the caecilians evolved from the lepospondyl lineage.

KEY EVENTS

300 Mya	290 Mya	280–248 Mya	250 Mya	200 Mya	200 Mya
Amphibamus inhabits the swamplands of Europe and North America. It may be one of the earliest ancestors of modern amphibians.	Commonly known as the "frogmander," *Gerobatrachus* exhibits characteristics of both modern frogs and salamanders.	Amphibians and reptiles dominate in the Age of Amphibians. Phytoplankton and plants oxygenate Earth's atmosphere to close to modern levels.	The small, froglike amphibian *Triadobatrachus* appears on the island of Madagascar.	*Prosalirus*, 2½ inches (6 cm) long, exhibits the three-pronged pelvic structure that is common to all modern frogs.	*Vieraella* appears, the earliest known example of the modern frog, with long, powerful hind legs adapted for jumping.

Another candidate for the common ancestor of modern lissamphibians is the froglike *Triadobatrachus* (see p.222), which lived 250 million years ago during the early Triassic. *Triadobatrachus* did not possess all of the characteristics of a modern frog. For example, it had a short tail to accommodate its longer backbone (modern frogs do not have tails). However, it is undeniably froglike in shape and form.

Despite such discoveries, the evolution of modern lissamphibians continues to be the subject of much debate. Unfortunately, there is a gap in the amphibian fossil record until the middle of the Jurassic period, when fossils show that the first true frogs started to appear. The earliest example is *Vieraella* from Argentina, dated to around 200 million years ago. This small frog measured only 1¼ inches (3 cm) in length, but possessed long, muscular legs for jumping and the shortened backbone and latticelike skull characteristic of modern species. The Jurassic also gave rise to *Enneabatrachus* from the family Alytidae, which includes living groups commonly known as midwife toads and painted frogs. However, most modern frog families do not appear in the fossil record until after the mass extinction at the end of the Cretaceous period, 66 million years ago (see p.364). It is likely that the evolutionary picture will change as experts uncover more fossils.

The ancestor of modern salamanders and newts (see image 3) is best represented by *Karaurus*, a tiny amphibian with a large head that lived in Asia during the late Jurassic. *Karaurus* measured about 8 inches (20 cm) in length and probably spent most of its life in fresh water, feeding on crustaceans, snails, and aquatic insects. Fossils appear throughout the remainder of the Mesozoic era but do not appear to undergo much evolutionary change. In fact, salamanders are occasionally referred to as living fossils because they have kept the same basic body plan for millions of years. Like their close frog and toad relatives, modern salamander families do not appear until the last 66 million years, during the Cenozoic era. The caecilians (see image 2) are the least understood of all modern amphibians, partly due to their underground lifestyle and dwindling population. These unusual creatures have no limbs and look similar to large earthworms or even snakes; some reach more than 5 feet (1.5 m) in length. The evolutionary history of caecilians is very unclear because there are so few examples in the fossil record. One of these, *Eocaecilia*, dates from the early Jurassic and retains small vestigial legs—similar to the evolutionary process that prehistoric snakes went through during the Cretaceous.

Modern amphibians are the descendants of some of the earliest tetrapods (four-limbed vertebrates). They have persisted through mass extinction events and severe climate change. It is therefore ironic that modern amphibians are some of the most endangered animals. The exact cause is unknown, but climate change, habitat destruction, pollution, and disease are all likely candidates. If the dwindling populations are not preserved, amphibians may be one of the first major vertebrate groups to disappear from our planet. **LG**

199 Mya	150 Mya	150 Mya	145 Mya	66 Mya	Present
Eocaecilia, the earliest known ancestor of modern caecilians, about 6 inches (15 cm) long, appears in Arizona.	*Enneabatrachus* is the earliest documented ancestor of the extant frog family Alytidae, which includes modern painted frogs and midwife toads.	*Karaurus* appears in the fossil record—a small salamanderlike amphibian with a triangular head and upward-pointing eyes.	The evolution of lissamphibians that resemble today's main groups is largely complete by the end of the Jurassic.	Many lissamphibians survive the end-of-Cretaceous mass extinction event that wiped out larger vertebrates, such as the great dinosaurs.	Lissamphibians are some of the most endangered vertebrates: up to 50 percent of all species are threatened with extinction.

Triadobatrachus
EARLY TRIASSIC PERIOD

SPECIES
Triadobatrachus massinoti
GROUP Protobatrachidae
SIZE 4 inches (10 cm)
LOCATION Madagascar

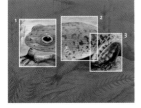

The species *Triadobatrachus massinoti* is one of the earliest known ancestors of modern frogs. This amphibian lived during the early Triassic, 250 million years ago. By the time of the early Jurassic, 50 million years later, relatives of *Triadobatrachus* had developed into frogs in the modern group of amphibians, lissamphibians, represented by the species *Prosalirus bitis* (see panel). A fossil *Triadobatrachus* was discovered in northern Madagascar in 1930. Six years later, French paleontologist Jean Piveteau (1899–1991) examined the specimen and found that the skeleton was almost complete, with all the bones in the correct anatomical position. Only the front skull and ends of the limbs were missing. Piveteau concluded that it was an early transitional form of the modern frog. He named the creature *Protobatrachus massinoti* after the man who found the fossil, Adrien Massinot (1887–1948). Later, the genus name changed to *Triadobatrachus*. The fossil shows that *Triadobatrachus* was about 4 inches (10 cm) in length. Like all amphibians, it developed in water as a larva and gradually metamorphosed into its adult body shape. Unlike modern frogs, however, the adults kept a very short tail. The species had the characteristic longer hind legs and shorter front legs of the frog, but it is unlikely that it could jump like modern frogs. It probably spent most of its time in water, using its hind legs to propel its body. Massinot found the fossil in marine-based rocks, even though the skeleton structure shows that it could also survive on land and breathe air. **LG**

⦿ FOCAL POINTS

1 SKULL

The skull was similar to that of a modern frog. It consisted of a narrow roof, made up of fused frontal and parietal bones, and large orbital sockets for the eyes. A lattice of thin bones was located at the base of the skull around the jaws. The lower jaw had no teeth. As the front part of the fossil's skull is missing, its exact structure is unknown.

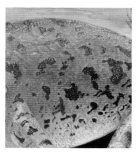

2 VERTEBRAE

Another skeletal feature that relates *Triadobatrachus* to modern frogs is the reduced number of vertebrae in the spine. The skeleton of *Triadobatrachus* has 14 vertebrae, six of which form a very short tail. Modern frogs have up to nine vertebrae, although none of them extend into a tail.

3 HIND LEGS

The long hind-leg bones and elongated ankle bones were very froglike. *Triadobatrachus*'s hind legs were at least as long as its body and represent a step toward modern frogs, which use them to jump on land as well as swim strongly. The tips of the limbs were missing so the structure of the hands and feet are unknown.

THE FIRST LEAPING FROG

In 1981, Farish Jenkins (1940–2012), a professor at Harvard University, discovered a new fossil frog in Arizona, called *Prosalirus bitis* (below). Dating from the early Jurassic, 200 million years ago, its most noticeable characteristic was the structure of the pelvic (hip) girdle. The largest bone of the pelvis—the ilium—was longer than that of *Triadobatrachus*, and the vertebrae that formed the short tail in *Triadobatrachus* had fused into one solid piece in *Prosalirus*, termed the urostyle. Together, the ilium and urostyle formed a three-pronged pelvic structure common to all modern frogs. This structure is very strong and resistant to bending. Together with the long hind legs and ankle bones, this suggests that *Prosalirus* was effective at leaping, like modern frogs. In fact, the genus name *Prosalirus* means "jump forward."

◄ The sole fossil of *Triadobatrachus* shows the skull to the far right, the shorter fore limbs just to the left of it, and the vertebrae extending to the larger hind limbs. To the far left, all feet are missing. Bones in such lifelike (articulated) positions are a very rare occurrence and indicate that the creature was preserved rapidly, before it was scavenged or it rotted away.

THE PERMIAN MASS EXTINCTION

1 The Araguainha Crater is 25 miles (40 km) across, in central Brazil—the suggested site of a meteorite impact at the end of the Permian.

2 *Lycaenops* was one of many therapsids that disappeared around 252 million years ago.

3 Paleodictyopterida, evidenced by a fossil wing of *Lithomantis*, were common early flying insects that resembled dragonflies, but all perished at the Permian's close.

Mass extinction events are scattered throughout Earth's turbulent geological history. The most cataclysmic occurrence of all took place 252 million years ago and set the boundary between the end of the Permian period and the start of the Triassic period. The event closed the Paleozoic era, the great time span of "ancient life," and opened the Mesozoic era, the span of "middle life." The end-of-Permian (P/T or P/Tr [Permian–Triassic]) extinction was probably Earth's most devastating in terms of its impact on biodiversity. Some studies estimate that more than 80 percent of Earth's species disappeared, including 96 percent of life in the oceans and 70 percent of vertebrates on land. The occurrence is commonly known as the "Great Dying," and it took 30 million years for life on Earth to recover. The event overshadows the better-known end-of-Cretaceous (K-Pg) mass extinction (see p.364) of 66 million years ago that killed the nonbird dinosaurs and many other species.

No one really knows the precise cause of the Great Dying. Given that the event occurred more than 250 million years ago, almost four times more distant than the end-of-Cretaceous one, much of the geological evidence is destroyed or buried deep beneath Earth's surface. Lacking evidence, most experts speculate that a global catastrophic event, such as an asteroid or comet collision, or a series of huge volcanic eruptions, was the primary cause. Such a short-term but cataclysmic event may then have triggered a secondary wave of longer-term events, such as acid rain and global warming, that

KEY EVENTS

299 Mya	280 Mya	270 Mya	259¾ Mya	255–250 Mya	254–251 Mya
The Permian begins, following the Carboniferous Age of Coal when life on land flourished in warm, moist conditions.	Earth's landmasses have moved together to form the supercontinent of Pangaea, surrounded by the super-ocean Panthalassa.	The interior of Pangaea becomes progressively drier and harsher. Olson's Extinction occurs, an early but still considerable life-extinction event.	The last epoch of the Permian, the Lopingian begins, during which species turnover at first increases.	Higher toxic levels of carbon dioxide in marine habitats penetrate the shallows and kill much life.	Many trees die, probably as a result of drastically changed atmospheric conditions. Fossils later indicate much higher levels of fungal spores.

combined to extinguish life on an unimaginable scale. An asteroid collision probably caused the end-of-Cretaceous mass extinction, and it may be reasonable to suggest that a similar incident brought about the even more damaging end-of-Permian event. Experts have identified several possible impact sites. The Araguainha Crater in Brazil (see image 1) is the most likely candidate, although this has not been proved. In Siberia, evidence has been found of violent volcanic activity at the end of the Permian, and the volcanoes would have caused widespread fires, choking Earth's atmosphere with toxic volcanic ash and carbon dioxide gas. Pollution on this scale would have brought acid rain and global warming for years, threatening any species lucky enough to survive the primary destructive event.

The precise impact of the end-of-Permian event on Earth's biodiversity is difficult to assess. The causes are complex and interrelated, and it is likely that the loss of species took place in stages, perhaps over millions of years. Marine species suffered the greatest losses, and hardest hit by the changes in atmospheric carbon dioxide were various brachiopods, corals (see p.104), echinoderms (see p.162), and sponges (see p.100); killed lichens formed stromatolites (see p.42). All these animals relied on stable carbon dioxide levels to maintain their skeletons, and the vast increases in atmospheric carbon dioxide would have devastated them. The event also wiped out up to nine major insect groups and greatly reduced the populations of 10 others. The extinction included *Lithomantis bohemica* (see image 3) and all the order Paleodictyopterida, which comprised 50 percent of all the known insect species of the Paleozoic. Experts link the decline with the devastation of Permian flora, although the overall impact on plants is poorly understood. The Permian is known as the Age of Amphibians (see p.214), but even before the end-of-Permian event the dominant amphibians were in decline. The climate had become increasingly dry, making it difficult for these water-dependent creatures to survive. Other land vertebrates suffered, too, with therapsids (mammal-like reptiles), such as *Lycaenops* (see image 2), going extinct or seriously depleted. The enormous loss of life was sufficient to open the way for a new era—the Age of the Dinosaurs.

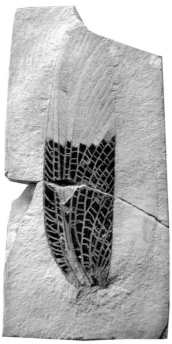

After the end-of-Permian event, life on Earth needed millions of years to recover. Experts believe that, before the recovery, successive waves of extinction continued into the early Triassic, as much as 5 million years after the earlier catastrophes. A study in 2012 by Chinese scientists concluded that the surface temperature of early Triassic oceans may have exceeded 104°F (40°C), creating conditions that were much too hot for most marine life to thrive. Land vertebrates also took millions of years to recover. The smaller temnospondyl amphibians (see p.214) fared much better than their larger lepospondyl cousins, while therapsid vertebrates (see p.236), including the ancestors of present-day mammals, and the archosaurs (see p.248), ancestors of dinosaurs and modern-day crocodylians, went on to dominate life in the Triassic. **LG**

254 Mya	252¼ Mya	252–250 Mya	252–250 Mya	245 Mya	125 Mya
Vastly increased volcanic activity at the end of the Permian results in climate change that is already making many species extinct.	The flashpoint of the "Great Dying" brings both the Permian and the Paleozoic to an abrupt end.	The predatory therapsid *Dinogorgon* is only one of the thousands of genera that disappear within 1 million years of the main extinction event.	Ash and particles from huge volumes of volcanic gases pour into the sky and block sunlight as the planet cools. Acid rain destroys plants.	Biodiversity shows signs of recovering from the Great Dying. Life begins to adapt to the new conditions. Some plants develop greater acid tolerance.	Global marine family numbers are restored to their late Permian levels, 125 million years after the Permian–Triassic mass extinction occurred.

The Luoping Biota Fossil Beds
EARLY–MIDDLE TRIASSIC PERIOD

LOCATION Luoping county, Yunnan province, China
AGE 243 million to 240 million years
GROUPS See right
BED DEPTH Up to 50 feet (16 m)

The Luoping fossil site in Yunnan province, China, was first investigated in 2007 when experts noticed fossils in the flat stone slabs that were used to roof nearby dwellings. The fossils were traced to a hilltop that, by 2010, had been excavated to reveal more than 20,000 fossils. The rocks, limestones, and similar sedimentary types, were once part of a seabed during the middle Triassic, around 243 million to 240 million years ago, 10 million years after the end-of-Permian extinction event.

The beautifully preserved Luoping remains were found in flat rock layers stacked neatly, which enabled them to be dated accurately in a sequence through time. They depict a whole marine community of plants, animals, and other organisms (biota) that appeared to have bounced back from the mass extinction. Biodiversity almost matched pre-extinction-event levels. The communities were organized in a similar way to those of the Permian, with plants at the base of the food chains, then herbivores, first-level carnivores, second carnivores (which ate the first-level ones), and so on. Planktonic filter-feeders, scavengers, and detritus-feeders sifted through mulch on the seabed. The vast majority of these organisms had newly evolved; only 1 in 10 Permian species had survived the extinction. Luoping animals included new kinds of planktonic foraminifera (single-celled organisms); shrimp, water fleas and other crustaceans; xiphosurans (horseshoe crabs; see p.140); bivalve shellfish, gastropod sea snails, belemnoids, ammonoids and other mollusks (see p.124); brachiopods (lampshells); conodonts; crinoids, starfish, urchins and other echinoderms (see p.162); many fish, including coelacanths (see p.200); and marine reptiles such as ichthyosaurs (see p.248). These creatures established lineages that would continue to evolve during the Triassic. Luoping presents a detailed snapshot of how life recovered and evolved afresh within a few million years of the world's greatest mass extinction event, yet continued the same ecological organization. **SP**

ISOPOD

Isopods are in the Crustacea group, along with the larger and better-known shrimp, prawn, crabs, and lobsters. Land isopods include woodlice, such as *Asellus*. *Protamphisopus baii*, about 1 inch (2.5 cm) long, is common in the Luoping biota. As one of the small creatures at the beginning of food chains, it sieved tiny pieces of nutrient from sea mud and was consumed by first-level predators.

PREDATORY FISH

Of the first 20,000 Luoping fossils analyzed, less than 4 percent were fish. Among the largest was the ray-fin *Saurichthys yunnanensis*, a slim, sleek 3 feet (1 m) long predator. With jaws one third of its total length, and the dorsal (back) and anal fins positioned well toward the rear, near the tail, it probably hunted with bursts of speed, much like the modern pike.

NEW TOP PREDATORS

One of the greatest changes seen at Triassic Luoping involved the carnivores. Innovative kinds formed a new level of larger, fiercer top predators that fed on smaller predatory fish and reptiles. They included the long bony fish *Saurichthys* and three new kinds of reptile: *Nothosaurus* (see p.242), which evolved to exceed 7 feet (2 m) in length; *Dinocephalosaurus*, more than 7 feet (2 m) long; and the ichthyosaur *Mixosaurus*, of similar length. These, and other top predator groups, diversified and raced to gigantic size during the Triassic. Then much larger marine reptile predators appeared, such as nothosaur *Ceresiosaurus* (above)—16 feet (5 m) long—in only a few million years.

HUNTING ICHTHYOSAUR

The genus *Mixosaurus* was the main representative of the ichthyosaur marine reptile group at Luoping. It was one of the largest and fastest of the new top predators. This genus later diversified into several species that roamed the Triassic seas, some exceeding 7 feet (2 m) in length. Fossils are widespread, from North America across Europe, to East and Southeast Asia.

5 | REPTILES

Evolution of the "waterproof egg" freed some tetrapods from reliance on water for breeding, and innovations such as non-moist skin furthered this independence. The resulting reptiles were able to colonize habitats on dry land and diversify enormously, although some lineages, such as ichthyosaurs and plesiosaurs, returned to fully aquatic lifestyles. A suite of features involving posture, bone shapes, and muscle attachments led to one of the most prolific and famous of all animal groups: the dinosaurs.

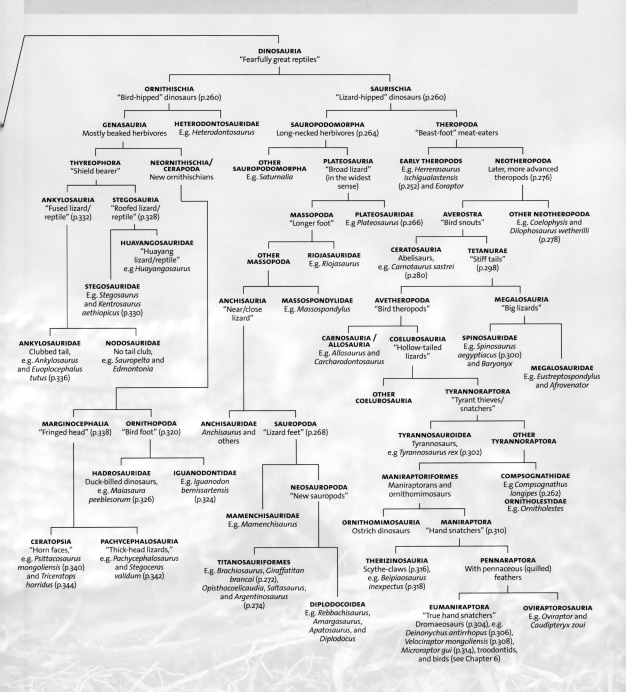

EARLY REPTILES

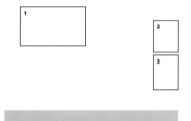

1 Around 10 feet (3 m) in length, *Scutosaurus* was a squat, heavy, herbivore pareiasaur that lived in ancient Russia.

2 Amphisbaenids (worm lizards) are legless members of the lizard-snake group Squamata. Having evolved separately from snakes, they are highly adapted to underground life.

3 *Pleurosaurus* was a 2-foot (60 cm) aquatic rhynchocephalian, a tuatara cousin from 150 million years ago.

R eptiles are one of the major land-living vertebrate groups and include turtles, terrapins and tortoises, lizards and snakes, crocodiles, alligators, and caimans. Birds and other feathered dinosaurs are also termed reptiles, as are furry flying pterosaurs. In common parlance, a "reptile" is thought of as a scaly-skinned, cold-blooded animal with sprawling limbs. However, fossils show that many extinct animals on the same evolutionary branch as lizards and crocodiles had feathers or hairlike structures, walked with erect limbs, and were capable of generating and storing body heat. The scientific term "reptile" does not, therefore, have the same meaning as the one commonly used.

The earliest reptiles were small, sprawling animals, closely related to synapsids, the group that gave rise to mammals (see p.424). Both shared a common ancestor, and both are included within a major tetrapod (four-limbed vertebrate) group called amniotes. Unlike earlier limbed vertebrates, amniotes protected their developing young with a membrane, termed an amniotic sac, and an external shell. They also shared several anatomical features, including claws on their fingers and toes, an ankle bone called the astragalus, and the complete loss of the lateral line system used by other vertebrates (especially fish) as a sensory mechanism for life in water. Combined, these features show that amniotes were dedicated to life on land since their first appearance, no longer needing water for the development of their eggs or young, unlike fish

KEY EVENTS

315 Mya	315–312 Mya	305 Mya	300 Mya	295 Mya	265 Mya
The earliest amniotes appear, possessing a set of anatomical features that show they are fully terrestrial and do not need water for reproduction.	Eureptiles evolve; the best known is *Hylonomus* from Nova Scotia. They are small, lightly built insectivores less than 8 inches (20 cm) long.	Diapsids appear, lightly built with long, slender legs, indicating that they are swift, agile predators of insects. *Petrolacosaurus* is the classic example.	Parareptiles and synapsids live alongside early diapsids. Synapsids soon evolve to become the most important predatory group.	Parareptiles evolve a diversity of new body shapes and lifestyles. They include herbivores and insectivores, some with specialized teeth.	The two major diapsid groups—archosaurs and lepidosaurs—evolve, both starting their history as agile, lizardlike insectivores.

and amphibians. The evolution of robust jaws and teeth gave early amniotes a competitive edge on the land. Members of several amniote groups became specialized predators while others became omnivores and even herbivores. It is often thought that early amniotes had a drier, more watertight skin than other tetrapods, which also gave them an advantage for the invasion of dry habitats on land. However, the groups most closely related to the earliest amniotes would have already had thick, dry skin.

Based on the earliest occurrences of reptiles and synapsids in the fossil record, the amniote common ancestor lived about 315 million years ago, during the late Carboniferous period. Early synapsids and reptiles looked alike and led similar lifestyles. Consequently, experts have confused fossils of the two on several occasions. However, they became increasingly distinct toward the end of the Carboniferous. A burst of reptile evolution late in the Carboniferous, resulted in new insectivorous, omnivorous, and herbivorous reptile groups such as the parareptiles and eureptiles. Stout, blunt-tipped teeth and serrated, leaf-shaped teeth show that some of these parareptiles, such as *Scutosaurus* (see image 1), were herbivores. Bony armor was common in this group, perhaps as a self-defence adaptation against the predatory synapsids, such as *Dimetrodon* (see p.238), that lived at the time. Some of these parareptiles were large, reaching 10 feet (3 m) in length.

The eureptiles—a major new reptile group—originated at the same time as the parareptiles. They had longer, slimmer limbs than synapsids and parareptiles. The earliest eureptiles were swift predators of insects. The classic example is *Hylonomus* (see p.232), which is a small predator that, together with several related forms, was the first to have the lizardlike body shape that persisted throughout history. The majority of lizard-shaped reptiles belong to a group that evolved two large bony openings behind the eye to allow for larger jaw-closing muscles. Termed the diapsids, this group gave rise to lizards, snakes, crocodiles, dinosaurs, birds, and many others. The skulls and teeth of these animals were predominantly adapted for feeding on insects.

Two of the most important diapsid groups—archosauromorphs and lepidosaurs—evolved around 265 million years ago. The former includes a large number of superficially lizardlike species and also the archosaurs—the group that includes crocodiles and dinosaurs. Archosaurs are one of the great success stories of vertebrate life on land: it is the group that evolved gargantuan size, feathers, powered flight, and numerous other innovations. Lepidosaurs, meanwhile, remained lizardlike throughout their history, and the vast majority of living species belong to the lizard-snake group Squamata (see image 2). Another lepidosaur group, the rhynchocephalians (see image 3), underwent a burst of evolution during the Triassic and Jurassic periods, with many forms thriving around 200 million years ago. Today, they are represented by only two surviving species: the tuataras of New Zealand (see p.234). **DN**

255 Mya	230 Mya	225 Mya	200 Mya	140 Mya	66 Mya
Armored parareptiles, the pareiasaurs, inhabit much of the Pangaean supercontinent. These bulky herbivores are among the biggest land animals yet.	An extinction event at the end of the middle Triassic causes the disappearance of many parareptile lineages, although some persist into the late Triassic.	Lepidosaur diversity increases. Little is known of squamates but rhynchocephalians, including tuataras, occupy roles later taken by lizards.	The diversity of rhynchocephalians flourishes to include insectivorous, terrestrial, and herbivorous species and marine predators.	Rhynchocephalians decline, but new squamate groups— including snakes and modern lizards— become important in ecosystems worldwide.	One rhynchocephalian lineage survives beyond the Cretaceous period. Its descendants persist in New Zealand, and become specialized for life in cool climates.

Hylonomus
LATE CARBONIFEROUS PERIOD

SPECIES *Hylonomus lyelli*
GROUP Protorothyrididae
SIZE 8 inches (20 cm)
LOCATION Canada

A classic early reptile often imagined as an ancestral, prototypelike reptilian form, *Hylonomus* (forest mouse) is part of a group called Protorothyrididae. Its fossil was discovered by William Dawson (1820–99) in 1852 from 312-million-year-old rocks in Joggins, Nova Scotia, Canada. The species, *Hylonomus lyelli*, was named after the geologist's colleague, Charles Lyell (1797–1875).

Only 8 inches (20 cm) in total length, *Hylonomus* was a long-tailed quadruped (walking on four legs) that was superficially lizardlike overall, but far more primitive in detail. The body was proportionally very long and the skull was solid behind the eye, in contrast to the skulls of true lizards, which are perforated with large openings. Furthermore, *Hylonomus* and its relatives lacked the proportionally longer neck and longer limbs that would evolve in later reptiles. Small, pointed teeth lined its jaws and indicated that it preyed on small, lightly armored insects and other arthropods. During the late Carboniferous and Permian periods, reptiles such as *Hylonomus* gave rise to diapsids, and led to the enormous diversity of reptiles that was to dominate Earth for the millions of years that followed. *Hylonomus* was one of several small fossil animals preserved within deep hollows left by the rotten stumps of gigantic trees. One theory is that the animals fell into these hollows while foraging for prey and eventually starved and died. This is supported by the fact that the fossils of insects have been found alongside those of *Hylonomus* and other animals. Another possibility is that the hollows were used as nesting sites and were visited routinely by *Hylonomus* and its kin. Yet another idea is that the animals perished in the hollows after seeking refuge from forest fires. The high oxygen content of the Carboniferous atmosphere made fires common, and small animals would have regularly needed to take shelter from them. **DN**

✣ NAVIGATOR

1 HEAD
The thumb-sized, slightly boxlike head had nostril openings at the front tip and relatively large eye sockets. Inside the mouth, in addition to jaw teeth, there were also teeth on the palate, the bony roof of the mouth cavity. These small, spiky palatal teeth are a typical feature of early reptiles and other early land-living vertebrate groups.

2 SKIN
Scales help to reduce moisture loss from skin, and fossils of early forms show that scaly skin was definitely present on at least the bellies of these animals, although skin impressions for *Hylonomus* are unknown. It is likely that it was covered in scales, but these would not have been the specialized overlapping scales of snakes and lizards.

3 LIMBS
Long hands and feet suggest that *Hylonomus* was faster than earlier four-footed vertebrates, and capable of climbing on low branches. Poor preservation has obscured details of its anatomy, but blocklike ankle bones suggest that its feet were not as flexible as those of later reptiles. It would have been clumsy compared to modern lizards.

1797–1829
Charles Lyell was born in Scotland to a botanist father who influenced his son's interest in nature. After studying classics and graduating from the University of Oxford in 1819, Lyell became a lawyer, but by 1827 was a full-time geologist.

1830–40
His landmark work, *Principles of Geology*, was published in 1830 to 1833 in three volumes. It championed the principle of uniformitarianism: the idea that processes and events seen today in the natural world have always occurred. For example, volcanoes have been erupting not only very recently, but also through great periods of geological time. Scottish geologist James Hutton (1726–97) had elucidated the principle in the 1780s, but Lyell brought it to a much wider audience.

1841–51
Lyell visited the United States and Canada in 1841, and again in 1845 to 1846, when he worked with William Dawson (1820–99), an energetic Canadian geologist, educational reformer, and principal at McGill University, Quebec.

1852–55
When Dawson discovered the fossils of *Hylonomus* at Joggins in 1852, he decided that he should honor his colleague and mentor, Lyell, by naming one of the three species after him: *Hylonomus lyelli*. At the time, he was unaware of the animal's true significance as one of the earliest known reptiles. The other two species—*Hylerpeton curtidentatum* and *Fritschia curtidentata*—were later combined with *Hylonomus*.

1856–59
Lyell's uniformitarianism implied that Earth was much older than the 6,000 years then commonly accepted, which gave Charles Darwin (1809–82) confidence that there had been enough time for evolution to occur. Lyell supported and encouraged Darwin as the latter formulated his theory of evolution by natural selection. Darwin was influenced greatly by Lyell's *Principles of Geology* and drew on its conclusions for his *Origin of the Species*, published in 1859.

1860–75
Lyell was made a baronet for his achievements in 1864. *Principles of Geology* was updated 11 times, and Lyell died in 1875 while working on the 12th edition.

◄ The jaw teeth were long and pointed, ideal to puncture and rip the bodies of insects and other arthropods eaten by *Hylonomus*. Each jaw carried 35 to 40 teeth. A few individual teeth toward the front of the mouth were elongated and fanglike, most likely to make the first grab and hold onto the prey while the reptile prepared to take a second, more thorough, bite.

Tuatara
PLEISTOCENE EPOCH—PRESENT

SPECIES *Sphenodon punctatus*
GROUP Rhynchocephalia
SIZE 30 inches (76 cm)
LOCATION New Zealand
IUCN Least Concern

✧ NAVIGATOR

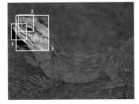

The two living tuatara species—*Sphenodon punctatus* and *S. guntheri*—are unique to offshore islands that surround New Zealand. They are the only surviving remnants of a once huge and diverse group of reptiles related to snakes and lizards: the rhynchocephalians. Members of this group lived worldwide during the Mesozoic era and evolved several different lifestyles and body shapes. The species alive today are among the world's most remarkable animals: they have slow metabolisms and lifespans of more than 100 years, they take up to 20 years to reach sexual maturity and their eggs take 2 years to hatch. Modern tuataras reach 30 inches (76 cm) in length and eat insects, snails, frogs, lizards, and seabird chicks, but some extinct rhynchocephalians were herbivores, omnivores, and even sea-going fish predators. During the Cretaceous, the largest rhynchocephalians exceeded 3 feet (1 m) in length, and there were armor-plated and sea-going herbivorous kinds. In fact, this group was so diverse during the Mesozoic that they occupied many of the roles later taken by squamates—lizards and snakes. Squamates originated early in the Mesozoic, but virtually all early types known were small, generalized animals. Only during the Cretaceous, as rhynchocephalian diversity declined, did snakes and lizards begin to resemble modern kinds in shape and lifestyle. Tuataras were thought to be primitive relict cousins of the more adaptable squamates, but this was not the case. A bony bar in the skull that connects the cheek to the jaw joint is present in tuataras and lacking in squamates, but it was also absent in early rhynchocephalians. It seems to have evolved again within the group, perhaps due to skull and jaw motion and the physical stresses of eating. **DN**

FOCAL POINTS

1 TEETH
The teeth are remarkable. Two massive incisors at the tip of the upper jaw form a projecting "beak." There are long, canine-like teeth at the front of the jaw and triangular teeth farther back, fused to the jawbones. The lower jaw slides backward and forward, and bladelike edges on the teeth shear food.

2 THIRD EYE
On top of the head is the parietal eye, named after the parietal skull bone. It has the structures of a lens, cornea, and retina, but is unable to see images. It could be connected to the sleep–wake cycle and hormonal system. It is obvious in newly hatched tuataras when semi-translucent skin covers it for 3 months.

3 SKULL
The skull looks archaic compared to that of lizards and snakes, because tuataras have a bony bar between the cheek and jaw joint. For this reason, tuataras were long thought to be living fossils that hardly changed in 100 million years. Fossils show that their ancestors lacked this bar, but it re-evolved.

TUATARA RELATIONSHIPS

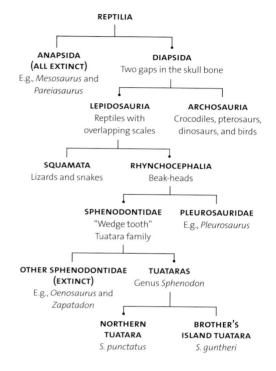

▲ The rhynchocephalians, to which the tuataras belong, are usually considered to be a "sister" or equivalent group to lizards and snakes. Pleurosaurs, a sister group to sphenodonts, were lizard-shaped aquatic rhynchocephalians about 2 feet (60 cm) in length.

FOSSIL COUSINS

Several fossil rhynchocephalians from the Jurassic and Cretaceous were especially similar to tuataras in overall appearance. Among these was *Cynosphenodon* from middle Jurassic period Mexico. It resembled the tuatara because it had caninelike teeth at the front of the lower jaw; the growth patterns of its teeth were also similar to the living species. Another is *Zapatadon*, also from Mexico, the smallest rhynchocephalian yet discovered. Its skull is only ½ inch (11 mm) long, so the whole animal would have been less than 6 inches (15 cm) in length. Living toward the end of the Jurassic was *Homoeosaurus* (right). Its fossils come from the well-known Solnhofen limestones of Bavaria, Germany, and again show remarkable skeletal similarities to today's tuatara. Oddly, despite such conservative anatomy over hundreds of millions of years, recent research shows that at the molecular level, the tuatara is evolving faster than almost any other animal yet studied.

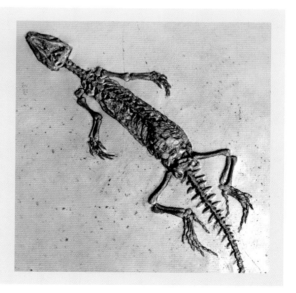

EARLY SYNAPSIDS

Approximately 315 million years ago, the earliest amniotes—the group of vertebrates that protected their young with shelled eggs and amniotic membranes—gave rise to two lineages. One amniotic branch led to reptiles: animals such as lizards, snakes, dinosaurs and birds; the other led to mammals. The mammal-branch amniotes are termed synapsids, and they share a set of anatomical features, the most notable of which is a single opening in the skull behind the eye. Early synapsids look superficially reptilelike, so have traditionally been called "mammal-like reptiles." In recent years, it has become accepted that the term "reptile" should be restricted to the group that includes turtles, lizards, snakes, dinosaurs, and crocodiles. Therefore, synapsids are not part of the reptile lineage and should not be called "mammal-like reptiles."

Two main groups of synapsids are recognized. The earlier, more reptilelike lineages are informally termed pelycosaurs. The later species were very similar to early mammals in appearance, anatomy, and biology and are included with early mammals in a group called therapsids. The earliest of the pelycosaur-type synapsids were small and superficially lizardlike. Teeth and skull shapes show that they were predators of small animals. They include a widespread group known as the varanopids, the fossils of which have been mistaken for those of early reptiles. Other subgroups—notably the ophiacodontids (see image 1)— have long skulls typical of animals that grab fish and other aquatic prey in shallow water. Members of another early synapsid group—the caseasaurs— were among the earliest large-bodied herbivores to evolve on land. They started their evolutionary journey at less than 3 feet (1 m) in length, but eventually

KEY EVENTS

315 Mya	310 Mya	305 Mya	300 Mya	300 Mya	290 Mya
The earliest synapsids appear. They are very similar to the reptiles and are closely related, but both groups soon evolve in different directions.	Early ophiacodontid *Archaeothyris* inhabits North America. Members of the varanopid and caseasaur groups also exist at this time.	Sphenacodontids—the group that includes *Dimetrodon*—undergo a burst of evolution. All have large canines but are otherwise variable in tooth shape.	*Ophiacodon*, the largest synapsid, inhabits North America. It is a long-skulled, amphibious predator 10 feet (3 m) long.	Herbivorous *Edaphosaurus* appears in Europe and North America—one of the first large-bodied herbivores. It inherits a dorsal (back) sail.	Members of the omnivorous-herbivorous caseasaur group appear, but older fossils from the Carboniferous period likely await discovery.

reached 20 feet (6 m) long. Their tiny heads, muscular fore limbs, and enormous claws indicate a way of life that involved digging. This lifestyle was successful, as caseasaurs survived beyond many other early synapsid lineages.

The best-known herbivorous early synapsid—with fossils found in Europe and North America—is *Edaphosaurus*, a wide-bodied animal. It was part of a larger group of omnivores and herbivores called edaphosaurids. *Edaphosaurus* had a specialized dentition adapted for gathering and chewing high-fiber plants. Short, peglike cropping teeth lined the outside of its jaws while small, rounded teeth covered the palate and parts of the lower jaw. *Edaphosaurus* was similar to its predatory cousin *Dimetrodon* (see p.238) because it possessed a giant back sail. However, the shapes of the rodlike bony processes that supported its sail were different, indicating that the two animals evolved separately. The reason why these two sizeable synapsids—both giants of their time—had sails is not known. Were the sails evolutionary solutions to temperature control, were they for sexual and social display, or could one have been mimicking the other? Edaphosaurids with sails appeared several million years before *Dimetrodon*.

Numerous trends can be seen within synapsid evolution. The teeth became increasingly heterodont, meaning they differentiated within the mouth, evolving incisors, canines, premolars, and molars. The number of bones in the skull and jaws decreased as they fused together or diminished. And, while early synapsids walked with sprawling limbs that projected sideways from the body, the later, more mammal-like forms had erect limbs. The fossil record provides little information concerning the behavior and lifestyle of early synapsids. However, there are hints of intriguing behavior among even the oldest members of this group. Five articulated fossil skeletons of the varanopid *Heleosaurus* (see image 2) were preserved in a burrow. This suggests parental care because only one of these individuals was an adult; the others were all similar-sized juveniles and almost certainly siblings. **DN**

1 The seemingly oversized skull of *Ophiacodon* (snake tooth) from the early Permian period, adapted for grabbing slippery victims, was more than 20 inches (50 cm) long in some specimens.

2 *Heleosaurus* was discovered in 270-million-year-old sediments in South Africa.

280 Mya	275 Mya	270 Mya	270 Mya	265 Mya	252 Mya
Several species of the predator *Dimetrodon* occur across Europe and North America. It is the first big, land-living top predator in prehistory.	*Edaphosaurus* becomes extinct after 25 million years—an unusually long span of time for any one genus of animal to persist.	A diverse set of caseasaurs inhabits North America, Europe and Russia. Some are small; others are the largest land animals of the time.	The first therapsids appear, probably evolving from sphenacodontids.	Most nontherapsid synapsids die out in a mass extinction event. Caseasaurs and therapsids survive. The latter eventually give rise to mammals.	An enormous extinction event brings a close to the Permian (see p.224). The therapsids are affected badly.

Dimetrodon
EARLY PERMIAN PERIOD

👁 FOCAL POINTS

1 TEETH
Dimetrodon grandis had prominent heterodonty (different shaped teeth) and its name means "two sets of tooth." At the front of the jaws were long, caninelike teeth, while shorter, blunter teeth sat farther back, some with cusps at the edges. Tiny saw teeth along the tooth edges sliced through flesh by forcing muscle fibers into the gaps between the serrations. Such serrated teeth evolved independently in several synapsid groups.

2 DORSAL SAIL
The sail was supported by long, bony structures that grew upward from the vertebrae. These spines are not always circular in cross section as in *D. grandis*: in some species they were shaped like a figure eight. It is probable that thin, translucent skin stretched between the spines, which were connected at their bases by muscles and ligament sheets. Fossils have shown that the sail may not have been entirely straight.

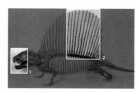
SPECIES *Dimetrodon grandis*
GROUP Sphenacodontinae
SIZE 5 feet (1.5 m)
LOCATION Europe and North America

The best-known and most familiar of the early synapsids are the nonreptiles that eventually gave rise to mammals. *Dimetrodon* was a large and successful predator known from fossils found in both Europe and North America. There are many species—the largest reached 10 feet (3 m) long. This powerful predator is instantly recognizable thanks to the giant bone-and-skin "sail" on its back—a feature that appeared much later in some dinosaurs, including the giant hunter *Spinosaurus* (see p.300).

Dimetrodon, which lived between 290 million and 270 million years ago, is significant for being the world's first truly big top land predator. This means it was the first predator to evolve a large enough body size, combined with various other features, to kill large animals. These features included an especially strong skull and laterally narrow, bladelike teeth with serrated or sawlike edges. These top predator characteristics continued to appear as vertebrates evolved. However, the reason for the evolution of *Dimetrodon*'s incredible dorsal (back) sail remains as contentious an issue today as it was when the fossils were first studied in the 1870s. A popular theory is that the sail provided the ability to control body temperature (see panel). However, the anatomy of the bones supporting the sail is not consistent with this. Furthermore, growth data for *Dimetrodon* shows that the sail grew much faster than would be expected for a role in temperature control. An alternative idea is that the sail evolved for reasons of sexual and social display. *Dimetrodon* is a member of a larger synapsid group called the sphenacodontids, only some of which possessed sails. Most were small predators, less than 5 feet (1.5 m) long, with pointed teeth. All had especially noticeable canines in their upper and lower jaws. *Dimetrodon* and other early synapsids are often depicted as sprawling animals, with arms and legs extending out to the side, bellies close to the ground, and tails dragging behind. However, fossil tracks show that they in fact walked in a semi-sprawling, more erect-limbed pose, keeping the body and tail above the ground. **DN**

THERMOREGULATION

The reasons behind the evolution of the tall, sail-like structure on *Dimetrodon* has generated endless discussion. One major theory centers on body temperature control, also known as thermoregulation. In hot conditions the sail acted as a radiator (right), dispersing excess heat into the atmosphere. When cool, it would have worked as a solar panel, absorbing heat from the sun and passing it to the body. In both cases the circulating blood carried out thermal transfer between the body and the sail. It meant that *Dimetrodon* could change its body temperature using its behavior, for example, standing side-on to the sun when chilled, or moving to the shade and standing side-on toward the breeze when overheated.

THE FIRST SEA REPTILES

Between 340 million and 250 million years ago, reptiles proliferated on land. Numerous different groups adapted to insectivorous, carnivorous, and herbivorous lifestyles in the deserts, swamps, uplands, and treetops. During this burst of evolution, it did not take long for certain reptiles to join their distant fish ancestors and readapt to life in the seas. The very first of these were the small, slender-jawed mesosaurs (see image 1), a poorly known group that lived 280 million years ago. During the Triassic period, from 252 million to 201 million years ago, several additional reptile groups became specialized for marine life. Some of these were short-lived and left no descendants, but others spawned great dynasties.

The transition to life in the water would have been relatively easy for the reptile body shape, especially in situations where a warm climate helped the animals to maintain a constant body temperature. Key anatomical changes occurred several times as reptiles became specialized sea-goers. Longer, more flexible necks evolved, as did jaws and throats that allowed swimming prey to be grabbed quickly or sucked into the mouth. Enhanced senses of sight and touch also evolved, enabling these animals to better find and capture their prey. Furthermore, the fossil record reveals that the limbs of aquatic reptiles were transformed into paddles or flippers, and that bodies and tails changed to enable efficient movement through the water.

The ability to give birth to live young (viviparity) was also a major evolutionary advantage for marine reptiles. However, it is not essential,

KEY EVENTS

290–270 Mya	250 Mya	245 Mya	245 Mya	245 Mya	245 Mya
The first aquatic reptiles, Mesosaurs exist. They returned to the seas from terrestrial ancestors.	Sauropterygians evolve from earlier diapsid reptiles, but it is uncertain which sort. They may have been members of the bird and crocodile branch.	Placodonts diverge from the common ancestor that leads to nothosaurs and plesiosaurs. Early species are a similar size and shape.	Turtlelike placodonts, cyamodontoids, evolve and spread along the Tethys Sea margins. They have shallow, wide bodies with bony armor plates.	Nothosaurs diversify to exploit several different lifestyles. Some are fish-grabbers with long jaws, while suction-feeders evolve short, wide jaws.	Giant nothosaurs—among the biggest marine reptiles yet to evolve—inhabit regions from Europe to China.

because turtles have persisted as sea-going reptiles for more than 100 million years and rely on a strategy that involves leaving the sea to dig nests and bury eggs along the coast. Fossils show that some major aquatic reptile groups have been giving birth to live young from very early on in their evolutionary history.

Among the most successful and diverse groups of marine reptiles were the sauropterygians, the group best known for the flippered, long-necked plesiosaurs (see p.282). These lived worldwide throughout the Triassic, Jurassic, and Cretaceous periods. Early on in its history, the Sauropterygia aquatic reptile group split into two branches. The branch of sauropterygians that includes plesiosaurs also contains several less-familiar groups—the small pachypleurosaurs, the larger, longer-skulled nothosaurs (see p.242), and the highly aquatic, long-necked pistosaurs—none of which survived beyond the wave of extinctions 200 million years ago, at the end of the Triassic. Members of these groups were amphibious reptiles that used both their paddlelike limbs and laterally compressed (narrow) tails to swim. Virtually all the key events of their evolution occurred along the fringes of the great Tethys Sea, a huge marine realm that extended from present-day eastern Asia to western Europe and consisted of warm, shallow lagoons and vast coastal zones as well as deep basins.

The members of the second branch fed on a diet of hard-shelled marine animals. They evolved unusual teeth that were variously peglike and used to pluck food from the seabed, or were rounded, flattened, and used to crush or break open shells. The bodies of these reptiles became wider, flatter, heavier, and increasingly more armor-plated. This was partly because a diet of bottom-dwelling prey caused predators to become heavier over time; being heavier meant less energy was expended in reaching the seabed to eat. These were the placodonts, a group of Triassic sauropterygians named for their distinctive teeth—"placodont" means "flat tooth" (see image 2).

Early placodonts were similar in shape to other early sauropterygians and were essentially lizardlike. However, the evolution of broader, increasingly armor-plated bodies eventually gave them a turtlelike appearance. A classic example of this was *Henodus* (see image 3), a lagoon-dwelling placodont unique to present-day southern Germany. The size of *Henodus*'s crushing teeth was reduced and they seem not to have played a major role in feeding or foraging. Small, toothlike denticles along the leading edge of its broad mouth could have been used to pick up small animals or scrape algae off submerged rocks, while hairlike bristles, tightly packed in gutterlike grooves lining the edges of its jaws, were presumably used to filter small animals or plants out of the water. *Henodus* remains enigmatic and experts have struggled to place it within the placodont family tree on account of its atypical features. It is one of the most unusual sauropterygians, and one of the strangest of all reptiles. **DN**

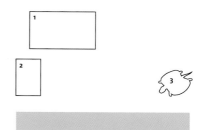

1 One of the earliest aquatic reptiles, slim *Mesosaurus* grew to 3 feet (1 m) in length. Its long tail provided the main propulsion; it may have been partly amphibious.

2 Shellfish-crunching *Placodus*, from around 240 million years ago, reached 7 feet (2 m) in length, and probably swam, paddled, and walked in shallow water.

3 At 3 feet (1 m) long, and bearing a striking resemblance to a turtle, *Henodus* was an enigmatic, highly evolved placodont.

240 Mya	235 Mya	235 Mya	230 Mya	225 Mya	201 Mya
Nothosaurus and related Triassic sauropterygians spread along the north Tethys Sea. Species of nothosaur now inhabit different environments.	Pachypleurosaurs found in China prove that even these early sauropterygians are viviparous. Viviparity is later inherited by all sauropterygians.	Pistosaurs, the most strongly adapted sauropterygians for life at sea yet, evolve. They can be imagined as prototypes of plesiosaurs.	The placodont *Henodus* evolves. Virtually all of its evolutionary history occurs in a giant lagoon located in Germany.	Early plesiosaurs evolve from pistosaur-like ancestors late in the Triassic. Fragmentary fossils of these early plesiosaurs are later found.	Mass extinction events at the end of the Triassic eliminate most sauropterygians; plesiosaurs are the only group to survive into the Jurassic.

Nothosaurus
MIDDLE–LATE TRIASSIC PERIOD

SPECIES *Nothosaurus giganteus*
GROUP Sauropterygia
SIZE 13 feet (4 m)
LOCATION Germany

One of the best known of the Triassic sea reptiles, *Nothosaurus* belongs to the sauropterygian group and inhabited the fringes of the great Tethys Sea that covered what is now western Europe all the way to eastern Asia. *Nothosaurus* was highly variable: very small, mid-sized, and gigantic species occurred in different areas and at different times, between around 240 million and 220 million years ago. The long, narrow shape of the *Nothosaurus* skull suggests that it grabbed prey by swiping its head sideways through the water, its jaws swiftly closing down on passing animals. The larger species would have been among the biggest and most powerful predators in their habitats and they preyed on other marine reptiles. Stomach contents from other nothosaurs show that they sometimes ate smaller sea reptiles, such as placodonts. Some related Triassic sauropterygians hunted in the same way as *Nothosaurus*, but others did not. *Simosaurus*—a close relative of *Nothosaurus*—had a flatter, broader, rounder snout and jaws, and was probably a suction-feeder, rapidly opening its mouth to engulf water and the prey it contained.

A large number of *Nothosaurus giganteus* fossils have been uncovered—mostly in Germany, dating back to the 1830s, providing plenty of opportunity to study this animal. Among the bigger species of *Nothosaurus*, the *N. giganteus* from Germany exceeded 13 feet (4 m) in total length and had a skull more than 28 inches (70 cm) long. In contrast, the tiny *N. haasi* from Israel had a skull only 5 inches (12 cm) long and the whole animal was less than 3 feet (1 m) in length. Small species such as this evolved alongside larger ones and descended from a *N. giganteus*-like ancestor. It is likely that they became dwarfed during their evolution as a response to a newly developing habitat or reduced food sources. **DN**

⦿ NAVIGATOR

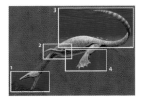

FOCAL POINTS

1 SKULL AND TEETH

The skull was long, narrow, and shallow, with interlocking teeth lining the jaws in front of the eyes. Large and long spaces for jaw-closing muscles show that *N. giganteus* must have had a very swift, snapping bite, which it would have used to catch passing prey. Skull and teeth shape indicate that nothosaurs such as *N. giganteus* mainly ate fish and soft-bodied squid.

3 TAIL

The *N. giganteus* tail was proportionally shorter than that of earlier sauropterygians. It certainly had a role in steering while swimming, but was less important than in its ancestor groups. Little study of nothosaur swimming behavior has occurred, but the paddlelike shapes of the limbs, combined with evidence from swimming tracks, indicate the use of fore limbs in a rowing motion.

2 SKIN AND SCALES

All sauropterygians were diapsid reptiles, and almost certainly started their evolutionary history with scaly skin. However, it is unknown whether scales were retained throughout the whole of their history. It is possible that in some groups the scales were lost and exchanged for a smoother skin. This could have reduced water resistance, thereby increasing swimming speed and saving energy.

4 PADDLELIKE LIMBS

Studies of limb, shoulder, and hip bones suggest *N. giganteus* could walk on land in conventional reptilian fashion. However, the bones of its hands and feet suggest that they formed more paddlelike structures, rather than separate-digit hands and feet shown here. Hundreds of fossil tracks from China made by nothosaurs "punting" along the seabed appear to show these paddlelike limbs.

NAMING SPECIES

As these organisms are known only from fossil remains, new finds can reveal that what were thought of as distinct species are in fact variations of the same species. The first specimen named, the reference "model" for subsequent classifications, was *N. mirabilis*, as described in 1834 by paleontologist Georg zu Münster (1776–1844). Following this, similar creatures were named *Chondriosaurus*, *Dracontosaurus*, *Elmosaurus*, *Kolposaurus*, *Menodon*, *Oligolycus*, *Opeosaurus*, *Paranothosaurus*, and *Shingyisaurus*. As more evidence emerged, experts reassigned names, so 50 species from those genera are now regarded as 12 to 14 species of *Nothosaurus*. Related genera include *Keichousaurus* (right), up to 10 feet (3 m) long, from China.

ICHTHYOSAURS

Of all the reptile groups that took to life at sea, by far the most specialized were the ichthyosaurs—the so-called "fish lizards" that lived between 250 million and 100 million years ago. Complete ichthyosaur fossils (see image 1) from 180-million-year-old rocks in Germany show that the ichthyosaurs of the Jurassic and Cretaceous periods were similar in body shape to modern sharks or dolphins. The body was torpedo-shaped and streamlined, with a triangular dorsal (back) fin and a vertical, two-lobed tail, although only the lower lobe was supported by bone. This body shape was very unusual compared to other reptiles, making it difficult to ascertain where ichthyosaurs belong in the reptile family tree.

Although the earliest known ichthyosaur fossils show that these reptiles were paddle-bearing species adapted for swimming, a definite ichthyosaur "prototype" remains elusive. However, it is possible that ichthyosaurs were closely related to the small, long-jawed hupehsuchians, known from rocks in China from the Triassic period. If this were proven, it would demonstrate that ichthyosaurs descended from aquatic ancestors and that many adaptations for life in water were already present in ichthyosaur ancestors before the evolution of the ichthyosaurs themselves. Analyzing the area behind the eye of the ichthyosaur skull has shown that it is highly evolved from most of the typical reptile patterns, but it seems to have originally possessed the set of skull openings considered typical for diapsids: the great group of reptiles that includes lizards and archosaurs such as dinosaurs and pterosaurs.

Advanced ichthyosaurs were "thunniform," meaning that they possessed the streamlined body shape seen in modern fast swimmers of the open ocean, such as tuna, white sharks, and dolphins. The thunniform body shape is associated with a predatory lifestyle and with the ability to produce and retain body heat. Thunniform ichthyosaurs were among the first animals to evolve this behavior and body shape, since they appeared some considerable time prior to dolphins, and they also long predated tuna and thunniform sharks. However, not all

KEY EVENTS

250 Mya	245 Mya	245 Mya	230 Mya	200 Mya	200 Mya
The oldest ichthyosaurs, including *Parvinatator* from British Columbia, appear. They have paddles and bodies suited for swimming.	Mixosaurs, a group of archaic ichthyosaurs with blunt-tipped or mound-shaped teeth, occur across Europe and Asia. They are less than 10 feet (3 m) long.	Predatory ichthyosaurs, such as *Thalattoarchon* from North America, with strong jaws and large, bladelike teeth, are the top predators of the oceans.	Parvipelvians appear: they are the first ichthyosaurs to have paddle-shaped hind limbs and reduced hip girdles. The oldest is *Hudsonelpidia*.	Advanced thunniform ichthyosaurs evolve from temnodontosaur-like ancestors. The oldest, such as *Ichthyosaurus*, are from Europe.	The majority of ichthyosaurs are large temnodontosaurs and leptonectids, with streamlined bodies and slimmer paddles than earlier groups.

ichthyosaurs were like this. Prior to the evolution of the thunniform group, ichthyosaurs, such as shastasaurs and cymbospondylids, were long-bodied and long-tailed and species existed with blunt-tipped, flat, or rounded tooth crowns. Shastasaurs and cymbospondylids were large predators. One of them, *Shastasaurus sikanniensis* from British Columbia, exceeded 66 feet (20 m) in length, which made it one of the largest animals ever to have evolved. It appears to have been toothless, so would have grabbed fish or squidlike prey with the toughened edges of its jaws. Other ichthyosaurs of this sort had huge, bladelike teeth and were predators of large marine animals. *Thalattoarchon saurophagis* (see image 2) was more than 28 feet (8.5 m) long and possessed a strongly built skull and large teeth with cutting ridges called keels.

Toward the end of the Triassic, 230 million to 220 million years ago, a shorter-bodied ichthyosaur group appeared. The shoulder blade changed from being large and fan-shaped to being smaller and slimmer; the pelvic girdle became smaller; and the hind limb bones became shorter and smaller, too. These ichthyosaurs are called parvipelvians and started their history as small, nimble predators of fish and other fast-swimming prey. By 200 million years ago, parvipelvians had increased substantially in diversity, while other ichthyosaur groups had become extinct. *Temnodontosaurus* was an especially large, long-snouted early parvipelvian, exceeding 40 feet (12 m) in length. Sizable, conical teeth and robust jaws suggest that it occupied the role of top predator, previously taken by the shastasaur-type ichthyosaurs. However, it was more streamlined and shark-shaped than earlier kinds. In some specimens, the eyeballs were more than 10 inches (25 cm) wide, which made them among the largest eyes ever to have evolved. However, later ichthyosaur eyeballs evolved that were even larger in relation to the size of the animal. A swordfishlike group, the leptonectids, appeared at the same time as the temnodontosaurs. In some cases the upper jaw was several times longer than the lower one.

A final parvipelvian group, termed Thunnosauria, had also appeared by this time. Early members include the familiar thunniform *Ichthyosaurus* of western Europe (see p.246) and the similar but larger *Stenopterygius*. The thunnosaurians were the only ichthyosaurs to survive into the late Jurassic and beyond, 150 million years ago, and they were by far the most strongly adapted for long-distance, open-water swimming and diving. Proportionally gigantic eyeballs show that some—such as *Ophthalmosaurus* of the Jurassic seas (see image 3)—could see in dim light at great depth. They would have been deep-diving predators of squid. Other *Ophthalmosaurus*-like ichthyosaurs were generalist predators of surface waters. Preserved stomach contents show that they ate fish, turtles, and seabirds as well as other mid-sized prey. They remained major players in many habitats of the world's oceans until their final extinction about 100 million years ago. **DN**

1 Many species of ichthyosaur, including this *Ichthyosaurus*, were preserved in marine muds and similar sediments, giving great details of their anatomy.

2 The bladelike teeth of *Thalattoarchon saurophagis* resembled those of predatory dinosaurs rather than other ichthyosaurs, indicating that this was a super-predator that tackled the largest prey.

3 The huge eyes of *Ophthalmosaurus* were protected and supported by spokelike circles of bone called sclerotic rings.

175 Mya	160 Mya	145 Mya	130 Mya	110 Mya	100 Mya
Ophthalmosaurids, with huge eyes and broad paddles, evolve from *Stenopterygius*-like ancestors. Their key adaptations relate to a deep-diving lifestyle.	*Ophthalmosaurus*, the most successful of its group, lives virtually worldwide, wherever there is deep water inhabited by squid and related mollusks.	The Jurassic ends with an extinction event that kills off many animals, but ichthyosaurs remain unaffected.	*Ichthyosaurus*-like *Malawania* survives in what is now Iraq, and is the only Cretaceous ichthyosaur that is not a member of the group Ophthalmosauridae.	Members of several ophthalmosaurid groups—generalist predators with broad fore paddles and robust teeth—occupy tropical to temperate seas.	The last ichthyosaurs—members of the group Ophthalmosauridae—disappear, possibly due to a major change in marine environments at this time.

Ichthyosaurus
LATE TRIASSIC PERIOD—EARLY JURASSIC PERIOD

SPECIES *Ichthyosaurus communis*
GROUP Ichthyosauria
SIZE 6½ feet (2 m)
LOCATION Europe, especially England, Switzerland, Belgium, and Germany

NAVIGATOR

One of the world's best-known ichthyosaurs is the namesake member of the group, *Ichthyosaurus communis*, from the Jurassic rocks of Dorset, England. It is the quintessential thunniform ichthyosaur: a shark-shaped reptile with long pointed jaws, huge eyes, fore paddles composed of numerous bones, small hind paddles, a triangular dorsal fin, and a vertical tail fin. Several species have been named, differing in length, shape, proportions of the skull, and anatomy of the paddles. None is longer than 8 feet (2.5 m), and remains date back to more than 200 million years ago. Part of the reason for this reptile's notoriety is that it was discovered very early in the history of paleontology. Members of the Anning family from Dorset excavated the first specimens in the early 1800s. Scientists initially misidentified these fossils as ancient fish, crocodiles, or fish-crocodile intermediates, but by the 1830s it was clear that they represented a unique, extinct group of sea-going reptiles. Numerous additional specimens were found in southwest England as well as in Belgium and Switzerland. This uniquely European range is surprising for an animal adapted for open ocean life, which suggests that it was specialized for a particular resource or environment that occurred only within this region. *Ichthyosaurus* was not the only ichthyosaur of this time and place, and remains are often found alongside those of the larger, more predatory *Temnodontosaurus*. Several smaller ichthyosaurs have been found preserved within the mouths and stomachs of *Temnodontosaurus* specimens and *I. communis* would have been prey for its larger cousin. Its slender, pointed jaws and numerous small, conical teeth suggest a diet of fish and squidlike mollusks. **DN**

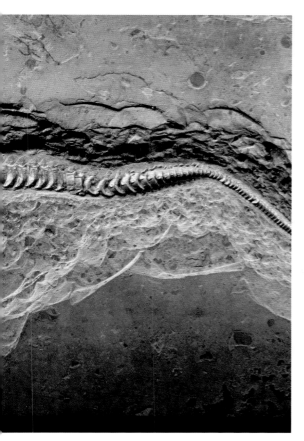

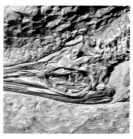 FOCAL POINTS

1 EYE SOCKET AND EAR BONES
The eye socket was occupied by a sclerotic ring formed of overlapping bony plates, embedded in the outer layers of the eyeball to prevent it from deforming when the muscles contracted during focusing. The ear bones do not indicate sensitive hearing, but large nostrils suggest a good sense of smell.

2 PADDLES
I. communis's fore paddles were formed from mostly rectangular, tightly fitting bones. The paddle was a hydrodynamic blade, used in steering and braking. The two lower arm bones in land reptiles were larger than the other paddle bones, but even they were flattened blocks. There were no elbow or wrist joints.

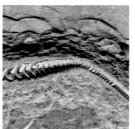

3 FORKED TAIL
With only bones as evidence, it was initially assumed that thunniform ichthyosaurs such as *I. communis* had straight tails. However, the bend at the tip was unusual. Ichthyosaur specimens from Germany revealed that the tail was forked, like that of a shark, with the tail vertebrae angled down into the lower lobe.

EVIDENCE OF LIVE BIRTH

Ichthyosaurs were so highly evolved for life in the sea that they could not come onto land to lay eggs in the way that today's marine reptiles, the sea turtles, do. It also seemed unlikely that they laid eggs in water because the reptile egg shell is not designed or optimized for salt-water immersion. It was suspected that instead they gave birth to fully formed babies, in the same way that mammals do. Fossils discovered during the 1850s proved this assumption, revealing preserved ichthyosaur mothers and babies who died at birth. In eastern China in 2014, the fossils of *Chaohusaurus geishanensis*, one of the first ichthyosaur species, were uncovered with three babies. One baby ichthyosaur was still inside the mother, another was exiting the pelvis, in the process of being born (right), and the third was outside. These remains date back to 248 million years ago, which reveals that ichthyosaurs were living much earlier than was previously thought.

DINOSAUROMORPHS

Dinosaurs were archosaurs—diapsid reptiles that also include the crocodile group and pterosaurs. Around 250 million years ago archosaurs divided into two lineages: one led to crocodylomorphs, the other to pterosaurs and dinosaurs. This latter group, ornithodirans, was characterized by the presence of a hingelike ankle joint, an especially long lower leg and a proportionally long neck. Pterosaurs—the membranous-winged archosaurs of the Mesozoic era—evolved from the earliest ornithodirans, as did the group that would eventually give rise to dinosaurs. These are known as the dinosauromorphs, a name meaning "dinosaur-shaped" or "of dinosaur form." Dinosaurs were the dominant land animals throughout the Jurassic and Cretaceous periods, between 200 million and 66 million years ago. For a long time, a lack of fossils inhibited the understanding of their origins. Until recently, the oldest known species were large, advanced members of groups typical of the Jurassic. However, it is now known that they appeared as far back as 245 million years ago during the middle Triassic period.

Early dinosaurs were one of many archosaur groups present during the Triassic, and both dinosaurs and other closely related dinosauromorph groups were minor players in ecosystems controlled by crocodile-line archosaurs. Among the earliest dinosauromorphs was *Marasuchus* from Argentina (see p.250),

KEY EVENTS

245 Mya	245 Mya	245 Mya	240–235 Mya	230 Mya	230 Mya
The earliest known dinosauromorphs (dinosaur-shaped) evolve a few million years after the "Great Dying" at the end of the Permian period.	Tracks from Poland show that small dinosauromorphs lived alongside other archosaurs. Archosaurs split into crocodile- and dinosaur-line lineages.	The oldest silesaurids, such as *Asilisaurus*, appear. Probable early members of the dinosaur lineage, such as *Nyasasaurus*, live at this time.	Lagerpetids and *Marasuchus* inhabit Argentina. These small, bipedal, predatory dinosauromorphs are often thought of as dinosaur prototypes.	Early members of all three main dinosaur lineages—theropods, sauropodomorphs, and ornithischians—are present in Argentina, alongside silesaurids.	The first sizable dinosaurs, such as *Herrerasaurus*, appear in South America. However, crocodile-line archosaurs remain the top predators.

a small, bipedal predator (walking on two legs). A group of similar, long-legged dinosauromorphs—the lagerpetids—is known from Argentina and the United States. They had unusual feet in which the first and second toes were shorter than the third and fourth. This was almost certainly an adaptation for running and possibly for leaping. For a long time, *Marasuchus* was the only well-known potential dinosaur "prototype," and it was assumed that bipedal, carnivorous habits were ancestral for dinosaurs and that they evolved from *Marasuchus*-type animals. However, bipedal habits and a carnivorous lifestyle were by no means typical for early dinosauromorphs.

An entirely new dinosauromorph group was recognized in 2010: the silesaurids. In contrast to *Marasuchus*, the proportions of their long, slender fore limbs show that they were quadrupedal (walking on four legs). Also, the toothless, beaklike lower jaw tip, with peglike teeth farther back, indicates that silesaurids were not carnivorous predators, but herbivores or omnivores. However, their anatomical features show that they were more closely related to dinosaurs than were lagerpetids or *Marasuchus*. Silesaurids are now known from Poland, Brazil, the United States, and Tanzania, indicating that they occurred at diverse latitudes during the middle Triassic period. Not only were they geographically widespread, but the whole group lived for many millions of years. The oldest members come from rocks 245 million years old, while the youngest are from 212-million-year-old sediments.

As close relatives of dinosaurs, silesaurids indicate that dinosaurs did not emerge as bipedal predators, but as quadrupedal omnivores or herbivores. This idea is supported by the diversity of dinosaurs themselves, because two of the three major dinosaur lineages—the long-necked sauropodomorphs and the ornithischians (bird-hipped dinosaurs)—consist of herbivores. In addition, dinosauromorphs were relatively rare during their time. For much of the Triassic, the role of large-bodied herbivores and carnivores was taken by archosaurs of the crocodile lineage (see image 2), most of which were larger than early dinosauromorphs or dinosaurs.

The first known dinosaur may have been *Nyasasaurus* from Tanzania (see image 1), which dates to 245 million years ago and was 10 feet (3 m) in length. Whether it was a true dinosaur is unknown because fossil remains are scant, but *Herrerasaurus* (see p.252), from 231 million years ago, was certainly a dinosaur. It was millions of years before dinosaurs diversified and became important animals in ecological terms. It would appear that the eventual success of dinosaurs was due to the elimination of crocodile-line archosaurs during the mass extinction events 220 million and 200 million years ago. Once these competitors had died out, new, large dinosaurs appeared. Theropods (meat-eating dinosaurs) became specialized predators of animals, while sauropodomorphs and ornithischians became successful herbivores. After a slow and protracted early history, the Age of Dinosaurs had finally begun. **DN**

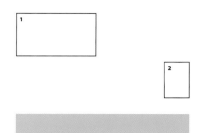

1 Assuming that further finds show *Nyasasaurus*, named from its discovery site near Lake Nyasa in Kenya, is a dinosaur, this pushes back the origins of the group 15 million years before the next well-known representatives, such as *Herrerasaurus* and *Eoraptor*.

2 Fossils of *Asilisaurus*, a silesaurid archosaur from Tanzania 245 million years ago, show that it was a slim, long-legged, 10-foot (3 m) quadruped. A relatively newly recognized group, silesaurids are seen as a "sister" group to dinosaurs.

220 Mya	220 Mya	212 Mya	210 Mya	204 Mya	201 Mya
Theropods appear with slim, shallow snouts and a set of features that make them more advanced than *Herrerasaurus*. They are termed neotheropods.	A mass extinction kills off various reptile groups and many synapsid lineages. Dinosauromorphs and early dinosaurs survive unscathed.	Nondinosaurian dinosauromorphs, such as lagerpetids and silesaurids, thrive.	Crocodile-line archosaurs—the main predators and competitors of dinosauromorphs—are at the peak of their diversity.	Theropods and sauropodomorphs that strongly resemble those of the Jurassic are known from North America, Europe and elsewhere.	A mass extinction ends the Triassic, removing most crocodile-line archosaurs. Dinosaurs are now the only major group of big, land-living animals.

Marasuchus
MIDDLE TRIASSIC PERIOD

SPECIES *Marasuchus lilloensis*
GROUP Dinosauromorpha
SIZE 28 inches (70 cm)
LOCATION Argentina

The species *Marasuchus lilloensis* is one of several Triassic archosaurs that was frequently imagined as a protodinosaur: a close relative of, and similar to, the dinosaur group's original ancestors. All dinosaur-line archosaurs are included within the Dinosauromorpha, a group that emerged 250 million years ago during the phase of evolution following the massive end-of-Permian extinction. *Marasuchus* was small at around 20 to 28 inches (50–70 cm) long, lightly built and bipedal, walking if not racing on its long, powerful rear legs. It had short slender fore limbs and an S-shaped neck. Its name means "mara crocodile" after the mara: a furry, harelike, fast-moving rodent from present-day South America. *Marasuchus* lived around 235 million years ago and its remains have provided substantial information on the evolutionary changes that preceded dinosaurs. However, now that many early dinosauromorphs are known, it is no longer clear that dinosaurs evolved from *Marasuchus*-like animals. They may have originated from quadrupedal herbivores, such as the silesaurids.

Compared to *Marasuchus*, dinosaurs had longer necks, larger muscle attachment sites on the arm and thigh bone, and hips better suited to take body weight and the stresses of movement. In addition, dinosaurs had a simple, hingelike ankle joint, very different from the more complex ankle present in crocodile-line archosaurs. These features evolved in lightweight animals and proved advantageous at small size, but more importantly they were crucial in allowing dinosaurs to reach large and even giant size. Erect limbs held beneath the body, strong hip joints that allowed the transfer of body weight to these limbs, and ankle joints that prevented the twisting of the foot all proved significant. It was these features, among others, that enabled dinosaurs to exceed other land-living animals in size. Several other anatomical traits that were also carried to extremes in later dinosaur groups, such as proportionally long necks and lightweight fore limbs, also had their origins in small protodinosaurs such as *Marasuchus*. **DN**

✦ NAVIGATOR

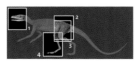

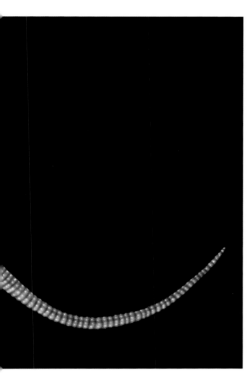

CLOSE RELATIVE

Named in 1971, *Lagosuchus talampayensis* (below) was very similar to *Marasuchus* and lived around the same time, 230 million years ago, in the same region of Argentina. Its fossils are fragmentary, however, therefore some experts believe it to be *Marasuchus*. In fact, one species, *Lagosuchus lilloensis*, was restudied and reclassified in 1994 by US paleontologist Paul Sereno (b.1957) as *Marasuchus*, leaving the other species, *L. talampayensis*, as the sole representative of the genus. Until more fossils are found, it will remain unknown whether *Lagosuchus* and *Marasuchus* were one and the same, and how closely related they were to the very first dinosaurs.

👁 FOCAL POINTS

1 SKULL AND TEETH
Only parts of the skull have been found, from the braincase—the upper jaw region just behind the nostril (maxillary area)—and a few teeth. The curved, slightly serrated teeth have not been well preserved, but they seem most suited to grabbing bugs, worms, and similar small prey, and would have coped with a more omnivorous diet.

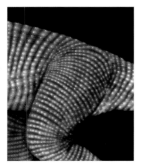

2 HIP JOINT
On the side of the hip bone, the socket for the thigh bone is an oval-shaped concavity; only a small gap in its lower half perforates the bone. This gap grew larger during dinosauromorph evolution, eventually occupying the entire thigh bone. Consequently, the dinosaur thigh bone is a large opening, not just a shallow concavity.

3 THIGH BONE
In dinosaurs, the head of the thigh bone (femur) is turned inward at about 90 degrees relative to the bone's shaft. *Marasuchus* had an "incipient" version of this condition: the head of the femur was turned inward to a degree, but not as strongly as in dinosaurs. This change occurred in step with the opening of the thigh bone in the pelvis.

4 FEET
Marasuchus had extremely long, narrow feet. All five of the metatarsals—the rodlike bones that make up the sole of the foot—were especially slender and close together. The fifth toe was lacking, but the remaining four were long, slim and arranged in parallel. Long, narrow feet of this sort evolve as animals become specialized for fast running or leaping.

Herrerasaurus
MIDDLE–LATE TRIASSIC PERIOD

SPECIES *Herrerasaurus ischigualastensis*
GROUP Saurischia
SIZE 14 feet (4.5 m)
LOCATION South America

More than 230 million years ago dinosaurs were small and insignificant. Their teeth and jaws indicate that they were omnivores or predators of small, lizard-sized prey. These dinosaur pioneers included the 3-foot (1 m) *Eoraptor,* the slightly larger 4-foot (1.2 m) theropod *Eodromaeus,* and the similar-sized early sauropodomorph *Panphagia,* meaning "eat-all." For much of the Triassic these tentative beginners were outnumbered by the large, dangerous, crocodile-line archosaurs that dominated life on land at that time. However, 230 million years ago, a new kind of dinosaur evolved in South America: *Herrerasaurus ischigualastensis.* It was the first known sizable dinosaur, although it was not particularly large compared to later kinds, and in total length it reached 14 feet (4.5 m) and weighed only 440 pounds (200 kg), similar to a small pony. *Herrerasaurus* is also significant for being the first sizable dinosaurian predator. Long, curved teeth grew from its upper jaw and three large, strongly curved claws tipped its fingers. A joint in the middle of the lower jaw suggests that it could widen its jaws slightly when swallowing large objects.

The first discoveries of *Herrerasaurus* were too restricted to permit a secure analysis of its place in the reptile and dinosaur family tree. Various experts suggested it might have been a predinosaur, a cousin, an early meat-eating dinosaur or even an early plant-eater. In 1988, more remains were uncovered, including a relatively complete skull. From these and other skeletal features, *Herrerasaurus* is now usually placed in the saurischian group (lizard-hipped dinosaurs; see p.260), which includes the predatory theropods and long-necked sauropodomorphs. Long finger bones and an enlarged angled ending, or "boot," on the pubic bone show that *Herrerasaurus* was most probably an early theropod, but was not in the group that contained the advanced, more birdlike theropods. **DN**

✪ NAVIGATOR

👁 FOCAL POINTS

1 PUBIC BONE

Herrerasaurus had an unusual pubic bone, pointing down from the hip socket and slightly backward. Ancestrally, dinosaurs had pubic bones that pointed downward and forward. This was true of predinosaur dinosauromorphs and early saurischians. Elsewhere within theropods, this is seen in birdlike coelurosaurs, a group that evolved millions of years after *Herrerasaurus* was extinct.

3 FINGERS

Herrerasaurus possessed the remnants of five fingers, but the three inner ones were large with enormous claws, while the outer two had withered away to nothing, as evolutionary relics or vestiges. This reduction and loss of the outer two fingers are part of an evolutionary trend seen repeatedly in archosaurs, and *Herrerasaurus* provides a good example of this.

2 SNOUT

The snout of *Herrerasaurus* is deep and rectangular, whereas early theropods (and other early dinosaurs) had shallow, pointed snouts. Similarly shaped snouts evolved several times in predatory archosaurs because they transferred bite force through the teeth into the bodies of prey animals. *Herrerasaurus* had a unique profile that made its skull very different from that of other early theropods.

4 CLAWS

The claws of *Herrerasaurus* present clues about its predatory behavior. It would have grabbed or raked prey with its hand claws, possibly pinning the victim to the ground with its feet. The long upper jaw teeth would have then been used to slash through or pierce tissue. Like most large predators, the majority of its prey was comparatively small—not much larger than its head.

LOCAL DISCOVERY

The first fossil specimens of *Herrerasaurus* were recognized by a local herder and fossil enthusiast, Victorino Herrera, in northwest Argentina in 1959. The dinosaur was officially named after him in 1963. The species name derives from the Ischigualasto-Talampaya Formation, a series of rocks in that area that date from 231 million to 225 million years ago. The scenery there is bare and rocky with strange eroded formations of stone, hence its name "Valley of the Moon" (right). In addition to *Herrerasaurus*, *Eoraptor*, and other early dinosaurs, findings here include rhynchosaurs, cynodonts, dicynodonts, and temnospondyls, which give an unparalleled view of evolution in the Triassic.

CROCODILES AND KIN

The 25 living species of crocodile, alligator, and gharial (see p.258) are amphibious predators of lakes, rivers, and tropical seas. Successful and adaptable, they are the top predators of many waterways—their conical teeth, lightning-fast snapping jaws, and excellent senses of touch, smell, hearing, and sight making them supreme hunters of fish, mammals, and other animals. Modern crocodiles and kin—grouped together as the crocodylians—are more diverse and successful than is often appreciated. However, they are the surviving remnants of a previously far more diverse group, the crocodylomorphs (crocodile-shaped). The members of several extinct crocodylomorph groups differed from living species. There were land-living omnivorous or herbivorous kinds, terrestrial predatory types, and sea-going versions with flippers and tail flukes. Several major evolutionary trends can be seen in the crocodylomorph fossil record. The bony palate becomes larger and more important in helping the skull resist stress; the way the vertebrae lock together changes over time, and the armor plates along the neck, back, and tail alter in shape and complexity.

The oldest crocodylomorphs are called sphenosuchians (see image 3) and the oldest have been found in 235-million-year-old rocks. They possess the key features unique to crocodylomorphs, such as elongated wrist bones but they were very different from modern crocodylians. They were small, long-limbed, terrestrial predators resembling lizards, and the shape of their teeth and jaws indicates that they ate insects and small animals. Other early crocodylomorphs were similar but they evolved wider, more reinforced skulls, a more complicated ear, and broader armor plates along the neck, back, and tail. These animals are called protosuchians, and they would have been better than the sphenosuchians at subduing larger prey. They also would have been predominantly terrestrial and tended to be less than 6 feet (2 m) long.

A burst of crocodylomorph evolution occurred approximately 210 million years ago, late in the Triassic period.

KEY EVENTS

235 Mya	230 Mya	210 Mya	200 Mya	125 Mya	85 Mya
The oldest crocodylomorphs, the sphenosuchians, live in Europe and the Americas. They are long-limbed predators.	Protosuchians evolve from sphenosuchians as small, terrestrial predators of small animals, but their skulls resemble those of modern crocodiles.	A burst of evolution sees the evolution of thalattosuchians, the notosuchians of the southern continents, and neosuchians, which survive today.	Sea-going thalattosuchians equipped with paddles, vertical tail fins and other specializations inhabit Europe and South America.	The last sea-going thalattosuchians become extinct, coinciding with a global cooling, argued by some as a Cretaceous period ice age.	The earliest members of modern crocodylian lineages—ancestors of living crocodiles, alligators, and gharials—occur in North America.

Protosuchian-like ancestors gave rise to three major branches, all of which evolved in very different directions. One branch, the thalattosuchians (see image 1), became larger and longer-snouted and were adapted for life at sea. They reduced their armor plating, evolved paddlelike limbs and changed the tail to a broad sculling organ. Eventually the tail became a more specialized structure with a vertical fin at its tip. Many were slender-snouted predators of small fish, but one group, the geosaurines, evolved deep skulls, stronger jaws, and more bladelike teeth, presumably to tackle larger prey.

The second major crocodylomorph branch, the notosuchians, comprised a variety of omnivores, carnivores, and herbivores. Many retained the long limbs of ancestral sphenosuchians and protosuchians and they were capable walkers and runners. One interesting feature concerns their teeth and jaws. The teeth in the jaws of a modern crocodylian are all alike: conical, suited for grabbing and dismembering animal prey. However, some notosuchians had incisorlike, caninelike, and molarlike teeth, similar to mammals, which must have been used for nipping, tearing, and slicing. These are sometimes called mammal-toothed crocodylomorphs. Other notosuchians could slide their jaws backward and forward in a chewing motion to break up hard-shelled animals or plants.

The notosuchian group also included a number of large terrestrial predators: the sebecids and baurusuchids. The teeth of these were laterally compressed (narrow) and bladelike—similar to those of theropods (meat-eating dinosaurs). In fact, teeth of the South American sebecid *Sebecus* discovered in rocks 50 million years old were misidentified initially as those of a dinosaur that had survived the extinction event 66 million years ago. Much of notosuchian evolution occurred in South America, where these diverse crocodylomorphs formed a major part of the terrestrial fauna and shared the environment with herbivorous and carnivorous dinosaurs. Several groups survived the extinction event 66 million years ago, although none persisted to modern times.

The third major crocodylomorph branch, the neosuchians, included a variety of short- and long-jawed predators. Among these were the ancestors of modern crocodiles, alligators, and gharials. Early neosuchians were less than 6 feet (2 m) long. However, giant size evolved in an especially long-jawed group—the tethysuchians—associated with Africa, Europe, and South America. Some tethysuchians were coast-dwelling and ate shellfish, turtles, and fish, but *Sarcosuchus* (see p.256) was a dinosaur killer. The alligator group also gave rise to some huge species. *Deinosuchus* (see image 2) from North America also preyed on dinosaurs, while *Purussaurus*, from 8-million-year-old South American rocks, was a broad-skulled caiman that would have eaten large mammals and turtles. Both were 36 feet (11 m) or more in length. **DN**

1 *Teleosaurus*, from 160 million years ago, was a 10-foot (3 m) thalattosuchian highly adapted to water, with slim gharial-like jaws and small, sharp, fish-grabbing teeth.

2 A fossil skull of the extinct crocodilian reptile, *Deinosuchus hatcheri*. Growing to around 30 feet (9 m) long, this crocodile lived during the Late Cretaceous period, around 80 million years ago.

3 Adult specimens of the Triassic *Terrestrisuchus* approached 3 feet (1 m) in length and weighed less than 33 pounds (15 kg). It was a small lizardlike creature with long legs.

80 Mya	75 Mya	66 Mya	50 Mya	8 Mya	5 Mya
Diverse notosuchians live in South America. There are omnivorous and herbivorous kinds as well as large, deep-skulled predators such as the baurusuchids.	The giant alligator *Deinosuchus* inhabits southern North America, occupying both sides of the seaway that divides the continent.	Mass extinction reduces crocodylomorph diversity. Many South American notosuchian groups disappear but tethysuchians survive.	*Sebecus* persists in South America, as do related forms. They prey on the newly evolving mammals of the continent.	Giant members of the alligator family evolve in South America, such as broad-skulled *Purussaurus* and "duck-faced" *Mourasuchus*.	Gharials disappear from South America, leaving tropical southern Asia as the last place inhabited by this once wide-ranging group.

Sarcosuchus
EARLY CRETACEOUS PERIOD

SPECIES *Sarcosuchus imperator*
GROUP Pholidosauridae
SIZE 39 feet (12 m)
LOCATION North Africa
and Brazil

Much of crocodile evolution occurred at small body size. Virtually all of the early sphenosuchians and protosuchians, as well as the early members of the several groups that evolved from them were less than 7 feet (2 m) long. However, giant size evolved on several occasions, mostly among aquatic lineages in which water helped buoy up their weight. Perhaps the largest member of all time among the whole crocodile group, the crocodylomorphs, was *Sarcosuchus imperator*. This long-jawed, aquatic predator is best known from fossils in 110-million-year-old rocks in North African countries such as Algeria, Morocco, Libya, and Nigeria. The biggest *S. imperator* specimens were almost 39 feet (12 m) long and weighed about 9 tons (8 tonnes). The largest skulls were around 5 feet (1.5 m) long, making *S. imperator* more than twice as long and more than eight times as heavy as living crocodylians, including the saltwater crocodile. *S. imperator* was so powerful that its diet would not have been restricted to fish and aquatic prey; it seems likely that it tackled dinosaurs on occasion. A variety of dinosaurs lived alongside it, and the small- and mid-sized species could have been targets. *S. imperator* belonged to a long-jawed group of tethysuchians called pholidosaurs. However, it was unlike many others in the group because it was a freshwater animal. It is likely that the group was originally marine and later moved to freshwater. This transition from marine to freshwater life is a pattern seen repeatedly in crocodylomorph evolution. Brazilian fossils indicate that *S. imperator* was not unique to Africa. A transcontinental distribution is expected from animals of this time, because Africa and South America were still connected and the South Atlantic was yet to form. Several other animals of the middle Cretaceous, including dinosaurs and freshwater fish have been found in western Africa and also in eastern Brazil. **DN**

✸ NAVIGATOR

FOCAL POINTS

1 SNOUT MOUND

The outer edges of *S. imperator*'s upper jaw were almost parallel, the tip of the upper jaw was longer than the lower one and noticeably overhung it, and a giant, hollow mound was located over the nostrils. This mound, or bulla, looks similar to the ghara present in modern gharials. Gharials use this structure to enhance their mating calls, and it is possible that *S. imperator* did too.

3 GASTRALIA

Living and extinct crocodiles, as well as certain members of other reptile groups such as theropods, pterosaurs, and plesiosaurs possess long, slim bony elements in the abdominal wall known as gastralia. Sometimes called "belly" ribs, they do not connect to the main skeleton. Their functions seem to be to anchor abdominal muscles, to aid respiration, and to provide a degree of physical protection.

2 SNOUT AND JAW PROPORTIONS

Compared to others in the group, the snout and jaws were—while very long—among the shortest and broadest in proportion to the whole body. This suggests that *S. imperator* was adapted for tackling large, powerful animals. The same evolutionary pattern is seen in other crocodylomorph groups: giant species with stocky jaws and robust teeth evolved from smaller, slimmer-jawed ancestors.

4 LEGS

S. imperator was probably too big and too bulky to lift its body up off the ground, with limbs too small relative to its overall size. Most modern crocodiles are capable of walking on land when the need arises, and the body is held up in what is called the "high walk" or even "gallop." However, it is unlikely that *S. imperator* would have walked or run in this position.

SAHARAN WETLANDS

The Sahara today is the world's biggest dryland desert where only a few specially evolved plants and animals survive. Through prehistory and even recorded history, the climate was moist, and plants and creatures thrived. Fossil remains of water-dwelling plants, fish, and turtles, and various rocks laid down in aquatic environments show that 100 million years ago the region was crossed by rivers, with shallow pools and marshy wetlands (left). Here *Sarcosuchus* lurked, probably camouflaged, ready to grab its victim in vicelike jaws. Many dinosaurs lived here at the same time, including the 39-foot (12 m), 9-ton (8 tonne) predator *Carcharodontosaurus*, the even bigger *Spinosaurus* (see p.300), and the sail-backed, 23-foot (7 m) *Iguanodon*-like herbivore *Ouranosaurus*.

Gharial
NEOGENE PERIOD—PRESENT

SPECIES *Gavialis gangeticus*
GROUP Crocodyloidea
SIZE 20 feet (6 m)
LOCATION India and Nepal
IUCN Critically Endangered

⟨✛⟩ **NAVIGATOR**

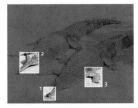

U nlike most living crocodiles and alligators, the gharial of the Indian subcontinent, *Gavialis gangeticus*, is descended from a line of crocodylians better adapted to life in water than a mixed aquatic and terrestrial existence. As a specialized fish-eater, it does not possess the broad snout, strong, slightly blunted teeth, or immensely powerful jaw muscles evolved by cousins that pursue much more sizable prey—the zebra of the Nile crocodile and the deer of the American alligator. The gharial's family can be traced back to the late Cretaceous, 70 million years ago, before the great dinosaurs became extinct. Well-preserved fossils from the late Eocene, 35 million years ago, in North Africa show that an ancestor, *Eogavalias*, had already established the trend of a slim, long snout. A later representative was *Gryposuchus* from South America, which attained the impressive length of 33 feet (10 m) and a weight approaching 2¼ tons (2 tonnes). Both prehistoric types became common and widespread. Remains of another relative, *Aktiogavialis*, from Puerto Rico in the Caribbean, suggest that this crocodylian lived in salt water. Fossils of *Rhamphosuchus*, a Miocene epoch representative from South Asia, show a gharial-type crocodylian perhaps 36 feet (11 m) long. Today's gharial is not quite as huge as this antecedent but, at up to 20 feet (6 m) in length, it rivals its fellow crocodylian, the saltwater crocodile, as the biggest living reptile. However, with its slim body, reduced limbs, and relatively long tail, gharial rarely weighs more than 440 pounds (200 kg), while its saltwater cousin can be almost 10 times as heavy. **SP**

FOCAL POINTS

1 POT NOSE
The bulbous "pot" at the end of the snout is partly hollow and is most pronounced in mature males. It likely serves as a visual sign of maturity and also amplifies the hisses, rasps, and low rumbles made by the male as it gathers a group of females and defends its territory against rivals. *Sarcosuchus* has a similar structure.

2 SLIM JAWS
The gharial has the long, narrow jaws of its ancient ancestors. Lightweight and streamlined, they enable it to swipe its head sideways at speed, snapping at fish to wound them before grabbing and swallowing. The jaws are lined with more than 100 interlocking teeth that grip and hold slippery prey.

3 REDUCED FORE LIMBS
The gharial's fore limbs are feeble and small in size, which contrasts with the ancestral long, powerful tail, large hind limbs and webbed feet that propel it at great speed as it lunges at prey in water. Consequently, on land the reptile is clumsy, tending to slide and wriggle its body rather than walk.

GHARIAL RELATIONSHIPS

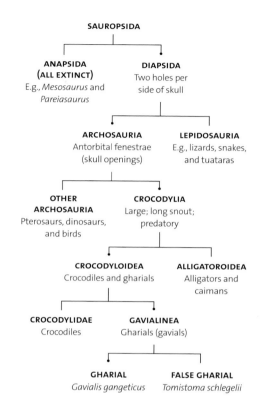

SAUROPSIDA

ANAPSIDA (ALL EXTINCT)
E.g., *Mesosaurus* and *Pareiasaurus*

DIAPSIDA
Two holes per side of skull

ARCHOSAURIA
Antorbital fenestrae (skull openings)

LEPIDOSAURIA
E.g., lizards, snakes, and tuataras

OTHER ARCHOSAURIA
Pterosaurs, dinosaurs, and birds

CROCODYLIA
Large; long snout; predatory

CROCODYLOIDEA
Crocodiles and gharials

ALLIGATOROIDEA
Alligators and caimans

CROCODYLIDAE
Crocodiles

GAVIALINEA
Gharials (gavials)

GHARIAL
Gavialis gangeticus

FALSE GHARIAL
Tomistoma schlegelii

▲ The gharial's evolutionary relationship with the similar false gharial is unclear. Physical features, especially bones, place the false gharial closer to the crocodile subgroup; molecular studies show a closer relationship with the gharial, as shown here.

SHARED HABITAT

The gharial's range overlaps with two other crocodylian species: the saltwater crocodile, *Crocodylus porosus* (right), and the mugger or marsh crocodile, *C. palustris*. The former is the bigger of the two—males can reach 23 feet (7 m) in length—and the more powerful. It is also able to take larger prey, up to the size of sambar deer and water buffalo. The reptile frequents coastal areas and rarely penetrates far inland. By contrast, the 10-foot (3 m) mugger crocodile favors swamps and marshes and pursues a wide range of small- to medium-sized prey, such as turtles and monkeys. The two species have evolved in terms of habitat and prey preferences to coexist with the freshwater, fish-specialized gharial. This natural balance has been disturbed by a reduction of the gharial's numbers by 95 percent due to egg collection, trapping in fish nets, and the degradation of its habitat.

THE IMPORTANCE OF HIPS

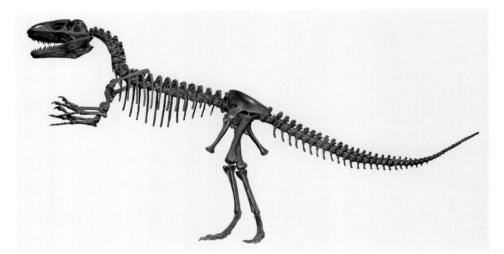

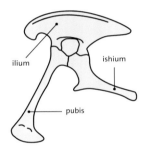

ilium

ishium

pubis

Lizard-hipped dinosaurs

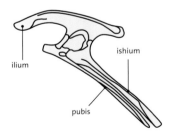

ilium

ishium

pubis

Bird-hipped dinosaurs

The first dinosaurs had evolved by about 230 million years ago, in the Carnian age of the late Triassic period. The group survives to the present day in the form of birds, but all nonavian dinosaurs went extinct in the great mass extinction at the end of the Cretaceous period 66 million years ago (see p.364). Some of the early dinosaurs were slender chicken-sized animals. Dinosaurs then grew to dominate Earth's fauna for 150 million years, diversifying into an array of different forms and attaining by far the largest sizes of all terrestrial animals.

Dinosaurs and their immediate ancestors are characterized by several features, most of which are rather esoteric details of individual bones. Historically, two important features, both related to the hips, were used to recognize members of the group. One was a fused sacrum of three or more vertebrae—the sacrum is the part of the spinal column that forms the upper (rear) part of the pelvis, with the lower backbones connected to it at the front, and the tail bones connected to the rear. The other feature is a bowl-shaped socket in the pelvis with a hole at its center, as opposed to a continuous inner surface, which held the head of the femur (thigh bone). These features contributed to the dinosaurs' ability to stand erect, with their straight legs propping up the body from beneath rather than sprawling out at the sides in the manner of other reptiles such as lizards. The erect leg arrangement required better balance, but allowed faster, more efficient movement when walking and running, which contributed to the success of the dinosaurs.

However, the discovery and study of early dinosaurs and their close relatives have shown that the evolution of these features did not coincide with the

KEY EVENTS

231 Mya	220 Mya	157–145 Mya	150 Mya	150 Mya	145 Mya
Eoraptor, one of the earliest known dinosaurs, lives in northwest Argentina. It is a primitive saurischian.	*Pisanosaurus*, a very early ornithischian, lives in northwest Argentina.	The first great peak of dinosaur diversity in the Kimmeridgian to Tithonian ages occurs at the end of the Jurassic period.	The early saurischian *Compsognathus* lives in Germany. It is a small predator with several birdlike features.	Early birds such as *Archaeopteryx* (see p.376) are evolving a birdlike hip arrangement within the lizard-hipped theropod group.	The end-of-Jurassic extinction reduces dinosaur diversity and the Cretaceous begins.

emergence of dinosaurs. This complexity is an inevitable by-product of gaining a better understanding of the fossil record: when a gap between two species is filled, the changes previously observed between those two species are divided between the new, intermediate species and the species on either side of it.

The group Dinosauria is composed of two main subgroups. One is the Saurischia (lizard-hipped dinosaurs; see image 1); the other is the Ornithischia (bird-hipped dinosaurs; see image 3). The names of the two subgroups refer to the orientation of the pubis, which is the frontmost bone in the lower pelvis. In most saurischians, the pubis retained the ancestral condition and pointed forward, as in lizards (see image 2, upper). In ornithischians, it tilted to the rear and pointed backward, as in birds (see image 2, lower). Despite the name meaning "bird-hips," the resemblance to the bird pelvis in ornithischians is misleading. In ornithischians, the backward orientation was achieved by the development of a backward-pointing portion of the pubis and the reduction of the forward-pointing part, whereas in birds, the whole forward-pointing pubis is rotated backward. These two arrangements were presumably alternative adaptations for stabilizing the rear body and anchoring the muscles that moved the upper legs.

Saurischians include the bipedal (walking on two legs) theropods (meat-eating dinosaurs) such as *Compsognathus* (see p.262), and the long-necked, herbivorous sauropodomorphs, the latter group including the giant sauropods (see p.268). Ornithischians comprise a panoply of herbivorous groups: the spiked stegosaurs and armored ankylosaurs (thyreophorans), the horned ceratopsians and thick-skulled pachycephalosaurs (marginocephalians), and the ornithopods (a diverse group that includes the duck-billed hadrosaurs).

When Richard Owen (1804–92) created the name Dinosauria in 1842, he recognized three genera as belonging to the group of "gigantic crocodile-lizards of the dry land": the theropod *Megalosaurus* (a lizard-hip) herbivorous, bird-hipped *Iguanodon* (see p.324); and *Hylaeosaurus*. Coincidentally, these three species constitute a good sample of dinosaur morphology. The early attempts by paleontologists to define the dinosaur group and their uniting features have partly stood the test of time. Hips are still used to demarcate the two great subgroups that followed separate evolutionary paths. The classification between saurischians and ornithischians was proposed in 1877 by London-born paleontologist Harry Seeley (1839–1909). For much of the 20th century it was thought that the two groups had separate origins, meaning that Dinosauria was an unnatural group and the two separate types of hips were not closely linked. However, it has been established that the two were evolutionary sister groups; they had the same common ancestor, which is regarded as one of the earliest dinosaurs—potentially the very earliest. The Saurischia and Ornithischia groups therefore represent an early—possibly the first—branching point in dinosaur evolution. **MT**

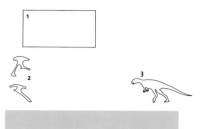

1 This skeleton of *Megalosaurus* ("great lizard"), from the middle Jurassic in Europe, clearly shows the hip configration shared by all saurischians.

2 Diagrams of the two dinosaur hip conditions demonstrate how the pubis bone angles down and forwards in "lizard-hipped" saurischians (see upper image). In "bird-hip" ornithischians it angles down but rearwards, parallel with another pelvic bone, the ischium (see lower image).

3 *Pisanosaurus*, the oldest-known ornithiscian, is known from a single skeleton from Argentina. It lived during the late Triassic.

126 Mya	124 Mya	110 Mya	84–66 Mya	66 Mya	Present
The early therizinosaur *Falcarius* shows a change in pubis angle, rotating from forward-pointing to rearward in this strange group.	Another therizinosaur, *Beipiaosaurus* (see p.318), shows that this saurischian group is adopting an ornithischian-type hip pattern.	Therizinosaur *Alxasaurus* continues the trend and now has an ornithischian-like hip, derived from saurischian origins.	The second and last great peak of dinosaurs diversity in the Campanian to Maastrichtian ages occurs at the end of the Cretaceous.	All nonavian dinosaurs become extinct at the end of the Cretaceous (see p.364).	The 10,000 species of living dinosaurs—birds—currently outnumber mammal species by a ratio of two to one.

Compsognathus
LATE JURASSIC PERIOD

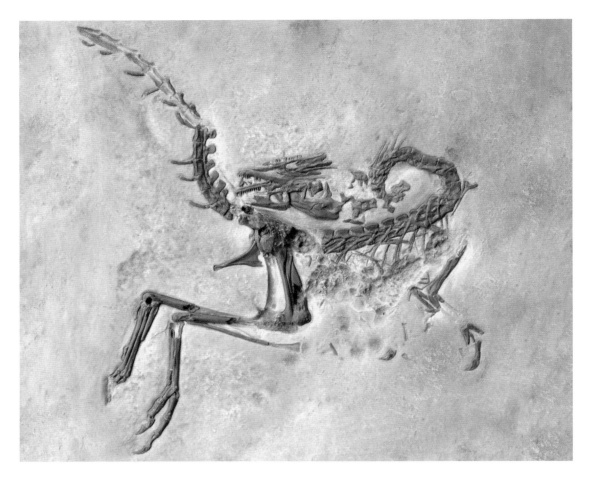

SPECIES *Compsognathus longipes*
GROUP Compsognathidae
SIZE 4 feet (120 cm)
LOCATION Germany

✦ NAVIGATOR

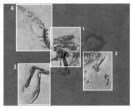

Miniature *Compsognathus* lived 150 million years ago during the late Jurassic and was initially thought to have been a relative of *Archaeopteryx* due to its birdlike characteristics, but it was in fact a saurischian dinosaur. Its fossils were found in 1859 in limestone deposits near *Archaeopteryx* fossils in Germany. It was the first near-complete theropod discovered. A second fossil specimen was found in France in 1972. At only 4 feet (120 cm) long and 4 to 6 pounds (2–3 kg) in weight, *Compsognathus longipes* was a small predator that used its hind legs to run swiftly. It had a long tail, long neck, narrow head with large eyes, and birdlike feet. It is evident that *Compsognathus* was a predator, not only from its anatomy (sharp teeth and clawed front limbs) but also because lizard remains were found inside the belly of a fossilized specimen. *Compsognathus* was sufficiently birdlike to cause speculation about it representing a stage in avian evolution, although fossils show no evidence of feathers and it could not fly. The similarities with birds were reinforced by the fact that two fossils now classified as *Archaeopteryx* were first identified as *Compsognathus*. In the 1860s, Thomas Henry Huxley (1825–95) noticed similarities between birds and reptiles. However, *Compsognathus* is not regarded as being on the evolutionary line to birds, because *Archaeopteryx* lived at the same time. Rather, the *Compsognathus* family Compsognathidae, which includes *Juravenator* and *Scipionyx*, is a sister group to the maniraptoran dinosaurs (see p.304) that gave rise to birds. **MW**

FOCAL POINTS

1 REPTILIAN SKULL

The skull of *Compsognathus* was elongated and the jaws set with sharp teeth, indicating a carnivorous diet that would have included insects and small lizards. The lower jaw was slender, hence the genus name, which means "elegant jaw." Large eye sockets suggest that *Compsognathus* had good vision, which could indicate a lifestyle among undergrowth or activity at dawn and dusk.

2 CLAWED FRONT LIMBS

The front limbs were relatively small and *Compsognathus* relied mainly on its powerful hind legs to walk and run. However, the fore limbs, which had three digits, carried sharp claws, indicative of a predatory lifestyle. The original specimen is displayed in the Bavarian State Institute for Paleontology and Historical Geology in Munich, and there are many copies worldwide.

3 HIND LEGS

The long, robust hind limbs would have enabled *Compsognathus* to run at considerable speed, both in search of its prey and to escape capture by larger predators. However, it was the only dinosaur found in its habitat at the time. Although tiny, it shares its body plan with most other theropods, from medium-sized raptors to giants such as *Tyrannosaurus* (see p.302).

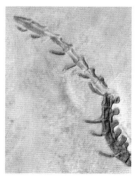

4 LONG TAIL

The long, lizardlike tail would have served a balancing role when running. For example, *Compsognathus* could swing it to one side, like a counterweight rudder, thereby giving its body momentum to turn and dart in the opposite direction. From the preserved caudal vertebrae (tail bones), estimates of the complete tail indicate that it formed almost half the dinosaur's total length.

COMPSOGNATHUS RELATIONSHIPS

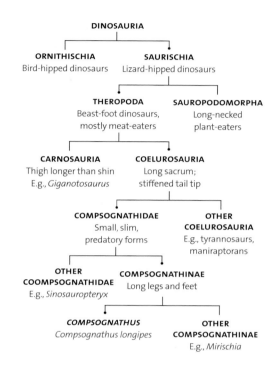

▲ Despite the similarities of *Compsognathus* to early birds, its group, Compsognathidae, is considered to be on a separate branch of evolution from that leading to the raptors and birds.

SCIENTIST PROFILE

1825–61

Thomas Henry Huxley was born in Middlesex, England, and received only 2 years of formal education. From 1835 he lived in Coventry, where he developed theological views that would lead to agnosticism, a term he coined. Huxley examined the fossil *Compsognathus* in 1859, and the first *Archaeopteryx* skeleton in 1861, along with other fossil and living reptiles and birds. He saw similarities in the body plans and showed how the anatomy of birds could be derived from that of dinosaurs. The first-found skeleton of *Archaeopteryx* was missing the head. Huxley stated that it could have a reptilian jaw with teeth, in keeping with its reptilian tail. Biologist Richard Owen disagreed, predicting that the head would have a bird beak.

1862–95

Huxley's book on human evolution, *Evidence As to Man's Place in Nature* (1863), was published four years after *On the Origin of Species* (1859) by Charles Darwin (1809–82). Huxley's support of the theory of evolution, and debates with evolution deniers, earned him the nickname "Darwin's bulldog." In 1868 Huxley published his theory that birds evolved from small meat-eating dinosaurs. Later fossil finds, such as the Berlin specimen of 1875, verified his prediction about *Archaeopteryx*.

PROSAUROPODS

The first dinosaurs split into the saurischians (lizard-hipped dinosaurs) and the ornithischians (bird-hipped dinosaurs), with the earliest saurischians including *Eoraptor* (see image 1) and *Herrerasaurus* (see p.252). Soon after, the saurischian lineage split into two subgroups: the theropods (meat-eating dinosaurs) and the sauropods (long-necked herbivores). The best-known species are the giant sauropods (see p.268), but the early sauropodomorphs, known informally as prosauropods, preceded the sauropods and had their own rich evolutionary history. More than 70 genera are recognized, and they have been discovered on every continent—including Antarctica. In most rocks where they are located, they are the most abundant large animals.

The earliest known prosauropods have been found in the Ischigualasto Formation in the Valley of the Moon, northwest Argentina. Some of the fossils here date from much of the Triassic period, but the layers containing prosauropod remains date from the Carnian age, 237 million to 227 million years ago. The early prosauropods *Panphagia* and *Chromogisaurus* have been found here. They were most probably omnivorous—like their saurischian ancestors—and they were small: the size of a dog or sheep. A reliable example of this stage in prosauropod evolution is *Thecodontosaurus* (socket-tooth lizard; see image 2), from the late Triassic in England.

KEY EVENTS

231 Mya	231 Mya	218–201 Mya	215 Mya	215–201 Mya	210 Mya
The saurischian lineage divides into Theropoda and Sauropodomorpha, represented by *Eoraptor* and *Panphagia*.	Early sauropodomorphs, such as *Chromogisaurus* and *Panphagia*, are known from Argentina.	An early large sauropodomorph, with a 3-foot-long (1 m) upper arm bone (humerus), appears in Thailand.	The earliest quadrupedal prosauropods evolve—walking habitually on all four limbs rather than just the rear two.	*Plateosaurus* is an abundant prosauropod during this time.	Prosauropods are taking to herbivory rather than their previous, more omnivorous, diets.

The first important evolutionary transition within the prosauropod group was the rise of herbivores that had teeth and jaws specialized for plant processing, long necks for high browsing, and large body sizes for digestion. Prosauropods of this kind included *Plateosaurus* (see p.266) from the late Triassic in northern Europe, and its relatives: the plateosaurid group. Plateosaurids retained habitually bipedal posture (walking on two legs) despite their size of up to 33 feet (10 m) in length and up to 4½ tons (4 tonnes) in weight. At this size, not only would they have inhabited their environment, they also would have shaped it.

A second important transition happened 220 million to 200 million years ago. Prosauropods such as *Melanorosaurus* and *Riojasaurus* had evolved to become obligate quadrupeds (walking on four legs), as is evident by the similar length of the fore limbs and hind limbs, as well as the structure of the fore foot. Prosauropods increased in size, heading toward a style of bulk feeding. It was from this group that true sauropods would emerge, taking the trends of quadrupedality, big size and bulk feeding to extremes not seen before nor since. Sauropods originated within the prosauropod group; recent studies established that there was a significant period of prosauropod evolution before sauropods.

Some lineages of prosauropods survived into the latest part of the early Jurassic period, as sauropods were becoming dominant. These include *Massospondylus* (see image 3), preserved in the Elliot Formation rocks of South Africa 210 million to 190 million years ago, and *Yunnanosaurus* from China, with some specimens exceeding 33 feet (10 m). Prosauropods were a disparate group with variation in size, posture, and diet, but they did have some common features. They all had long necks containing 10 or more cervical (neck) vertebrae, with each individual vertebra elongated. With the long neck came a proportionally small head with large bony nostrils. They also had thumb claws and hind limbs that were short in relation to the torso. **MT**

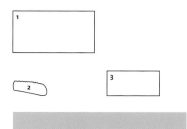

1 *Eoraptor* lived in the late Triassic and was among the earliest dinosaurs. No more than 1⅓ feet (1 m) long, the tiny carnivore is believed to be close in appearance to the common ancestor of all dinosaurs.

2 The jaw of *Thecodontosaurus* bears leaf-shaped serrated teeth, a common design for chopping plant foods.

3 *Massospondylus* (longer vertebra) lived during the early Jurassic. It was around 13 feet (4 m) long, with a long neck and a small head in relation to its slender body. These features make it a typical prosauropod.

210 Mya	203 Mya	200 Mya	200 Mya	200 Mya	168 Mya
Efraasia inhabits Germany. It is medium-sized, about 20 feet (6 m) long and weighs 5½ tons (5 tonnes).	*Thecodontosaurus*, one of Britain's early dinosaurs, lives in the south of England and is typically 6½ feet (2 m) long.	*Massospondylus* exists in South Africa. It was probably a bipedal herbivore, and had a sharp claw on each thumb, used in defence or feeding.	*Melanorosaurus*, a near-sauropod from the late Triassic lower Elliot Formation of South Africa, is represented by a near-complete skeleton.	The earliest true sauropods emerge from the prosauropod lines. They later become Earth's most ecologically important vertebrates.	*Yunnanosaurus*, one of the last surviving known prosauropods, from Yunnan, China, may have survived until this time.

Plateosaurus
LATE TRIASSIC PERIOD

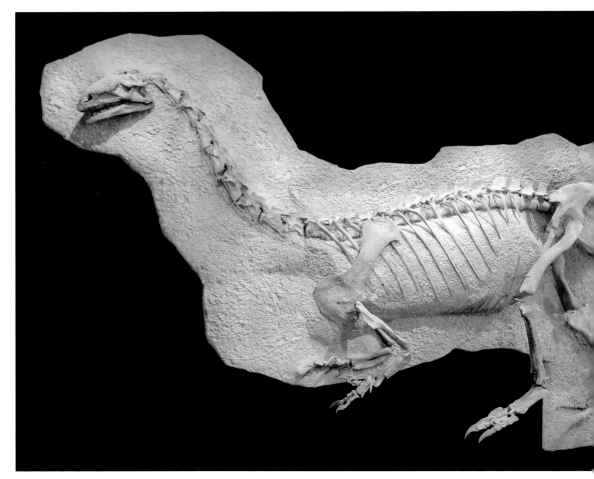

SPECIES *Plateosaurus engelhardti*
GROUP Plateosauridae
SIZE 26 feet (8 m)
LOCATION Europe

A typical prosauropod, *Plateosaurus* had a small head, long neck and tail, and a bulky body. Its rear legs were longer and stronger than the front, but less so than in earlier prosauropods, demonstrating an evolutionary trend toward four equal limbs that would continue in the sauropods. The taxonomy of *Plateosaurus* is complicated: at least eight species have been named, but the modern consensus is that there is only one valid species: *Plateosaurus engelhardti*. Some paleontologists consider *Sellosaurus gracilis* to be *Plateosaurus*. In the past, scientists were much more ready to name new genera, and as a result more than half a dozen other dinosaurs have been reallocated as *Plateosaurus*.

In addition to being one of the most familiar of the long-necked herbivores, *Plateosaurus* is among the best known of all dinosaurs. It was the sixth nonavian dinosaur to be named, in 1837, and only the second from outside England. Numerous complete fossil specimens from Germany, Switzerland, and elsewhere provide excellent evidence of its anatomy. Skeletons represent animals up to 26 feet (8 m) in length and 2¼ tons (2 tonnes) in weight. Other fragmentary specimens, belonging either to *Plateosaurus* or a close relative, are elephant-sized at 33 feet (10 m) and 4½ tons (4 tonnes). The cross section of sizes—and ages—has made it possible to determine that growth stopped at different ages in different individuals: some as early as 12 years old, others nearer to 20. **MT**

✦ NAVIGATOR

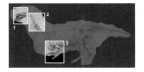

1 SKULL AND TEETH
The skull was long with large nostrils and the front of the mouth turned down. When viewed from either above or the front, it was narrow. The jaw joint and muscles were arranged for a very powerful bite, and there were up to 60 teeth: leaf-shaped, sharp, and sturdy indicating a plant-dominated omnivorous diet.

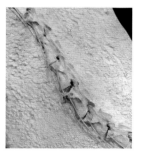

2 NECK
Plateosaurus's neck had only 10 vertebrae, compared with up to 19 for some sauropods and 25 in swans. However, the body of each individual vertebra was elongated, so the neck was long. The neck and front of the torso were flexible up and down and side to side, but were not capable of twisting.

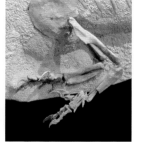

3 ARMS
The arms of *Plateosaurus* were short, stocky, and powerful. Their palms faced inward, with hands optimized for grasping and perhaps digging. The thumb was shorter and more robust than the other digits, and it carried a powerful claw, which could have been useful in defence or during competition for mates.

HABITUAL POSTURE

Plateosaurus has been studied for many years and has been depicted in almost every posture, including lizardlike sprawling, kangaroo-style bipedal sit-up-and-beg, and a horizontal-backed, all-fours quadruped. Despite the existence of many good-quality specimens, its habitual posture remains controversial. In recent years, computers have been used to model complete articulated skeletons into possible postures. *Plateosaurus* is unlikely to have been quadrupedal because of its unusually short arms and inward-facing hands, although the hands were capable of rotating palm downward. Computer simulations showed that the arc movement on the fore limb was limited, so the upper arm could only move backward from the vertical, making it unsuitable for striding. Therefore, a two-legged gait is most likely (right).

GIANT SAUROPODS

Sauropods were the terrestrial superlatives, dwarfing all other land animals before and since. Near-complete skeletons indicate that some sauropods weighed 33 tons (30 tonnes), and there is evidence of giants of more than 110 tons (100 tonnes). Fragmentary fossils suggest even bigger sizes: Indian titanosaur *Bruhathkayosaurus* may have massed 165 tons (150 tonnes), which puts it on a par with the heaviest blue whales. It is hard to image such creatures walking on land, but evidence shows that sauropods were heavier than the next biggest land animals: hadrosaurian dinosaurs and rhinolike mammals, such as *Paraceratherium* (see p.484). Modern elephants are puny by comparison.

Not only were sauropods enormous herbivores, they also had a unique body plan, with a tall, slab-sided torso carried on long legs—one at each corner—a small head on a long neck and a long tail. The necks of sauropods are astonishing: a single cervical (neck) vertebra of *Supersaurus* measures 4½ feet (1.5 m), which would imply that the complete neck was 49 feet (15 m) in length. This is more than six times as long as the 8-foot (2.5 m) neck of the modern giraffe, and five times as long as the longest nonsauropod necks on land: those of the giant azhdarchid pterosaurs (see p.290). With regard to the causes of sauropods' extraordinary dimensions, it has been suggested that environmental factors, such as high temperatures or increased atmospheric oxygen content, played an important role in their gigantism, but these theories are not supported by evidence. In fact, sauropods reached huge sizes throughout the

Jurassic and Cretaceous periods, including at times when temperatures, oxygen levels, sea levels, and other factors were both high and low.

Several aspects of their anatomy would have contributed to their great mass. First, the sauropod approach to obtaining nutrition was not to process food in their mouths. Unlike elephants, which spend most of each day chewing, sauropods simply cropped vegetation and swallowed each mouthful whole, to be digested over time in the massive torso. This enabled them to eat more quickly than mouth-processing herbivores, such as hadrosaurs and elephants. Second, the long necks of sauropods enabled food to be taken in quickly by increasing the feeding range, which enabled them to gain weight quickly. Finally, the reproductive strategy of mammals—live births—alters with size. Broods become smaller in larger animals: modern elephants, for example, give birth to a single baby after a gestation of nearly 2 years, which makes recovery after a population crash slow. However, a sauropod's large egg clutches could repopulate a devastated area very quickly, making them less vulnerable to extinction.

Necks exceeding 33 feet (10 m) evolved in at least four different sauropod groups that were not closely related: mamenchisaurs, diplodocids (see image 1), brachiosaurs, and titanosaurs. In fact, a suite of adaptations that was in place near the root of the sauropod evolutionary tree enabled neck elongation. First, sauropods had many cervical vertebrae—as many as 19 in *Mamenchisaurus*. Second, the individual backbones were elongated to a degree seen, among mammals, only in the giraffe. Third, the no-chewing strategy allowed sauropod teeth, jawbones, and jaw muscles to be minimal, thereby reducing the size of the head and lessening stress on the neck (see image 2). Fourth, having a large torso squarely supported by four limbs gave the neck a very stable platform. Fifth, an efficient respiratory system like that of modern birds overcame the problems of breathing through a long neck. Sixth, pneumatic (air-filled) vertebrae reduced the weight of the neck.

The vertebrae of sauropods were extremely complex and elaborate structures, so they are just as useful as skulls when studying sauropod lifestyles. Air spaces inside the vertebrae allowed these enormous bones to be extremely light: for example, in the 4½-foot-long (1.5 m) mid-neck vertebra of the brachiosaur *Sauroposeidon*, the bone is not more than a few millimeters thick, and much of it is eggshell-thin. One cross section through this vertebra shows that it was 85 percent air and only 15 percent bone. Air entered the sauropod vertebrae through openings called foramina, which were penetrated by offshoots of the lungs and air sacs called diverticula. The evidence of foramina suggests

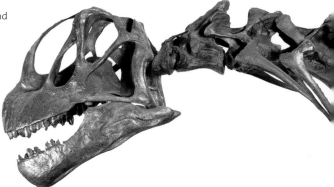

1 *Diplodocus* and a human shows the former's vast size, yet at 10 to 20 tons (9–18 tonnes) it was one of the slimmer sauropods.

2 The skull of the late Jurassic North American sauropod *Camarasaurus* was equipped with 8-inch (20 cm) chisel-like teeth to harvest vegetation rapidly.

150 Mya	150 Mya	140 Mya	100 Mya	95 Mya	66 Mya
The brachiosaurs *Brachiosaurus* and *Giraffatitan* live in North America and Africa respectively. They are among the earliest of their group.	*Janenschia* lives in the same time and place as *Giraffatitan*. It is likely to be a titanosaur: if so, it is the earliest known.	*Rebbachisaurus*, around 60 feet (18 m) long and the earliest known rebbachisaur "ridge-back," occupies Niger, Africa.	*Abydosaurus* is the latest-surviving known brachiosaur, in the earliest part of the late Cretaceous in North America.	The rebbachisaur *Limaysaurus* of Argentina is the latest surviving diplodocid: only macronarian sauropods survived hereafter.	Giant titanosaurs, such as *Alamosaurus* of southern North America, continue to thrive until the end-of-Cretaceous extinction.

3

4

5

3 Charles R. Knight (1874–1953) painted many prehistoric scenes. In this work from 1897, *Apatosaurus* looks up with an almost straight neck, while *Diplodocus* behind grazes at ground level.

4 This *Cetiosaurus* vertebrae shows the main body (centrum) below, the opening for the spinal cord within the neural arch, pronglike side (transverse) processes and the neural spine at top.

5 This *Giraffatitan* skeleton towers over a human, showing the vast scale of these dinosaurs. This specimen resides in the Natural History Museum of Humboldt University in Berlin.

the presence of a bird-style respiratory system with a rigid flow-through lung ventilated by pliant air sacs in front and behind. This system allows birds to extract twice as much oxygen as mammals, enabling them to fuel a high metabolism that can maintain energetic flapping flight. Sauropods probably had a similarly high metabolism, which meant that they could grow very quickly. Evidence from microscopic bone structure suggests that sauropods became reproductively mature by their mid-teens and reached full adult body size by the age of 30.

The habitual neck posture of sauropods has long been controversial. Early images showed them with raised, flexible, swanlike necks, although some skeletal reconstructions depicted the necks lower, horizontal, and more rigid. The pioneering US paleoartist Charles R. Knight's classic painting of 1897 (see image 3) is a compromise: it depicts an *Apatosaurus* in the foreground with an all but vertical neck and a *Diplodocus* in the background with its neck horizontal. Computer-modeling work at the turn of the millennium gave support to the horizontal neck hypothesis, but these models omitted intervertebral cartilage, the gristly "washers" between the vertebral bones. X-rays of living animals show that most habitually hold up their necks—even those such as rabbits, which are not usually imagined as having necks. Sauropods would probably have followed this rule: their necks were held high unless drinking or low browsing. Having a very long up-sloping neck was not without problems. It is not clear how food transferred down the gullet, nor how blood circulated to the high brain. The evolutionary heritage of all vertebrates includes a nerve from the larynx that is connected to the brain by a path that loops under the major blood vessel, the aorta, just above the heart. In *Supersaurus*, this nerve would have been 100 feet (30 m) long to traverse a path less than 39 inches (1 m) in a straight line.

In the early days of sauropod paleontology, it was assumed that they were aquatic, or at least amphibious, because the idea of such large animals on land was hard to accept. From the 1970s, multiple pieces of evidence showed that sauropods were terrestrial. Their torsos were slab-sided rather than barrel-shaped, as in aquatic mammals such as the hippopotamus; their limbs were proportionally long; their fore feet were compact and would have been unable to traverse swampland without getting mired; the extensive air-filling of their vertebrae was a weight-saving adaptation not needed in amphibious animals; and sauropod fossils are found in rocks laid down in seasonally dry environments.

Although the basic sauropod body plan is well understood, the majority of individual sauropod genera are known from imperfect remains: usually a single fossil specimen that is only a small part of the whole skeleton. Several sauropods are named from single bones—usually vertebrae, because the air-filling complexities make them very distinctive. Huge animals fossilize with difficulty. Fossils form when a carcass is covered by sediments such as sand, mud, or silt, before scavenging, weathering, or decay destroys the skeleton. This rapid burial can happen easily to small animals, which may be buried completely; but for an animal the size of a sauropod, it is more common for only part of the skeleton to be preserved. As a result, most sauropods are reconstructed by comparison and "borrowing" from related forms. For example, the giant *Argentinosaurus* (see p.274) is known only from a few vertebrae and a tibia (shin bone), but it is reconstructed along the lines of related sauropods from its group, the titanosaurs.

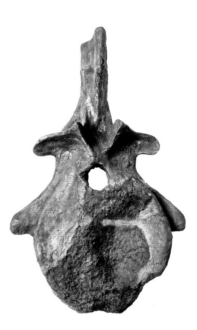

The imperfect nature of most sauropod fossils inhibited knowledge of their family tree, but a consensus has now emerged. The most basal (original) sauropods resembled their ancestors, the prosauropods (see p.264). All more derived (evolved) sauropods are included in the Eusauropoda (true sauropods), which had more characteristically sauropodlike limbs and feet. Within this group are the long-necked mamenchisaurs from China, several groups such as cetiosaurs (see image 4) whose exact positions are unclear, and the great clade (group with one common ancestor) Neosauropoda (new sauropods), which contains most of the best-known kinds.

Neosauropoda in turn evolved and branched into the Diplodocoidea and Macronaria. The former contains the bizarre, goofy-skulled rebbachisaurs and the sister groups of diplodocids, such as *Diplodocus*, *Apatosaurus*, and kin as well as the Dicraeosauridae, a group of relatively small sauropods with non-pneumatic vertebrae and tall bony rods (neural spines) projecting along the back. Macronaria contains the camarasaurs—the most abundant dinosaurs in the late Jurassic of North America—along with the long-armed brachiosaurs such as *Brachiosaurus*, *Giraffatitan* (see image 5; see p.272) and allies, and the titanosaurs. This last group remains the least well understood of the sauropods, encompassing numerous genera of widely different kinds.

Despite common assumptions, sauropods were far from similar. For example, the tail of *Diplodocus* was well over twice as long as its neck, but in *Mamenchisaurus*, the converse is true. *Apatosaurus* was built much like *Diplodocus*, but was far heavier and had a much fatter neck, triangular in cross section. *Opisthocoelicaudia* had even more robust limbs and an unusual barrel-shaped torso that suggests that it spent more time in water than its relatives. *Limaysaurus* had tall neural spines that might have formed a sail on its back. *Agustinia* had spikes (although the fossils are not complete enough to know where on its body) and *Bonitasaura* had a beak. *Nigersaurus* had a unique skull, very lightly constructed and with a mouth much wider than the main part of the skull. It had a broad, flat row of teeth, which numbered about 500 and were replaced every 2 weeks. Sauropods reached peak diversity at the end of the Jurassic, 145 million years ago. However, they did not merely survive into the Cretaceous; they thrived. Rebbachisaurs, brachiosaurs, and titanosaurs were all prevalent until the end of the early Cretaceous; rebbachisaurs continued into the late Cretaceous; and titanosaurs survived until the end-of-Cretaceous extinction. Furthermore, diversity was fairly high, and many of the biggest titanosaurs are from the last ages of the Cretaceous. **MT**

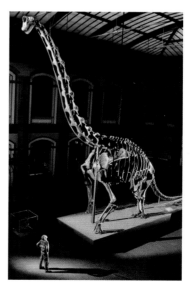

Giraffatitan
LATE JURASSIC PERIOD

SPECIES *Giraffatitan brancai*
GROUP Titanosauriformes
SIZE 73 feet (22 m)
LOCATION Africa

One of the biggest sauropods for which there exists a reasonably complete skeleton is *Giraffatitan* (giant giraffe), which weighed around 33 tons (30 tonnes). This genus was an offshoot of the more familiar *Brachiosaurus* (arm lizard/reptile) from North America. When *Brachiosaurus* was described in 1903, it was recognized immediately as the biggest sauropod and therefore dinosaur, yet described. However, the remains were tantalizingly incomplete, as there was no skull or neck. Similar but much more complete specimens were found by the German-organized expeditions to Tendaguru in what is now Tanzania in 1909 to 1912 (see panel). German paleontologist Werner Janensch (1878–1969) named these specimens as a new species of *Brachiosaurus*, *B. brancai*. It was from this connection that the concept of *Brachiosaurus* mostly derived. Reconstructions of the dinosaur pounding through and browsing in North American late Jurassic forests, 150 million to 145 million years ago, were largely based on these African fossils. Numerous skeletal differences between the North American and African versions became apparent. US writer, illustrator, and paleontologist Gregory S. Paul (b. 1954) noticed the overall lighter or "gracile" build of the African form compared to the North American one, and also observed distinctions between their backbones. Further studies took place and confirmed that, although related, the African and North American versions were not similar enough to be regarded as the same genus. Accordingly, in 1988 the African species was recognized as its own genus, *Giraffatitan*. In 2009 a study by British expert Mike Taylor (b. 1968), comparing many individual bones, supported this.

Janensch had supervised reconstruction of a composite fossil skeleton of *Giraffatitan* —then still called *Brachiosaurus*—between the two World Wars. It was taken down during World War II for safekeeping, reconstructed afterward and finally remounted in 2007 in a pose reflecting modern understanding of sauropod anatomy. It has pride of place in Berlin's Museum für Naturkunde (Museum of Natural History) and stands an immense 41 feet (12.5 m) high, twice as tall as the largest living giraffe, and has a length of 73 feet (22 m). **MT**

✪ NAVIGATOR

FOCAL POINTS

1 SKULL
The skull was very distinctive, with a "helmet" formed by a single narrow curving strip of bone along the midline. Although the large openings either side of this strip form the bony nostrils, the fleshy nostrils are likely to have been toward the front of the snout. A similar skull, with a less extreme helmet, has been found in the United States and is considered to belong to *Brachiosaurus*.

3 NECK
The most arresting feature of *Giraffatitan* was the neck. At 28 feet (8.5 m), it was three and half times as long as the longest giraffe neck—although little more than half the likely 49-foot (15 m) neck length of *Supersaurus*. The skeleton in Berlin shows the neck in an elevated pose that was probably habitual in life, which raised the head more than 43 feet (13 m) above ground level.

2 TEETH
The small teeth resembled widened chisels. As in all sauropods, the teeth were simply food gatherers, slicing off and collecting vegetation that was swallowed instantly down the lengthy esophagus to the immense stomach. Here, swallowed stomach stones called gastroliths pummeled and mashed the food into a more easily digestible pulp.

4 FORE LIMBS
Giraffatitan had fore limbs longer than its hind limbs, like its relative *Brachiosaurus* and a distinctive profile with the back sloping upward to the shoulder. Although the humerus (upper arm bone) and thigh bone are roughly equal in length, the "hand" is tall with digits pointing straight down, as in all sauropods. The shoulder articulates with the torso at a lower point than the hip.

NAMING THE FIRST FOSSILS

Werner Janensch (right) led extensive fossil-hunting expeditions to the Tendaguru region of what was then German East Africa, now Tanzania. He recognized a similarity between his largest fossil sauropod specimens in Tendaguru and North America's *Brachiosaurus*, and in 1914 he named his new find *B. brancai*. The fossils included those from several partial skeletons, three skulls, and many smaller remains, including teeth, parts of skulls, vertebrae, and limbs. These huge and heavy finds, many in their original rocks, were then transported back to Berlin where Janensch organized a fully reconstructed skeleton. Janensch was honored in 1991 by the naming of another large sauropod, *Janenschia*, also from late Jurassic Tanzania.

Argentinosaurus
LATE CRETACEOUS PERIOD

SPECIES *Argentinosaurus huinculensis*
GROUP Titanosauria
SIZE 130 feet (40 m)
LOCATION Argentina

U sually cited as the biggest sauropod yet discovered and the largest land creature ever, *Argentinosaurus huinculensis* is known from dorsal (back) vertebrae, some ribs, a damaged sacrum (where the vertebrae contact the pelvis), and a lower leg bone, as well as possibly a partial femur (thigh bone). The sparse fossils are enough to identify *Argentinosaurus* as a member of the titanosaur group, but this group encompasses a wide range of body shapes and it is not possible to give very precise size estimates. Thought to exceed 130 feet (40 m) long, it was at first calculated to weigh more than 110 tons (100 tonnes); recalculations suggest a more likely range of between 66 and 88 tons (60–80 tonnes). Probably a herd-dweller, *Argentinosaurus* lived in the late Cretaceous, 95 million years ago. Like other well-known titanosaurs, it would crane its neck over a huge area to crop and swallow plant matter as it tramped through openly wooded landscape. Computer modeling of its movements suggest a maximum speed of 5 miles per hour (8 km/h). Even at its colossal size, it probably had enemies. The huge predatory theropod (meat-eating dinosaur) *Giganotosaurus*, for example, similar in size to *Tyrannosaurus* (see p.302), is known from fossils only a few million years older.

Other big sauropods for which good evidence exists include the African *Giraffatitan*, and the Argentinean titanosaur *Dreadnoughtus*, the latter 85 feet (26 m) long and weighing 66 tons (60 tonnes). Evidence for larger sauropods is scarce. *Puertasaurus*, another titanosaur from Argentina, is known from only four vertebrae; it could have been slightly smaller than *Argentinosaurus*. **MT**

✪ NAVIGATOR

▲ Discovered in 1987, the remains of *Argentinosaurus* were excavated from rocks known as the Huincul Formation in Neuquén province, Argentina. The species was officially named in 1993.

👁 FOCAL POINTS

1 SKULL
Sauropod skulls are rare and difficult to assign to the body, and no skull of *Argentinosaurus* is yet known. The skull and "tooth fringe" are based on remains borrowed from more complete, better-known finds of other titanosaur sauropods, such as the smaller 39-foot (12 m) *Saltasaurus*.

2 VERTEBRAE
The complex air chambers in *Argentinosaurus* vertebrae made these parts very light. They aid the diagnosis and identification of specimens from very limited remains. However, their fragile thin-walled nature makes them vulnerable to disintegration as they are fossilized or excavated.

3 TAIL BONES
The tail here is arranged in the elevated position, in line with the body, as is suggested by most current expert opinions. However, a common problem is that the many caudal vertebrae (tail bones) are so similar, with only a few isolated finds it is difficult to be sure of their exact positions.

4 LIMBS
The meager remains of *Argentinosaurus* include the tibia—the larger of the two rear shin bones—and part of a femur that may be from this creature. These finds are valuable for estimating overall height and weight, and also the type of gait and speed of movement.

EARLY CARNIVOROUS DINOSAURS

Theropods (meat-eating dinosaurs) were the dominant land-living hunting creatures between approximately 225 million years ago and the mass extinction event 66 million years ago at the end of the Cretaceous period. The earliest theropods include *Herrerasaurus* (see p.252), notable for its deep, rectangular skull, and the far smaller, less predatory *Eodromaeus*. A new group evolved from ancestors similar to these, characterized by the loss of the fifth finger and fifth toe. These were the neotheropods, and they comprise three major groups. The first group was the coelophysoids, a slim-bodied, shallow-skulled group with a distinctive notch in the upper jaw, termed the subnarial gap. The second was the ceratosaurs, a more robust, bigger-bodied group, many of which had horns or head crests. The third was the tetanurans (stiff-tails), the enormous group that possessed birdlike features in the hands, spinal columns, and hind limbs. Many familiar big predatory dinosaurs, such as megalosaurs, spinosaurs, allosaurs, tyrannosaurs—and the birds—were part of the Tetanurae group, making it by far the most successful theropod group.

Coelophysoids also existed throughout the Mesozoic era; they were mostly 7 to 10 feet (2–3 m) long. The best known are the North American *Coelophysis* (see image 1) and the predominantly African *Syntarsus*, two dinosaurs that were so similar that many experts regard them as different species of the same genus. They were widespread and inhabited Europe and China. One of the most noticeable coelophysoid features concerned their jaw anatomy. The conical teeth located in the shallow, pointed snout-tip were separated from those farther back by the subnarial gap. The evolutionary reasons behind this remain unknown. It may have helped the

KEY EVENTS

225 Mya	220 Mya	215 Mya	210 Mya	200 Mya	200 Mya
Theropods, a group of bipedal saurischians, appear in the late Triassic and eventually lead to modern birds.	Neotheropods—with narrower, more birdlike feet than earlier theropod groups— evolve from ancestors similar to the broad-footed *Herrerasaurus*.	Early North American *Tawa* shows that the subnarial gap typical of *Coelophysis* and its relatives was probably widespread in early neotheropod groups.	Mesozoic dinosaur *Coelophysis* inhabits North America. Huge numbers die at Ghost Ranch, New Mexico, possibly killed by a flash flood.	*Coelophysis*-like theropods, such as *Syntarsus*, occur across the Pangaean supercontinent in deserts and other habitats.	Dilophosaur-type theropods give rise to the first ceratosaurs and tetanurans. They are medium-sized dinosaurs about the size of horses.

animals to grab and manipulate small prey items, perhaps pulling animals out of crevices or burrows. Alternatively, it might have been used as a fish-catching tool because these theropods may have foraged in shallow water.

The three groups—coelophysoids, ceratosaurs, and tetanurans—could have had separate evolutionary paths; alternatively, neotheropods might have gone through a phase of being *Coelophysis*-like in form before the ceratosaurs and tetanurans evolved. The fossil record shows that coelophysoids of several lineages evolved bigger species exceeding 20 feet (6 m) in length. Among these was *Dilophosaurus* (see p.278), well known for its twin platelike skull crests. Many ceratosaurs and tetanurans were substantially larger than *Dilophosaurus*, but the earliest members of their groups were small in size, which shows that the evolution of large body size occurred repeatedly in parallel within these groups. This is a persistent theme in theropod evolution: small ancestors gave rise to large or gigantic descendants. In fact, groups of small-bodied species often evolved their own giant-sized species only when the previously dominant group became extinct.

Approximately 200 million years ago, the ceratosaurs evolved from a branch of dilophosaur-like theropods. The oldest ceratosaurs were from Africa, but evidently they became widespread before continental fragmentation occurred, because younger species occur worldwide. The deep-snouted, horned ceratosaur *Ceratosaurus* (see image 2)—the namesake of the group—occurred in North America, western Europe, and possibly eastern Africa. *Ceratosaurus* shared anatomical features with another group, the abelisaurs, and this group forms the bulk of nontetanuran theropod diversity. Abelisaurs are known from rocks worldwide and more than 25 different kinds are recognized at present. Some are known from spectacularly well-preserved complete specimens, such as *Majungasaurus* from Madagascar (see image 3). *Ceratosaurus* was large and deep-skulled, and its long curved teeth show that it was a predator of other large animals. This lifestyle and skull shape were also present among abelisaurs, such as *Carnotaurus* (see p.280), which evolved especially short-snouted, absurdly deep skulls.

However, not all ceratosaurs were large hunters like *Ceratosaurus*. A variety of small-bodied abelisaurs termed noasaurids inhabited South America, Madagascar, and India. Shallow skulls and protruding teeth suggest that they caught fish, insects, and other small animals instead of large prey. A remarkable ceratosaur from China, *Limusaurus*, was in fact toothless, with reduced hands and long, slender hind limbs. Small stones were found in its belly, suggesting that it was a herbivore and swallowed the stones to help grind up its plant food. Noasaurs and *Limusaurus* show that ceratosaurs were not limited to evolutionary paths that involved hunting, killing, and eating large animals. **DN**

1 This well known 1947 specimen of *Coelophysis* from Ghost Ranch was thought to contain young of its own kind in the stomach region, indicating cannibalism. In fact the bones are from an early crocodile-line reptile, *Hesperosuchus*.

2 *Ceratosaurus* grew to a length of more than 20 feet (6 m).

3 Plentiful fossils of sturdy, well-built, late Cretaceous *Majungasaurus* allow size estimates in excess of 24 feet (7.3 m).

190 Mya	185 Mya	155 Mya	150 Mya	120 Mya	75 Mya
Dilophosaurus inhabits North America and China. Its probable relatives *Dracovenator* and *Cryolophosaurus* occur in South Africa and Antarctica.	Early ceratosaur *Berberosaurus* inhabits Morocco. Early African ceratosaurs indicate they originated in the center of the southern supercontinent.	*Ceratosaurus*, one of the first large, robust-bodied theropods, inhabits North America, Europe, and perhaps east Africa.	*Limusaurus*, a toothless, short-armed, long-legged herbivore, lives in China. Ceratosaurs evolve a wide range of forms and lifestyles.	*Genyodectes* lives in Argentina, a possible relative of Jurassic *Ceratosaurus*. The ceratosaur group survive into the Cretaceous period.	Abelisaurs are known from France. It is probable that they migrate from Africa when low sea levels allow them to use land bridges.

Dilophosaurus
EARLY JURASSIC PERIOD

SPECIES *Dilophosaurus wetherilli*
GROUP Neotheropoda
SIZE 23 feet (7 m)
LOCATION North America

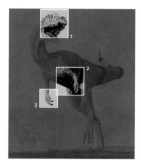
One of the best-known early theropods was *Dilophosaurus*, an inhabitant of North America that lived 190 million years ago. *Dilophosaurus wetherilli* is often imagined as a human-sized theropod with an expandable neck frill and an ability to spit venom. In reality, *Dilophosaurus* was 23 feet (7 m) long, weighed up to half a ton (0.5 tonne) and was lightly built with a shallow skull and slender jaws. Its large size relative to *Coelophysis* and earlier theropods reflects a trend seen in all dinosaur lineages of the time: they were becoming larger, partly because the archosaurs that previously controlled the top positions in terrestrial ecosystems were now extinct.

Like its close relatives, *Coelophysis* and *Syntarsus*, *Dilophosaurus* had an interesting jaw shape, suggesting that it regularly preyed on small animals. However, it also overpowered animals similar in size to itself. Its best-known feature is a pair of thin, semicircular skull crests, hence its name, which means "two-crested lizard." Similar but smaller crests of this sort appear in other *Coelophysis*-like theropods, thus suggesting that they were widespread in early members of the group. Quite why these crests evolved remains unknown. A role in predation is unlikely. It is more likely that they served a sexual or social signaling function, to attract partners or to subdue rivals. The evolution of bony display structures on the head is a pervasive theme for dinosaurs, as is the increase in size. **DN**

👁 FOCAL POINTS

1 HEAD CRESTS
The striking twin head crests of *Dilophosaurus* were formed by thin sheets of bone, buttressed with vertical pillars. Seen from the front they would have diverged to form a V shape. Short, thin spikes projected from their rear margins and extended backward, parallel to the skull roof.

2 ARMS
Large muscle attachment sites and ligament evidence indicate that *Dilophosaurus* had strong, flexible arms and clawed hands, suited for grabbing prey. A well-preserved hand shows three clawed fingers. A fourth finger was also present but it was only a tiny stump and lacked a claw —an evolutionary leftover.

3 HANDS
It is likely that the palms of *Dilophosaurus* naturally faced inward. A fossil track from Utah shows that the outer surfaces of a dilophosaur's fingers contacted the ground as it squatted, which supports this theory. This configuration is typical of theropods and is retained in birds today.

GLOBAL DISTRIBUTION

Dilophosaurus was one of the first big predatory dinosaurs. Chinese fossils discovered in 1987 were suggested by some experts to be a second species of *Dilophosaurus* (above). Broadly similar to the US version, the creature was at first named *Dilophosaurus sinensis*. Further studies have since assigned a new name, *Sinosaurus* (Chinese lizard/reptile) although it remains in the dilophosaur group. Another crested theropod from the same time period but from the region that is now Antarctica is called *Cryolophosaurus* (cold crested lizard/reptile). Slightly smaller than the US *Dilophosaurus*, it also seems to be a close relative. Consequently, these dinosaurs might have been a globally distributed group, adapting to varied conditions in widely separated regions.

Carnotaurus
LATE CRETACEOUS PERIOD

SPECIES *Carnotaurus sastrei*
GROUP Abelisauridae
SIZE 30 feet (9 m)
LOCATION Argentina

The abelisaur group of predatory dinosaurs was first recognized in 1985 following the discovery in Argentina of *Abelisaurus*, a large, blunt-snouted predator that lived around 80 million years ago. Since then, numerous additional members of the group have been identified. *Carnotaurus sastrei* (flesh-eating bull) from Argentina, the best-known member of the group, is one of the most "extreme" abelisaurs from an anatomical point of view. It had small, blunt-tipped horns that projected sideways from above the eyes and several competing theories have been proposed to explain why these horn structures evolved. A social or sexual role is plausible, and their shape implies that they were used in headbutting or shoving matches. The horns were not the only unusual feature of *Carnotaurus*: it also had a remarkably short, deep snout. Several abelisaurs—grouped together as the brachyrostrans—were similar. Large theropods of several groups had short arms, presumably because the skull was their main weapon. This trend was apparent in *Carnotaurus*, but its arm-shortening would have occurred in an unusual manner: the upper arm bone (humerus) was long and straight, but the lower arm oddly short. A mobile elbow was absent and the lower arm bones (radius and ulna) were similar in length to the hand. The hand was also unusual: the four fingers were reduced to clawless stumps that could not have had a role in predation or fighting. **DN**

FOCAL POINTS

1 MOBILE ARMS
The arms of *Carnotaurus* were unusual. The long, straight humerus had a ball-like head, which suggests that the arm was extremely mobile and that *Carnotaurus* could rotate its arms in circles. One theory is that these dinosaurs whirled their arms about, either when fighting or for display.

2 VISION
The design of *Carnotaurus*'s skull allowed for limited binocular vision, which suggests that its sense of smell was more important. Most theropods had some degree of binocular vision: there was a zone where the views from both eyes overlapped, thereby giving them depth perception.

3 FLAT BACK
The neural spines were short and the bony processes on either side were as tall as the neural spines. As a result, *Carnotaurus* must have had a peculiarly flattened back. In the majority of theropods, the bony spines that grew upward from the vertebrae were tall and formed a ridge along the back.

▲ Expedition leader and Argentinean paleontologist José Bonaparte (b. 1928) recovered this *Carnotaurus* skeleton in 1984 and it was named officially in 1985. The fossil is remarkably complete; even skin patches have been preserved.

PLESIOSAURS AND PLIOSAURS

A mong the most familiar and successful of extinct reptiles are the plesiosaurs. Members of this group inhabited seas and oceans worldwide, evolved a diversity of body shapes including species that ranged in length from less than 6 feet (2 m) to more than 40 feet (12 m), and some even moved into freshwater rivers and lakes. Plesiosaurs were a perpetual presence in the world's water between 225 million and 66 million years ago, and they included some of the greatest predators ever to roam the oceans. They were in the sauropterygian group and were related to the nothosaur and pistosaur groups of the Triassic period, neither of which survived the mass extinction 201 million years ago. Plesiosaurs alone made it through to the Jurassic period, and their dedicated open-ocean lifestyle may have contributed to this because their ecology and behavior were no longer dependent on the events and climate of the land. These reptiles faced several extinction events during their history, but proved remarkably resilient and only the devastating

KEY EVENTS

225 Mya	200 Mya	195 Mya	180 Mya	160 Mya	160 Mya
Plesiosaurs descend from a pistosaurlike ancestor. These earliest plesiosaurs are small, slender-necked predators of coastal waters.	Several small, long-necked plesiosaurs survive the Triassic mass extinction. These groups evolve in different directions.	The first plesiosaur to be scientifically named, *Plesiosaurus*, inhabits the seas of Europe. It is a long-necked generalist.	Giant, big-skulled rhomaleosaurs, up to 23 feet (7 m) long, are the most important large marine predators. Most species live in western Europe.	Older rhomaleosaurs become extinct. The small, slender-skulled pliosaurs that have long lived alongside them take over the role of top predator.	Multitoothed, big-eyed cryptoclidids live in the seas covering Europe and North America. They are mid-sized, long-necked plesiosaurs.

mass extinction at the end of the Cretaceous period 66 million years ago finally caused their disappearance.

Several innovative features enabled plesiosaurs to become successful marine animals. Paddlelike limbs (see image 2), a stiff body, and a short tail were inherited from their ancestors, and all of these features became exaggerated. Their limbs changed from paddles to slender-tipped "underwater wings," which were almost certainly used in an up–down flying motion rather than a forward–back rowing action. The body became stiffer and the tail shorter and the shoulder and hip bones evolved into flattened plates, often located on the body's underside and serving as attachment sites for the muscles that powered the limbs. Plesiosaurs had two pairs of "wings;" their fore limbs and hind limbs were often near-identical in size and shape, so experts remain uncertain about the details of how they swam. Robots and computer simulations have shown which styles would have worked best, and it is thought that plesiosaurs used different gaits at different times.

The evolution of especially long, flexible necks, long snouts and jaws, retracted nostrils, and very large eyes made plesiosaurs better at finding prey than their earlier sauropterygian cousins. Once they had the ability to give birth to live young—a feature inherited from earlier sauropterygians—plesiosaurs no longer needed to visit the land. They became larger over time and eventually exceeded 33 feet (10 m) in length and 5½ tons (5 tonnes) in weight. The anatomy of plesiosaur vertebrae, their neck length and the form of their shoulders, hips, and limbs all show major variation among the species. They do not fall neatly into short-necked and long-necked types, because several groups have necks of intermediate length. This variation is such that experts believe that the multiple short-necked groups were not closely related to one another, because short-necked kinds evolved independently from long-necked ancestors at least three times.

The short-necked rhomaleosaurs, such as *Rhomaleosaurus cramptoni* (see image 1), occupied the top predator role early in plesiosaur evolution, but they were almost entirely extinct by 160 million years ago. In 1828 the first articulated plesiosaur skeleton—with the bones preserved in lifelike positions relative to one another—was found by Mary Anning (1799–1847; see image 3). Its appearance was so remarkable and unexpected that some doubted its authenticity. An unfortunate break across the base of the neck caused French anatomist Georges Cuvier (1769–1832) to suggest that it could be a fake, that combined parts of several different fossils, but subsequent analyses proved this was not the case. Mary Anning was one of the most significant contributors to the discovery and study of ancient life. She was a self-taught, working class woman from a small, provincial town on the southwest coast of England: Lyme Regis. Her father, Richard Anning, a carpenter and furniture maker, familiarized her with fossils and both she and her brother Joseph were finding and

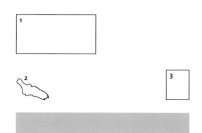

1 *Rhomaleosaurus cramptoni*, 23 feet (7 m) long, was an early plesiosaur dating to the early Jurassic, 180 million years ago.

2 The front right limb of the plesiosaur *Cryptoclidus* shows the short, broad forearm and wrist bones, as well as close-knit finger bones, all forming one large, firm, water-pushing surface.

3 In the 1820s renowned fossil hunter Mary Anning discovered two excellent plesiosaur skeletons, the first described for science.

150 Mya	145 Mya	145 Mya	105 Mya	100 Mya	66 Mya
Leptocleidids appear: a group with short and mid-length necks, they are mostly small predators of estuaries, deltas, rivers, and lakes.	The Jurassic mass extinction event causes the extinction of most cryptoclidids and pliosaurs. Only a few plesiosaur groups survive.	A group of long-necked plesiosaurs evolves: elasmosaurs. They occupy roles once held by other long-necked groups and evolve into open-ocean giants.	A peculiar new group of elasmosaurs, the aristonectines, evolves. They possess more than 200 teeth and superficially resemble earlier cryptoclidids.	Giant pliosaurs are the top ocean predators worldwide. A distantly related group, the polycotylids, also evolves big predators.	A major extinction event kills the last plesiosaurs. The groups affected are already low in diversity due to a changing climate.

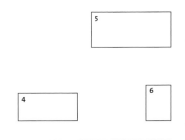

4 Like all marine reptiles, *Liopleurodon* had to surface to breathe. Its nostrils had moved to a higher position on the skull, close to the eyes, so that only this upper part needed to surface to see and breathe.

5 *Kronosaurus* displayed the long, slim jaws and sharp, conical teeth of a top predator adapted to tackling large aquatic prey items, in the manner of the dinosaur *Spinosaurus* (see p.300) and big crocodiles.

6 After 8 years of excavations in the far north, "Predator X" was officially described in 2012 as a species of *Pliosaurus*, *P. funkei*. Its length is estimated at 39 feet (12 m).

assembling ichthyosaur fossils by 1811 (see p.244). Their mother, Molly, was also involved in what became a burgeoning business. After Richard's death in 1811, the family sold an ichthyosaur fossil for a large sum, and by 1820 the Annings had become international sellers of quality fossils. Mary was the most important member of the family business: she wrote to and met with many important names in the 19th-century world of paleontology and geology. She became a local celebrity and her shop was a noted tourist attraction. Between 1818 and 1830 she discovered, in addition to the first articulated plesiosaur fossil, the skeleton of the early pterosaur *Dimorphodon*, several key ichthyosaur specimens, and the early cartilaginous fish *Squaloraja*.

A second short-necked group, the pliosaurs, had lived alongside the rhomaleosaurs for 40 million years, remaining small and slender-snouted throughout that time. With rhomaleosaurs extinct, pliosaurs quickly became the new archpredators. *Liopleurodon* (see image 4) and *Pliosaurus* from the Jurassic of Europe, along with *Kronosaurus* (see image 5) of Australia and South America, and *Brachauchenius* of North America were all members of this big-bodied group of superpredators. They were the ultimate hunters in the world's oceans until about 80 million years ago.

Liopleurodon (smooth-sided teeth) has been the subject of several movies and television programs. Living 160 million to 150 million years ago, it was a large-headed, huge-mouthed, long-fanged predator with all the characteristics of a creature adapted to tackling the biggest prey. As is the case with other oversized predators, from *Tyrannosaurus* (see p.302) and *Sarcosuchus* (see p.256) to the living great white shark, its dimensions have often been exaggerated. One method used to estimate the size of pliosaurs involves estimating the length of the whole body, nose to tail, based on the proportions of the skull and jaws. Consequently, if only fossil jaws and teeth are known for a certain species, its total length can be calculated. Using this technique, upper-end estimates for the length of *Liopleurodon* are 49 feet (15 m). However, these dimensions are now considered too great, and 23 feet (7 m) appears to have been typical—the maximum length of the largest great whites, and equating to a weight of 2¼ to 3¼ tons (2–3 tonnes). Exceptional cases reached up to 30 feet (9 m) long.

Kronosaurus (Kronos lizard/reptile) was even larger. Fossils from 120 million to 100 million years ago are known from Australia and from Colombia. Similar in overall appearance to *Liopleurodon*, it could have reached 33 feet (10 m) in length. Another great pliosaur was *Pliosaurus* (more lizard/reptile). Between five and 10 species are currently described and many more have been merged or transferred to other genera in the past. This broad range means that assigning partial remains to a particular species is often difficult. The first-named species, *Pliosaurus brachydeirus*, was labeled initially *Plesiosaurus brachydeirus* in 1841 by Richard Owen (1805–92), eminent paleontologist, anatomist, and inventor of the term "Dinosauria." As a result of a misspelling, it then became *Pleiosaurus brachydeirus* before, finally, *Pliosaurus brachydeirus*.

"Predator X," recently discovered in Svalbard, Norway (see image 6) has been assigned to another species, *Pliosaurus funkei*. This specimen took archeologists several expeditions to excavate, finishing in 2012. This pliosaur was first estimated to have been more than 49 feet (15 m) in length, but revisions have since reduced it to a "mere" 39 feet (12 m) long. All species of *Pliosaurus* lived toward the end of the Jurassic, 160 million to 145 million years ago. Experts have questioned why long-necked plesiosaurs, which mostly specialized on small prey such as fast-swimming fish, gave rise to these huge-headed, short-necked descendants. It is possible that plesiosaurs adapted their ecology and body shape to take advantage of new foods or other resources.

Long-necked plesiosaurs were often preyed upon by their rhomaleosaur and pliosaur relations. Many long-necked plesiosaurs were generalists that could survive on all kinds of animal prey. A group called the cryptoclidids evolved numerous teeth and very big eyes, and presumably took mouthfuls of myriad small prey. They were mostly wiped out in an extinction that occurred 145 million years ago. Around 80 million years ago, an extraordinary new plesiosaur group, the elasmosaurs, evolved, such as *Elasmosaurus* (see p.286). They subsequently diversified to occupy many roles in the world's oceans. In several ways, elasmosaurs were the most extreme plesiosaurs because they had the longest necks, the biggest paddles, and several other unusual features, such as a low skull, an almost Y-shaped lower jaw with a pointed front, and a bony bar in the shoulder to strengthen the motion of the front flippers. **DN/MT**

Elasmosaurus
LATE CRETACEOUS PERIOD

SPECIES *Elasmosaurus
platyurus*
GROUP Elasmosauridae
SIZE 46 feet (14 m)
LOCATION North America

All long-necked plesiosaurs were animals of biological extremes, and their remarkable necks have raised all kinds of questions about the evolution of physiology, behavior, and lifestyle. Elasmosaurids—a group of Cretaceous plesiosaurs with necks longer than their bodies and tails combined—were the most extreme of all. One of the best-known members of this group is the North American *Elasmosaurus platyurus* from around 80 million years ago, mistakenly reconstructed in 1868 as if its neck were its tail (see panel). Its neck, almost 23 feet (7 m) long and comprising 72 vertebrae, is one of the longest in history. However, *Elasmosaurus* was outclassed by another member of its group: *Albertonectes vanderveldei*, also from North America and named in 2012, possessed three or four more neck (cervical) vertebrae and had a neck length about 2 feet (60 cm) greater. But what evolutionary pressures led to the development of this extraordinary structure? This question is difficult to answer because the role and function of the elasmosaurid's neck remain uncertain. Studies suggest that the neck was reasonably flexible, especially with regard to sideways and downward movement, but it is doubtful whether it would have bent into an S-shape. Furthermore, the neck was so heavy that lifting it out of the water completely would have been virtually impossible.

Tooth anatomy and stomach contents indicate that elasmosaurids exploited the diverse prey available at various depths of the ocean. They would have grabbed fish with upward, downward, or sideways swipes, but also reached down to capture bottom-dwelling creatures such as crustaceans and mollusks. Their heavy bones and deliberately swallowed stones (gastroliths) would have enabled them to partially control their weight and buoyancy in order to remain at depth while foraging. The picture that emerges is one of supreme marine generalists that used their necks to approach prey of many different kinds. It seems that the evolution of ever-longer necks provided a real advantage in the food-rich marine environments that elasmosaurids inhabited. **DN**

✦ NAVIGATOR

👁 FOCAL POINTS

1 SHORT TAIL

All plesiosaurs had short tails, but *Elasmosaurus* especially so. However, the assumption that plesiosaur tails were essentially useless has been challenged by evidence that certain members of the group possessed a vertical fin at the tail's tip. This could have been used either for steering or to maintain stability while swimming.

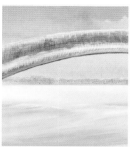

2 LONG NECK

Elasmosaur fossils have been discovered worldwide, but the species with the especially long necks, such as *Elasmosaurus*, were unique to the Western Interior Seaway—the sea that covered much of central North America from 110 to 60 million years ago. Certain conditions or a unique resource could have encouraged the "extreme" necks.

3 SENSES

Elasmosaurus would have been on constant alert for the giant predators that shared its habitats. Acute eyesight and a good sense of smell were therefore important. Its massive paddles and heavy body were best suited to a slow swimming style, but it would have increased its speed when danger threatened.

BONE WARS

In 1868, US fossil collector and anatomist Edward Drinker Cope (1840–97) received some strange fossil bones from Kansas. "Dinosaur fever" was becoming popular, and Cope assembled the bones to produce a new kind of sea reptile that he named *Elasmosaurus*, with a shorter neck than tail (above). Another equally well-known fossil hunter, Othniel Charles Marsh (1831–99), saw Cope's reconstruction in a scientific journal; he realized that the skull was at the wrong end—on the tail rather than the neck. To Cope's embarrassment, Marsh publicly pointed out this mistake. So began a feud between the two that developed into the infamous "Bone Wars" of the 1870s to 1890s, as their respective teams raced to unearth and name more dinosaurs and other prehistoric beasts than their rivals.

◀ Specialized feeding techniques are likely to have been a primary reason for evolution of the immense neck of *Elasmosaurus*. However, such extravagance may also be linked to sexual display, such as courtship, as depicted here. This has contributed to the evolution of extraordinary features such as the huge antlers of the extinct giant elk and the impressive plumage of living birds of paradise.

WINGED PTEROSAURS

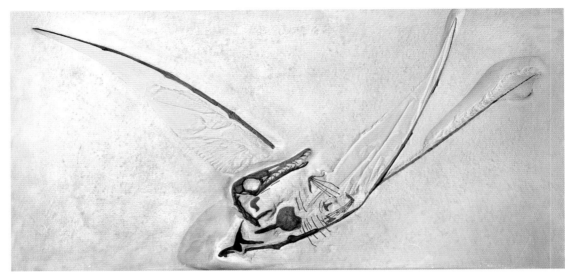

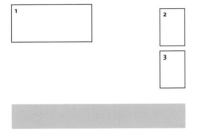

1 *Rhamphorhynchus*, an early pterosaur, had a long tail characteristic of its group.

2 A *Rhamphorhynchus* wing shows the upper arm and forearm bones as a "V," upper left; the wrist with clawed digits, upper center; and the fourth digit with membrane folds, curving down to the right.

3 Transitional *Darwinopterus* had a snout crest that varied among individuals, perhaps with age.

Pterosaurs (wing lizards/reptiles), a name preferred to the less accurate pterodactyls (wing fingers), were a group of flying reptiles that existed between 225 million and 66 million years ago. They occurred worldwide in a range of sizes and all possessed the same body form: a large head, a small and compact torso, slim hind limbs and huge fore limbs that had evolved as wings and dominated their shape. The smallest pterosaurs were similar in size to modern gulls, while the biggest were the largest flying animals that ever lived, with wing-spans of 33 feet (10 m). The details of pterosaur ancestry remain uncertain, largely because prototype pterosaurs, part-evolved from their predecessors, remain undiscovered. However, pterosaur skulls and limbs indicate that they belonged to the diapsid reptile group archosaurs, as did the crocodiles and dinosaurs. Their long necks, curved hand claws, and long legs suggest that they had ancestors in common with dinosaurs, which would have lived 245 million years ago.

The pterosaur wing was unique. Its key support was formed by a gigantic fourth finger and its associated hand bone (see image 2). The main wing membrane, the brachiopatagium, was attached to this "wing finger" and formed the principal aerodynamic surface. The membrane was broad near the body but became narrower as it reached the wing tip. Another membrane, the propatagium, extended from the shoulder to the wrist, its position controlled by the rodlike pteroid bone projecting from the wrist. A third membrane, the uropatagium, stretched between the legs. Well-preserved fossils show the complexity of these membranes: stiffening fibers, blood vessels, and a layer of

KEY EVENTS

225 Mya	210 Mya	180 Mya	160 Mya	160 Mya	152 Mya
The oldest known pterosaurs inhabit the skies. They are already well adapted for flight, with large wings and all the key features of the group.	Dimorphodontids and campylognathoidids inhabit Europe, Greenland, and North America. Some have multicusped teeth and bony horns.	Rhamphorhynchids, a new group of slim-snouted, long-toothed pterosaurs, inhabit marine and freshwater habitats in Europe.	*Darwinopterus* and similar transitional pterosaurs appear. They have the features of early, long-tailed kinds as well as later pterodactyloids.	In Asia, the first pterodactyloids appear. They are members of the short-tailed pterosaur group that later evolve successively larger species.	The gull-sized pterosaur *Rhamphorhynchus* inhabits western Europe. It lives and hunts around lagoons and beaches.

thin muscles were all embedded within the wing structure. The membranes could be folded away when the animal was grounded or perched. Fossil tracks and the anatomy of limbs show that many pterosaurs were capable terrestrial walkers, and some would have foraged on foot instead of in the air. Fossils also indicate that pterosaurs were covered in a pelt of hairlike structures, pycnofibers. Air-filled sacs ran throughout their bodies—including into the bones—and were connected to the lungs. This air-sac system also featured in many dinosaurs and could have been present in the common ancestor of pterosaurs, and of the wider group: archosaurs. Alternatively, it could have evolved independently several times. The heat-generating flight muscles, active lifestyle, insulating outer covering, and air-sac system suggest that pterosaurs generated and retained heat internally and therefore were "warm-blooded."

The rhamphorhynchoids, the name taken from *Rhamphorhynchus* (see image 1), were a group of early pterosaurs that lived between 225 million and 125 million years ago. They were long-tailed insectivores or carnivores and their elongated jaws contained long, pointed teeth at the front and shorter, stouter, sometimes multicusped teeth at the back. The largest rhamphorhynchoids had wingspans of 8 feet (2.5 m), but most were less than half this size. Stomach contents show that some ate fish while others foraged for insects and small vertebrates on land. The tails of these early pterosaurs often had diamond-shaped structures at their tips, or had long, serrated lobes made from soft tissue that extended along the tail. Some rhamphorhynchoids from Germany are believed to have flown along coastal environments grabbing prey from the water. However, certain species, such as *Dimorphodon* from England, had deep, narrow skulls and fanglike teeth, so would have hunted prey on land, including in treetops. Their large, hooked claws are evidence of good climbing abilities.

The anurognathids—a group with short faces, enormous eyes, and widely spaced teeth—had short tails in contrast to other early pterosaurs, so it is possible that they were nocturnal and pursued insects in flight, similar to modern bats. The majority of pterosaurs belonged to a later group, the pterodactyloids, all of which were short-tailed. Several pterosaurs from China, notably *Darwinopterus* (see image 3), appear to bridge the anatomical gap between early, long-tailed pterosaurs, such as *Rhamphorhynchus*, and the first pterodactyloids. In fact, *Darwinopterus* seems to have had a pterodactyloid skull on a *Rhamphorhynchus*-like body, making it a near-perfect evolutionary intermediate. Early pterodactyloids had wingspans of less than 7 feet (2 m) and mostly lived in freshwater and marine environments. Some evolved hundreds of fine teeth to strain small animals from the water. Several groups, however, were toothless, including *Pteranodon* (see p.290). Azhdarchids were gigantic, storklike pterosaurs that picked up prey from the ground, while tapejarids were hornbill-like forest-dwellers that ate fruits, seeds, and small animals. Meanwhile, giant, albatrosslike, crested pteranodontids hunted out at sea. **DN**

140 Mya	125 Mya	100 Mya	85 Mya	68 Mya	66 Mya
Azhdarchids evolve. Early kinds have wingspans less than 10 feet (3 m), but have the long necks and long, slender jaws typical of this group.	The last non-pterodactyloid pterosaurs disappear. The short-faced anurognathid group survive due to a nocturnal lifestyle.	Large pterodactyloids with long skulls, termed ornithocheirids or anhanguerids, hunt in lakes and seas across South America, Europe, and Asia.	New toothless pterodactyloids—giant pteranodontids and nyctosaurids—become specialized for life at sea. Their large head crests show status.	Azhdarchids with wingspans of more than 33 feet (10 m) inhabit Europe and North America. They are the largest flying animals of all time.	The last of the pterosaurs—members of the azhdarchid group—become extinct, dying out in the mass extinction of the Cretaceous period.

Pteranodon
LATE CRETACEOUS PERIOD

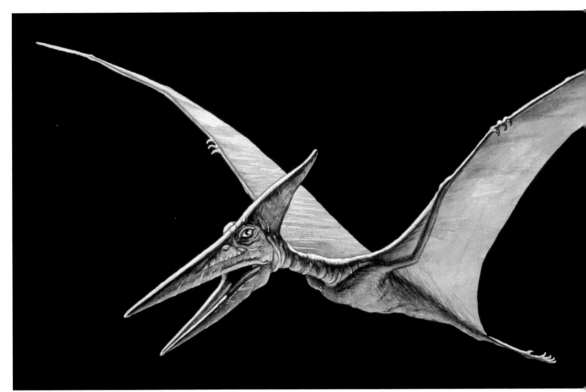

👁 FOCAL POINTS

1 JAWS AND DIET
P. longiceps's marine lifestyle and slender jaws suggest a diet of fish and marine prey. This is confirmed by a specimen that contained fish bones in its jaws, perhaps regurgitated during death. Its method for catching prey is uncertain: it could have grabbed prey while flying or floating, or diving from the surface.

2 NOSTRILS
During evolution, pterosaur nostril openings moved backward and eventually fused with another opening in the skull. This could mean that the nostrils were small, were shifted backward, or were even absent altogether and covered by skin or beak tissue. Furthermore, the parts of the pterosaur brain devoted to smell were tiny.

3 EYES
P. longiceps had enormous eyes and excellent eyesight. The parts of the brain devoted to sight were very large, confirming the importance of their vision. Like reptiles, pterosaurs presumably had sensitive color vision, and it is likely that the head crest of *P. longiceps* was brightly colored and used in display.

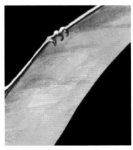

4 WINGS
There were four fingers on each hand: the gigantic wing (fourth) finger and three tiny clawed fingers located close to the wing finger's base. Tracks show that these were used to walk, and they presumably had roles in climbing and grooming too. Nyctosaurid pterosaurs lacked these fingers, perhaps spending most of their time in flight.

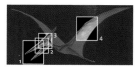
SPECIES *Pteranodon longiceps*
GROUP Pterosauria
SIZE 21 feet (6.5 m) wingspan
LOCATION North America

One of the best-known pterosaurs is *Pteranodon longiceps*, a giant, long-winged member of the group that flew over the Western Interior Seaway: the sea that split North America during the latter part of the Cretaceous. *Pteranodon* was not the largest pterosaur—the azhdarchids *Hatzegopteryx* from Romania and *Quetzalcoatlus* from North America were larger—but the biggest specimens of *Pteranodon* had wingspans that reached an impressive 21 feet (6.5 m). *Pteranodon* belonged to a toothless group that evolved from toothed ancestors around 100 million years ago. From smaller, more terrestrial ancestors, several different lineages independently evolved large size and a marine lifestyle. The reasons behind this are unlikely to have been differences in the world's atmosphere or climate. Their unique wing anatomy and take-off style possibly allowed them to become bigger and heavier than other flying animals. Furthermore, the anatomy of their fore limbs indicates that pterosaurs vaulted into the air by pushing off with their strong wing muscles.

Pteranodon had a bony head crest that varied in shape; some had long, slim crests while others sported short, rounded ones. It seems likely that this variation represents sexual dimorphism: the larger, longer-crested individuals were male and the smaller, shorter-crested animals female. If correct, the crests would have had a role in sexual display, perhaps demonstrating genetic quality. Bony crests are widespread in pterosaurs, perhaps indicating that sexual display was a major driving force behind their diversity and anatomy. Illustrations often show *Pteranodon* flying over dinosaurs *Tyrannosaurus* (see p.302) and *Triceratops* (see p.344) but this is inaccurate, partly because *Pteranodon* was oceanic, but also because it lived around 85 million years ago and became extinct more than 10 million years prior to the evolution of *Tyrannosaurus* and *Triceratops*. **DN**

TOOTHLESS BEAKS

Pterosaurs were known from fossil finds in Europe when, in the early 1870s, North American fossil hunters began to find the first specimens of *Pteranodon*. Rivals Othniel Charles Marsh (1831–99) and Edward Drinker Cope (1840–97) both identified and named several kinds from the skeletal remains, although preserved skulls had not been found. Then, in 1876, biologist and fossil excavator Samuel Wendell Williston (1851–1918) uncovered a pterosaur skull in Kansas. Marsh examined the find and saw at once that it had great relevance, because unlike the European pterosaurs known to that time it lacked teeth (left). Marsh coined the name *Pteranodon* (toothless beak) and reclassified some of his earlier finds, which he had named *Pterodactylus*, as this new genus.

TURTLES AND TORTOISES

T urtles, terrapins, and tortoises—the testudines or chelonians—are one of the strangest and most enigmatic groups of reptile. Their key feature, the turtle shell, is formed by a domed upper part, termed the carapace, and a lower section that covers the chest and belly, termed the plastron. Remarkably, the carapace is a modified ribcage, formed from expanded, platelike ribs covered in horny scutes (waterproof scales made of keratin). During turtle evolution, the ribcage expanded to enclose the shoulder and hip bones. As turtle embryos grow, their bodies replay this evolutionary event: their ribs grow down from the spinal column before expanding outward and downward to "capture" the shoulders and hips. Turtles also became toothless during their evolution and all modern varieties have a toothless beak.

The oldest known turtles lived 220 million years ago during the late Triassic period. For a long time a *Proganochelys* fossil specimen (see image 1) from Germany provided evidence of the oldest known turtle. It was broad-bodied with a spiky, club-tipped tail, had small teeth on its palate, and its limb proportions suggested a terrestrial lifestyle. This *Proganochelys* had a fully developed shell, so the fossil did not provide new information on the earliest stages of turtle evolution. A more recently discovered Triassic turtle, *Odontochelys* from China, may represent an earlier stage in evolution because it seemingly lacked a carapace and had teeth along its jaw edges and also on the palate. However, its lack of a carapace is controversial—the shape of the ribs suggests that a carapace may have been present but was not preserved. Alternatively, its aquatic lifestyle could have led to the reduction or loss of the carapace.

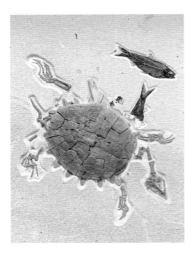

KEY EVENTS

220 Mya	210 Mya	150 Mya	115 Mya	112 Mya	110 Mya
Odontochelys, the oldest known turtle, emerges. A small reptile, it has limb proportions consistent with life in shallow ponds.	*Proganochelys* and similar archaic Triassic turtles inhabit the land. They have wide bodies and armor-plated tails, and retain teeth on the palate.	Pleurodires and cryptodires—the two main modern turtle groups—evolve by the late Jurassic period. Both groups diversify during the Cretaceous.	Pleurodires undergo a burst of evolution, developing new short-skulled and long-skulled groups in North and South America, Africa and Europe.	Aquatic cryptodires take to marine life, evolving flippers and streamlined shells. They become sea turtles and evolve larger sizes and lighter shells.	Softshell turtles evolve among cryptodires and are soon numerous in freshwater habitats, especially across Asia. They evolve unusual skulls and carapaces.

Proganochelys-like turtles with bulky, armored tails persisted well beyond the Triassic. Termed meiolaniids, they inhabited South America, Australia, and a few Southwest Pacific islands. The youngest meiolaniid remains are from Vanuatu, Southwest Pacific, and are 3,000 years old. Meiolaniids reached more than 7 feet (2 m) in length and had horns and bony collars on the back of their skulls, perhaps used for mating displays, battles, digging, or self-defence.

The majority of turtles and tortoises belong to one of two major groups: the side-necked turtle (pleurodires) or the hidden-necks (cryptodires, see image 2). The former retracts the neck by pulling it horizontally beneath the carapace, whereas the latter retracts it vertically. Ball-and-socket joints allow both of these actions. Modern living pleurodires are limited to South America, Africa, Madagascar, and Australasia; they are generally less than 3 feet (1 m) long and are amphibious omnivores and carnivores of lakes and rivers. However, the fossil record reveals a far richer history. Members of several pleurodire groups occurred in the northern continents; some were marine, and a diversity of skull shapes indicates varied diets and lifestyles. Some pleurodires were giants, exceeding 10 feet (3 m) in length. Long necks evolved at least twice in the group.

Fully aquatic cryptodires with flippers evolved from amphibious ancestors that lived in shallow pools and lakes. It it likely that turtles of this sort originated in estuaries and deltas before moving to the open sea, where they then evolved gigantic winglike paddles and skulls specialized to eat marine prey, including sea-jellies, sponges, and swimming mollusks. The horn content of the shell (but not its size) was reduced substantially in two marine cryptodire groups, both of which evolved into giants: the leatherbacks (see image 3) and the protostegids of the Cretaceous period, such as *Archelon* (see p.294). Presumably, this made them lighter, faster swimmers. A separate cryptodire group moved from marshes and ponds onto land, becoming fully terrestrial tortoises. In recent geological history, South American and African tortoises colonized islands in the Pacific and Indian Oceans and became giants. Different shell shapes and neck lengths evolved to match the local climate and vegetation, and many species resulted, making giant island-dwelling tortoises a major ecological presence across the world's tropical island groups. Sadly, such animals were highly vulnerable to sailors, and many of these tortoises have been hunted to extinction in the last few hundred years.

The exact place of turtles in the reptile family tree is controversial. Turtle skulls lack bony openings behind the eye, so they are anapsids and have often been regarded as archaic reptiles closely related to or descended from the most ancient groups. However, it has also been recognized that turtles have changed a lot in relation to other reptiles; consequently their apparently primitive skull anatomy could be deceiving. In fact, genetic studies consistently place turtles in the diapsid group, which indicates that they are perhaps closely related to dinosaurs and other archosaurs. **DN**

1 The skull of *Proganochelys*, a Triassic form about 3 feet (1 m) long, shows the typically turtlelike toothless beak for snipping and cutting up vegetation.

2 This softshell turtle fossil dates to the Eocene epoch, around 50 million years ago. The skull and neck are lower right, between the front limbs; two co-preserved fish are in the middle and upper right.

3 The leatherback *Dermochelys coriacea*, the largest living chelonian, has a much reduced shell.

80 Mya	60 Mya	55 Mya	4 Mya	3 Mya	3,000 Ya
Gigantic leatherback and protostegid turtles inhabit the seas of Europe, Asia, North America, and Australia. Some are more than 13 feet (4 m) long.	Land turtles with stout limbs appear: tortoises. Early kinds are small and live near marshes and lakes. Grassland and desert-dwellers evolve later.	The many diverse pleurodires of the northern continents mostly disappear, possibly as a result of climatic cooling.	Global cooling causes the loss of several turtle groups from northern continents. Snapping turtles and giant tortoises both disappear from Europe.	Tortoises in South America and Africa travel to the Galapagos and Mascarene Islands respectively, and evolve into long-necked, island-dwelling giants.	The last meiolaniids— an archaic group of horned, land-living turtles that evolved during the Jurassic— are hunted to extinction by humans.

Archelon
LATE CRETACEOUS PERIOD

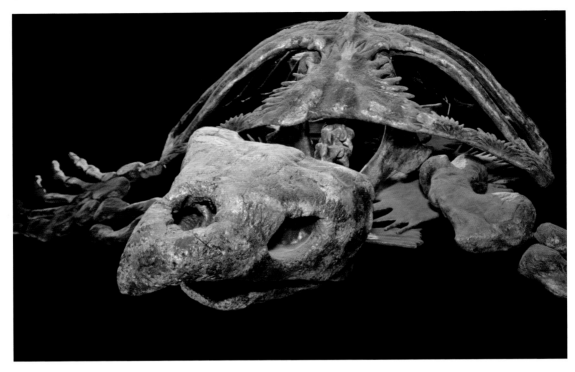

SPECIES *Archelon ischyros*
GROUP Protostegidae
SIZE 13 feet (4 m)
LOCATION North America

It is likely that *Archelon* belonged to the cryptodire, or hidden-neck, group, and within this the protostegids, all of which are now extinct. Their closest relations were once thought to be the dermochelyids, another group that is extinct apart from the leatherback turtle, *Dermochelys coriacea*. However, recent studies have questioned their evolutionary position, postulating that *Archelon* and other protostegids may in fact represent a separate group of turtles with much earlier origins.

Both the fossil record and the diversity of living turtles show that the average ancestral body size for turtles is 2 feet (60 cm) long. However, giant-sized turtles evolved several times within the group, in land-living tortoises and sea turtles, as well as in a number of amphibious turtle groups. Among the largest and most spectacular turtles is the extinct marine *Archelon ischyros* of ancient North America. The total length of this aptly named "turtle ruler" was 13 feet (4 m). Its front flippers spanned an immense 16 feet (5 m) and its total mass would have exceeded 2¼ tons (2 tonnes). However, the leatherback turtle is the living record-holder for size. It has been documented at almost 10 feet (3 m) from nose to tail, with a similar flipper span and a weight of almost 1¹⁄₁₀ tons (1 tonne). Fossils of *Archelon* have been found mainly in central North America and date to the late Cretaceous, 80 million years ago. At this time, vast expanses of a shallow salty sea with numerous lagoons covered the area known as the Western Interior Seaway. This sea was populated by numerous other extra-large representatives of sea reptiles, such as the long-necked plesiosaur *Elasmosaurus* (see p.286) and the enormous 49-foot (15 m) mosasaur *Tylosaurus*, as well as the powerful hunting bony fish *Xiphactinus* (see p.204) and the great-white-sized shark *Squalicorax*. Clearly, the Western Interior Seaway contained some very productive habitats that allowed diverse plant life and small creatures to thrive in large enough quantities to feed these various kinds of top predators. **SP/DN**

⬡ **NAVIGATOR**

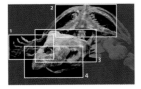

FOCAL POINTS

1 FLIPPERS
All four limbs of *Archelon* were adapted as broad water-thrusting flippers. The front pair were larger and it was likely that they moved up and down, to "fly" through the water, rather than forward and backward, to "row." The limb skeleton follows the standard vertebrate pattern, with five digits, a wrist or ankle, and lower and upper limb bones. All were modified into the broad paddle shape.

3 BODY FORM
When compared to terrestrial and semi-aquatic species, the head of *Archelon* was very large and long in proportion to the main body—in some specimens the head was 3⅓ feet (1 m) in length. The neck was relatively short, so head movement was fairly limited. The protostegids set the pattern for marine turtles, with a wide bony shell more flattened top-to-bottom than in other families.

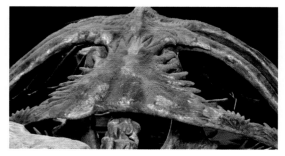

2 SHELL
Archelon evolved a very large shell with bony rib struts, which were probably covered by a very thin carapace or leathery skin. This light covering would have saved weight in order to increase speed and agility in the water. It would also have made the turtle more effective in catching prey and evading predators efficiently, thereby self-spiraling this evolutionary trend.

4 BEAK
The horny beak is reminiscent of that of the living leatherback turtle, which consumes mainly jellyfish. *Archelon* is likely to have eaten a wider variety of food given its size and abilities. Its diet would have included small-to-medium floating and slow-swimming prey, such as squid, salps, fish, and ammonites. It may have targeted the weak or dead because its jaws were not very strong.

DARWIN'S GIANT TORTOISES

In the absence of mammal competitors for plant food on some small islands and island groups, several kinds of giant tortoises evolved. This happened on the Seychelles, Mascarene, and Galapagos Islands (right). Naturalist Charles Darwin (1809–82) visited the Galapagos in 1835 and noticed that tortoises had colonized this remote island group, growing to great size. He was also intrigued by the differences in shell shapes and body proportions between tortoises from different islands. On drier islands they were smaller, with a "saddleback" shell shape and longer necks and limbs. Local people could identify a tortoise's island of origin from these features. Along with mockingbirds and finches, the giant tortoises played a pivotal role in Darwin's theory of evolution.

Snapping Turtle
NEOGENE PERIOD—PRESENT

SPECIES *Macrochelys temminckii*
GROUP Chelydridae
SIZE 3 feet (1 m)
LOCATION United States
IUCN Vulnerable

◈ NAVIGATOR

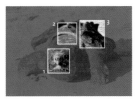

Among the most distinctive and formidable of the world's approximately 300 living turtle species are the snappers: the three species of snapping turtle—genus *Chelydra*—and the three species of alligator snapping turtle—genus *Macrochelys*. All are aquatic freshwater predators with large heads, enormous hooked jaws, and long, chunky tails that in the three *Chelydra* species have a series of serrations along the upper surface. Powerful claws are used to walk, swim, and kill prey. The alligator snapper *Marochelys temminckii* has three ridges running the length of its carapace formed from pyramidal projections. It is the largest of the living snapping turtle species; exceptional specimens reach 3 feet (90 cm) in length and 220 pounds (100 kg) in weight.

Snapping turtles are part of the cryptodire or hidden-necked turtle group termed Americhelydia, meaning "turtles of America." Genetic studies indicate that they are closely related to the small, amphibious mud and musk turtles as well as to sea turtles. Today, all members of the Americhelydia group are unique to the Americas, occurring from Nova Scotia to Ecuador. However, fossils show that snapping turtles were more widespread in the past. Numerous extinct species are known from Europe, Asia, and possibly Africa, the oldest dating back to more than 66 million years ago. Most modern turtle groups originated at this time (the late Cretaceous) and subsequently survived the extinction event at the end of the Cretaceous before evolving rapidly in the following Cenozoic era. By 40 million years ago, ancient snapping turtles had spread across much of the northern continents, and the group was especially species-rich between about 30 million and 5 million years ago. Apart from the six surviving species mentioned above, these had all become extinct by 3 million years ago, predominantly because of the global cooling at this time. **DN**

 FOCAL POINTS

1 BITE POWER

The skull of *M. temminckii* is adapted for biting, and the jaw muscles occupy a large region at the back. A long, bony, fingerlike crest projects backward from the rear edge, providing areas for muscle attachment. The tips of the jaws are hooked, supporting a curved beak that digs into the tissues of prey.

2 SECRETIONS

Rathke's glands or musk glands line the outside edge of the shell and produce strong-smelling secretions. Their function could be social, sexual, or a defence from predators. The secretions may deter small organisms that would otherwise attach themselves to the shell of *M. temminckii*.

3 CAMOUFLAGE

Due to its bumpy skin, large scales, soft spines, and rough areas on the shell, the alligator snapping turtle is often partially covered in algae and mud. This makes it near invisible on the lake or pond floor. It gives it a prehistoric appearance, and presumably its extinct relations covered themselves in this way, too.

▲ Snapping turtles, such as this *Chelydra serpentina*, hunt prey including fish, frogs, swimming snakes, mammals, and even other turtles, by ambushing with their fast lunging bite. This mode of feeding has been known in turtles since their origins, although the teeth in early types have now disappeared.

FOSSILIZED RELATIVES

Fossilized remains show that snapping turtles have been in existence for at least 66 million years. One of the earliest known types is *Emarginachelys*, whose remains come from Montana. *Protochelydra*, from North Dakota, dates to several million years later. This snapper evolved a taller, more rounded shell than its modern cousins. It is suggested that the higher dome of the carapace would have discouraged predators of the time, such as crocodiles and alligators, from attacking it. The predators would have had problems opening their mouths wide enough to encompass the turtle, and with their jaws gaping so much they would have been unable to bite very hard. Snapping turtle remains (right) dating from 50 million years ago, during the Eocene, occur in Wyoming, while more recent discoveries of *Protochelydra* extend its range far to the north, to Alaska.

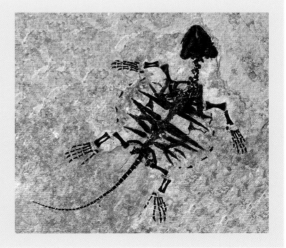

GIANT CARNIVOROUS DINOSAURS

1 *Giganotosaurus*, from Argentina, had
one of the largest known dinosaur skulls,
measuring more than 5 feet (1.5 m).
In life, this theropod would have weighed
8 to 9 tons (7.5–8 tonnes).

2 Teeth of the North African
Carcharodontosaurus grew to 9 inches
(22 cm) long and show a curved shape
with serrated edges.

A mong the most spectacular creatures of all time, giant carnivores—
Spinosaurus (see p.300), *Tyrannosaurus* (see p.302), *Allosaurus,* and their
relatives—had the ability to kill and dismember large prey items. These
giant theropod (meat-eating dinosaurs) were the top terrestrial predators of
the Jurassic and Cretaceous periods, from 200 million to 66 million years ago.
Despite their similarities, they were members of only distantly related branches
of the theropod family tree, which independently evolved similar solutions to
their ways of life. For most of their Mesozoic era history, theropods weighed
in the region of 220 pounds (100 kg), large enough to tackle prey but not
super-predators capable of killing huge animals. Giant theropods, in contrast,
exceeded several tons and the record holders, among them *Tyrannosaurus*,
Giganotosaurus (see image 1), *Carcharodontosaurus*, and *Spinosaurus*, were
more than 33 feet (10 m) in length and weighed upward of 5 tons (4.5 tonnes).
These giant theropods were members of Tetanurae, the "stiff-tailed" group that
evolved from dilophosaur-like ancestors 195 million years ago. Until the 1990s,
it was thought that all big tetanurans belonged to the group Carnosauria
(flesh-eating lizard/reptile). This view stated that allosaurs evolved from
megalosaurs and that short-armed, two-fingered tyrannosaurs evolved from
bigger-armed, three-fingered allosaurs. However, recent scientific analyses and
newly discovered fossils show that carnosaurs are not a natural group.

KEY EVENTS

195 Mya	170 Mya	165 Mya	150 Mya	150 Mya	145 Mya
Tetanurans evolve from dilophosaur-like ancestors, diversifying into megalosaurs and birdlike allosaurs and coelurosaurs. All three groups become larger.	Early allosaurs— members of the metriacanthosaurid group—appear in Asia and Europe. They have tall ridges along their backs and nasal crests.	Europe is inhabited by a diversity of megalosaurs. Many are small compared to the giant tetanurans that evolve later.	*Allosaurus* lives in North America and Europe, alongside large megalosaurs and big predatory non-tetanurans, such as *Ceratosaurus*.	Early tyrannosaurs inhabit Europe, North America, and Asia. They are small, long-armed, and not like the gigantic species that evolves much later.	Megalosaurs are mostly extinct at the end of the Jurassic, but the crocodile-headed spinosaurs survive in both northern and southern continents.

Megalosaurs represent one branch and allosaurs another. Also, tyrannosaurs have features showing they belong to the birdlike group Coelurosauria. Tyrannosaurs were therefore not closely related to the other big tetanurans, and the coelurosaurs were more closely related to allosaurs than to megalosaurs. This means that the several tetanuran groups evolved from different ancestors.

In each case, the availability of large prey drove the evolution of large body sizes. Furthermore, species belonging to these three groups (megalosaurs, allosaurs, and tyrannosaurs) lived in different places at different times. The majority of large allosaurs existed after megalosaurs were extinct and they were more strongly associated with Asia than the European and North American megalosaurs. Meanwhile, the species of tyrannosaur that lived alongside these two groups were small-bodied, only becoming huge once the others disappeared. This indicates that most of the "takeover events" observed in the theropod fossil record do not reflect protracted, aggressive confrontations between groups. Instead, a big-bodied group became extinct and a previously small-bodied group evolved to fill the void. Tyrannosaurs existed for tens of millions of years as small predators, in danger of being eaten by larger megalosaurs and allosaurs, before evolving into giant super-predators.

The most archaic or "primitive" of the big-bodied theropod groups, the megalosaurs included species with long muzzles, recurved (back-curved) teeth, and powerful arms and hands. The largest of them—such as *Torvosaurus* from North America and Portugal—were 33 feet (10 m) long. These became extinct 145 million years ago. However, a group of especially long-jawed megalosaurs evolved prior to this; their crocodilelike jaws and conical teeth indicate that they were specialized for fish hunting. These were the spinosaurs, the only megalosaur group that lived into the Cretaceous.

Some spinosaurs and other megalosaurs lived alongside allosaurs. This last group includes the Asian and European metriacanthosaurids, *Allosaurus* of North America and Portugal, and the widespread carcharodontosaurs, named for the similarity between their teeth and those of *Carcharodon*, the great white shark. South American and African carcharodontosaurs became enormous, similar in size to *Tyrannosaurus*. However, while tyrannosaurs evolved thick, massively constructed skull bones, spikelike teeth, and specializations to increase their bite power, carcharodontosaurs retained the lightly built skulls and bladelike teeth (see image 2) typical of allosaurs. These features suggest that they took quick, slashing bites from prey animals, weakening them through trauma and blood loss. Tyrannosaurs, in contrast, were capable of biting chunks of flesh and breaking through bones. Although they overlapped in time, giant members of these two different groups never met. Giant tyrannosaurs were animals of the northern continents while giant carcharodontosaurs were restricted to the south. **DN**

135 Mya	125 Mya	110 Mya	100 Mya	94 Mya	68–66 Mya
Spinosaurs—fish-eating megalosaurs—live in Europe. *Baryonyx* is one of the oldest in the group and probably originated earlier.	Carcharodontosaurs, a group of allosaurs, appear in Europe. *Neovenator* is one of the first. Later species occur in Africa and the Americas.	Giant allosaurs disappear from Asia and North America, but persist in South America. There is no giant predator in the northern continents.	Carcharodontosaurs occupy Africa and South America: *Carcharodontosaurus* in Africa and *Giganotosaurus* in Argentina.	The last spinosaurs disappear, ending the megalosaur dynasty. Spinosaur extinction may be linked to an event that affects freshwater fish groups.	Top predator roles in Asia and North America are dominated by advanced tyrannosaurs, including slender-snouted alioramins and *Tyrannosaurus*.

Spinosaurus
EARLY CRETACEOUS PERIOD

👁 FOCAL POINTS

1 TEETH
The teeth were conical or semiconical, not back-curved and bladelike as is typical for other theropods. Teeth of this sort are typical of animals that grab fish. Stomach contents from the related *Baryonyx* prove that fish formed part of its diet, but it is likely that dinosaurs and other prey were eaten too.

2 JAWBONES
The jawbones were covered in tiny openings that housed abundant pressure-sensitive organs. These indicate that the jaws were sensitive to vibration and touch, suggesting it hunted by holding its jaws motionless in the water. The movements of swimming fish could be sensed, and prey located.

3 BACK SPINES
The long, rodlike bony processes that supported the sail of *Spinosaurus* grew up from the vertebrae and are known as neural spines. The longest were more than 5 feet (1.5 m), strap-shaped and thin from side to side. They tapered only slightly and dwarfed the vertebrae at their bases.

4 FORE LIMBS
The fore limbs of *Spinosaurus* were robust and powerfully muscled, and the claws on its three-fingered hands were huge and intensely curved. This is reflected in the name of the related *Baryonyx* (heavy claw). The giant hand claws might have been used as gaffs that pulled large fish onto land.

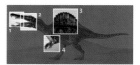
SPECIES *Spinosaurus aegyptiacus*
GROUP Spinosauridae
SIZE 49 feet (15 m)
LOCATION North Africa

From 110 million to 100 million years ago in North Africa lived *Spinosaurus aegyptiacus*, the best known of the spinosaurids. It partly owes its high public profile to the long bony spines that grew upward from its back and part of its tail, which show that it had a giant sail. The spines on the sail somewhat resemble those of mammals such as bison, and some experts have argued that *Spinosaurus* was hump-backed. However, the spines probably looked more like a sail because skin and ligaments would have spanned the upper parts. For a long time, little else was known of *Spinosaurus*, and consequently old reconstructions show this spinosaurid looking like a sail-backed *Tyrannosaurus*.

Thanks to the discovery of the spinosaurid *Baryonyx* in Dorking, England, and to new remains of *Spinosaurus* uncovered in North Africa (the originals were destroyed during World War II), it is known that spinosaurids were among the most unusual of meat-eating dinosaurs. Their fore limbs were strong and muscular, and spinosaurid skulls were highly modified compared to those of other megalosaurs. They are elongate, giving the snout and jaws a superficially crocodilelike appearance, and the nostrils are some distance back from the snout tip. Combined, these features indicate that spinosaurids evolved to be fish-eaters that plucked prey from shallow waters while they waded or swam. It seems likely that spinosaurid evolution was driven by an abundance of large, accessible fish prey, and fossils found alongside these dinosaurs show that such animals were common in their habitat. *Spinosaurus* inhabited mangroves and floodplains, while other spinosaurids such as *Baryonyx* occurred in environments where marshes, giant lakes, and meandering rivers were inhabited by large, carplike fish. This lifestyle was so successful that spinosaurids became larger over time. *Spinosaurus*, one of the very last members of the spinosaurid group, was a gargantuan beast around 49 feet (15 m) long and weighing more than 11 tons (10 tonnes). **DN**

THE PURPOSE OF THE SAIL

Back sails have evolved several times in large creatures, including the synapsids *Dimetrodon* (see p.238) and *Edaphosaurus*, and dinosaurs such as *Spinosaurus* and the *Iguanodon*-like plant eater *Ouranosaurus*. Possible reasons include thermoregulation and control of body temperature. However, this ignores the fact that, as shown by reptiles and insects today, cold-blooded animals use numerous behavioral tricks to control body temperature. Furthermore, giant theropods from the same environment as *Spinosaurus*, and even from hotter ones, did not have specialized heat-regulation structures such as a sail. A more plausible idea is that the sails were flamboyant display structures, like the peacock's tail or anole lizard's chin flap (right), to show off to the pack at breeding time, to territorial intruders, or to other meat-eaters attempting to usurp prey.

Tyrannosaurus
LATE CRETACEOUS PERIOD

SPECIES *Tyrannosaurus rex*
GROUP Tyrannosauridae
SIZE 39 feet (12 m)
LOCATION North America

One of the best-known animals of all time, *Tyrannosaurus rex* is also one of the most intensively studied and best understood of all fossil dinosaurs. Gigantic, broad-chested, and hugely muscular, it combined enormously thick skull and jawbones with stout, spikelike teeth and long, powerful hind limbs. Its total length exceeded 39 feet (12 m) and it weighed more than 6½ tons (6 tonnes). Some herbivorous dinosaur bones bear *Tyrannosaurus* bite marks, confirming that it could bite deep into thick bones. One *Triceratops* (see p.344) specimen shows that a *Tyrannosaurus* bit off one of its horns. Studies indicate that *Tyrannosaurus* had one of the strongest bites that has ever evolved, exceeding powerful modern biters such as alligators by a considerable margin.

As the last and geologically youngest member of the entire tyrannosaur lineage, living around 68 million years ago, *Tyrannosaurus* (king tyrant lizard/reptile) can be seen as the "end member" of a trend toward increasing body size and firepower. A reliance on the biting intensity of the jaws and teeth resulted in a dwindling of the fore limbs. All advanced tyrannosaurs—grouped together as the tyrannosaurids—exhibit the same big-headed, short-armed body plan. In contrast, the hind limbs of *Tyrannosaurus* were especially long and powerful compared to other big theropods. The elongate sole bones, forming the long part of the foot, were locked tightly together for strength and the shin bones were lengthened. While theropods such as megalosaurs would have used a short dash to ambush prey, tyrannosaurids would have been faster sprinters, swiftly covering distance. It is not known exactly how fast *Tyrannosaurus* could run, but speeds exceeding those of a running human are likely. This super-predator was the evolutionary end product of increasing specialization toward hunting large herbivorous dinosaurs, but despite its size, power, and varied ways of feeding, *Tyrannosaurus* perished in the Cretaceous mass extinction event 66 million years ago, along with all other nonbird dinosaurs. **DN**

NAVIGATOR

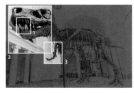

1 TEETH

Short teeth that were D-shaped in cross section were present at the front of the upper jaw and would have served a nipping or scraping role. Much longer teeth, round in cross section, line the middle of the upper jaw and would have been the main weapons used in predation. Shorter, blunter teeth were present at the back of the jaws.

2 SKULL

The muzzle was broader than was typical for theropods. The skull was wide across the cheeks, so the eye sockets pointed partially forward. Reconstructions of the field of vision suggests that *Tyrannosaurus* had a large area of overlap in its line of sight, which points to good binocular vision and depth perception.

3 SHORT ARMS

The purpose of the arms is unknown, but some have suggested that they latched on to prey, carried carcasses, were mating claspers or props enabling *Tyrannosaurus* to raise itself from the ground. The arms were not useless, but are unlikely to have been specialized for a role. They were relict structures on their way to loss.

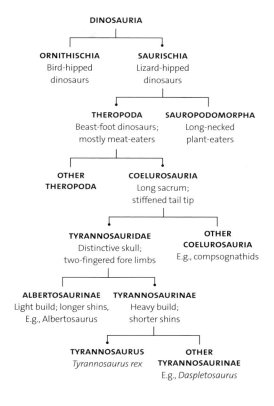

DINOSAURIA

ORNITHISCHIA
Bird-hipped dinosaurs

SAURISCHIA
Lizard-hipped dinosaurs

THEROPODA
Beast-foot dinosaurs; mostly meat-eaters

SAUROPODOMORPHA
Long-necked plant-eaters

OTHER THEROPODA

COELUROSAURIA
Long sacrum; stiffened tail tip

TYRANNOSAURIDAE
Distinctive skull; two-fingered fore limbs

OTHER COELUROSAURIA
E.g., compsognathids

ALBERTOSAURINAE
Light build; longer shins, E.g., Albertosaurus

TYRANNOSAURINAE
Heavy build; shorter shins

TYRANNOSAURUS
Tyrannosaurus rex

OTHER TYRANNOSAURINAE
E.g., *Daspletosaurus*

▲ *Tyrannosaurus* was used to establish its family, Tyrannosauridae: very large predators with tiny two-fingered arms. It was on a different branch of the theropod lineage to carnosaurs such as *Giganotosaurus* and *Carcharodontosaurus*.

ACTIVE HUNTER

One view of *Tyrannosaurus* is of a dedicated scavenger, with eyes too small and legs too short to make an effective predator. However, the eyes are larger and the legs longer than is expected in a theropod of this size, and its body build and proportions suggest it was an active hunter. Although, similar to modern predators, *Tyrannosaurus* would have scavenged when the opportunity arose. CT (computer tomography) scans have revealed that, similar to many animals, the head of *Tyrannosaurus* was riddled with 10 sets of air cavities, or sinuses (colored areas, right). This meant that the head was one-fifth lighter than if it had contained simple tubelike breathing passages. This lightness would have facilitated sudden head movements, as the dinosaur tackled live prey, while retaining the structural integrity of the bone.

DROMAEOSAURS (RAPTORS)

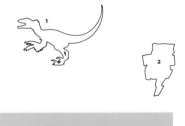

1 North American *Utahraptor* hunted in packs during the Cretaceous period. It weighed 660 pounds (300 kg) and was 17 feet (5 m) long.

2 *Sinornithosaurus* from about 123 million years ago was 3 feet (1 m) long. It had two main kinds of feathers, with micropreservation indicating their colors were brown-red, gray, yellow-orange, and black.

Today, birds are a unique and distinctive group of animals. At the start of their history, however, birds were merely one among several lineages of small, feathered coelurosaurian dinosaurs, all of which pursued similar lifestyles and were similar in body shape and appearance. The more birdlike coelurosaur groups—those with large, crescent-shaped wrist bones and large, platelike breastbones—are known as the maniraptorans. Among the best-known maniraptorans are the dromaeosaurs, also known as dromaeosaurids or raptors. This subgroup appeared 165 million years ago and persisted until the extinction at the end of the Cretaceous period, 66 million years ago. Two dromaeosaurs—*Deinonychus* from North America (see p.306) and *Velociraptor* from Mongolia and China (see p.308)—are among the most important dinosaurs of all, in terms of shaping scientific and popular ideas about dinosaur diversity, biology, and appearance. A huge number of recent discoveries show how dromaeosaurs evolved from small, birdlike ancestors into wolf-sized and even bear-sized predators, while others became long-snouted fish-eaters.

After the discovery of their fossils in the 1920s, dromaeosaurs were thought to be fairly nondescript. However, during the 1960s fossils of their remarkable feet were found. The second toe was hyperextendable, meaning that it could be raised off the ground. The rollerlike shapes of the joints showed that this was its normal pose. At the end of this toe was a giant, sickle-shaped claw. Experts have debated

KEY EVENTS

165 Mya	130 Mya	125–120 Mya	120 Mya	113–109 Mya	110 Mya
Dromaeosaurs evolve from a maniraptoran ancestor, a crow-sized, feathery omnivore with slim jaws, that later gives rise to birds and troodontids.	The long-snouted dromaeosaur unenlagiine evolves in South America and the Velociraptorinae group appears in Asia.	*Utahraptor* inhabits North America. It is one of several dromaeosaurs that evolve huge size from smaller ancestors.	Crow-sized dromaeosaurid and generalist predator *Microraptor* inhabits China, as do other small, feathered members of the group.	*Deinonychus* occurs across North America, in habitats from semi-deserts to humid swamps. It is an adaptable, mid-sized dromaeosaur.	Continental break-up causes dromaeosaurs (and others) on the North American–Asian block to pursue distinct evolutionary paths from those elsewhere.

whether this was a slashing, ripping, or stabbing tool, and it has even been suggested that these dinosaurs leaped onto prey before using their sickle claw to rip open the belly. However, this idea is now seen as unlikely because tests have demonstrated how such claws interact with skin and muscle. Similar enlarged claws are present on the second toes of modern hawks, and their role is to pin the prey to the ground while the bird attacks with the beak. It seems most likely that dromaeosaurs also used this strategy, pinning mid-sized animals to the ground before attacking them with teeth. This strategy could explain why dromaeosaurs have proportionally short, stocky legs compared to other coelurosaurs.

The size of an average dromaeosaur was between a large rooster and a human, although this varied a lot. Shallow jaws lined with back-curved, serrated teeth were typical, as were long, slender, three-fingered hands equipped with large, strongly curved claws. Typically for theropods (meat-eating dinosaurs), the hands were held with their palms facing inward, which suggests a specialized grabbing function. Specimens with their feathering preserved show that feathers grew from the upper surface of the second finger as well as the lower arm bone (ulna), meaning that much of the fore limb was obscured by large feathers. Exceptionally well-preserved Chinese dromaeosaurs such as *Microraptor* (see p.314) and *Sinornithosaurus* (see image 2) show that these dinosaurs—seemingly like all maniraptorans—were covered in feathers.

The dromaeosaur pubic bone often projected backward. This unusual feature is typical of modern birds and ornithischians (bird-hipped dinosaurs), but not of the group to which dromaeosaurs belonged: the saurischians (lizard-hipped dinosaurs; see p.260). However, it is now known to have been present in several birdlike coelurosaurian groups. In fact, the earliest and most primitive dromaeosaurs were extremely similar in size and form to the earliest birds. It seems that dromaeosaurs, birds, and related maniraptorans all descended from a small-bodied ancestor, somewhat similar to *Archaeopteryx* (see p.376). While some of these evolutionary lines remained small, dromaeosaurs evolved to a large size on at least two occasions. Stomach contents from the crow-sized early dromaeosaur *Microraptor* show that it was a generalist predator of birds, fish, and other small vertebrates. In contrast, the larger size of forms such as 155-pound (70 kg) *Deinonychus* and 660-pound (300 kg) *Utahraptor* (see image 1), combined with information from the herbivorous dinosaurs they were preserved with, suggests that they were predators of large animals, perhaps cooperating in packs to kill victims larger than themselves.

A mostly South American dromaeosaur group called the unenlagiines had especially long, slender snouts and numerous teeth, features that suggest they grabbed fish, frogs, and similar prey while wading. One member of the group, *Austroraptor* from Argentina, was a giant, reaching 20 feet (6 m) in length and with a shallow skull more than 32 inches (80 cm) in length. **DN**

90 Mya	77 Mya	75 Mya	75 Mya	70 Mya	66 Mya
Unenlagiines of several different kinds coexist in South America. Their long, slender jaws suggest they are wading fish-eaters.	*Hesperonychus* inhabits forests and swamps of Canada. It is the last member of the dromaeosaurid group that includes *Microraptor*.	*Velociraptor* inhabits Mongolia and China. It is mostly associated with deserts but also occurs in dry woodlands and scrublands.	Small and medium-sized dromaeosaurs occur in North America, for example, *Dromaeosaurus*, *Bambiraptor*, and *Saurornitholestes*.	*Austroraptor*, a huge member of the unenlagiine group, lives in Argentina. It has evolved its size independently of other big dromaeosaurs.	The last dromaeosaurs inhabit North America and become extinct at the end-of-Cretaceous extinction event. Their close cousins, the birds, survive.

Deinonychus
EARLY CRETACEOUS PERIOD

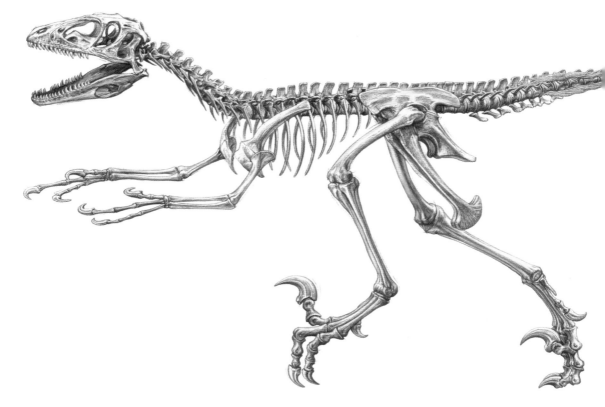

SPECIES
Deinonychus antirrhopus
GROUP Dromaeosauridae
SIZE 10 feet (3 m)
LOCATION North America

Among the best known of the dromaeosaurs (raptors), is *Deinonychus antirrhopus* (awesome/terrible claw) from North America—a wolf-sized theropod with only one described species known from numerous remains. In the anatomy of its hands, wrists, vertebrae, and other parts, *Deinonychus* was similar to the early bird *Archaeopteryx*, albeit several times larger. (*Deinonychus* was around 10 feet (3 m) long and 155 pounds (70 kg) in weight.) This discovery led to a series of studies in which US paleontologist John Ostrom (1928–2005) showed that birds were members of the theropod group and had evolved from *Deinonychus*-like ancestors—an idea that has since been supported by numerous additional birdlike dinosaur species, many preserved with feathers in position. The suggestion that birds might be dinosaur descendants arose in the 1860s but fell into disfavor in the intervening decades. Ostrom resurrected it, and it is now widely accepted. The agile, dynamic nature of *Deinonychus* also inspired Ostrom to propose that it and similar dinosaurs were more birdlike than reptilelike in their physiology (body chemistry). The curved claw on the second toe of *Deinonychus* appears to have been its main weapon. Ostrom concluded that it could be deployed only if the dinosaur balanced on one leg while kicking. Accordingly, he reasoned that *Deinonychus* had lightning reflexes and a sophisticated sense of balance. His reinvigorated view of dinosaurs—that they were birdlike, sophisticated, and even warm-blooded—began a surge of scientific interest in the 1960s (see panel). *Deinonychus*, which lived between 115 million and 108 million years ago, was named in 1969 and consequently was a relative latecomer, as another of its group, *Velociraptor*, had been named in 1924. However, the completeness of its fossils, combined with thorough research by Ostrom, resulted in the realization that dinosaurs were more agile, dynamic, and birdlike than had been appreciated. **DN**

◆ **NAVIGATOR**

FOCAL POINTS

1 SKULL AND JAW

The skull was about 16 inches (40 cm) long and intermediate in form between the shallower, delicate shape of *Velociraptor* and the deeper, rectangular form of *Dromaeosaurus*. These differences suggest that the animals were specialized for different prey. *Deinonychus* had a strong bite and its powerful jaws were lined with around 70 curved teeth.

2 SHOULDER AND CHEST

Deinonychus's shoulder and chest was very birdlike. The shoulder joint was higher than was typical for theropods, and large, platelike breastbones featured along the bottom of the chest. A V-shaped wishbone was at the front of the ribcage. The wishbone was not exclusive to the maniraptorans and was present in the earliest theropods.

3 STRAIGHT TAIL

Long, flexible rods ran along *Deinonychus*'s tail, keeping it straight. The rods on the upper surface were super-elongate versions of the pronglike bony structures (zygapophyses) that help vertebrae articulate with one another. The rods on the lower surface were extended versions of the chevron bones that grew downward and backward from the vertebrae.

4 SICKLE CLAW

Each second toe had hyperextendable joints and carried a large, sickle-shaped claw. This is a trademark of the raptor group and was probably held up off the ground during normal walking and running to keep it protected. It is likely that the toe and claw held down medium-sized prey while *Deinonychus* tore at it with its 70 slicing teeth.

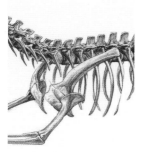

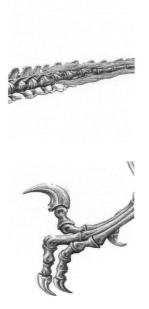

DINOSAUR RENAISSANCE

From the 1960s, US fossil expert John Ostrom (left) was central to the changing views of dinosaurs. Previously they had been regarded as clumsy, dim-witted sluggards that lived slow and simple lives. Ostrom's work on *Deinonychus* challenged this. He methodically analyzed all features of this dinosaur and how it had evolved to suit the various parts of its life. The results, showing *Deinonychus* as intelligent and hunting in coordinated packs, caught the public imagination and began the Dinosaur Renaissance, which gained worldwide exposure in the *Jurassic Park* (1993–2015) films and continues today. Enormous numbers of scientists have since been attracted to the study of dinosaurs, many encouraged by Ostrom's conclusions and his inspired image of dinosaurs as clever, cunning, fast, and furious.

Velociraptor
LATE CRETACEOUS PERIOD

SPECIES *Velociraptor mongoliensis*
GROUP Dromaeosauridae
SIZE 7 feet (2 m)
LOCATION Mongolia and China

The well-known dromaeosaur *Velociraptor mongoliensis* lived in Mongolia and China approximately 75 million years ago in the late Cretaceous period. *Velociraptor* is often confused with its larger, North American cousin *Deinonychus*. However, it was not a tough, muscular, mid-sized predator, but a slender dinosaur only 7 feet (2 m) long and similar in weight to a domestic dog, such as a labrador.

Velociraptor was extremely birdlike. Among its skeletal avian features were large breast bones, curved bony prongs that project from four of its rib shafts, and back-turned pubic bones. Like all dromaeosaurs and related maniraptorans, it had air-filled pockets in its ear region, pneumatic (air filled) vertebrae, a large wishbone, and long, slender, three-fingered hands with foldable wrists—all of which are birdlike features. *Velociraptor* fossils were discovered in sandstone rocks that were once dune fields, indicating that it was a desert-dweller at least part of the time. However, not all of the environments present in the area would have been as arid as is often thought, because large rivers, seasonal rainfall, and areas of greenery covered part of the region at times. Nevertheless, *Velociraptor* was well-adapted for life in hot, desert environments and presumably had behavioral features associated with this lifestyle. Some studies have found that *Velociraptor* is closely linked to similar Asian dromaeosaurs, such as *Tsaagan* and *Adasaurus*. This small, localized Velociraptorinae group is surrounded in the family tree by North American species, and the Asian species almost certainly descended from ancestors that moved west across a land bridge from North America. However, the skeletal anatomy of *Velociraptor* and its close relatives is not especially different from that of other dromaeosaurs who inhabited cooler, less arid habitats. It seems that dromaeosaur lifestyle and body shape were highly adaptable, and members of the group could transfer from forested habitats to swampy places, floodplains, semi-deserts, or deserts. **DN**

⚙ NAVIGATOR

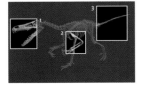

FOCAL POINTS

1 SNOUT
The snout was shallow and slightly concave on its upper surface. It was also less boxlike and rectangular than related dromaeosaurs. This slender snout could have evolved because *Velociraptor* was adapted to tackle tiny animals or to probe into small spaces. The skull shape was distinctive and quite slight.

2 LOWER ARMS
Velociraptor's lower arm bone (ulna) showed regularly spaced bony lumps along its lower edge. These are identical in shape and position to structures present in living birds that are attachment points for wing feathers. However, *Velociraptor* fossils come from sandstone rocks where feathers are rarely preserved.

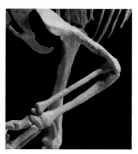

3 TAIL BONES
The tail bones had rows of rodlike projections. Fossils have shown that these rods were most probably flexible, allowing tail movement that would have been necessary for balance when running. It was initially thought that the tail was stiff and could only be moved from side to side by waving the hips.

▲ *Velociraptor* is one of many dinosaurs discovered during the American Museum of Natural History's Central Asia Expeditions of the 1920s, led by Henry Fairfield Osborn (1837–1935, pictured), one of the most accomplished paleontologists of the 20th century. Osborn arranged these expeditions because he regarded Mongolia as the place where humans most probably originated.

LOCKED IN COMBAT

The best-known *Velociraptor* specimen is a near-complete skeleton seemingly locked in combat with the ceratopsian (horn-face) *Protoceratops* (right). The animals were fighting at the point of death. *Velociraptor* grabbed *Protoceratops*'s frill and kicked it in the neck region, while its hind claw was embedded in the belly of *Protoceratops*. At the same moment, *Protoceratops* bit *Velociraptor*'s right arm. It is unknown whether *Velociraptor* routinely tackled such large, dangerous prey. The fossil was discovered in 1971 in Mongolia by a joint team of experts from Mongolia and Poland and it is housed at the Mongolian Academy of Sciences based in Ulaan-Baatar.

FEATHERED DINOSAURS

1 *Oviraptor* is often pictured with feathers
  because of its close relatives, but fossils
  lack good evidence of feathers.

2 The primitive bird *Jeholornis*, 120 million
  years old from China, with its skull
  curved back over the shoulders, had
  feathers up to 8 inches (20 cm) long.

3 The small meat-eating theropod
  Sciurumimus (squirrel mimic) had long
  tail filaments similar to a squirrel's tail.

Thanks to the discovery of *Archaeopteryx* (see p.376), the first known feathered dinosaur, it has been known since the 1860s that feathers of modern-type birds evolved long prior to the appearance of modern-looking, toothless, short-tailed birds. But was this the very first instance of feathers, or had more structurally primitive feathers evolved earlier in animals ancestral to *Archaeopteryx*? The similarity between *Archaeopteryx* and the small, active, predatory dinosaurs called maniraptorans (hand snatcher) led scientists to propose that feathers may have been present in nonbird dinosaurs. Reconstructions from the 1980s show *Deinonychus* (see p.306) with feathers, even though they had not yet been discovered. This theory was confirmed during the 1990s, when feathered nonbird dinosaurs were reported from the rocks of the lower Cretaceous period in Liaoning in northeast China. Numerous feathered dinosaurs are now known from these rocks, and feathered species have also been discovered in North America and Germany.

Notably, the earliest complex feathers were restricted to the fore limbs and tails of maniraptorans, a body distribution that suggests large feathers on these parts of the body proved advantageous. One theory is that the large surfaces formed by these feathers helped maniraptorans to steer, brake, and accelerate when running, and also helped them to control their direction when leaping. Eventually two theories for the origins of true flight were proposed: "ground-up,"

KEY EVENTS

165 Mya	165 Mya	150 Mya	130 Mya	130 Mya	125 Mya
Early tyrannosaurs inhabit Asia and North America. They evolve from a compsognathid -like ancestor. Chinese fossils show early kinds covered in furlike pelt.	The common ancestor of birds, troodontids and dromaeosaurs evolves. Based on its descendants' anatomy, it is a small, fully feathered omnivore.	*Archaeopteryx* inhabits Germany. It will later become the first feathered dinosaur to be discovered for science.	Early ostrich dinosaurs, or ornithomimosaurs, occur in Europe and Asia. Later species from Canada reveal large fore limb feathers.	Therizinosaurs live in North America. The earliest kinds, such as *Falcarius*, are human-sized. They are omnivorous or herbivorous.	*Microraptor* appears in China and survives for 5 million years. It had long flight feathers on its arms, hands, legs, and feet. These are later seen in birds.

with fast runners evolving wings to help them jump higher, especially to feed or escape enemies; and "tree-down," when tree-dwellers gained wings to help them glide, then fly. However, early maniraptorans were not the swift-running, leaping predators that scientists envisaged when proposing these ideas: most were omnivores or herbivores that did little or no climbing and had no need to leap or chase prey, but they may still have needed to run fast to escape predators.

An alternative idea is that evolution of large, complex feathers was driven by a role in display. There is evidence from the way in which the feathers of early maniraptorans were arranged that they were display structures, as was the case for *Caudipteryx* (see p.312) and *Microraptor* (see p.314). Early birds, such as the long-tailed *Jeholornis* (see image 2), had curved feathers in the tail fan that do not appear suited for an aerodynamic role. In many living birds, such as peacocks, individuals, usually males, exhibit their patterned plumage to outdo rivals, to repel intruders from their territory, and to impress females. If feathers did evolve for insulation, terrestrial maneuverability, or display, their later use in flight represents a re-use for another function. The process by which structures that evolve for one role, but later become co-opted for another, is exaptation.

Complex feathers, with the barbs and barbules that maintain their shape, are also present in many dinosaurs found in Liaoning. They show that true feathers were present in the omnivorous or herbivorous oviraptorosaurs (see image 1) and in maniraptorans, including the dromaeosaurs or raptor dinosaurs (see p.304), featuring large, complex feathers, while groups only distantly linked to birds had simple, hairlike filaments. Feathers were widespread across the coelurosaurian dinosaurs before birds evolved, so it seems that birds inherited a plumage that had been "invented" much earlier. Because simple, filamentlike feathers evolved before complex ones, it seems that feathers did not evolve for reasons related to parachuting, gliding, or flapping. It is more likely that they initially served an insulatory function. The feathered dinosaurs from Liaoning lived in a cool or temperate climate, not a tropical one, which supports this theory. However, this does not explain why feathers became increasingly elaborate over time.

The main factor determining the preservation of feathers is geology—fine-grained volcanic ashes and lake-floor sediments, such as those present in the Liaoning beds, are better at retaining fine details compared to the coarse sands that typically preserve dinosaurs. These fossils prove that coelurosaurs, such as the theropod (meat-eating dinosaur) *Sciurumimus* (see image 3), were covered in hairlike filaments that were almost certainly the structural ancestors of the feathers present in the more birdlike groups. The filaments formed a thick pelt across the whole body. They were not limited to small dinosaurs, and the fearsome tyrannosaur *Yutyrannus* was covered in them and reached 30 feet (9 m) in length. Members of the long-necked therizinosaur group, known as the scythe-claw dinosaurs, also preserve structures of this sort. **DN**

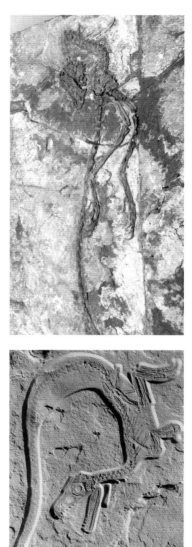

125 Mya	125 Mya	125 Mya	115 Mya	115 Mya	95 Mya
Caudipteryx, one of the first fully feathered nonbirds discovered, inhabits Liaoning in China. It is a long-legged, short-tailed omnivore.	The giant tyrannosaur *Yutyrannus* lives in China. It shows that some tyrannosaurs possess a thick pelt of filamentlike feathers.	Several early bird groups possess peculiar feathers at the ends or bases of their tails, which are almost certainly specialized for a role in display.	Oviraptorosaurs diversify; toothless lineages evolve from small, *Caudipteryx*-like ancestors. Human-sized and giant species (*Gigantoraptor*) evolve.	Advanced troodontids evolve from small ancestral forms. The advanced ones are larger, longer-legged, and have proportionally shorter arms.	The small, short-armed alvarezsaurids evolve. They are fuzzy-coated, and their skulls and arms indicate that they are ant- or termite-eating theropods.

Caudipteryx
EARY CRETACEOUS PERIOD

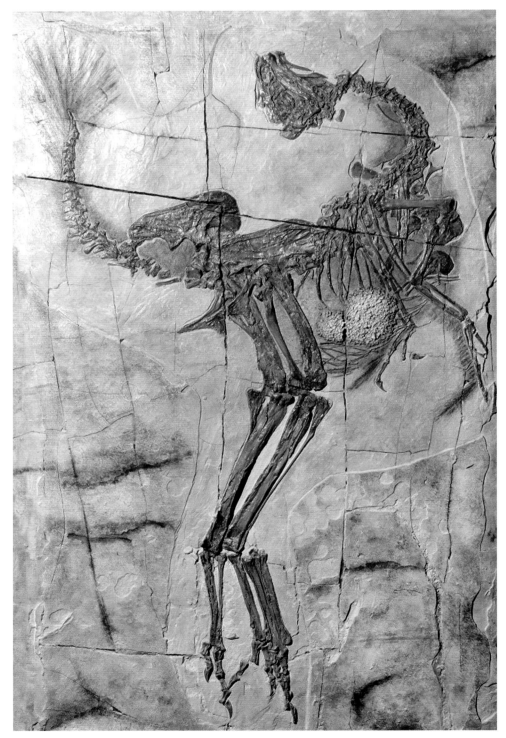

Named by scientists in 1998, *Caudipteryx* (tail feather) was one of the first nonbird dinosaurs to confirm that feathers were not unique to birds, but were present in theropod dinosaurs. A turkey-sized, short-tailed maniraptoran, *Caudipteryx* was long-legged with proportionally short fore limbs where only two of the three fingers were clawed. Its relatively small skull, short, blunt snout and peglike teeth suggest an omnivorous diet of fruits, leaves, and seeds. This is supported by the presence of small rounded stones (gastroliths) in the stomach, which are typical in herbivores and used to break down plant material for digestion. The best known feature of this dinosaur is its feathering. Similar to a modern bird, *Caudipteryx* was fully covered in feathers except around its mouth, lower legs, and feet. Long, narrow feathers grew from the upper surfaces of its second finger and hand, but not from its arms. Twelve long feathers, six on each side, also grew from the end of its tail, forming a fan. Juveniles had a smaller fan with ribbon-shaped feathers, which supports the idea that tail fans were used in display—only mature adults would need showy fans. Another theory is that they were waved to cool the body. The long, slender legs indicate that *Caudipteryx* was a fast runner, and the proportions of its fore limbs and toes show that it was not a climber, nor was it able to glide or fly. Scientists initially suggested that *Caudipteryx* could have been a flightless bird, or bird ancestor, but the anatomy of its hips and skull shows that it was an early member of Oviraptorosauria—a group of theropods best known for the human-sized, toothless, crested species that inhabited Asia during the late Cretaceous. Compared to *Caudipteryx*, most later oviraptorosaurs were larger, stockier and had shorter, deeper snouts and jaws. **DN**

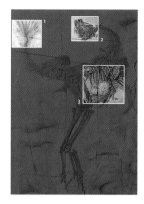

SPECIES *Caudipteryx zoui*
GROUP Oviraptorosauria
SIZE 2 feet (60 cm)
LOCATION China

FOCAL POINTS

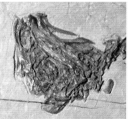

1 TAIL TIP
The vertebrae that formed the tail tip were massed together, forming a stiff, rodlike structure similar to that of modern birds. This may have supported the fan at the tip. The bone reduction saved weight and nutrient use, providing a base for the muscles that moved the feathers.

2 BEAK
The edges of *C. zoui*'s fossil jaws show that beak tissue grew around its mouth. Beaks evolved several times in omnivorous and herbivorous dinosaurs, presumably because they made jaws better at withstanding wear and tear caused by biting plants.

3 DOWNY COAT
The body of *C. zoui* was covered in downy feathers, each consisting of unbranched filaments sprouting from a base. They are thought to have helped maintain a constant body temperature. The skin could have given protection or camouflage.

EGG THIEF

The dinosaur that gave its name to *Caudipteryx*'s clade, Oviraptorosauria, was *Oviraptor*, officially described in 1924. Its fossils were recovered from Mongolia by a joint United States–Mongolian expedition that in fact hoped to find early human fossils. *Oviraptor* had a deep toothless beak like a parrot's, possibly a head crest, too, and was about 6½ feet (2 m) in length. It was found with fossilized eggs (above), and the conclusion at the time was that it was raiding another's nest and cracking open the eggs with its powerful beak, hence its name "egg thief/raider." Further finds of related dinosaurs in the 1990s revealed that *Oviraptor* looked after or brooded its eggs. Therefore, it seems the "egg thief" was actually at its own nest, caring for its own offspring.

Microraptor
EARLY CRETACEOUS PERIOD

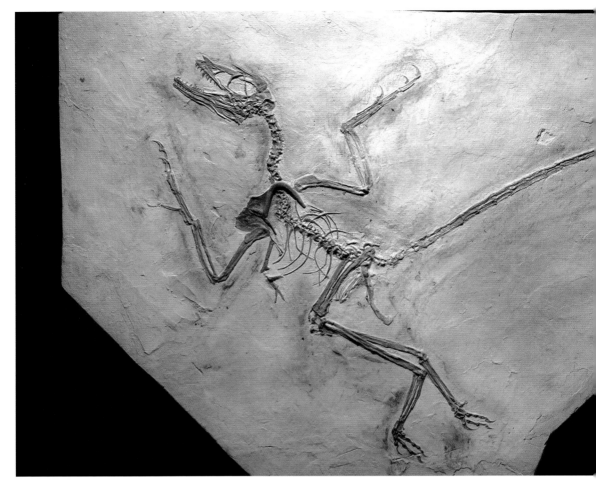

SPECIES *Microraptor gui*
GROUP Dromaeosauridae
SIZE 28 inches (70 cm)
LOCATION China

NAVIGATOR

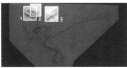

As suggested by its name, *Microraptor* was a small relative of the dromaeosaurs (raptors). Only 28 inches (70 cm) in length, it is one of the smallest-known dinosaurs. There are three species: *M. zhaoianus*, *M. gui*, and *M. hanqingi*. The first of the three species, *M. zhaoianus*, named in 2000, was one of the most remarkable of all maniraptoran dinosaurs. Like other raptors, it possessed a large, raised, strongly curved claw on its second toe, serrated teeth set in shallow jaws, elongated three-fingered hands and fingertips with large, needle-sharp claws. *Microraptor* lacked the anatomical features present in the majority of dromaeosaurs and seems to have been an unusual early branch of this dinosaur group. It were primitive in size, shape, and lifestyle compared to the big dromaeosaurs that later evolved; instead, it was similar to the ancestor that also gave rise to troodontids and birds. One of the remarkable things about *Microraptor* was its plumage. Long aerodynamic feathers grew from the arms, hands, and end of the tail. Even more noteworthy were the long feathers that also grew from its legs and feet, giving it what looked like four wings. This configuration led to the idea that *Microraptor* was a specialized glider, able to spread its limbs, launch from height and glide. However, there are problems with this theory: the hip-joint shape does not allow the leg to be sprawled out to the side. At best, it might have been held directed outward by about 45 degrees. **DN**

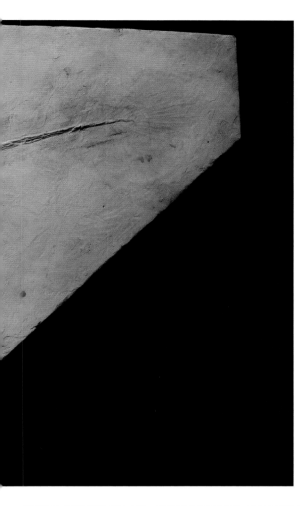

1 LARGE EYES
The eyes of *Microraptor gui* were so large that they suggest a nocturnal lifestyle, based on data from living animals. However, this theory is inconsistent with the idea that *M. gui* had iridescent feathers, because iridescence would have been of little use in the dark: today's iridescent birds are day-active.

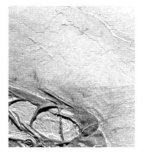

2 HEAD CREST
Some *M. gui* specimens appear to have possessed a bushy head crest. However, it seems that these apparent crests were in fact caused by the crushing and displacement of head feathers. Fossil evidence shows feathering on the face and most of the snout, which contradicts some early reconstructions that gave them naked, scaly faces.

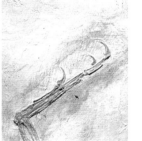

3 PLUMAGE AND COLOR
Fossils of several feathered dinosaurs have been found in Liaoning, China, and it has been argued that microscopic, pigment-holding structures in the tissues—melanosomes—are preserved on *M. gui*'s feathers. The size and shape of these melanosomes suggest it had a black, iridescent plumage, but this is yet to be confirmed.

SUSTAINED FLIGHT

More than a dozen scientific studies have examined whether *Microraptor* could glide a little, glide well, fly a little, or was capable of sustained flight (right). Some of the investigations involved physical models with simulated feathers, whereas others had simpler models that were installed in wind tunnels. Several studies focused on movements of the shoulder, hip, and limb joints, while others examined whether the animal held its arms higher than its legs in "biplane" mode. Virtual versions of this mini-dinosaur were studied in computerized aerodynamic simulations. The results varied hugely. At least one investigation supported each of the fly/glide options, but overall the question remains unanswered.

SCYTHE-CLAWED THERIZINOSAURS

A mong the most unusual dinosaurs were the therizinosaurs, a group with a peculiar evolutionary history and a surprising set of anatomical features that include scythelike claws. They are now accepted as members of the maniraptoran group (hand snatcher; see p.304) because the detailed anatomy of their skeletons indicates a close affinity with oviraptors, dromaeosaurs, and birds. Therizinosaurs were birdlike in appearance and were covered with feathery pelts of the sort known to be widespread within the oviraptor, dromaeosaur, and bird groups. Therizinosaurs were first recognized during the 1950s when *Therizinosaurus* (scythe lizard/reptile) was named from fore limb bones discovered in 70-million-year-old rocks in Mongolia. Its name refers to the fact that its hand claws were scythe-shaped (see image 2), the largest of them an incredible 28 inches (70 cm) long. These remain the biggest known claws of any animal. Early reconstructions of *Therizinosaurus* mistakenly portrayed it as broad-bodied and turtlelike and, in the years that followed, it was falsely suggested that *Therizinosaurus* could have been a slothlike dinosaur.

The position of the therizinosaurs within the maniraptoran group has interesting ramifications for the evolution of lifestyles and diets within both these dinosaur groups. Maniraptorans, being theropods (meat-eating dinosaurs), have always been regarded as mainly predatory, therefore it was assumed previously that a predatory lifestyle was present in the maniraptoran common ancestor. However, as therizinosaurs were most likely herbivorous, and as they are one of the oldest maniraptoran groups, it can be inferred that herbivory was the ancestral condition of the maniraptoran group. This theory is

KEY EVENTS

165 Mya	130 Mya	125 Mya	125 Mya	125 Mya	112 Mya
Therizinosaurs evolve from a common ancestor that later gives rise to all other maniraptorans.	*Falcarius* inhabits North America. It is a long-necked, long-tailed therizinosaur with a small skull.	New therizinosaurs evolve in Asia and North America, such as *Beipiaosaurus* (see p.318). Specimens with intact feathering are preserved in China.	*Martharaptor*, from Utah, shows that early therizinosaurs occur in North America as well as in Asia. It lives alongside the huge *Utahraptor*.	Land links eastern Asia and North America and therizinosaurs occur in both regions. Lineages within the group move between these areas.	Mid-sized *Alxasaurus* is known in China. It is intermediate between early kinds, such as *Beipiaosaurus*, and more advanced kinds, such as *Nothronychus*.

supported by the fact that the earliest members of certain other maniraptoran groups were herbivorous or omnivorous. Before the 1990s, therizinosaurs were misidentified as long-necked, plant-eating sauropod dinosaurs, or as an evolutionary link between sauropods and other herbivorous dinosaurs from a different line of evolution: the ornithischians (bird-hipped dinosaurs).

More complete remains of the Mongolian therizinosaurs *Segnosaurus* (see image 3) and *Erlikosaurus* were discovered, showing that therizinosaurs were wide-hipped theropods. They had stout bodies, heavily built hind limbs, broad feet, and skulls that combined a toothless beak with rows of small, leaf-shaped teeth. They were not predators; their features suggested that they were slow-moving herbivores that used long arms and large, curved hand claws to manipulate branches. These claws could also have been used in self-defence against tyrannosaurs and other dangerous predators.

A large number of additional therizinosaurs has now been discovered in eastern Asia and North America. One of the oldest members of the group is *Falcarius* (see image 1), discovered in 130-million-year-old rocks in Utah. *Falcarius* possessed key features that identify it as a therizinosaur, including an outwardly flaring front part on the ilium, the large, platelike bone that forms the upper part of the pelvis. Its teeth were leaf-shaped and coarsely serrated, and based on comparison with living reptiles would have been suited to a diet of plants. Furthermore, the tooth pair at the tip of the lower jaw was larger than the other teeth in the lower jaw, which suggests that herbivory was present from the start of the group's history. Overall, *Falcarius* was far more lightly built and slender than later therizinosaurs, with a longer tail and longer, slimmer hind limbs. It would have been similar in appearance to early members of the other maniraptoran groups.

Although *Falcarius*'s hips possessed key therizinosaur features they were not as wide as those of later species. Furthermore, the pubis—the long, rodlike bone at the front of the pelvis—pointed downward and forward, which is the arrangement typical in saurischians (lizard-hipped dinosaurs; see p.260). However, in later therizinosaurs, the pubis projected backward. Therefore, during therizinosaur evolution, the pubis rotated while the hind limbs became shorter and stockier, the hips wider, and the tail shorter. The reversed pubis and shortened tail would have resulted in a more posteriorly positioned center of gravity, and later, short-tailed therizinosaurs would have been heavier in the hip region. This suggests that they walked with a diagonal posture, which would have kept the head and neck raised, making it easier to reach tall foliage, and the arms and clawed hands would have been better placed for self-defence. This trend culminated in *Therizinosaurus* itself in the late Cretaceous period. It was a huge dinosaur, 33 feet (10 m) long, standing 16 feet (5 m) tall, and with its enormously wide, bulky body, weighing in excess of 5½ tons (5 tonnes)— measurements equivalent to the great *Tyrannosaurus* (see p.302). **DN**

1 *Falcarius* shows the characteristic "flared" ilium, visible upper left, to the rear of the ribs.

2 Three claws, each the length of a human arm and hand were located on the fore limb of *Therizinosaurus*.

3 The teeth of *Segnosaurus* were small, leaf-shaped and serrated, suited to chopping vegetation.

90 Mya	90 Mya	90 Mya	75 Mya	70 Mya	66 Mya
Large *Nothronychus* and *Suzhousaurus* occur across North America and Asia. They are heavier and more diagonal in posture than earlier kinds.	The North American therizinosaurs become extinct for unknown reasons. The group is now unique to eastern Asia, where members of several lines persist.	Advanced *Segnosaurus* and *Erlikosaurus* inhabit Mongolia and China. These are larger than earlier kinds, with longer arms and bigger claws.	A diverse set of Asian therizinosaurs disappears, leaving two lineages. One leads to giant *Therizinosaurus* and the other to *Nanshiungosaurus*.	Giant *Therizinosaurus* —the last of the group —inhabits Mongolia. It has the longest, straightest hand claws of all therizinosaurs.	The end-of-Cretaceous extinction event occurs. All maniraptorans except birds die out. It is unknown whether therizinosaurs survived until the event.

Beipiaosaurus
EARLY CRETACEOUS PERIOD

SPECIES *Beipiaosaurus inexpectus*
GROUP Therizinosauria
SIZE 6½ feet (2 m)
LOCATION China

⚙ NAVIGATOR

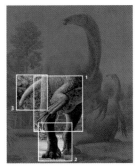

One of the oldest but best-known therizinosaurs is *Beipiaosaurus inexpectus*. Remains were found in the fossil beds of Liaoning province, China, from 125 milion years ago. *Beipiaosaurus* had preserved tightly packed, hairlike filaments across its body, which confirmed that therizinosaurs were clothed in a furry pelt, like an emu. However, the long, broad, spinelike structures on the top and sides of the neck and also on its tail were unexpected. These could be leftovers from an earlier stage in integument (skin and related structures) evolution, but it is likely that they were unique defensive structures. *Beipiaosaurus* had a large skull compared to later therizinosaurs. It was long and shallow-snouted with enormous eye sockets and a slender lower jaw. Numerous small, coarsely serrated, leaf-shaped teeth were present—teeth typical of leaf-eating reptiles. In subsequent therizinosaur evolution, the number of teeth at the front of the jaws reduced, and the beak-covered region enlarged. With a total length of 6½ feet (2 m), *Beipiaosaurus* was large in relation to the maniraptorans that shared its environment, but small compared to later therizinosaurs, most of which were twice as long and several times its weight. **DN**

1 FORE LIMBS
The three-fingered hands of *Biepiaosaurus* were long, slim and similarly proportioned to the hands of predatory maniraptorans, such as *Deinonychus* (see p.306). In later therizinosaurs, the arm and hand bones became thicker and more robust, and the hand and its claws were longer relative to the rest of the fore limb.

2 FEET
Biepiaosaurus had a narrower, longer coelurosaurian foot, as is typical for early members of the group. Therizinosaurs have short, broad feet compared to other coelurosaurian theropods. In the most specialized therizinosaurs, the first toe—usually positioned high up on the side of the foot—was in contact with the ground.

3 TAIL TIP
Five vertebrae at the tip of the tail were fused into a straight, rodlike structure. In other maniraptorans, similar structures have been regarded as a strengthening device, the tail supporting the fan-shaped arrangement of feathers. But as *Biepiaosaurus* did not possess a tail fan, the reasons behind the fusing remain unknown.

⊢ BEIPIAOSAURUS RELATIONSHIPS

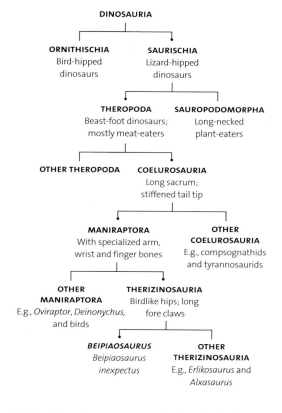

DINOSAURIA

ORNITHISCHIA
Bird-hipped dinosaurs

SAURISCHIA
Lizard-hipped dinosaurs

THEROPODA
Beast-foot dinosaurs; mostly meat-eaters

SAUROPODOMORPHA
Long-necked plant-eaters

OTHER THEROPODA

COELUROSAURIA
Long sacrum; stiffened tail tip

MANIRAPTORA
With specialized arm, wrist and finger bones

OTHER COELUROSAURIA
E.g., compsognathids and tyrannosaurids

OTHER MANIRAPTORA
E.g., *Oviraptor, Deinonychus,* and birds

THERIZINOSAURIA
Birdlike hips; long fore claws

BEIPIAOSAURUS
Beipiaosaurus inexpectus

OTHER THERIZINOSAURIA
E.g., *Erlikosaurus* and *Alxasaurus*

▲ Despite their outward appearance of a plump body and an ungainly posture, therizinosaurs such as *Beipiaosaurus* are in the "hand snatcher" maniraptoran group (eumaniraptorans), from which the agile, predatory raptor dinosaurs and birds evolved.

TWO KINDS OF FEATHER

The quality of fossil preservation for *Beipiaosaurus* has revealed that this dinosaur had two kinds of feathery skin structures (right). One was a pelt of fuzzy down, with each feather consisting of several filaments branching from the same short shaft in the skin, like the down feathers of birds. This type of filamentous covering is found in many feathered dinosaurs. The other set of feathers was elliptical in cross section, unbranched, and ribbonlike at ⅛ inch (3 mm) wide and up to 6 inches (15 cm) long. These grew from among the soft downy feathers and may have been quite stiff. The reasons for this two-stage feathery coat are unclear. Flight was out of the question. The downy set could have insulated the body, while the ribbonlike set could have been for display during breeding time, or even for protection.

ORNITHOPODS: BIRD-FEET

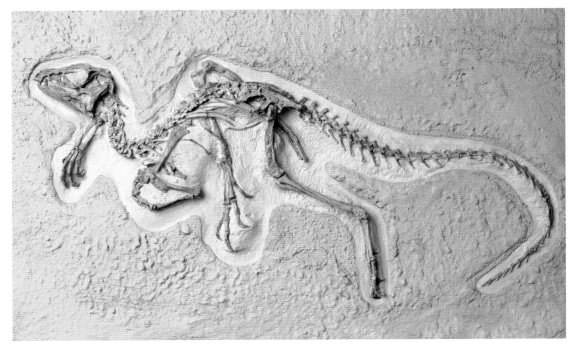

The ornithopods (bird-feet dinosaurs) belong to the Cerapoda group, which also include the ceratopsians (horn-faces), such as *Triceratops* (see p.344), and pachycephalosaurs (thick-heads), such as *Stegoceras* (see p.342). The sister group to the Cerapoda is the Thyreophora (armor bearers), which include the ankylosaurs and stegosaurs (see p.328). Both the Cerapoda and Thyreophora were ornithischians (bird-hipped dinosaurs), and the great majority of ornithischians fall into one of these two groups. However, the name ornithopod is slightly misleading, because the feet of the theropods (meat-eating dinosaurs, whose name means "beast feet") more closely resembled those of birds, and birds evolved from theropods not ornithopods. Both ceratopsians and ornithopods flourished and diversified until the end of the Mesozoic era.

A few early ornithischians, such as the heterodontosaurs, fall outside the two groups mentioned above. *Heterodontosaurus* (see image 1) was a small dinosaur only 3 feet (1 m) in length whose fossils were found in South Africa. The name means "different toothed lizard/reptile." In most dinosaurs the teeth differed only in size: they were sharp and pointed in meat-eaters, flat for chewing in plant-eaters or peglike for raking vegetation. However, *Heterodontosaurus* had

KEY EVENTS

165 Mya	155 Mya	125 Mya	125 Mya	98 Mya	84 Mya
Kulindadromeus, one of the first ornithopods, has feathers on its torso and neck.	Early ornithopods become more established, such as *Callovosaurus* in England, known only from a single partial thigh bone.	The well-known *Iguanodon* thrives in various locations throughout Europe.	*Mantellisaurus* (once thought to be a species of *Iguanodon*) in England is one of the first hadrosauriformes: the group that gave rise to hadrosaurids.	*Eolambia* lives in western North America. It is thought to be a near relation of the earliest hadrosaur.	The earliest known hadrosaur, *Jaxartosaurus*, roams Kazakhstan and China.

three different designs of teeth. At the front were small incisorlike teeth for nipping and chopping; at the rear were broad, more robust teeth for chewing; and in between, near the snout, was a pair of tusks in each jaw, like the canines of mammal meat-eaters. These variations might not have been connected to diet; they might have been for display, to threaten predators, or to attack rivals when mating. The dentition of *Heterodontosaurus* was more like that of a mammal than a reptile, and it was unique among dinosaurs.

The first ornithopods were small runners that were bipedal (walking on two legs), such as *Hypsilophodon* from the early Cretaceous period. One found on the Isle of Wight was less than 6½ feet (2 m) in length. Although some small species persisted, the main ornithopod evolutionary trend was toward larger body size, quadrupedality (walking on four legs), and more elaborate chewing equipment. *Tenontosaurus*, from the middle Cretaceous of North America, 110 million years ago, reached 26 feet (8 m) in length, including its disproportionately elongated tail; while *Iguanodon* (see p.324) attained 39 feet (12 m). In Africa, 120 million to 110 million years ago, the 23-foot (7 m) ornithopod *Ouranosaurus* (see image 2) may have sported a large sail of skin on its back—similar to the sail of the tremendous carnivore *Spinosaurus* (see p.300) from the same region and time span. The reasons for the existence of back sails on dinosaurs and other prehistoric vertebrates, such as *Dimetrodon* (see p.238) and *Edaphosaurus*, have been debated exhaustively. Possibilities range from thermoregulation to colorful displays to outshine rivals or to impress potential mates. A different interpretation is that the bony supports projecting upward from the backbone supported a hump of flesh that was a food store.

In 2014, fossils from Siberia were described that shed new light on, and posed further questions about, dinosaurs with feathers (see p.310). They were from *Kulindadromeus* (meaning "Kulinda runner" after its region of discovery and its body build) from the late Jurassic period, 170 million to 145 million years ago. Around 5 feet (1.5 m) long, it was a fairly typical small ornithopod with arms shorter than legs and a very long tail. The extremely well-preserved fossils show that *Kulindadromeus* had feathers on its torso and neck, but not on its face, lower arms, lower legs, or tail. The feathers were of several kinds: thready 1⅓-inch (3 cm) fuzzlike filaments over much of the head and body; multifilament designs, with several threads coming from one base, on the upper arm and upper leg; and ribbon-shaped ⅘-inch (2 cm) versions on the upper thigh. The significance of this is that *Kulindadromeus* is the first known well-feathered dinosaur not only from the ornithopods, but also from the larger ornithischian group containing many different plant-eaters. Other feathered dinosaurs, from tiny *Mei* and *Microraptor* (see p.314) to huge *Yutyrannus* and *Therizinosaurus*, were from the theropod branch of the dinosaur family tree, which is on the saurischian (lizard-hipped dinosaurs) side. One deduction from the find of *Kulindadromeus* is that feathers evolved independently in ornithopods and theropods. An alternative proposal is

1 This specimen of *Heterodontosaurus* from 197 million years ago has large "fangs" near the front of the jaws, one fore limb displaced to the neck area and a right knee above the backbone.

2 *Ouranosaurus* was an early iguanodont-like dinosaur that lived during the Cretaceous. It had strong hind limbs that supported the body for a bipedal posture.

80 Mya	77 Mya	73 Mya	70 Mya	68 Mya	66 Mya
Hadrosaurus lives in New Jersey. Later it becomes the first North American dinosaur named from good remains, and the first mounted dinosaur skeleton.	*Maiasaura* nests in huge colonies, and parents care for their young after hatching.	At up to 49 feet (15 m) in length, the largest known North American hadrosaur is *Magnapaulia* in Mexico.	*Shantungosaurus*, the longest and heaviest of all bipedal animals, lives in China. It is 16 feet (5 m) at the hip, taller than most giraffes.	*Olorotitan* from eastern Russia is notable for having 16 fused vertebrae in its sacrum—more than any other dinosaur.	Although most hadrosaurs live in North America and Asia, some are also found in Europe, such as *Arenysaurus* in Spain.

that *Kulindadromeus*—an ornithopod and ornithischian—and theropods, which were saurischians—inherited feathers (or the potential to evolve them) from a common ancestor that was in existence before the saurischian–ornithischian evolutionary lines separated very early in the dinosaur group. The extreme stance for this idea is that all dinosaurs, from their beginnings, were feathered by default to some degree. Those that were not feathered had lost their feathers as evolution progressed. The lack of widespread evidence for feathers in the fossil record could be due to difficulties in preservation. However, as time goes by, more feathered dinosaurs from different groups are discovered, or already-known fossil specimens are re-examined and experts find signs or traces of feathers in them. In the future, the popular vision that all dinosaurs were scaly could be replaced by a feathery-scaly representation.

It was in the hadrosaurs (duck-bills) of the late Cretaceous that ornithopods reached their apex in size, specialization and ecological dominance, becoming the most numerous large animals in their habitats. A typical hadrosaur was a large-bodied animal, habitually quadrupedal but capable of walking only on its hind limbs, which were much larger than its fore limbs. Its backbone curved downward toward the front of the torso before arching up into the neck. Its tail was vertically tall and often stiffened with rods of bony cartilage. The front of the mouth was formed into a broad, flat, toothless beak, ideal for cropping low-growing vegetation, complemented by a complex battery of grinding cheek teeth. There were two main groups of hadrosaurs: the saurolophines (sometimes referred to as the hadrosaurines) and the lambeosaurines. The former had unornamented heads and can be difficult to tell apart. *Maiasaura* (see p.326) is a typical representative of this group. However, the lambeosaurines are easily distinguished by their characteristic head crests: the single unicorn horn of *Tsintaosaurus* (see image 3), the helmet shape of *Corythosaurus* (see image 4), the hatchet design of *Olorotitan*, the forward-leaning crest of *Lambeosaurus* (see image 6) and the huge, curved, backward-pointing crest of *Parasaurolophus* (see image 5), which was longer than the rest of the skull.

It was once thought that the hollow crests of lambeosaurines were snorkels that allowed them to breathe underwater. However, this hypothesis cannot be

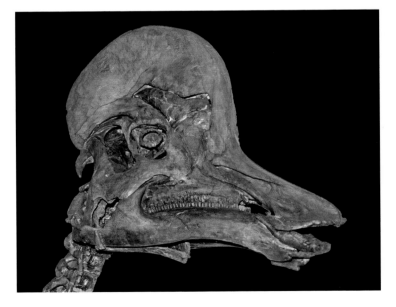

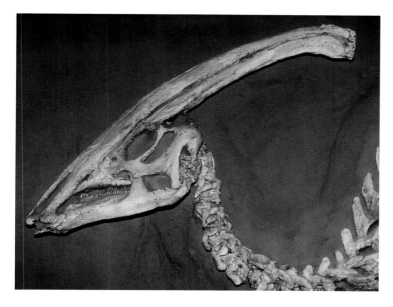

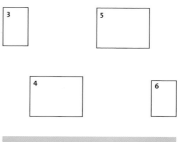

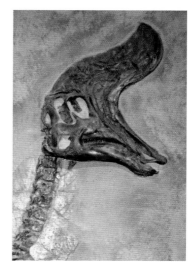

correct because there is no opening at the top of the crest to admit air. Furthermore, there is no evidence that hadrosaurs spent more time in water than other dinosaurs. The tall tail, which moves from side to side to power it through water, may resemble that of a crocodile but such a tail shape could have evolved for other reasons. The assumption that hadrosaurs lived partly in water may have arisen simply due to the similarity of their beak-fronted mouths to the bills of ducks, but in other respects their anatomy is quite unsuited for an aquatic lifestyle. Instead, their massive dental batteries are strong evidence that they ate tough terrestrial plants. Some hadrosaurs had more than 1,000 teeth, arranged not in a single row but in arrays that formed grinding plates. The plates in the upper jaws faced downward and inward; those in lower jaws, upward and outward. As the mouth closed for a chewing stroke, the lower jaw unit was forced upward in-between the upper jaws, which would bow apart to receive it. The upper and lower tooth plates on both sides of the mouth ground past one another, reducing vegetation to a pulp in the most efficient chewing set-up of any animal. However, the detailed diet of hadrosaurs is not well established. Studies of dental microwear (tiny scratches on teeth) indicate that these ornithopods probably grazed at ground level; however, some fossilized stomach contents have suggested browsing above, in bushes and tree branches. Different species no doubt had particular feeding habits, just as different modern hoofed mammals prefer diverse foods and feed from low to high.

Hadrosaurus, whose description established the group, was named in 1858, and it was the first North American dinosaur known from reasonable skeletal remains. (It was preceded by the sauropod *Astrodon*, but that name was based on a few nondiagnostic teeth.) A decade later, its skeleton—rather creatively completed because many of the fossil bones had not been found—was mounted at the Philadelphia Academy of Natural Sciences and became the world's first mounted dinosaur skeleton.

Once an animal group becomes established, the evolutionary trend to larger body size is common. The largest known hadrosaur and the biggest biped of all time was *Shantungosaurus*, from the late Cretaceous of China. At more than 52 feet (16 m) in length and heavily built, it may have weighed more than the long-necked sauropod *Diplodocus*. **MT/SP**

3 About 33 feet (10 m) long and from China, the hadrosaur *Tsintaosaurus* sported a unicornlike crest up to 20 inches (50 cm) tall. This could have been the rearmost part of a larger crestlike structure.

4 *Corythosaurus* (helmet lizard/reptile) shows the hadrosaurian toothless, flattened, beaklike front of the mouth, with batteries of chewing teeth behind.

5 Like other hadrosaur head crests, that of *Parasaurolophus*—some 5 feet (1.5 m) long—was hollow but had no openings to the outside air.

6 Numerous specimens of the "hatchet head" hadrosaur *Lambeosaurus* were once named as separate genera and species, but are now regarded as females and males of different ages from only a few species.

Iguanodon
EARLY CRETACEOUS PERIOD

SPECIES *Iguanodon bernissartensis*
GROUP Iguanodontia
SIZE 39 feet (12 m)
LOCATION Belgium, possibly other sites in Europe

The genus *Iguanodon* has a complex history. After *Megalosaurus* was named in 1824 by English geologist and paleontologist William Buckland (1784–1856), *Iguanodon* became the second nonavian dinosaur to receive a formal name, given in 1825, by English doctor and geologist Gideon Mantell (1790–1852). However, it was based on very incomplete remains that gave no real idea of what the animal would have looked like. Mantell's original conception was of a gigantic scaled-up iguana, 59 feet (18 m) in length. On the discovery of better fossil material, he reconceived it as an elephant-sized reptilian rhinoceros. It was this idea of *Iguanodon* that informed the well-known life-size sculptures by Benjamin Waterhouse Hawkins (1807–94) for the Crystal Palace Park in London in 1854. They are still on display today. Useful *Iguanodon* fossils were found nearly a quarter of a century after the public unveiling of this sculpture. About 38 essentially complete skeletons were excavated from a coal mine at Bernissart in Belgium, more than 1,000 feet (300 m) underground. These skeletons were placed in a new species, *Iguanodon bernissartensis*, and it is this species that has exemplified *Iguanodon* ever since. This was formally recognized with the establishment of *I. bernissartensis* as the new "type" or original specimen of the species. The name Iguanodontia has been adopted for its larger group, including the hadrosaurs.

The Bernissart skeletons demonstrated *Iguanodon* to be a very substantial animal, reaching lengths of 33 feet (10 m) and weighing as much as 3⅓ tons (3 tonnes). It had the ability to stand on its hind limbs and to walk on all fours—although the kangaroolike posture used for numerous museum mounts would not have been possible, because the tail was too stiff to bend at ground level. *Iguanodon*, which lived around 125 million years ago, was herbivorous, as shown by its teeth, which closely resemble those of living leaf-eating iguana lizards. Like other ornithopod dinosaurs, it was able to chew its food efficiently—not as mammals do, by back-and-forth sideways movements of the lower jaw, but by means of pleurokinesis, which is sideways bowing of the upper jaws as the lower jaws push upward within them. **MT**

✪ NAVIGATOR

FOCAL POINTS

1 FIFTH FINGER
The fifth finger was unusually flexible and would have been used to manipulate food while its thumb was still. This is the opposite of humans and is an example of the concepts of homology and analogy: the human thumb is homologous with *I. bernissartensis*'s thumb spike but analogous with its fifth finger.

2 THUMB SPIKE
The thumb of *I. bernissartensis* was a distinctive spike and could have been used for courtship displays, intraspecies fights, or defence against predators. It was interpreted as a nose horn in early reconstructions based on incomplete remains, which explains the rhinoceroslike aspect of the Crystal Palace Park model.

3 TAIL
The tail, like that of numerous duck-billed hadrosaurs, was stiffened by many thin bony rods (ossified tendons) running at angles to the axis of the tail and forming a complex lattice. The resulting inflexibility would have prevented the tail from being useful in swimming: evidence that helped to dispel the idea of aquatic hadrosaurs.

▲ At first, *Iguanodon* was posed on all fours, but in the 1880s Belgian paleontologist Louis Dollo (1857–1931) reconstructed the Bernissart skeletons in a kangaroolike posture: body semi-erect, head forward, front limbs dangling, and tail supporting the body. Modern images of *Iguanodon* on the move show the body horizontal, head in line with the torso and the tail balancing the head and torso over the rear legs.

SCIENTIST PROFILE

1790–1819
Gideon Mantell was a doctor by profession and an amateur geologist who lived in Sussex, England. By the age of 26, he had married Mary Woodhouse and was a partner in a profitable practice and a fellow of the Linnean Society of London.

1820–23
According to popular accounts, it was Mary, accompanying her husband in 1822, who found the teeth that would be named *Iguanodon*. French anatomist Georges Cuvier (1769–1832) dismissed them as belonging to a rhinoceros. Another blow was the opinion of William Buckland, who later named the fossil reptile *Megalosaurus*. Buckland said that the teeth were from a fish. However, Mantell maintained they were new to science and published his book *The Fossils of the South Downs* in 1822.

1824–31
In 1824, English naturalist Samuel Stutchbury (1798–1859) noted the teeth's resemblance to iguana teeth. Mantell proposed the name *Iguanasaurus*, but geologist and naturalist William Conybeare (1787–1857) said this could apply to the living iguana and instead advised *Iguanodon*, which Mantell adopted in 1825.

1832–52
In 1832 Mantell acquired another set of fossils and recognized their reptilian nature. The following year he proposed the name *Hylaeosaurus* (since identified as an ankylosaurian dinosaur). Thus, when anatomist and paleontologist Richard Owen (1804–92) coined the term Dinosauria in 1842, based mainly on three genera, Mantell had named two of them—*Iguanodon* and *Hylaeosaurus*—with the third being Buckland's *Megalosaurus*.

Maiasaura
LATE CRETACEOUS PERIOD

SPECIES *Maiasaura*
peeblesorum
GROUP Saurolophinae
SIZE 30 feet (9 m)
LOCATION Montana

✦ NAVIGATOR

In many respects, *Maiasaura* is a rather uninspiring duck-billed hadrosaur: not especially large or small and a typical representative of the saurolophine subgroup, with no special bony head crest. What sets it apart is its spectacular fossil record—or more specifically, a single extraordinary fossil site: Egg Mountain in western Montana, dating to 77 million years ago near the end of the Cretaceous. This site has yielded more than 200 individuals of *Maiasaura peeblesorum* of all growth stages, offering a unique insight into how a species grew, as well as the evolution of dinosaur social structures and breeding methods. One of the first excavations revealed that juvenile *Maiasaura* stayed near the nest after emerging: signs of eggshells that had been trampled, presumably by the hatchlings, supported this. The discovery of numerous additional nests demonstrated that *Maiasaura* bred in colonies, similar to seabirds. Adjacent nests were relatively near in proximity— closer than the length of an adult animal. Each nest held a clutch of 30 to 40 eggs, similar in size to those of the ostrich, even though the adult animal weighed about 20 times that of an ostrich. Eggs cannot exceed a certain size due to the difficulty of allowing enough oxygen through their shells by diffusion for the embryos to survive. *Maiasaura* hatchlings were about 16 inches (40 cm) long and quickly grew to 5 feet (1.5 m) in their first year, eventually reaching 30 feet (9 m) in length. *Maiasaura* lived in herds of thousands of individuals. Since it was contemporaneous with tyrannosaurs (see p.302) and lacked obvious means of defence, the herd lifestyle was important for protection, especially for juveniles—an evolutionary strategy that continues today in hoofed mammals defending themselves from big cats on the African savanna. **MT**

👁 FOCAL POINTS

1 ATTENTIVE CARER

Maiasaura means "good mother lizard," because one of the first finds revealed a nest containing eggshells from hatchlings and juveniles too large to be newly hatched. This was evidence that posthatchling *Maiasaura* lived around the nest, implying the growing juveniles were directly cared for by their parents.

2 NEST

Nests were simply constructed from earth. Vegetation was placed inside the nest so that the heat from its rotting would incubate the eggs, since the adults were too large to sit on the eggs. Crocodiles have a similar incubation strategy, as do some birds, including the turkeylike megapodes.

3 CLUTCH

Multiple hatchlings shared the nest—here, seven or eight. This could have been the average amount of juveniles to survive from a single clutch. The nest would have held more eggs, but an amount of loss is expected in animals with large clutch sizes. In the blue tit, it is common for only half a clutch to survive.

⊢ MAIASAURA RELATIONSHIPS

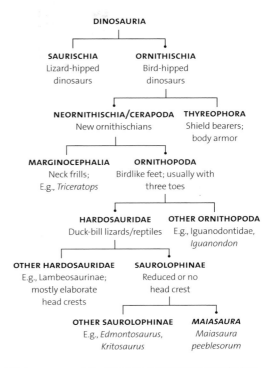

DINOSAURIA

SAURISCHIA
Lizard-hipped dinosaurs

ORNITHISCHIA
Bird-hipped dinosaurs

NEORNITHISCHIA/CERAPODA
New ornithischians

THYREOPHORA
Shield bearers; body armor

MARGINOCEPHALIA
Neck frills;
E.g., *Triceratops*

ORNITHOPODA
Birdlike feet; usually with three toes

HARDOSAURIDAE
Duck-bill lizards/reptiles

OTHER ORNITHOPODA
E.g., Iguanodontidae, *Iguanondon*

OTHER HARDOSAURIDAE
E.g., Lambeosaurinae; mostly elaborate head crests

SAUROLOPHINAE
Reduced or no head crest

OTHER SAUROLOPHINAE
E.g., *Edmontosaurus, Kritosaurus*

MAIASAURA
Maiasaura peeblesorum

▲ *Maiasaura* was a fairly standard member of the saurolophine duck-bill family, the hadrosaurids, more than 20 species of which are known from fossils. This is the case for all hadrosaurs that lived toward the end of the Mesozoic.

🕐 SCIENTIST PROFILE

1946–81

John "Jack" Horner was born in Montana, and had an eye for fossils since childhood, discovering his first fossil at only 8 years old. After studying geology and zoology at the University of Montana, in 1973 Horner secured a position as technician at Princeton University's Natural History Museum. In the mid-1970s he discovered *Maiasaura* nests in Montana, which contained the first dinosaur eggs in the western hemisphere. This groundbreaking discovery provided the first evidence of parental care among dinosaurs.

1982–present

Horner has been curator of paleontology at the Museum of the Rockies, Montana, since 1982, and has access to the world's largest collection of *Tyrannosaurus* fossils (see p.302), many of which he uncovered. His research supports the theory that *Tyrannosaurus* was an obligate scavenger, not a hunter. His main areas of expertise are dinosaur evolution, growth, and behavior, including the study of the microscopic structure of bone and tissue, known as histology. Horner has technically advised on all the *Jurassic Park* (1993–2015) movies, and is the inspiration for the films' protagonist, Dr. Alan Grant. He is widely regarded as one of the world's foremost paleontologists.

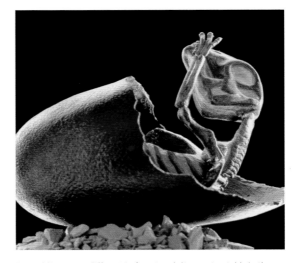

▲ Hatchlings were different in form to adults, most notably in the skull: those of the hatchlings were larger, with shorter snouts and bigger eyes. These differences are common in juvenile tetrapods (four-limbed vertebrates), including in mammals, reptiles, and birds.

PLATED STEGOSAURS

1 *Stegosaurus* is well known from almost 100 specimens in North America, like this example; recent finds have also been uncovered in Europe.

2 *Dacentrus*, 25 feet (7.6 m) long from late Jurassic Europe, had paired rather than alternating back plates.

The first major split in the ornithischians (bird-hipped dinosaurs) was between Thyreophora (armor bearers) and Cerapoda. Thyreophora consists of the stegosaurs (roofed lizard/reptiles) and ankylosaurs (see p.332) and the Cerapoda consists of ceratopsians (horn-faces), pachycephalosaurs (thick-heads; see p.338) and ornithopods (bird-feet; see p.320). Of the two armored thyreophoran groups, the stegosaurs were the first to spread widely, but also the first to lose prominence. Fragmentary fossils suggest that some stegosaurs may have survived to the end of the Cretaceous period, 66 million years ago, but they were a minor presence at most. The best-known early thyreophoran is *Scelidosaurus*, from the early Jurassic period of England. It was a 13-foot (4 m) quadrupedal (walking on all fours) herbivore with several rows of small osteoderms (skin bones) embedded in the skin of its back and sides.

Stegosaurs are among the most recognizable dinosaurs. Their unique body plan includes a large humped torso carried on four legs with the spine sloping down at the front to a small, narrow head held close to the ground and at the rear to a short tail. In basal (original) stegosaurs, such as *Huayangosaurus*, the fore limbs were long and slender, and the structure of the limbs suggests that it was a proficient runner. In later or more derived stegosaurs, the fore limbs were much shorter than the hind limbs, but robust and able to push sideways with force so that the stegosaur could pivot around its center of mass, above the hind limbs. The peak of the stegosaurs was 145 million years ago in the late Jurassic, when *Stegosaurus* (see image 1) and *Hesperosaurus* roamed western North America along with their fellow herbivores from other groups, such as

KEY EVENTS

165 Mya	160 Mya	157 Mya	156 Mya	155 Mya	152 Mya
The earliest known definitive stegosaur is *Huayangosaurus* from the middle Jurassic of Sichuan province, China. It is later known from 12 specimens.	Chinese *Tuojiangosaurus*'s back plates morph from rounded at the shoulders to pointed at the hips, like several other stegosaurs.	*Gigantspinosaurus* inhabits the earliest part of the late Jurassic, in China. Its shoulder spikes are twice as long as its shoulder blades.	*Hesperosaurus* is present in Wyoming. It will be named in 2001 from fossils in the well known Morrison Formation rocks.	*Dacentrurus* lives in England. It is later the first stegosaur known from more than fragmentary remains. It was initially known as *Omosaurus*.	The long-necked "sauropod- mimic" stegosaur *Miragaia* inhabits Portugal. It is named after the locality and means "wonderful goddess of Earth."

Diplodocus, *Camarasaurus*, and *Camptosaurus*. They were stalked by carnivores including *Allosaurus* and *Ceratosaurus*.

The most distinctive feature of stegosaurs was the ancestral osteoderms, that evolved into plates and spikes running along the neck, torso, and tail. The exact arrangement of plates and spikes varied greatly between different stegosaurs. In *Stegosaurus* itself, the neck, back, and most of the tail were ornamented with plates, and spikes were restricted to a clump of four at the tail end. The plates were large and roughly pentagonal. *Dacentrurus* (see image 2) had smaller plates than *Stegosaurus*, roughly rounded and arranged in symmetrical pairs rather than offset, alternating left and right as in *Stegosaurus*. *Kentrosaurus* (see p.330) had spikes instead of plates everywhere except on the neck and the foremost part of the torso. Increasing their armory yet further, many stegosaurs had additional spikes on their shoulders or hips. Shoulder spikes reached their most extreme form in the clumsily named *Gigantspinosaurus*, in which they were as long as the torso.

There is little doubt about the function of stegosaur spikes: they were used in combat. Renowned paleontologist Charles Gilmore (1874–1945) proposed in 1914 that they were display structures, but this is contradicted by the common occurrence of trauma-related damage in spikes, which supports the assumption that they were deployed in defence. The robust fore limbs and efficient hip-pivots of stegosaurs served to keep the deadly tail facing adversaries. The flexible neck allowed a stegosaur to look back over its shoulder to track the movements of an attacker. The role of the plates has been more controversial. They were assumed initially to have been defensive, but a bewildering selection of alternatives has been proposed, including species recognition, sexual display, threat display, thermoregulation (body temperature control), and even in the 1920s, as airfoils for gliding. For thermoregulation, the stegosaur warmed itself by standing with its plates side-on to the sun and cooled itself by standing with the plates edge-on to the sun or in the shade, in the manner of the back sails of *Dimetrodon* (see p.238) and *Spinosaurus* (see p.300). The bones of the plates were laced with blood vessels, so blood could have been warmed or cooled by circulating through them. Although this may have worked well for large-plated stegosaurs such as *Stegosaurus*, it would not have been practicable for spiked stegosaurs such as *Kentrosaurus*.

The modern consensus is that the plates and spikes of stegosaurs—like the much smaller osteoderms of modern crocodilians—actually provided a combination of benefits, differently optimized in different stegosaurs. The largest genera, including *Stegosaurus*, would have had less need for defence and more need of cooling, which accords with its having evolved the tallest, broadest and proportionally thinnest plates of any stegosaur. **MT**

152 Mya	150 Mya	150 Mya	125 Mya	125 Mya	113 Mya
The spiked stegosaur *Kentrosaurus* lives in Tanzania alongside the sauropods *Giraffatitan* and *Dicraeosaurus*.	*Stegosaurus* is the largest known stegosaur, at 30 feet (9 m) in length and perhaps 5½ tons (5 tonnes) in weight, in western America.	*Gargoyleosaurus* was one of the earliest, and smallest, plated dinosaurs to develop tail clubs. Short spikes covered its back.	A stegosaur possibly lived in Argentina at this time and is the only known South American group representative.	*Regnosaurus* is a plated dinosaur that inhabits woodlands of Western Europe.	*Wuerhosaurus*, the last known stegosaur, roams western China at the end of the early Cretaceous.

Kentrosaurus
LATE JURASSIC PERIOD

SPECIES *Kentrosaurus aethiopicus*
GROUP Stegosauridae
SIZE 15 feet (4.5 m)
LOCATION Tanzania, Africa

🧭 NAVIGATOR

Originating from Tanzania, Africa, 155 million to 150 million years ago at the end of the Jurassic, the stegosaur *Kentrosaurus aethiopicus* differed from the better known *Stegosaurus* in having spikes along its upper length, rather than plates. At 15 feet (4.5 m) in length—more than half of which was the tail—*Kentrosaurus* was around half the length of the largest *Stegosaurus*. It weighed less than 1 1⁄10 ton (1 tonne), but made up for its small size with impressive weaponry: not only the spikes along its back and tail but also an additional pair of long asymmetrical spikes. When German paleontologist Werner Janensch (1878–1969) first reconstructed the skeleton of *Kentrosaurus* in 1925, he placed these long spikes on the hips, pointing outward and backward. Subsequently discovered specimens of other stegosaurs revealed shoulder spikes, so the present version of the *Kentrosaurus* mount has had its long spikes moved to the shoulders. However, it is not clear how these spikes related to the shoulder, so the original hip placement may prove to be correct. All known specimens are from the Tendaguru excavations of the 1910s (see panel). Although no complete articulated skeletons exist, the expeditions found 50 partial specimens consisting of 1,200 bones, so the entire skeleton is known. A fine composite skeleton is mounted at the Museum für Naturkunde (Museum of Natural History) in Berlin, but many of the other collected fossil bones were destroyed during World War II. The name *Kentrosaurus* closely resembles that of the ceratopsian *Centrosaurus*, and both names mean "pointed or prickled reptile/lizard." For this reason, the original describer, German paleontologist Edwin Hennig (1882–1977), thought it necessary to rename this stegosaur *Kentrurosaurus*. However, because both the original spelling and pronunciation of *Kentrosaurus* differ from that of *Centrosaurus*, the original name has been retained. **MT**

1 TEETH

The teeth were small yet wide-based with vertical grooves and ridges, which was typically stegosaurian. These teeth would have torn off vegetation to be swallowed immediately. The size of the skull, flexible neck, and short fore limbs suggest that *Kentrosaurus* had evolved to browse at a height of 5 to 6½ feet (1.5–2 m) on low- to medium-height foliage.

2 NECK

The neck of *Kentrosaurus* was reasonably long and flexible. This contradicted earlier reconstructions of stegosaurs, which tended to show them with very short necks. One stegosaur, *Miragaia*, from the end of the Jurassic in Portugal, had a neck elongated to 17 vertebrae, giving it general proportions similar to those of a small, 20-foot (6 m) sauropod.

3 PLATES

There was a gradual transition from plates to spikes in *Kentrosaurus,* with a complete covering of spikes by the hip region. The spikes then remained roughly the same length from the hip to the tip of the tail. Contrastingly, in *Stegosaurus*, a series of plates ran along the neck and back and most of the tail, and spikes occurred only at the tip of the tail.

4 SPIKES

The shoulder (or hip) spikes of *Kentrosaurus* were very long—nearly as long as the fore limb—and had broad bony plates at their base for anchoring, although it is not known exactly how the spike articulated with the bones of the shoulder—or hip. The preserved bony tip of the spike is not particularly sharp, but its keratin-based covering would have come to a sharp point.

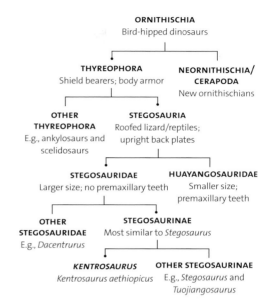

ORNITHISCHIA
Bird-hipped dinosaurs

THYREOPHORA
Shield bearers; body armor

NEORNITHISCHIA/ CERAPODA
New ornithischians

OTHER THYREOPHORA
E.g., ankylosaurs and scelidosaurs

STEGOSAURIA
Roofed lizard/reptiles; upright back plates

STEGOSAURIDAE
Larger size; no premaxillary teeth

HUAYANGOSAURIDAE
Smaller size; premaxillary teeth

OTHER STEGOSAURIDAE
E.g., *Dacentrurus*

STEGOSAURINAE
Most similar to *Stegosaurus*

KENTROSAURUS
Kentrosaurus aethiopicus

OTHER STEGOSAURINAE
E.g., *Stegosaurus* and *Tuojiangosaurus*

▲ *Kentrosaurus* was originally thought to be similar to the early stegosaurs (Stegosauria). However, recent studies indicate that it was more derived, or highly evolved, therefore was more closely related to *Stegosaurus* itself.

TENDAGURU EXPEDITIONS

Between 1909 and 1912, Werner Janensch headed a series of expeditions to Tendaguru in Tanzania (below). He recruited more than 500 local people to work in quarries under the supervision of trained local foremen, one of whom—Boheti bin Amrani—was honored in the name of the sauropod *Australodocus bohetii*. The expeditions are best known for their dinosaur fossils. In addition to *Kentrosaurus*, they found long-tailed ornithopod *Dysalotosaurus*; two sauropods, *Giraffatitan* (see p.272) and *Dicraeosaurus*; and theropod *Elaphrosaurus*, among many other animals.

ARMORED ANKYLOSAURS

The ankylosaurs (fused lizard/reptiles) were the dinosaurian tanks. These massive herbivores sacrificed speed and agility for strength and durability, and would have been formidable combatants. In addition to their bulky construction and short, powerful limbs they were heavily armored. Their heads were encased in bony armor; they had solid plates covering their necks; and there were numerous plates and nodules of armor embedded in the skin of their backs. These plates and nodules, not directly attached to any other bone, are known as osteoderms (skin bones). In life they would have been covered with keratin—the tough, horny protein that also covers the beaks of birds, the shells of tortoises, and forms the claws, horns, and hooves of many animals, as well as human fingernails. Ankylosaurs such as *Minmi* (see image 1) and their cousins the stegosaurs belonged to the main dinosaur group Thyreophora (armor bearers).

Ankylosaurs were obligate quadrupeds (walking on all fours), but they had shorter fore limbs than hind limbs. They also had short necks, and their heads would have been no higher than 3 feet (1 m) above ground level, so they would have been limited to low browsing.

KEY EVENTS

165 Mya	163 Mya	127 Mya	126 Mya	116 Mya	100 Mya
Tianchisaurus may be the earliest known ankylosaur. It lives in China during the middle Jurassic, 165 million to 160 million years ago.	*Sarcolestes*, from England, is based on a single lower jawbone, which suggests it is a nodosaurid. If so, it is the earliest known.	In the early Cretaceous, the polacanthid *Polacanthus* roams the Isle of Wight, off the south coast of England.	Large bone beds of *Gastonia* are laid down in the Cedar Mountain Formation of Utah. It is closely related to *Polacanthus*.	Early Australian ankylosaur *Minmi* lives during the early Cretaceous.	*Nodosaurus*, 16 feet (5 m) in length, roams what is now Wyoming; it will lend its name to one of the two major ankylosaur groups.

They had small, triangular teeth, most likely complemented by a long prehensile tongue like that of the giraffe. This is indicated by the hyoids—bones in the throat that function as anchors for the tongue—which are unusually large in ankylosaurs. They had proportionally broad torsos, which would have held an expanded stomach, suggesting that they retained food for a long time to extract maximum nutrients. This would have allowed them to be undiscriminating eaters, like elephants, pulling in and chewing a great range of vegetation, from tough roots and bark to soft buds and succulent stems. The biggest ankylosaur would have been *Ankylosaurus* itself, which was up to 26 feet (8 m) long and may have weighed as much as 5½ tons (5 tonnes). It lived in western North America 66 million years ago, at the end of the Cretaceous period. It is not the ankylosaur best known from fossils, because many more completely preserved remains of other genera have been described following its discovery. However, its description and naming in 1908, relatively early in the study of dinosaurs, led to the whole group being named after it. For such a large creature, *Ankylosaurus* was not well endowed with a brain: the brains of ankylosaurs were proportionally very small. Only those of the giant sauropods were smaller in relation to the whole body.

There were two main family branches of the ankylosaur group: ankylosaurids and nodosaurids. The second is named after *Nodosaurus*, another North American representative, from 100 million years ago. An ankylosaur was among the first nonbird dinosaurs to be recognized: the English nodosaurid *Hylaeosaurus* (see image 2) was a third of the trio found by biologist Richard Owen (1804–92), along with *Megalosaurus* and *Iguanodon* (see p.324), when he coined the term Dinosauria in 1842. Another ankylosaur—*Palaeoscincus*—was among the first named American dinosaurs. But it was based only on the remains of teeth and the name is no longer in use. Ankylosaurids and nodosaurids developed their armor in different ways. Ankylosaurids are most recognized by their tail clubs, which are almost unique among dinosaurs (see image 3). Some Chinese sauropods, including *Shunosaurus* and perhaps *Mamenchisaurus*, had small clubs at the ends of their tails, but they were insignificant compared to those of ankylosaurs, which weighed as much as 44 pounds (20 kg). The core of an ankylosaur club was formed from a sequence of around 10 fused caudal tail bones (vertebrae), but additional osteoderms on the side and the tip provided more mass and contributed to a formidable weapon. The shapes of the clubs varied between species, and appear to have changed throughout the life of an individual: clubs could be rounded, bluntly pointed, or elongate. Ankylosaur

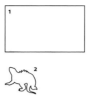

1 *Minmi* was an early ankylosaur, just 7 feet (2 m) long, from Queensland, Australia. Fossilized stomach contents show a mix of leaves, seeds, and ferns.

2 In this historic drawing made in 1862 by Benjamin Waterhouse Hawkins (1807–94), *Hylaeosaurus* is shown in a lizardlike squat pose that has since been heavily revised to a more upright stance.

3 The tail club of *Ankylosaurus* consisted of fused lumps and plates of bone called osteoderms, supported by several interlocked tail vertebrae.

90 Mya	84 Mya	72 Mya	72 Mya	66 Mya	66 Mya
Talarurus (basket tail), with a length of 20 feet (6 m), is one of the first ankylosaurids known from Central Asia.	Ankylosaur *Antarctopelta* lives in Antarctica. It will be the first nonbird dinosaur of any kind to be discovered for science on that continent.	The ankylosaurid *Tarchia* lives in the Nemegt Basin of Mongolia. It is known from several good specimens, including two complete skulls.	The nodosaur *Edmontonia*, with its enormous lateral spikes, roams North America.	The smallest known ankylosaur, *Struthiosaurus*, forages in Europe. It may have been no longer than 6½ feet (2 m).	*Ankylosaurus*, one of the last and biggest of the ankylosaurs, lives at the very end of the Cretaceous in North America.

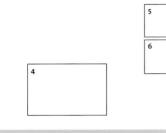

4 *Edmontonia* was a typically bulky, heavily armored ankylosaur almost 23 feet (7 m) long and weighing 3 tons (2.7 tonnes). It was named from the Edmonton Group formation of rocks in Canada.

5 American fossil hunter Othniel Charles Marsh, namer of *Nodosaurus*, directed a decades-long feud with ...

6 ... Edward Drinker Cope, who named more than 1,000 fossil species, although some are no longer recognized.

tails were optimized for wielding the club: their vertebrae had limited vertical mobility but allowed a sideways sweep of about 100 degrees. The foremost part, or base, of the tail was flexible, but the rear part was stiffened by ossified tendons (thin bony rods). The result of these features would have been that the club was swung just above ground level, where it could impact the relatively fragile metatarsals (foot and toe bones) of an attacker, including a theropod (meat-eating dinosaur) such as *Tyrannosaurus* (see p.302), rather than hit the more robust bones higher up the leg.

The neck armor of ankylosaurids consisted of two bands of bone wrapped around the top half of the neck, surmounted by osteoderms. That of nodosaurs was formed differently: there were no separate bony bands, but the osteoderms themselves were broader and fused to one another. Nodosaurs also lacked the tail clubs of their cousins, but compensated by evolving strong, sharp, sideways-pointing spikes along the sides of the torso. These were particularly well developed in the Canadian nodosaur *Edmontonia* (see image 4), which had murderous shoulder spikes that could have inflicted grievous wounds on predators.

A third family of ankylosaurs, Polacanthidae, is sometimes referred to, but this is now thought to be a subgroup of the nodosaurs. The polacanthids had lighter armor than the other ankylosaurs, and it could have been used as much for display as for defence. Ankylosaurs were a successful group, eventually spreading across the world. Some of the earliest examples are known from the middle Jurassic period in China, 165 million to 160 million years ago. One example is *Tianchisaurus* (heavenly pool/stream lizard/reptile) named in 1993 by leading Chinese paleontologist Dong Zhiming (b.1937). It has the extraordinary species name *T. nedegoapeferima*, devised with the help of film director Steven Spielberg. There are two letters from each family name of the

stars of *Jurassic Park* (1993–2015)—"ne" for Sam Neill, "de" for Laura Dern, "go" for Jeff Goldblum, and so on.

As the Jurassic gave way to the Cretaceous, ankylosaurs ranged into new areas, and their fossils have now been found on all continents except Africa. Even Antarctica has yielded an ankylosaur fossil, that of *Antarctopelta*. This was the first nonavian dinosaur discovered as fossils on the great southern continent, in 1986, although it was not formally named until 2006. Living 84 million years ago, *Antarctopelta* was not large, at around 13 feet (4 m) in length, and was somewhat intermediate between the ankylosaurid and nodosaurid groups, although it is usually placed in the latter. The "-pelta" suffix indicates a shield and is a common ending for ankylosaur names, just as "-raptor" is for birdlike theropods. Ankylosaurs continued to evolve and spread, especially from around 80 million years ago, and the group persisted until the end-of-Cretaceous extinction, which killed all nonavian dinosaurs.

Nodosaurus was one of the first nodosaurids found in North America and was named in 1889 by renowned fossil hunter Othniel Charles Marsh (1831–99; see image 5). This was the time of the North American "dinosaur rush" when fossil hunters competed to describe and name the most genera and species. Its two pre-eminent characters were Marsh and Edward Drinker Cope (1840–97; see image 6)—the latter still holds the record for the most published scientific papers, at more than 1,400. The rivalry between them is legendary and known as the "Bone Wars." The two fossil hunters came from very different backgrounds. Marsh's immediate family was not well off, but an enormously rich uncle, George Peabody, paid for his education and eventually endowed the Peabody Museum of Natural History at Yale University, with the condition that Marsh be a curator. Although Cope's family was fairly wealthy, he did not have the advantage of a rich benefactor—something that would later contribute to his resentment of Marsh.

Marsh and Cope were initially on good terms, and each even paid the other the honor of naming a species after him. However, their ambition and unscrupulousness led to a destructive rivalry. Relations began to deteriorate when Cope took Marsh to visit the marl pits where *Hadrosaurus*, the first American nonbird dinosaur to be named, had been discovered; Marsh secretly bribed the quarrymen to send all future finds to him instead of Cope. In 1869, Cope described and illustrated the long-necked plesiosaur *Elasmosaurus* (see p.286) but made the embarrassing mistake of putting the head on the wrong end; Marsh publicized the error. Soon both Cope and Marsh had teams of workers excavating dinosaurs and other fossils in the American West, and relations between the crews deteriorated. Spying to determine the best digging locations was commonplace, and Marsh and Cope frequently hired good workers away from one another. On occasion there were even physical fights between Marsh's and Cope's crews. Both teams, if they were unable to excavate interesting fossils for some reason, destroyed them rather than allow them to fall into the other's hands.

The conflict was destructive both professionally, as both men lost the respect of their peers, and scientifically, because in their haste to name more dinosaurs and more vertebrates than their rivals, shoddy work was done and specimens were described and illustrated inadequately. Many synonyms were created: the same genus or species was given different names. For example, Marsh's *Morosaurus* has long been considered a species of Cope's *Camarasaurus*. Despite these events, both men have been honored by having dinosaurs named after them. Drinker, described in 1990, was a small plant-eater, moving on two back legs, related to *Hypsilophodon*. *Othnielosaurus*, officially renamed in 2007 from some earlier specimens called *Othnielia*, was a similar kind of dinosaur. **MT/SP**

Euoplocephalus
LATE CRETACEOUS PERIOD

SPECIES *Euoplocephalus tutus*
GROUP Anklyosauridae
SIZE 20 feet (6 m)
LOCATION North America

One of the best represented of the ankylosaurs, *Euoplocephalus* is known from more than a dozen specimens, including some near-complete skeletons. It reached 20 feet (6 m) in length, 6½ feet (2 m) in height and the same in breadth. Its mass has been estimated at 2¼ tons (2 tonnes), equal to a large hippopotamus. The broad, stable torso; heavily armored head, neck, and back; and hefty tail club made *Euoplocephalus* a formidable creature. Its upper fore limb bone (humerus) had huge muscle attachment areas, and it has been suggested that this hefty, armored beast could gallop as well as a hippopotamus. This may seem unlikely based on the robust, broad shape of its skeleton, but the hippopotamus is even stockier and carries a great deal of soft tissue, yet is capable of running at 19 miles per hour (30 km/h)—a speed only exceptional humans can match.

Euoplocephalus lived in North America in the late Campanian age, the penultimate age of the Cretaceous, 75 million years ago. The tyrannosaur *Gorgosaurus* was among its contemporaries. The first specimen was discovered by Canadian paleontologist Lawrence Lambe (1863–1919) in Dinosaur Provincial Park, Canada, and named in 1910, but it is also known from the United States. However, multiple skeletons are yet to be found together, which suggests that *Euoplocephalus* led a solitary lifestyle, although numerous other ankylosaurs coexisted alongside it, such as *Dyoplosaurus* and *Anodontosaurus*. These other ankylosaurs were possibly more gregarious, just as tigers are solitary but their close relatives—lions—live in prides. At one point all contemporary ankylosaurs were considered to belong to the *Euoplocephalus* genus—even the Asian *Tarchia*. **MT**

✦ NAVIGATOR

FOCAL POINTS

1 TORSO

The proportions of the torso were typical for an ankylosaur, but not seen in any other dinosaurs: very broad, so that *Euoplocephalus* was as wide as it was tall. The wide sacrum and pelvis helped to support the torso. The limbs did not sprawl to the side, as was depicted in old artworks.

2 ARMOR

Bony armor ran along the back in lines. Each individual armor bone was an osteoderm: an oval with the long axis parallel to the animal and strong midline ridges. The armor was especially concentrated around the neck, with bony plates that formed protective half-rings around the upper neck.

3 TAIL CLUB

The bony tail club formed by the fusion of vertebrae and osteoderms is a distinctive feature of *Euoplocephalus*. The rear half of the tail was stiffened by ossified tendons, which the front half lacked. This allowed flexibility to wield the club, which could break bones.

TWISTED NASAL PASSAGES

In 2008, a computed tomography (CT) scan of a *Euoplocephalus* skull (right) uncovered a surprise: the airways from the nostrils to the trachea (windpipe) were twisted and repeatedly doubled back on themselves, making their total length more than twice that of the skull. The airways were also broad, with a total volume 10 times that of the brain. Similar, though less extreme, airway convolution has been observed in other ankylosaurs, including the nodosaurid *Panoplosaurus*. The airways of *Euoplocephalus* ran close to the canals that carried blood through the skull bones, so a possible explanation is that the airways acted as a heat exchanger—a cooling system for the blood. Furthermore, vocalizations would have been amplified and deepened as they made their way through these passages, similar to the air in a trombone. Perhaps the primary role of these enormous and complex nasal passages was to produce a unique call, recognized by other members of the species from a great distance.

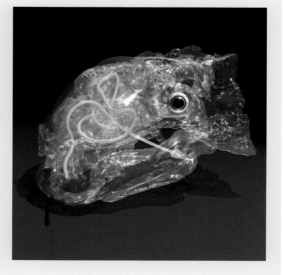

HORN-FACES AND DOME-HEADS

Ceratopsians (horn-faces) and pachycephalosaurs (thick-heads), including *Styracosaurus* (see image 1) and the well-known *Triceratops* (see p.344), belong to the group Marginocephalia (fringed-head). This was the sister evolutionary group to the ornithopods (bird-feet dinosaurs), such as *Iguanodon* (see p.324), and was known for the frill or ledge around the back of the skull.

The ceratopsian group was wildly successful. More than 70 genera are recognized, many from numerous specimens. Ceratopsians began as small bipedal herbivores, similar to their pachycephalosaur cousins, and they were distinguished mainly by increasingly large, solid skulls. *Psittacosaurus* (see p.340) exemplifies this early stage in their evolution. Later ceratopsians—called neoceratopsians—became larger, heavier, and quadrupedal (walking on four legs). Early neoceratopsians are exemplified by *Protoceratops* (see image 2), a sheep-sized animal whose fossils have been found in rocks 80 million to 70 million years old from Mongolia. *Udanoceratops*, from the same time and place, was the size of a cow and had an oversized lower jaw.

The most iconic ceratopsians belong to Ceratopsidae, the group of heavy, quadrupedal, rhinoceros-to-elephant-size dinosaurs with large bony frills projecting over the neck, and with horns, spikes, and other decorations on the head. The ceratopsids had the largest skulls of any terrestrial animal, approaching 10 feet (3 m) long in some species, including the

KEY EVENTS

160 Mya	113 Mya	90 Mya	90 Mya	89 Mya	77 Mya
Yinlong, from Xinjiang province, China, is the oldest known ceratopsian and the first marginocephalian.	*Archaeoceratops* is the oldest known neoceratopsian. It is only 3 feet (1 m) long and lives in north-central China.	*Amtocephale* lives in ancient Mongolia. It is most likely the earliest known pachycephalosaur.	*Turanoceratops* roams Uzbekistan early in the late Cretaceous. It is either the earliest known ceratopsid or a close relative.	*Zuniceratops*, the earliest ceratopsid from North America, looks like a small *Triceratops* at 11½ feet (3.5 m) long and weighing 330 pounds (150 kg).	Numerous *Styracosaurus* remains from this time, including horn-cores, frills, and jaws, are later found in a bone bed in Alberta, Canada.

frills. Ceratopsidae comprised centrosaurs and chasmosaurs. The centrosaurs had shorter faces, frills, and taller muzzles and developed their nose horns and frill decorations. Their remains are often found in large fossil accumulations called bone beds, suggesting that they lived in herds. Centrosaur skulls evolved a variety of shapes: *Styracosaurus* had seven horns (one on the nose and six along the frill margin); *Einiosaurus* had a single horn on the nose, bent forward and downward; *Pachyrhinosaurus* had no horn, only a thick, heavy pad on its nose.

Others in the Ceratopsidae group, the chasmosaurs, had longer skulls and frills, and instead developed their brow horns. *Triceratops* with its long, sharp beak, small nose-horn and long brow horns, is the classic chasmosaur, yet it differs from its relatives in having a solid frill. In *Torosaurus* and other chasmosaurs, two large holes developed in the frill. In *Chasmosaurus*, this is taken to an extreme, so that the frill is more like a frame consisting only of a narrow central bar, two side bars and a trailing margin. It is likely that these holes were covered over by skin. The roles of the frills and horns are debated. The robust bony frill of *Triceratops* would have been valuable armor, protecting the vulnerable neck. However, the fragile frill of *Chasmosaurus* would have been useless in this role, and, in some ceratopsids, the frill was a display structure and might have been brightly colored. Similarly, it has been assumed that the horns were defensive weapons used in combat with tyrannosaurs, but they may have been better suited for interspecies combat.

The sister group to the ceratopsians, the pachycephalosaurs, are known from significant finds dating from the late Cretaceous period, 90 million to 66 million years ago. They have not been found in great numbers; fewer than 20 genera have been recognized. Some specimens have flat heads, and rather than being a separate species, it is possible that these are the females or juveniles of dome-headed species, such as *Pachycephalosaurus* (see image 3) and *Stegoceras* (see p.342). If this is the case, the number of recognized genera would decrease further. The pachycephalosaurs were small-to-medium-sized herbivorous bipeds (walking on two legs), and most had high, domed heads, reinforced by bone up to 8 inches (20 cm) thick, which earned them the nicknames "bone-heads," "dome-heads," and "helmet-heads." It is likely that they engaged in headbutting contests and their spongy bone would have efficiently absorbed shocks. An alternative theory that the domes were for display is contradicted by their microscopic structure. Pachycephalosaurs also had proportionally broader torsos than any other bipedal dinosaurs. The hip bones were displaced outward to allow the stomach to continue past the pelvis, whereas in other dinosaurs the stomach was contained entirely within the main body cavity. This wide torso would have enabled more complete digestion of plants, but could also have provided protection for the internal organs if pachycephalosaurs used their domes for flank-butting (bashing the head into the opponent's side) rather than headbutting. **MT**

1 *Styracosaurus* was one of the larger ceratopsians, almost 20 feet (6 m) in length.

2 *Protoceratops* demonstrates the ceratopsian toothless beak at the mouth front and a nose lump, or boss, that may have supported a short horn.

3 The bone-dome of *Pachycephalosaurus* was up to 8 inches (20 cm) thick and had spiky cones around the rear; the teeth were tiny.

76 Mya	74 Mya	73 Mya	68 Mya	66 Mya	66 Mya
Centrosaurus lived in Alberta, Canada, at this time. The Hilda mega-bone bed later contains thousands of *Centrosaurus* fossils.	The chasmosaurine *Pentaceratops* from North America has the longest skull of any land animal at 8¾ feet (2.75 m).	Although its name implies that it is an ancestor, *Protoceratops* from Mongolia lives after many other ceratopsids.	*Eotriceratops* (dawn *Triceratops*) may be biggest of the horned dinosaurs, at 30 feet (9 m) in length and weighing 13 tons (12 tonnes).	*Pachycephalosaurus* is the largest known pachycephalosaur, at 15 feet (4.5 m) in length and weighing 1,000 pounds (450 kg).	*Triceratops*, one of the last chasmosaurine ceratopsids, roams North America until the end-of-Cretaceous mass extinction.

Psittacosaurus
EARLY—MIDDLE CRETACEOUS PERIOD

SPECIES *Psittacosaurus
mongoliensis*
GROUP Psittacosauridae
SIZE 6½ feet (2 m)
LOCATION China and Mongolia

This animal was a fine example of the early stages of evolution that led to the great horned dinosaurs, or ceratopsians, such as *Triceratops*. It was a small, dog-sized biped with a powerful beaked head that was large compared to similar-size ornithopod plant-eaters, but proportionally much smaller than those of ceratopsians. All species of *Psittacosaurus* were found in Asia, specifically China and Mongolia. One of the largest was *Psittacosaurus mongoliensis*, which reached up to 6½ feet (2 m) in length and 55 pounds (25 kg). The majority of dinosaur genera have only one single valid species, such as *Tyrannosaurus* (see p.302), but unusually, around 10 species of *Psittacosaurus* are considered valid, spanning a wide range of time from 123 million to 100 million years ago in the middle Cretaceous. It is possible that this complex of species may change after future analysis, with some species combined or synonymized, and others transferred to a new genus. *Psittacosaurus* is one of the best understood dinosaurs because of the large amount of fossil specimens available—more than 400 have been found and many of them are complete. This sizable sample has made it possible to compare animals of different ages and to understand how their body shape, or morphology, changed throughout their lives. For example, in juveniles the fore limbs were much longer in relation to the hind limbs than they were in mature adults. This could indicate that juveniles were quadrupedal and transitioned to bipedality as they matured.

Several fossils have been found in which multiple *Psittacosaurus* were preserved together, most notably at a site in Liaoning province, northeast China, which contains 34 babies. Another multiple find included six individuals of two different age classes, which is evidence that *Psittacosaurus* lived in groups beyond the nest. It is also the only dinosaur known from direct evidence to have been preyed upon by mammals. A specimen of *Repenomamus* (see p.426), the largest Mesozoic era mammal, was found with the partial skeleton of a juvenile *Psittacosaurus* in its stomach. **MT**

 NAVIGATOR

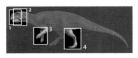

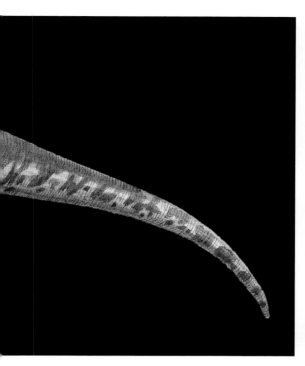

1 BEAK AND CHEEKS

It is likely that the toothless front of the mouth, or beak, was covered by a horny sheath. Its shape gave *Psittacosaurus* its name, meaning "parrot lizard." It would snip vegetation to chew using the small cheek teeth. At the tip of the lower jaw was an extra bone, the predentary—a distinctive feature found in all ornithischians (bird-hipped dinosaurs).

2 HEAD

When viewed from above, *P. mongoliensis*'s head was triangular in shape, with a narrow front to the beak and the cheeks widely drawn out toward the rear. Each cheek bone, or jugal, bore a rudimentary pointlike "horn" that projected sideways out of the face. The skull was tall and robust, firmly braced to deliver a powerful seed-cracking bite in the manner of modern parrots.

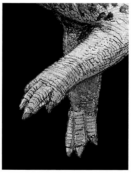

3 FORE LIMBS

The structure of the fore arm and wrist would have made it impossible for the hand of an adult *P. mongoliensis* to rotate or twist to face downward, to be placed palm or knuckles down on the ground, which is necessary for quadrupedal locomotion. Similarly, the limb was too short to reach the mouth. Possible fore limb functions include carrying objects and sparring with rivals.

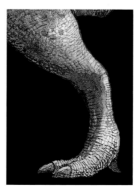

4 HIND LIMBS

In juveniles, the fore limbs were proportionally longer than the hind limbs. However, differential growth meant that by adulthood—6 to 10 years of age—the hind limbs were much longer than the fore limbs, necessitating bipedal movement. The wealth of fossils that has been discovered has allowed experts to study how *P. mongoliensis* grew and changed in size, shape, and proportion.

QUILLED PSITTACOSAURUS

In 2002, a particularly well-preserved specimen of *Psittacosaurus* was described. Its startling feature was a group of long bristles protruding above its tail. These bristles were well embedded in the flesh of the tail and seemingly hollow (below). The quills would have been the same length as the upper arm, and would have given the dinosaur a distinctive visual profile. The result would have been a porcupinelike effect. These quills have not yet been observed in other specimens, so it is unclear how widespread they were. They may have occurred only in one species, or perhaps only in males. Similar structures may also have existed in other ceratopsians.

Stegoceras
LATE CRETACEOUS PERIOD

FOCAL POINTS

1 TEETH
Gently curved, serrated teeth set in the front parts of S. validum's jaws were superficially similar to those of theropods (meat-eating dinosaurs). S. validum was long mistaken for the maniraptoran Troodon. These teeth could show that it did tackle animal prey.

2 SKULL FIBERS
The outer surface of the skull dome had a peculiar texture, with bundles of microscopic structures arranged throughout the bone, termed Sharpey's fibers. These indicate that a substantial amount of soft tissue—thick skin or horn—was attached to the outside.

3 NECK
The cervical (neck) vertebrae, adopted a natural curved profile, able to be flexed into a U- or an S-shape. This would be very difficult to straighten, so that the lowered head, neck, and body were all aligned, like a battering ram, perhaps for full-frontal clashes.

4 FEET
Evidence shows that S. validum had long toes and claws typical of animals that evolved in moist lowlands and woodlands. It was initially thought that S. validum and kin occupied upland environments, in which case small feet would be needed for climbing dry, rocky terrain.

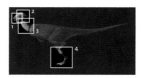
SPECIES *Stegoceras validum*
GROUP Pachycephalosauridae
SIZE 7 feet (2 m)
LOCATION North America

One of the best known pachycephalosaurs (thick heads), is *Stegoceras* from North America, which lived around 75 million years ago. With a length of 7 feet (2 m), it was a typical dome-skulled member of the group. The majority of pachycephalosaurs were small compared to other dinosaurs. They all shared a generalized body shape suited for a lifestyle of eating leaves, sprinting away from predators, and foraging in well-vegetated environments—a lifestyle that remained unchanged for ornithischians (bird-hipped dinosaurs) for more than 100 million years. However, in the anatomy of their skulls, hips, and tails, pachycephalosaurs were anything but typical.

Fossilized skull domes of *Stegoceras* show variation in size, shape, and thickness. Some suggest this reflects sexual dimorphism, with the males having more strongly domed, thicker skulls than the females. If correct, this supports the theory that the skull domes were used in sexual displays or battles. It may also support the view that these skulls evolved under sexual selection pressure, meaning their size and shape evolved based on the role they played in securing a mate, rather than, for example, for feeding. Several features of the *Stegoceras* skull could have evolved for visual display. A bony shelf projects from the back of the head, its upper surface covered in conical protrusions. A continuous curving ridge extends from the shelf to the nostril region, which could have supported soft tissue. Alternatively, the different skull shapes could represent different species. Some experts recognize the species *Hanssuessia* and *Colepiocephale*, while others class these as variants of *Stegoceras*. Experts also disagree over several aspects of pachycephalosaur biology, such as growth changes. Dome-skulled pachycephalosaurs could be adults and the flat-skulled kinds their young. Fossils of varying ages have been found of *Stegoceras*, which confirms that it started life with a flat skull roof that domed with age. **DN**

HEADBUTTING GROUP

The thick, sometimes dome-shaped skulls of pachycephalosaurs, such as *Stegoceras* (right), led to the idea that these dinosaurs engaged in butting or shoving matches. Based on the behavior of modern sheep and goats, it has been argued that they rammed heads for mating reasons. Since the 1990s, however, this idea has been questioned. First, using a "bowling ball" analogy, dome-shaped skulls do not make good headbutting weapons—unless clashing exactly head on, they would glance off at angles. Second, the microscopic anatomy of the pachycephalosaur skull contradicts the idea that they could withstand traumatic impact without serious injury. Perhaps the domes were only used in visual display. However, a notable percentage of fossilized domes preserve dents and other evidence of injury. Perhaps these dinosaurs headbutted one another's flanks, or objects, such as tree trunks, to show their prowess.

Triceratops
LATE CRETACEOUS PERIOD

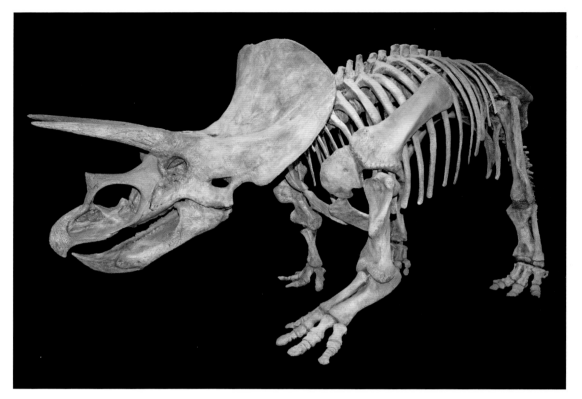

SPECIES *Triceratops horridus*
GROUP Chasmosaurinae
SIZE 30 feet (9 m)
LOCATION North America

NAVIGATOR

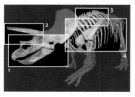

The best known of the ceratopsians is *Triceratops*, of which there are probably two species, *T. horridus* and *T. prorsus*. Its fossils are plentiful in the Hell Creek Formation rocks in North America, from the latest part of the Cretaceous, 66 million years ago. Around 50 skulls have been found since the turn of the millennium. *Triceratops* superficially resembled a rhinoceros, but had a larger head, a more substantial tail, and of course the weapons that gave it the name "three-horned face." *Triceratops* was one of the largest ceratopsians, measuring more than 26 feet (8 m) in total length, possibly reaching 30 feet (9 m), and weighing between 6½ and 11 tons (6–10 tonnes).

Triceratops is among the most iconic dinosaurs, second only to *Tyrannosaurus*, and its powerful body and intimidating weaponry make it worthy of this status. In fact, there is evidence that *Tyrannosaurus* ate *Triceratops* when it could—some *Triceratops* hip bones have distinctive tooth marks that no other predator could have left. However, these marks may have been caused by *Tyrannosaurus* scavenging an already-dead *Triceratops* carcass. Evidence of actual fighting is less conclusive, although partially healed bite marks on some *Triceratops* frills suggest a fight that the ceratopsian survived. Specimens of different ages show that the distinctive horns and frill changed dramatically through life. In juveniles, the brow horns were mere stubs and the edge of the frill was decorated with elaborate arrowhead bones called epoccipitals. The horns then became longer with growth, changed orientation—from curving backward to curving forward—and eventually may have shrunk in older individuals. Meanwhile, the epoccipitals were progressively fused into the frill and eventually obliterated completely, leaving a smooth trailing edge. It has been suggested that the similar *Torosaurus* is the most mature growth stage of *Triceratops*—an idea known informally as the Toroceratops Hypothesis—but this remains controversial. **MT**

👁 FOCAL POINTS

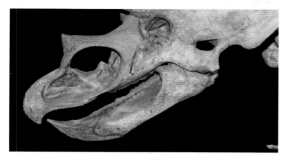

1 SKULL

The skull was robustly built and the eye sockets were well protected, directly under the brow horns. Unlike most other ceratopsians, *T. horridus*'s frill was solid with no holes. The skull was elongated, the lower jaw substantial, and the beak drawn out to a sharp point, specifically in the lower jaw.

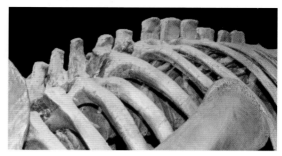

3 BACKBONE

The neck of *T. horridus* was reinforced by the fusion of the first four cervical (neck) vertebrae into a single unit known as a syncervical. Most tetrapods (four-limbed vertebrates), from amphibians to mammals, have a sacrum, which is a sequence of fused vertebrae at the rear of the pelvis, but the syncervical is unique to ceratopsians.

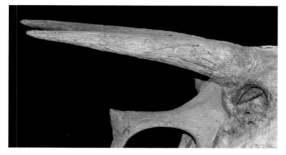

2 HORNS

The two main brow horns of *T. horridus* grew to 3 feet (1 m) in length and the nasal horn was shorter. They are preserved as bony "cores" and would probably have been covered in a horny, sharp-tipped sheath. Theories on their function include combat, self-defence, and a visual display to impress mates.

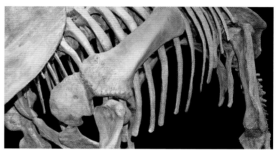

4 LIMB BONES

The upper arm bone (humerus) was huge and had a very prominent deltopectoral crest—the forward protrusion at the top of the bone was for muscle attachment. The limb bones were more robust than those of elephants. It has therefore been suggested that, despite its great size, *T. horridus* was capable of galloping like a rhinoceros.

DISCOVERING AND NAMING

The first fossils of *Triceratops* to be scientifically described and named were a pair of brow horns and the portion of skull attached to them (left) in 1887. Since the horned dinosaurs were not then known as a family, US paleontologist Othniel Charles Marsh (1831–99) misidentified them as belonging to an extinct bison and created a new species, *Bison alticornis*. It was only after fossil hunter John Bell Hatcher (1861–1904) discovered a more complete skull that Marsh understood that the horns belonged to an animal similar to *Ceratops*, which he had named a year previously. Over the next decades, 18 species of *Triceratops* were named, many of them based on skulls of different sizes and growth stages, subject to different degrees of crushing. Most of these have now been discarded: *T. horridus* remains, as does *T. prorsus*, but the latter's validity is disputed by some.

LIZARDS

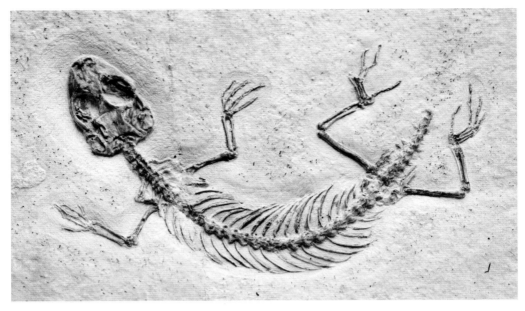

Lizards are one of the most diverse and successful reptile groups—excepting only snakes, which are essentially extremely modified, legless lizards. There are more than 5,900 living lizard species and they occur on all continents except Antarctica. Most share the same quadrupedal (walking on four legs) body plan—an arrangement so versatile that it is present in species ranging from 1 inch (2.5 cm) to almost 13 feet (4 m) in length. The majority of lizards are predators of insects and other small animals. Most combine conical, pointed teeth, slender limbs with clawed digits, and a long, slender tail. In some groups, the tail can be shed as a defensive mechanism. However, tricuspid (three-pointed), fanglike and flatter molarlike teeth have all evolved within the group, as have limbs and bodies specialized for climbing, digging, burrowing, swimming, and even gliding. Leglessness has evolved more than 60 times, predominantly in groups that have taken to a burrowing lifestyle. The ability to give birth to live young has also evolved numerous times.

Lizards originated around 200 million years ago from an ancestor shared with the rhynchocephalians, the group that includes tuataras (see p.234). However, events in lizard history are understood only since the middle Cretaceous period, around 100 million years ago, from fossil specimens such as *Yabeinosaurus* (see p.348). Lizards fall into four major groups: gekkotans (geckos and limbless flap-footed lizards), iguanians (iguanas), scincomorphs (skinks; see p.350), and anguimorphs (alligator lizards and kin). Early members of all four groups are known from sediments of the Cretaceous (see image 1). Some gekkotans are predators of other lizards that seemingly evolved to take the place of predatory snakes. Gecko toe pads—scansor pads—are among the most extraordinary evolutionary innovations in nature. Hundreds of thousands of hairlike structures, termed "setae," are arranged in rows and their adhesive properties allow geckos to cling to walls, smooth leaves, and even glass. The setae are simple in some geckos and complex in others, where the tips are either spatulate (flattened) or branched. The scansor pads have been reduced—and even lost altogether—in groups where normal digits have proved more useful for a terrestrial lifestyle.

One of the most distinctive lizard groups is the iguanians; within it are iguanas, basilisks, anoles, agamas, and chameleons. Historically, iguanians have been regarded as separate from other lizards on account of their large, fleshy tongue and the large upper opening at the back of the skull—features that changed substantially in other lizard groups. However, genetic studies contradict this distinction and have found iguanians to be a central line of lizard evolution. Iguanians commonly evolve herbivorous species, presumably because of their large size, big tongues, and strong skulls. Fossils show that iguanians with teeth and jaws suited for plant-eating have been in existence for at least 55 million years, and it seems that the especially warm conditions present at that time were favorable for some to become giants. *Barbaturex*, from 53-million-year-old rocks in Myanmar, was more than 6 feet (2 m) long, making it one of the biggest herbivores of its habitat.

Among the most successful iguanians are the anoles, a North American group containing more than 300 species, many of which possess toe pads, similar to those of geckos, which they evolved independently. Several fossil anoles have been preserved in amber. Elsewhere in iguanian evolution, agamas and chameleons (see image 2) form a lineage united by the presence of teeth that are fused to the jawbones and are not replaced. Chameleons are among the most bizarre of lizards, combining pincerlike hands and feet with prehensile tails, laterally flattened bodies, tall, bony head frills and occasional horns. The fossil record sheds little light on when and how these features evolved.

The third major lizard group, the scincomorphs, contains the majority of species. The skinks, whiptails, wall lizards, and many others belong here. Their small size, insectivorous lifestyle, and ability to breed and evolve quickly means that many are excellent colonizers of remote places, including islands. The final major lizard group, the anguimorphs, began as small predators of insects and snails but evolved into large predators of vertebrates. Slow-worms, monitors, and gila monsters are part of this group, and many possess big, bladelike teeth and venom-secreting glands that may help them to subdue prey. It is likely that the sea-going mosasaurs (see p.352) also belonged to this group. **DN**

1 This lizard, *Eichstaettisaurus*, was preserved 150 million years ago in the fine-grained limestone of Solnhofen, Germany, along with the early bird *Archaeopteryx* and many other fossils.

2 The Parson's chameleon, *Calumma parsonii*, is a native of the humid forest of Madagascar. Like all chameleons, it can change its skin color, not only for camouflage in varied terrain but also as a response to social stimuli and temperature changes.

50 Mya	18 Mya	15 Mya	10 Mya	4 Mya	4,000 Ya
Messelosaurines, which are similar to modern basilisks, inhabit Europe and North America.	Modern chameleons inhabit Africa. Some species are arboreal and occupy a range of forest habitats.	Ancient anoles inhabit Mexico and the Dominican Republic. They are essentially identical to modern anoles that climb trees.	Diverse anguimorphs, including alligator lizards, slow-worms, and glass lizards, live in Europe, Asia, and North America. Galliwasps inhabit the Caribbean.	Giant monitors evolve in Australia and also inhabit Timor. The biggest exceed 18 feet (5.5 m) in length.	The giant 5-foot-long (1.5 m) iguana *Lapitiguana* inhabits Fiji. Its closest relatives are South American. It is hunted to extinction by humans.

Yabeinosaurus
EARLY CRETACEOUS PERIOD

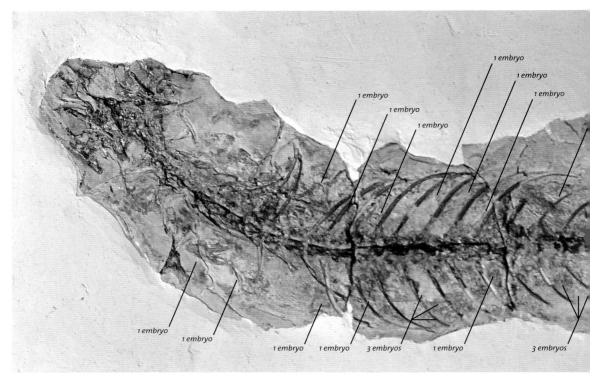

1 embryo
1 embryo
1 embryo
1 embryo
1 embryo
1 embryo
1 embryo
1 embryo
1 embryo
1 embryo
1 embryo
1 embryo
3 embryos
1 embryo
3 embryos

SPECIES *Yabeinosaurus tenuis*
GROUP Squamata
SIZE 2 feet (60 cm)
LOCATION Northeast China

In recent years, large numbers of new dinosaurs have been discovered, and numerous fossil lizards have been found alongside them. *Yabeinosaurus tenuis* is a fossil lizard from the northeast region of China, discovered in the same rocks as feathered dinosaurs. Its genus includes two species, the other being *Y. youngi*. *Y. tenuis* was first named in 1942, but the specimens found at that time were small, lightly built, and provided little indication that this lizard was especially interesting. However, new *Y. tenuis* fossils unearthed since 2005 have revealed several remarkable features about this lizard. Rather than being small and nondescript, it was large and robust: the head and body were around 1 foot (30 cm) in length, but including the tail the total length was more than twice this. The small specimens described during the 1940s appear to be juveniles. Taken together, the old and new fossils show a lengthy growth sequence from early juveniles to mature adults. The former had bones that were weakly ossified, meaning that they were only partially hardened with true bone tissue, while the latter had very strong bones, especially the skull, jaws, and vertebrae.

Lizards are typically regarded as egg-laying animals. However, the ability to give birth to live young, termed viviparity, has evolved more than a hundred times within the group, indicating that this is a relatively easy and useful transition for lizards to make. A preserved specimen of a pregnant *Y. tenuis* uncovered in 2011, the oldest one yet discovered, proves that this was yet another viviparous lizard. The specimen concerned has more than 15 embryos preserved in its body. It has been predicted that lizards from cool climates are most likely to evolve this trait, and the cool Cretaceous environment inhabited by *Y. tenuis* supports this. Eggs left to incubate in a cool climate may take a long time to hatch or fail to hatch entirely. Meanwhile, a mother carrying embryos within her body acts as a mobile incubator, ensuring that the developing young are kept at an optimum temperature. **DN**

 **NAVIGATOR**

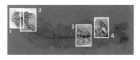

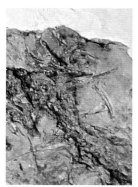

1 TEETH

Sharp conical teeth, together with the preserved stomach contents of *Y. tenuis*, show that it ate fish, which is unusual for a lizard. It could have been a swimmer that regularly caught aquatic prey, but other obvious aquatic adaptations are absent from the fossil skeleton. However, several living lizards, including the water monitor, swim and dive for food yet lack obviously aquatic features in the skeleton.

2 SKIN TEXTURE

A rough texture is present on the skull bones and small, rounded plates are scattered within the skin. Some specimens show variation in scale shape and size on different body parts, from small disks to large squares. Together, these features could indicate a rough skin surface. This could have made *Y. tenuis* a less palatable prey for dinosaurs and other predators, or could have provided camouflage.

- *1 embryo*

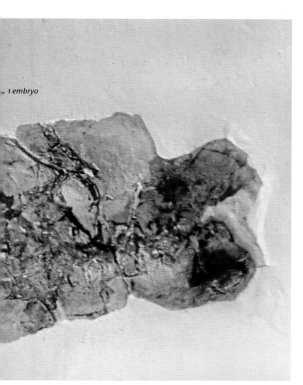

A RELICT SPECIES

Experts are uncertain about where *Yabeinosaurus* (below) fits in the lizard family tree. Some studies indicate that it is a primitive lizard, outside the group that includes modern kinds. During its time, the early Cretaceous, 120 million years ago, early members of all of the major lizard groups had already evolved. Yet *Yabeinosaurus* seems not to be a member of any of those lineages. It might be, therefore, that this lizard was already a relict—in effect, a living fossil of its time, left over from bursts of lizard evolution through the preceding Jurassic period.

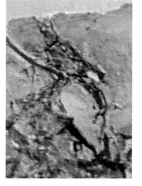

3 EMBRYOS

This 120-million-year-old specimen of a female *Y. tenuis* contains 18 well-developed embryos that were close to birth. The quality of preservation is so detailed that close examination reveals tiny teeth in the skulls of the offspring. This fossil was found in the fine-grained limestone rocks of the Jehol Group in northeast China.

4 SHORT LIMBS

Y. tenuis had relatively short limbs and a narrow, flexible body. These features are consistent with a variety of lifestyles, but could indicate that it was a ground-dwelling generalist instead of a specialized climber, burrower, or runner. Short limbs would enable movement through dense undergrowth, where long limbs would become tangled.

Florida Sand Skink
PLEISTOCENE EPOCH—PRESENT

👁 FOCAL POINTS

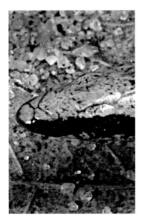

1 RECESSED JAW
The lower jaw is recessed beneath the upper jaw (set back and slightly upward) so that the upper jaw partly protects the lower jaw. This is typical of burrowing skinks. It improves streamlining and prevents sand and soil from entering the mouth as the reptile forces its way through sand or loose soil. The wedge-shaped snout is toughened to push between particles, but is sensitive, so the skink can locate the quickest path—useful when escaping predators.

2 RELICT LIMBS
The Florida sand skink is one of many burrowing skinks that appear to be in evolutionary transition. On each tiny fore limb the skink has a single digit, with two toes on each small hind limb. The limbs themselves are interpreted as evolutionary relics—useless appendages that will almost certainly disappear in future generations. Thus, while many animals worldwide use their legs for movement, the Florida sand skink tends to tuck away its limbs when moving.

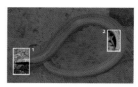
SPECIES *Neoseps reynoldsi*
GROUP Scincidae
SIZE 5 inches (12 cm)
LOCATION Florida
IUCN Vulnerable

The skinks, one the most successful and diverse lizard groups, are a globally distributed family of more than 1,400 species that occur in virtually all habitats that lizards could colonize. Variations include tree-climbing skinks, burrowing skinks, and skinks living in grassland, desert, and woodland environments. Possible fossil members of the group are known from European sediments that are around 130 million years old, but little reliable information is available on their early history. Anatomy and genetics show that skinks are part of the Scincomorpha; the same major lizard group as the armor-plated girdled lizards and armadillo lizards; the wall lizards and kin; and the whiptails, tegus, and kin. The majority are insectivores. This major group originated approximately 165 million years ago, and the earliest fossil members are known from locations in North America, Europe, Asia, and Africa. A key characteristic in skinks is the presence of smooth, glossy scales, which, combined with their streamlined shape, could explain why skinks have repeatedly adopted a burrowing lifestyle. When submerged in sand, loose soil, or sediments, skinks such as the Florida sand skink, *Neoseps reynoldsi*, breathe in the air trapped between the grains or particles. The shape and position of the nostrils, on either side of the snout facing sideways, each with a forward ridge, prevents sand grains from being inhaled. The Florida sand skink and several other burrowing lizards have evolved ridges on the sides of their bodies that prevent sand from filling in the spaces formed by their ribcages when they exhale.

The Florida sand skink and skinks of many groups, inhabiting all continents except Antarctica, and also oceanic islands, have independently evolved long bodies and reduced or absent limbs. Successive stages of limb loss can be seen in several groups. Within the Australian *Lerista* genus are species with short limbs and a full complement of digits, others with reduced limbs and three digits, two digits, one digit or no digits, and even limbless species. These skinks provide clues to the origin of snakes. It has been argued that snakes originated from lizards that became specialized for burrowing. This would explain why surface-dwelling snakes have lidless eyes, retained from their burrowing lizard ancestors. **DN**

SWIMMING IN SAND

It is thought that skinks such as the Florida sand skink have evolved a burrowing technique that involves "swimming through sand," like this burrowing skink (right). They propel forward with powerful lateral (side to side) undulations of their streamlined body and tail. The muscles along either side of the backbone are well developed for this purpose, throwing the backbone into S-shaped curves that travel from head to tail. Lizards with more typical proportions use their arms and legs to push and pull the soil or sand when burrowing. This method is successful, but is slower and more inefficient than lateral undulation alone.

MOSASAURS

Approximately 100 million years ago, the world's seas were occupied by fish of all sizes, from tiny to giant, along with vast numbers of swimming mollusks—notably the spiral-shelled ammonites (see p.157)—and a diversity of long-necked and short-necked marine reptiles called plesiosaurs (see p.282). The ichthyosaurs (see p.244), the once-numerous shark-shaped "fish lizards" of the Mesozoic era, were dwindling to extinction at this time, and a new group of marine reptiles evolved to take their place: mosasaurs. Members of the mosasaur group had existed for some time in shallow water habitats, and they quickly became gigantic dominant predators, occurring from the Arctic to the Antarctic and evolving a variety of skull and tooth shapes. By 70 million years ago, they included species that were among the largest sea-going reptiles of all time, such as *Prognathodon currii* and *Mosasaurus hoffmanni* (see p.354; see image 1).

The earliest mosasaurs were superficially similar to modern monitor lizards, such as the Nile monitor and Komodo dragon (see p.356), but more lightly built. Including coniasaurs, dolichosaurs, and aigialosaurs, the lizards were less than 7 feet (2 m) in length and were able to walk, rest, and forage on land as well as swim and dive. Some had pointed, conical teeth for grabbing fish; some had bulbous-crowned teeth suited for breaking open shelled prey; others had a mixture of tooth shapes. Mosasaur teeth, and their long, flexible necks and bodies, suggest that they were predators of reefs and other cluttered nearshore environments.

Over the course of their history, mosasaurs became larger and the clawed hands and feet of aigialosaurs evolved into paddles, or winglike flippers. They

1 *Mosasaurus hoffmanni* lived in shallow seas. It is thought to have relied upon ambush techniques rather than fast hunting to catch its prey.

2 With short and broad forelimb bones, *Tylosaurus* was able to thrust powerfully against water resistance.

3 Around 75 million years ago the Western Interior Seaway, covering much of central North America, was home to many oversized aquatic creatures, from ammonites to mosasaurs.

KEY EVENTS

140–100 Mya	100 Mya	92 Mya	90 Mya	88 Mya	87 Mya
Early ancestors of today's monitor lizards thrive. By 100 Mya, small marine lizards, the dolichosaurs, have descended from the early lizards.	Amphibious *Coniasaurus* in England and North America and *Aigialosaurus* in eastern Europe live in reefs and nearshore marine habitats.	Evolutionary intermediates between aigialosaurs and fully marine mosasaurs inhabit seas of North America and Morocco. They have clawed toes.	*Tethysaurus*, around 10 feet (3 m) long, has both primitive and more advanced mosasaur features, suggesting it was an intermediate form.	Mosasaurs with fully developed flippers evolve, perhaps several times from different ancestors. They are streamlined and larger than earlier kinds.	Vertical tail fins occur in some species. This either evolves once and is inherited by many species, or evolves independently more than once.

developed a sleek shape, their snouts became longer and their nostrils moved away from the snout tip. Similar changes had already occurred among several other reptile groups during the Mesozoic, with each repeating the same approximate pattern of evolutionary change. It was originally thought that this strong series of adaptations to a sea-going, rather than amphibious, life occurred once within mosasaurs, and that all marine, paddle-limbed species were the descendants of the same single ancestor. However, some experts now argue that different paddle-limbed mosasaur groups evolved more than once from different aigialosaur-type ancestors, meaning that mosasaurs were a mixed group with several lines of descent. Fossils show that the ability to give birth to live babies was present throughout the group, even in the aigialosaurs.

By 88 million years ago, advanced, fully flippered mosasaurs cruised the oceans. These included tylosaurs (see image 2), which lived in the Western Interior Seaway (see image 3). Fossil remains of their stomach contents show that they ate mollusks, fish, seabirds, plesiosaurs, and even other mosasaurs. Based on their general similarity to their cousins—living large predatory lizards—mosasaurs are expected to have had a generalized diet of marine animals, large and small. These living lizards are rarely specialists; they are instead supreme opportunists that forage in diverse habitats and eat anything they can subdue and swallow. Some mosasaurs, however, evolved specializations for more specific diets. Rounded or flattened teeth, probably used for breaking up or crushing animals with hard shells, such as crustaceans or mollusks, evolved more than once. There were also species with especially long, slender jaws and numerous teeth, and others that possessed bladelike teeth and also thicker teeth with widened bases. Thick, deep jawbones and blunt-tipped teeth in the giant *P. currii* from Israel suggest that it was a hunter of big, bony prey, and it has been described as a "marine tyrannosaur." Large eyes are evidence of excellent eyesight.

The mosasaur tail was narrow from side to side, so experts long thought that it was a sculling organ: side to side undulations provide the main thrust during swimming. However, the downturned shape of the tail's tip, the detailed anatomy of the tail tip bones, and the preserved skin impressions show that many advanced mosasaurs had vertical tail flukes. Instead of being used as a sculling organ, the tail's tip generated most of the thrust. The accepted view of the appearance of mosasaurs has changed over time. Illustrations from the middle of the 19th century show them as crocodilelike, with a serrated frill along the back and tail, and a rough, armored skin texture. The idea of the serrated frill was a mistake that stemmed from the misidentification of cartilage rings from the windpipe. Preserved patches of mosasaur skin show that these animals were covered in small, closely spaced scales, each covered in fine grooves and ridges. Similar ridges and grooves on shark body scales channel water around the animal's surface, improving its streamlining. **DN**

83 Mya	80 Mya	75 Mya	75 Mya	70 Mya	66 Mya
Globidens appears to crush shelled, bottom-dwelling prey with mound-shaped teeth. It occurs in Europe, Africa, and North and South America.	Mosasaurs occur in the Western Interior Seaway, including generalist predators *Clidastes*, *Platecarpus*, and *Tylosaurus*.	Several mosasaurs with long, slender jaws and numerous teeth evolve, notably *Pluridens* from Niger, with twice as many teeth as others.	*Mosasaurus*, one of the biggest marine reptiles of all time, evolves. Species swim in the seas over Europe, Africa, and North America.	*Plioplatecarpus* of North America and Europe lives in deltas, estuaries, and large rivers. It has big, broad paddles.	The last mosasaurs disappear in the mass extinction that ends the Cretaceous period. Of all the Cretaceous marine reptiles, only turtles survive.

Mosasaurus
LATE CRETACEOUS PERIOD

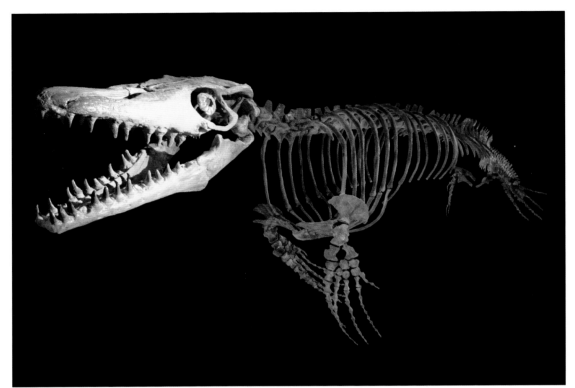

SPECIES *Mosasaurus hoffmanni*
GROUP Mosasauroidea
SIZE 60 feet (18 m)
LOCATION Western Europe and
eastern North America

Late in the 1700s, the gigantic partial skull of an ancient animal now known to be a mosasaur was discovered in a limestone mine at Maastricht, in the Netherlands. Thought by some to be a whale or a crocodile, the fossil was identified in 1799 by scientist Adriaan Camper (1759–1820) as that of a gigantic lizard, rather similar to modern monitor lizards such as the Nile monitor and Komodo dragon. This identification was confirmed by French comparative anatomist Georges Cuvier (1769–1832). The mosasaur from Maastricht went without a proper scientific name until 1829. In that year, English geologist and paleontologist Gideon Mantell (1790–1852; see p.325) named it *Mosasaurus hoffmanni*, a name that simply means "Hoffmann's lizard from the Meuse." Mantell is best known for his work on English dinosaurs.

Mosasaurus lived between 70 million and 66 million years ago. More complete remains found after the initial fossil confirm that *M. hoffmanni* was the largest mosasaur yet discovered, and could reach an incredible 60 feet (18 m) in length and up to 22 tons (20 tonnes) in weight. Only giant ichthyosaurs such as *Shastasaurus sikanniensis*, and some of the plesiosaurs, were larger. *M. hoffmanni* was also widespread, inhabiting the shallow seas that occurred during the late Cretaceous over western Europe and also eastern North America. In 2012, another mosasaur, possibly *M. hoffmanni*, was discovered near the site of the Maastricht find. The fossils include parts of the skull, jaws, backbones, and tail bones, and represent a 43-foot (13 m) individual dated at 67 million years old. *M. hoffmanni* evolved to be one of the top predators of its era, but its body and limbs seem unsuited for prolonged chasing after prey. It is likely to have lurked quietly, waiting for an opportunity to dash forward with a burst of speed to seize a passing victim. Fish, squid, and similar creatures would have been its prey. **DN**

NAVIGATOR

⊙ FOCAL POINTS

1 TEETH
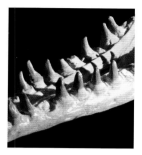
M. hoffmanni evolved large, curved teeth on its palatal bones, a feature that some have cited to link mosasaurs with snakes. However these teeth are viewed from an evolutionary perspective, it is clear they would have restrained and disabled prey. The main jaw teeth grew to 6 inches (15 cm) in length.

2 JAW JOINTS
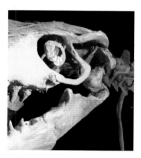
M. hoffmanni had a joint mid-way along the lower jaw and another at the chin, suggesting it could stretch the lower jaw to swallow large prey. Some argue this foreshadowed the jaw flexibility in snakes, and supports the view that snakes and mosasaurs are cousins. Others claim that the details of the creatures' jaws are different.

3 TAIL

The long, slim, high tail of *M. hoffmanni* would have provided forward propulsion force. It may have moved with S-like undulations, similar to an eel or a snake, rather than waving from one side to the other. The body was fairly rigid. The limbs acted as rudders when swimming fast, and as paddles when moving slowly.

⏲ SCIENTIST PROFILE

1769–96
Georges Cuvier was born in Montbéliard, then Germany, in 1769. After moving to Paris in 1795 he became a professor of animal anatomy at Musée National d'Histoire Naturelle (National Museum of Natural History). For decades, as Europe's leading authority, he studied fossil specimens from around the world.

1797–98
In his first book, *Tableau élémentaire de l'histoire naturelle des animaux* (*Elementary Survey of the Natural History of Animals*), published in 1797, he opposed the theory of evolution, believing that fossils were indicative of species that had appeared independently of one another, rather than gradually developing from pre-existing species.

1799–1811
Upon discovery of the "great fossil animal of Maastricht," Cuvier argued that it must represent an animal that had vanished from Earth. However, the notion that fossils might represent extinct animals was new and controversial. It meant that God's creations were not everlasting. The suggestion was a pivotal moment in scientific views on evolution.

1812–13
In *Discours préliminaire* (*Essay on the Theory of the Earth*, 1813), Cuvier disagreed with French naturalist Jean-Baptiste Lamarck (1744–1829), who believed that evolution was based on the inheritance of acquired characteristics. Cuvier instead supported catastrophism: that great natural disasters had destroyed existing life forms to be followed by the creation of a new set to populate the world.

1814–32
Cuvier was elected to the Council of State shortly before Napoleon's abdication in 1814. His theories on animal classifications were a departure from the 18th-century belief that all living things were arranged in a single linear system. He grouped similar animals together, insisting that the structure and function of organs were dependent on the environment, which later scientists developed further.

◄ *Prognathodon currii* was a huge mosasaur, up to 33 feet (10 m) long, that lived about 75 million years ago. The powerful jaw and sharp teeth of this fossilized head, found near Khouribga, Morocco, indicate that the water-dwelling mosasaur was as fearsome as *Tyrannosaurus* was on land.

Komodo Dragon
PLIOCENE EPOCH—PRESENT

SPECIES *Varanus komodoensis*
GROUP Varanidae
SIZE 10 feet (3 m)
LOCATION Indonesia
IUCN Vulnerable

Monitor lizards are a widespread group that includes the Australian goannas and the great Komodo dragon of the southern Indonesian islands, *Varanus komodoensis*. The theory that monitors and mosasaurs are close cousins in evolutionary terms has been popular for decades. This is because mosasaurs possess the relatively "open plan" rear skull anatomy typical of lizards; their long, slender jaws, slit-shaped nostril openings, and other features make them similar to monitors. It has also been suggested that gila monsters—venomous lizards of North America's southwest—are relatives. However, new studies show that monitor, gila monster, and mosasaur lineages separated 100 million years ago. Monitors have a complex and fascinating history. From Asia, the lizards spread to Africa and Australasia. Fossils show that the Komodo dragon originated in Australia before some populations moved westward and colonized islands that are now part of Indonesia. However, rather than being an island-dwelling giant, the Komodo dragon may be an island-dwelling dwarf, and modern populations are small-bodied relics of a larger continental predator.

Approximately 4 million years ago, *Varanus priscus*, the largest monitor lizard known to have existed, evolved. It is often known as *Megalania*, and remains from mainland Australia show that it was more than 11 feet (3.5 m) long and may have exceeded 20 feet (6 m). It became a true giant, its great size presumably an evolutionary response to the availability of new prey. This enormous lizard died out around 30,000 years ago—perhaps, coincidentally, as humans spread across Australia. **DN**

✦ NAVIGATOR

FOCAL POINTS

1 VENOM GLANDS
The komodo dragon possesses venom glands between its lower jaw teeth, and this venom causes septicemia-like (blood poisoning) symptoms in prey. It was once thought that prey were weakened in this way, mainly by being "infected" by rotting, bacteria-filled flesh, left from previous meals and carried in between the dragon's tooth serrations. However, it is uncertain to what extent either the tooth-trapped rotting flesh or the venom contribute to hunting success.

2 LIMBS AND TAIL
The komodo dragon has the typical sprawling, sideways-projecting legs of lizards. Each of the five toes per foot has a large, curved, sharp claw. The ungainly gait is due partly to the curvature of the body from side to side, and also the movement of the legs at the hip and shoulder joints. The tail is as long as the body and is muscular and powerful, acting as a fearsome weapon in self-defence. The tail is also used to deter competitive carnivores from a prized prey item.

BALANCED DIET

Monitor lizards are terrestrial predators, good at climbing, digging, and swimming, with teeth and jaws that allow them to prey on animals of all sorts, and sometimes eat plants, too. Komodo dragons are adaptable generalists that exploit a variety of resources. Prey items range from insects, snakes, and rodents to exhumed turtle eggs and even other Komodo dragons. They are attracted to carrion from miles away (right) and can kill large mammals, including deer, cattle, and humans. This flexibility has made monitors a successful group for 70 million years.

SNAKES

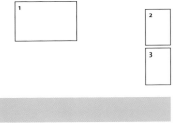

The evolutionary story of snakes is one of the most remarkable in the development of reptiles, and is also one of the most complex and controversial. Snakes range in length from threadsnakes of less than 4 inches (10 cm) long to boas and pythons, the biggest of which can reach 33 feet (10 m) in length and are capable of eating mammals similar in size to humans. The fossil record shows that even bigger snakes once lived. *Titanoboa*, a boa from Colombia that lived 60 million years ago, reached 42 feet (13 m) in length.

Snakes are part of the same group as lizards, termed squamates. Both snakes and lizards possess outgrowths of the skin, sheathed externally by a horny layer of slats like tiles on a roof. These structures are referred to as scales, even though they share no similarities with the detachable structures called scales present in fish and ancestral four-footed vertebrates. However, snakes have been modified by evolution far more than lizards, causing experts to debate the relationship between the two. The carnivorous habits and long retractile tongues of snakes recall those of monitor lizards, causing scientists to postulate that they share an ancestor with this group. Snakes are also similar in jaw anatomy to mosasaurs (see p.352), the enormous sea-going lizards of the Cretaceous period, causing some scientists to argue that both these groups descend from the same marine ancestor. Several ancient limbed snakes were sea-going animals with laterally compressed (narrow) tails, which would

1 The South American bushmaster *Lachesis muta* is the longest pit viper, exceeding 10 feet (3 m) in length. Its jaw joints stretch to swallow victims larger than its head.

2 Threadsnakes are part of the scolecophidian snake group and follow an underground lifestyle, preying on small soil creatures.

3 The royal python, *Python regius*, reaching 6 feet (1.75 m) in length, is from the family Pythonidae, with about 25 living species and more known from fossils.

KEY EVENTS

110 Mya	100 Mya	95–92 Mya	95 Mya	95 Mya	66 Mya
Snake fossils of fragmentary remains reveal that snakes emerged at this time.	Archaic snake *Najash* (see p.360) inhabits Argentina. It possesses hind limbs and a sacrum (where the vertebrae contact the pelvis).	Hind-limb-bearing snakes, *Pachyrhachis*, *Haasiophis* and *Eupodophis*, live in the seas covering eastern Europe and the Middle East, eating a fish diet.	Superficially pythonlike snakes, madtsoiids, inhabit Europe, Asia, Africa, and South America. They may be primitive snakes.	Early colubroids—the group of snakes that includes cobras, vipers, and colubrids—appear in Africa. They are rare and unimportant until the Miocene epoch.	The end-of-Cretaceous mass extinction has little impact on snake diversity. Some archaic snakes survive the event, including small, American *Coniophis*.

support this view, and snakes are similar to several limbless squamates, including limbless skinks and the obscure dibamids and amphisbaenians.

Snakes are not simply "stretched lizards." The neck and tail are generally proportionally short—much of a snake's length is formed by an incredibly long body. More than 200, sometimes more than 400, vertebrae are present in total. Specialized joints between the vertebrae, an unusual musculature, and large, platelike belly scales are all used for movement. Most snakes use lateral (side-to-side) undulations to move across the ground, throwing the long body into S-shaped curves—a form of movement clearly derived from that of lizard-shaped ancestors. However, a novel way of moving the body known as rectilinear locomotion has evolved in some heavy-bodied groups, which involves a back-and-forth motion of the large belly scales. Several key snake adaptations relate to the mobility of the bones of the palate and jaw. Distensible joints (see image 1) allow these bones to stretch apart so that large prey items can be swallowed. The joints also mean that many snakes can perform asymmetrical movements when handling prey, using the teeth on the left side of the jaws to drag prey into the mouth before detaching them and performing the same action on the right side. This is called unilateral feeding. It may be that an ability to take in large prey and use unilateral feeding were key factors in driving snake evolution, encouraging the development of their tubular body shape.

Snakes can be grouped into three major assemblages. The least familiar snakes are the wormsnakes, threadsnakes (see image 2), and blindsnakes, collectively termed scolecophidians. All are burrowers, mostly less than 12 inches (30 cm) long, with reduced eyes and an underslung mouth. They use their flexible sometimes toothless jaws to scoop up ants and termites. Certain aspects of their anatomy suggest that they are descended from larger ancestors that ate big prey and used unilateral feeding. The second major group, alethinophidians, includes the pythons (see p.362; see image 3) and boas, the consummate experts at constricting and swallowing large animals. Finally, the largest and most diverse group is the colubroids, which includes the venomous, fang-bearing vipers and cobras as well as the numerous water snakes, grass snakes, garter snakes, and rat snakes. There are some 3,000 living species of colubroids, making up almost nine-tenths of all snake species alive today, although ideas about their relationships are changing with ongoing genetic studies. Most colubroids are geologically young and much of their evolution can be linked to the evolution of their prey—rodents and other small mammals.

The majority of fossil snakes fit into these three main groups, but some do not. Several marine snakes from 95-million-year-old rocks of the Middle East and eastern Europe have tiny external hind limbs and even possess small feet. One theory is that these are primitive snakes, outside the main group. Another theory is that they are advanced snakes, closely related to pythons and colubroids, that re-evolved hind limbs for an unknown reason. **DN**

60 Mya	30 Mya	22 Mya	11 Mya	6 Mya	50,000 Ya
Gigantic boa *Titanoboa* inhabits Colombia. It is most likely an aquatic fish predator and lives at a time of extremely high temperatures worldwide.	The colubroid group that includes cobras—the elapids—appears in Asia, spreading to Africa and Australasia. Early elapids are rare and poorly known.	Early vipers appear in Europe and pit vipers in North America. They are similar to modern species, suggesting earlier evolutionary stages await discovery.	Small boas and similar snakes are diverse in Europe and elsewhere. They later decline to near-extinction when colubroids become more numerous.	Sea snakes, a group of elapids, evolve and spread throughout the Indo-Pacific region. This is a very young, rapidly evolving group of snakes.	The last madtsoiid—gigantic *Wonambi* from Australia—becomes extinct. Up to 30 feet (9 m) in length, it was a likely predator of large mammals.

Najash
LATE CRETACEOUS PERIOD

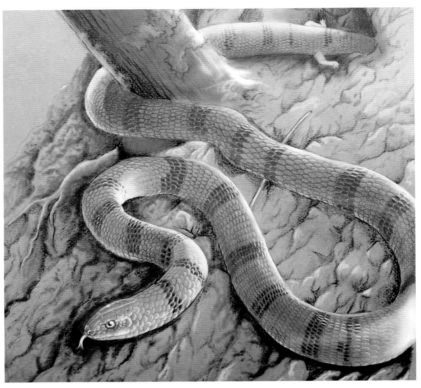

SPECIES *Najash rionegrina*
GROUP Ophidia
SIZE 5 feet (1.5 m)
LOCATION Argentina

For many years, ancient snakes were mostly unknown from the fossil record; even today the oldest known snakes are represented only by fragmentary remains. However, several semi-complete ancient fossil snakes have been described since 2000—and all have been the subject of controversy regarding the interpretation of their anatomy and evolutionary significance. Among these is *Najash rionegrina*, named in 2006 from the Cretaceous rocks of Argentina, which existed around 95 million years ago. The name *Najash* comes from the Hebrew word for the "legendary snake of the Bible." It was approximately 5 feet (1.5 m) long. It was clearly a snake because it possessed the extremely long body typical of this group: 122 articulated vertebrae (joined backbones) were discovered and more would have been present in the complete skeleton. Furthermore, the detailed anatomy of its vertebrae and skull are specifically snakelike.

Najash was especially interesting, and different from modern snakes, because it possessed hind limbs and a sacrum. This may mean it is one of the most primitive snakes yet discovered, and an early member of the group that still possessed a sacrum, pelvis, and hind limbs. All of these features were later lost in snake evolution. If *Najash* is such a primitive snake, it could have a bearing on our assumptions about snake origins. The skull and vertebrae indicate a terrestrial form, even an underground lifestyle. This complies with the major theory that snakes are lizards that lost their legs due to a subterranean lifestyle—although genetic studies may eventually prove otherwise. Regarding *Najash*'s diet and ecology, it is estimated to have hunted small reptiles and other such prey. However, some Cretaceous snakes were predators of reasonably large land-living animals. Proof of this comes from the fact that one of these snakes, *Sanajeh*, from India, was preserved within an egg-filled nest belonging to a sauropod dinosaur, its body coiled around a baby dinosaur as though ready to eat. **DN**

⚙ NAVIGATOR

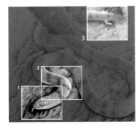

👁 FOCAL POINTS

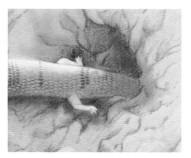

1 LOWER JAWS
Partial lower jaws were found close to the articulated skeleton of *Najash* and are thought to belong to it. The jaws found lacked teeth, but sockets show that the teeth would have had wide bases. Strongly curved, hook-shaped teeth with wide bases are typical of predatory snakes and help them to grab and swallow prey.

2 REAR SKULL AND BACKBONE
The broad, low vertebrae and the broad shape of the back of its skull indicate that this was a land-living snake, and perhaps a subterranean form that used burrows and hunted prey underground. This evidence could counteract the idea, suggested several times in the past 150 years, that snakes originated in the sea.

3 SACRUM, HIPS, AND LEGS
The sacrum is the part of the vertebral column that joins to the hips. Hip bones are present in several living snake groups, but none has a sacrum. The hind limbs here were relict features evolving toward absence. The original fossil description noted the presence of thigh and lower leg bones, but it is unclear whether foot bones were present.

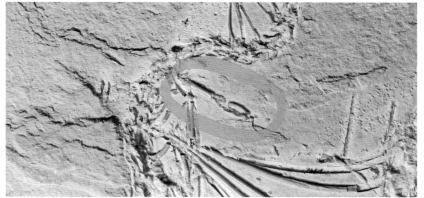

◄ Living 92 million years ago, slightly later than *Najash*, in Lebanon, *Eupodophis* was around 3 feet (1 m) long. It had no fore limb bones but did possess rear limb bones (circled). Although small, they have the same components as lizard leg bones, indicating the evolutionary pattern of snake limb reduction.

VENOM PRODUCTION

Venom is a substance poked or injected into a living creature that causes damage, harm, or even death. It is different to poison. Many living snakes are venomous (right), as are creatures as diverse as starfish, jellyfish, and shrews. Among lizards and snakes, a popular theory is that venom evolved just once and many snakes descended and diversified from this common ancestor, forming a clade known as the Toxicofera. During evolution, the amount of venom in some snakes was then reduced, or the ability to make it was lost. Another theory is that venom evolved separately in different snake groups. The former view, once dismissed by many experts, has recently gained credence, as genetic studies have found that genes for venom production are present in snakes that do not actually make venom or "toxic saliva."

Reticulated Python
PLEISTOCENE EPOCH—PRESENT

SPECIES *Python reticulatus*
GROUP Pythonidae
SIZE 30 feet (9 m)
LOCATION Southeast Asia
IUCN Not Yet Assessed

The oldest python fossils date to 45 million years ago in western Europe, and they would have certainly been constricting predators of vertebrates. Modern pythons appeared about 20 million years ago and were widespread, with remains found in Europe, the Middle East, southern Africa, and Australia. It seems likely that modern-type pythons, such as the reticulated python, *Python reticulatus*, originated in Africa and later spread to other regions, often crossing water barriers. However, some experts place python origins in Asia or Australia. The reticulated python is one of the largest pythons; a giant snake up to 30 feet (9 m) long that occurs in southern China and various islands around New Guinea. Pythons inhabit Africa, Asia, and Australasia, with the most diverse set of species living in Australia, New Guinea, and the Indonesian islands to the west.

Snakes such as pythons and boas are anatomically archaic when compared to the larger Colubroidea group, but are ecologically important for being large, globally distributed predators of diverse prey such as fish, alligators, lizards, birds and mammals. Reticulated pythons have been known to kill and eat large mammals including bears, pigs, deer, and sometimes adult humans. However, it is generally thought that the wide shoulders of humans prevents successful swallowing. An anatomical peculiarity of pythons and boas is the presence of tiny, simplified hind limbs embedded within the body wall. A small thigh bone supports a short, curved claw that projects from either side of the cloaca (the external opening for the gut and reproductive tract). Males use these claws (also known as "cloacal spurs") to stimulate females during mating, and may also use them in combat. **DN**

✦ NAVIGATOR

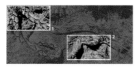

ISLAND COLONIES

The reticulated python is a supremely good colonist and inhabits numerous islands, large and small. It must have reached many of these islands either by swimming in the open sea or "rafting," which involves being carried on mats of vegetation or fallen trees that were swept from rivers out to sea, usually in flood conditions. Reticulated pythons are highly variable across their range, and members of some of the island-dwelling populations have evolved unusual traits. Those on Sulawesi are "island giants" (above) while those on other islands are "island dwarfs." Tales abound of pythons 50 feet (15 m) long, but accurate scientific measurements are rare. Even zoo specimens are difficult to measure because they may have been stretched unnaturally to get them into the record books.

👁 FOCAL POINTS

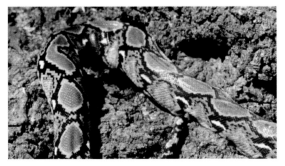

1 MUSCULATURE

The reticulated python has evolved to be heavy bodied with a very well-developed musculature. The snake grabs prey with its jaws before forming coils around the body to subdue the victim's movements. Gradually, with each out-breath, the snake tightens its grip so that the victim suffocates. This technique enables snakes to immobilize large prey items, and it is possible that the evolution of constriction occurred in step with the ability to swallow large prey.

2 HEAT PITS

Reticulated pythons possess heat-sensitive pits in the labial scales that line the edges of their jaws, like others in the python family. This is a form of infrared detection that enables the snake to sense warm-blooded prey and target vulnerable body parts. It has been suggested that heat pits are used to sense oncoming predators and for thermoregulation (body temperature control). They are also present in boas and some colubroids, such as pit vipers.

THE CRETACEOUS MASS EXTINCTION

1 Eliminated at the end-of-Cretaceous event, rudists were Jurassic and Cretaceous bivalve mollusks related to modern clams and oysters. Their varied shell shapes packed close together built up vast reefs, much as corals do today.

2 This image plots tiny variations in Earth's gravity around Mexico's Yucatán Peninsula (the coastline and the town of Chicxulub are marked in white). The lowest measurements, coded green and blue, are at the center. These indicate weak gravity due to looser, less dense impact debris and sediments that have replaced the denser, heavier, solid rock that was blasted away.

At the close of the Cretaceous period, 66 million years ago, a major extinction event occurred and a wide range of animals—famously nonbird dinosaurs, pterosaurs, and giant marine reptiles such as mosasaurs (see p.352) and plesiosaurs (see p.282)—disappeared forever. This extinction has been referred to by various names, including the end-of-Cretaceous event, the Cretaceous–Tertiary event ("Tertiary" is a name that is no longer used for the time span that now includes the Paleogene period) or the Cretaceous–Paleogene event. The geological symbols for the Cretaceous and Paleogene are "K" (for *kreta* or *creta*, chalk) and "Pg"; therefore, it is also referred to as the K–Pg or KPg event. Giant reptiles were not the only animals affected by the end-of-Cretaceous event; in fact, an estimated 75 percent of all species disappeared. Among these were an enormous number of invertebrates, including the shelled, swimming ammonites, the reef-building rudist mollusks (see image 1), and some major groups of plankton. What could have caused this catastrophic event? Many different ideas have been proposed, many of them without support from reliable evidence. However, the most popular ideas have centered around climate change, a surge of volcanism and its associated climatic effects, and the impact of an object from space.

Many large dinosaurs died out, such as *Tyrannosaurus* and *Triceratops* at Hell Creek (see p.368). However, other fossils, such as plankton, arguably provide better information on the climate and temperature of this time—even though it

KEY EVENTS

90 Mya	72 Mya	72 Mya	72–66 Mya	72–66 Mya	72–66 Mya
Ichthyosaurs (see p.244), fish-shaped marine reptiles have all but disappeared 30 million years before the end-of-Cretaceous mass extinction.	The Maastrichtian age begins; it is the final age of the Cretaceous.	Some dinosaur groups are already showing signs of decline, especially in North America, while others continue to flourish.	Plesiosaurs and mosasaurs become extinct some time during the Maastrichtian, together with the largest marine turtles.	At the beginning of the Maastrichtian only nine genera of ammonites remain. The end-of-Cretaceous mass extinction will extinguish them all.	In the Maastrichtian, the flying pterosaurs reach up to 50 feet (15 m) in wingspan, but none of these huge animals will survive the extinction.

is still puzzling. The dinosaur fossil record complicates the picture because, for example, some dinosaur groups in North America were already in decline before the event, while others, such as the Asian dinosaurs, were not. The fact that some groups were seemingly already in decline indicates that gradual changes associated with climate or habitat had been adversely impacting diversity for some time prior to the event itself. This could mean that dinosaurs were at a higher risk of being severely affected by an unusual event than they would otherwise have been. Indeed, dinosaurs and other reptiles suffered several declines during the Mesozoic era, but diversity always recovered. The fossil record of plankton and other nondinosaurian groups is similarly controversial. Many groups faded several million years prior to the event, while others were apparently flourishing just before it occurred.

Scientists have long speculated that an extraterrestrial event might have caused, or contributed to, the end-of-Cretaceous mass extinction. For many decades, the main problem with the theory was a lack of obvious evidence for an impact crater at a particular site. In fact, several sets of information had already been collected in the Yucatán area of Mexico—now known to be the impact site—but, at those times, they were not linked. Investigations during the 1950s had identified rock layers later seen as relevant to the crater's presence. In the 1960s and 1970s, during searches for oil-bearing rocks, geophysics data showed gravity anomalies—that is, variations in the pull of Earth's gravitational force (see image 2). These clues suggested some kind of craterlike remnants, publicly

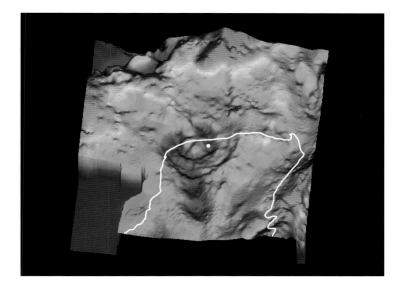

72–66 Mya	72–66 Mya	66 Mya	66 Mya	63 Mya	60 Mya
Marsupial mammals have proliferated into many different species. The extinction will drastically reduce the total compared to that of other mammals.	Crocodiles, lizards, iguanas, snakes, and slow-worms reach levels that, remarkably, are little affected by the event that destroys nonbird dinosaurs.	The end-of-Cretaceous mass extinction occurs, the end of the late Cretaceous, and with it the close of the Mesozoic era of "middle life."	The Danian age— first in the Paleocene epoch, in turn first in the Paleogene—and the Cenozoic era of "recent life" begin.	Already some mammal groups are showing rapid evolution to larger body sizes and more varied forms and lifestyles.	Mammals and birds are evolving. They spread rapidly to become the main land vertebrates in place of the nonbird dinosaurs.

announced in 1981, but again the significance was not realized. Meanwhile, in another line of evidence, unusually high concentrations of iridium, a platinum-group metal, were discovered in 1980 in a thin layer of sedimentary rock that had formed at the Cretaceous–Paleogene boundary in Italy, Denmark, and New Zealand. Iridium is rare on Earth but occurs in higher concentrations in meteorites and asteroids. Its presence seemed to indicate the impact of a giant extraterrestrial object, the particles of which were subsequently scattered across the world after it was blasted to pieces.

At last the clues were assembled, and in 1990 announcements were made of a crater of the right age and size, on and near the Yucatán Peninsula in Mexico. Named the Chicxulub crater (see image 4), after a town near its center, it preserves evidence that an enormous object—most likely an asteroid around 6 miles (10 km) in diameter—struck this region at the end of the Cretaceous. More recently, new computer programs that map gravity anomalies as three-dimensional images have allowed detailed simulations of the appearance of the crater. It is around 9 to 15 miles (15–25 km) deep, with a width of 110 miles (180 km), although ring-shaped gravity anomalies present around its main part suggest that it could be as much as 190 miles (300 km) wide.

It is estimated that the impact of an object this big would have emitted more than a billion times more energy than that caused by the detonation of the largest man-made nuclear explosion. An event of this magnitude was almost certainly sufficient to explain the extinctions of the time. Furthermore, dust thrown into the atmosphere would have reduced light levels and caused a cooling effect, resulting in the death of plants and the collapse of food chains: an "asteroid winter." Geological evidence shows that tsunamis caused by the impact crashed against adjacent shorelines and destroyed coastal environments. Theoretically, ash and similar ejecta from the blast, falling back to Earth, may have been hot enough to ignite forest fires and even animals themselves, but a lack of abundant charcoal in sediments of this age indicates that global fires were not prevalent, and the idea that the atmosphere was super-heated enough to kill animals has been challenged.

Obvious signs of the Chicxulub crater have become obscured by continental drift, local earth movements, erosion, and settling sediments, both on land and

in the sea. The area of the original impact now lies far below the visible surface. When an asteroid or similar body hits Earth, super-heated glasslike pellets are blasted into the atmosphere from the pressure and dispersed over a wide area. These are called tektites (see image 3), and tektites dated to the end of the Cretaceous have been discovered on Haiti and Mexico. Microfractured fragments of the mineral quartz also result from impact events, and "shocked quartz" has been discovered in numerous rocks dating to 66 million years ago.

Substantial debate has centered on whether the impact of the Chicxulub crater occurred at the right time to be linked reliably to the extinction event. Since 2009, several experts have argued that the age of the crater does not match the timing of the extinction event. This instigated a series of detailed studies, the newest of which shows that the crater is indeed the correct age: in terms of geological time, the impact and extinction event occurred almost simultaneously. However, the possibility remains that other events could have contributed to the extinction. The late Cretaceous was an extremely volcanic time. Active volcanoes erupted in western North America and the southeast Atlantic Ocean, and gargantuan outpourings of lava over the Deccan traps in India resulted in the extrusion of more than 240,000 cubic miles (1 million cu km) of basalt over an enormous area of land. The resultant climate change and acid rain seemingly made some habitats inhospitable to dinosaurs and other animals. However, numerous animal groups persisted after the mass extinction. It appears that small animals (see image 6) had a better chance of survival than larger creatures. On land, mammals and birds were the most conspicuous survivors, but certain amphibian, turtle (see image 5), lizard, snake, and crocodile groups also continued, as did vast numbers of fish and insects.

In 2010, a group of experts voted to agree that an asteroid impact caused the Chicxulub crater, and that this was a primary cause of the end-of-Cretaceous mass extinction event. However, evidence for additional impact craters has been reported from elsewhere in the world, including India and Ukraine. The Indian structure off the eastern coast, near Mumbai, has been named the Shiva crater and could have resulted from a strike by an even bigger object, up to 25 miles (40 km) wide. Given that impact events sometimes occur as showers, the possibility that multiple impacts occurred near-simultaneously with Chicxulub is plausible. However, firm evidence that these craters impacted Earth during the correct time span has yet to be confirmed. **DN**

3 Tektites are hardened, glassy remains of minerals instantly melted by an extraterrestrial impact and often hurled huge distances. Many can be linked to the K–Pg event.

4 The imagined immediate aftermath of the Chicxulub impact, partly on the Yucatán Peninsula and partly in the Gulf of Mexico.

5 Marine turtles, such as *Protostega*, fared well during the K–Pg event, with only about one-fifth of species lost.

6 With large predatory dinosaurs gone, mammal evolution produced hundreds of new forms in the few million years after the extinction, from squirrel-shaped multituberculate *Ptilodus* to huge oxlike herbivores.

Hell Creek Fossil Beds
CRETACEOUS PERIOD–PALEOGENE PERIOD

LOCATION Jordan, Montana
AGE 68 million to
65 million years ago
GROUPS See right
BED DEPTH 165 to 330 feet
(50–100 m)

At the end of the Cretaceous, a coastal floodplain occupied land that is now Montana, parts of North Dakota, South Dakota, and Wyoming. To the east was the Western Interior Seaway; to the west, flat plains led to the growing Rocky Mountains. The floodplains were interspersed with deltas, rivers, creeks, lakes, swamps, and dry bushland with clusters of small trees. Plants ranged from mosses and ferns to evergreens, such as cycads, ginkgoes, and conifers, as well as numerous flowers and trees, such as palms, magnolias, sycamores, laurels, and beeches.

Many creatures thrived within this rich tapestry of habitats: insects, clams, and the last ammonites; fish such as sharks, paddlefish, bowfins, and sturgeons; amphibians, including frogs and salamanders; many kinds of lizards, snakes, and turtles; and giant flying azhdarchid pterosaurs. Also present were various crocodylians and a number of dinosaurs, from giants such as *Tyrannosaurus* and *Triceratops*, groups of armored ankylosaurs, hadrosaurs (duck bills), pachycephalosaurs (thick heads), ornithomimids (ostrich dinosaurs), and predatory dromaeosaurs (raptors), to lesser-known dinosaurs such as hypsilophodonts. Many mammals lived here, too, and most of these were mouse- to rabbit-size, including multituberculates such as *Mesodema*, the opossumlike early marsupial *Didelphodon* and the intriguing early primate *Purgatorius*. Approximately 66 million years later, in 1902, celebrated fossil hunter Barnum Brown (1873–1963) and his team explored the region that is now Hell Creek, Montana. They found specimens from which *Tyrannosaurus* was named in 1905 by Henry Fairfield Osborn (1857–1935). Excavation continues today and Hell Creek is a world-renowned geological and paleontological site. The Hell Creek Project (1990–2010), a multi-institute program based at the Museum of the Rockies, Montana, formulated comprehensive databases and analyzed these immensely important finds. **DN**

FOCAL POINTS

GEOLOGY

The main rocks of the Hell Creek Formation, the geological entity laid down at the time, consist of clay, mudstone, and sandstone. They formed in the very late Cretaceous (Maastrichtian) and very early Paleogene (Danian). The rocks reveal a rich, continually changing mosaic of habitats, both freshwater and brackish (part salty), with an almost subtropical climate that lacked cold seasons.

SURFACE TOPOGRAPHY

Much of the Hell Creek area is steep, eroded bare rock where fossils can be seen exposed at the surface. There is local evidence of the iridium-rich layer noted around the world at the Cretaceous–Paleogene boundary—evidence of meteorite impact. It has been dated at close to 66 million years ago by various means, including the argon–argon method, fossils and magnetic features.

TYRANNOSAURUS

One surprise result from the Hell Creek fossil census is the number of *Tyrannosaurus* (see p.302) remains. In some sections these form a quarter of all fossils found. Usually, top predators such as *Tyrannosaurus* are much less common than their prey, but the large herbivorous hadrosaur *Edmontosaurus* has also been found in large quantities, and these results are still being studied.

TRICERATOPS

The most common large dinosaur fossils at Hell Creek are from *Triceratops* (see p.344). The remains show a gradual change over a period of 1 million to 2 million years. Specimens from the lower rocks have a long beak and a small nasal horn. Intermediate stages of evolution can then be seen, and the upper layers of rock show those with a shorter beak and a prominent nasal horn.

SCIENTIST PROFILE

1857–90

Born into a wealthy family in Fairfield, Connecticut, Henry Fairfield Osborn studied archeology and geology at Princeton University, under the guidance of paleontologist Edward Drinker Cope (1840–97). He studied in London and New York before returning to Princeton as an academic.

1891–1905

In 1891, he took dual posts of Columbia University professor of zoology and vertebrate paleontology curator at the American Museum of Natural History (AMNH). This began his long and distinguished paleontology career. As Hell Creek specimens arrived at AMNH, he realized their importance and named *Tyrannosaurus rex* in 1905 from Barnum Brown's finds.

1906–25

In 1908 Osborn became president of AMNH, and the following year president of the Zoological Society, holding both positions for a further 25 years. Predominantly interested in mammals, he arranged several expeditions to Mongolia during the 1920s to find early human remains, but succeeded in discovering a wealth of dinosaur and other fossils, such as *Protoceratops* and *Velociraptor* (see p.308), which he also named.

1926–35

Osborn continued to produce huge numbers of technical articles, popular books, and treatises on fossil elephants and the rhinoceroslike brontotheres. He championed the now-discredited idea that organisms progress toward a goal of being bigger and better, without reference to natural selection (orthogenesis). In paleontology he was a hyper-splitter: he named a new species for almost every variant among similar fossils, rather than seeing them as natural variations within one species.

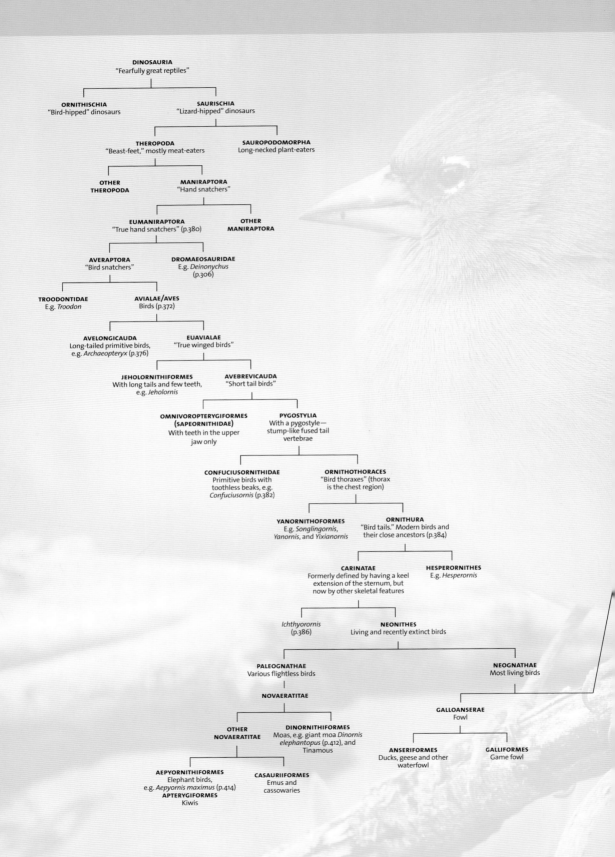

DINOSAURIA
"Fearfully great reptiles"

ORNITHISCHIA
"Bird-hipped" dinosaurs

SAURISCHIA
"Lizard-hipped" dinosaurs

THEROPODA
"Beast-feet," mostly meat-eaters

SAUROPODOMORPHA
Long-necked plant-eaters

OTHER THEROPODA

MANIRAPTORA
"Hand snatchers"

EUMANIRAPTORA
"True hand snatchers" (p.380)

OTHER MANIRAPTORA

AVERAPTORA
"Bird snatchers"

DROMAEOSAURIDAE
E.g. *Deinonychus*
(p.306)

TROODONTIDAE
E.g. *Troodon*

AVIALAE/AVES
Birds (p.372)

AVELONGICAUDA
Long-tailed primitive birds,
e.g. *Archaeopteryx* (p.376)

EUAVIALAE
"True winged birds"

JEHOLORNITHIFORMES
With long tails and few teeth,
e.g. *Jeholornis*

AVEBREVICAUDA
"Short tail birds"

**OMNIVOROPTERYGIFORMES
(SAPEORNITHIDAE)**
With teeth in the upper
jaw only

PYGOSTYLIA
With a pygostyle—
stump-like fused tail
vertebrae

CONFUCIUSORNITHIDAE
Primitive birds with
toothless beaks, e.g.
Confuciusornis (p.382)

ORNITHOTHORACES
"Bird thoraxes" (thorax
is the chest region)

YANORNITHOFORMES
E.g. *Songlingornis*,
Yanornis, and *Yixianornis*

ORNITHURA
"Bird tails." Modern birds and
their close ancestors (p.384)

CARINATAE
Formerly defined by having a keel
extension of the sternum, but
now by other skeletal features

HESPERORNITHES
E.g. *Hesperornis*

Ichthyorornis
(p.386)

NEONITHES
Living and recently extinct birds

PALEOGNATHAE
Various flightless birds

NEOGNATHAE
Most living birds

NOVAERATITAE

GALLOANSERAE
Fowl

**OTHER
NOVAERATITAE**

DINORNITHIFORMES
Moas, e.g. giant moa *Dinornis
elephantopus* (p.412), and
Tinamous

ANSERIFORMES
Ducks, geese and other
waterfowl

GALLIFORMES
Game fowl

AEPYORNITHIFORMES
Elephant birds,
e.g. *Aepyornis maximus* (p.414)

CASAURIIFORMES
Emus and
cassowaries

APTERYGIFORMES
Kiwis

6 | BIRDS

Feathers were long believed to be unique to birds, but in the 1990s the study of certain fossils revealed that they came directly from a dinosaurian ancestry, perhaps serving as insulation, or coloured and shaped for visual display—flight came later. The evolutionary connections are so intimate that birds are now regarded as living dinosaurs. The ways in which the 10,000 or so living bird species are related are discussed intensively, and new evidence from DNA and other sources is challenging traditional bird groupings.

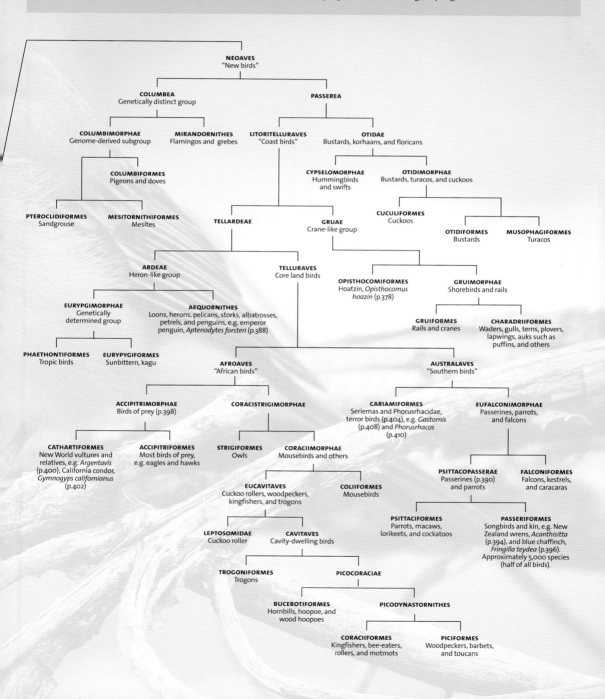

FROM DINOSAURS TO BIRDS

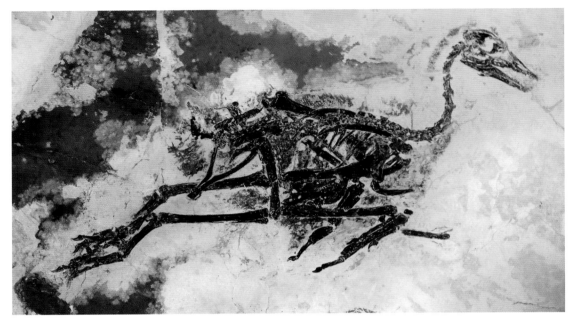

The origins of birds have been long debated. This is partly because bird fossils are generally rare: the creatures have thin, lightweight, often hollow bones that decay easily and also tend to be broken up by predators, and so are seldom preserved intact. However, since the late 1980s, evidence has accumulated that birds came from a group of small dinosaur meat-eaters whose arms, wrists, and fingers were "pre-adapted" to evolve into wings. The term "pre-adapted" refers to an existing feature that changes slightly to fulfill a new and different function. The new function is not predetermined, but evolution is given a head start when setting off in a new direction.

Together, the scientific studies of ornithology and paleontology have both contributed to the effort of charting the probable evolution of birds. Most famously, a link between birds and reptiles was made in 1861 with the discovery in limestone deposits in Bavaria, Germany, of several remarkably well-preserved specimens of a distinctly birdlike fossil, named *Archaeopteryx* (ancient wing; see image 3). Only 2 years before, in 1859, the same Bavarian limestone deposits had yielded another remarkable, birdlike dinosaur, named *Compsognathus* (see p.262). The two finds sparked renewed speculation about how dinosaurs and birds might be related, and since then the evolutionary path from reptile to bird has been extensively discussed.

KEY EVENTS

230–210 Mya	227–206 Mya	195 Mya	160 Mya	150 Mya	150 Mya
Dinosaurs give rise to the ornithischians (bird-hipped dinosaurs), and the saurischians (lizard-hipped dinosaurs).	Saurischians split into the mostly four-legged plant-eating sauropods and the bipedal (walking on two legs) meat-eating theropods.	Theropods give rise to various branches. One leads to *Tyrannosaurus* and its relatives. It is the eumaniraptoran branch that will lead to birds.	*Anchiornis* thrives in northeast China; it is usually regarded as a small, feathered maniraptoran dinosaur rather than a true bird.	*Rhamphorhynchus* (beak snout) is an early pterosaur, an extinct flying reptile closely related to dinosaurs.	The small, light, swift-running *Compsognathus* is a dinosaur with birdlike features.

Since the early 1990s, many important birdlike dinosaur fossils have been unearthed, most notably in China's northeastern Liaoning province, from rocks of the early Cretaceous period, between approximately 145 million and 100 million years ago. Local people and expert paleontologists soon realized their value, in both monetary and scientific terms, and a dinosaur fossil "gold rush" ensued. Many of those fossils have helped to fill in the gaps of possible evolutionary paths from dinosaur to bird, although the precise route is still the subject of much debate. Recent birdlike fossils found in China include the remarkable *Anchiornis huxleyi* (Huxley's near-bird, see image 1), a tiny dinosaur or dino-bird described in 2009 and named in honor of English biologist Thomas Henry Huxley (1825–95; see p.263). Dated at about 160 million years old, *A. huxleyi* is up to 10 million years older than the known finds of *Archaeopteryx*. It had well-developed wings and winglike feathers on its hind legs.

In the evolutionary history of birds, however, the most important fossils remain those of *Archaeopteryx*. This is one of the first known birds, or rather, transitional creatures between reptiles and birds. The first fossil of *Archaeopteryx* consisted of a single feather (see image 2) uncovered in 1860 in a limestone quarry near Solnhofen in Bavaria. The bird to which the feather belonged was named *A. lithographica*, in recognition of the fact that the fine-grained limestone in which the fossil was found was used for lithography, a printing process invented in 1798. The rock's tiny grain size helped to preserve the remains in extraordinary detail. The first more or less complete fossil, around 147 million years old, was discovered one year later, in 1861. This well-known fossil of *A. siemensii* was sold the following year to the British Museum, London, at the request of Richard Owen (1804–92), superintendent of the museum's natural history division, and it is now a star exhibit at London's Natural History Museum.

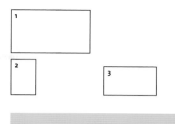

1 The exquisite fossils of *Anchiornis* reveal a creature that was less than 20 inches (50 cm) long—a tiny dinosaur within the nonavian rather than avian group.

2 The fossil feather of *Archaeopteryx* has a main shaft or quill (rachis) with vanes branching from it, similar to the flight feathers of modern birds.

3 Reconstructions of *Archaeopteryx* often show it chasing small creatures, snapping with its toothed jaws. This is in contrast to modern birds, which lack teeth.

150 Mya	130 Mya	125 Mya	124½ Mya	122 Mya	77 Mya
The first recognized birds, especially *Archaeopteryx*, are a mix of bird and reptile features, suggesting that birds had been long evolving.	*Sinosauropteryx* lives in China. Later fossil finds show clear evidence of feathers.	*Caudipteryx*, discovered in the Yixian Formation, China, is a flightless bird that combines reptilian and birdlike features.	*Mei long* (sleeping dragon), a tiny dinosaur, has features that place it close to the theoretical lineage of true birds.	The small dinosaur *Jinfengopteryx* grows feathers; classified as a bird in 2005, it is later reclassified as a maniraptoran dinosaur.	*Troodon* lives in North America. It gives its name to the troodontids: small eumaniraptoran dinosaurs from which birds may have evolved.

It was described by Owen in 1863 and is now considered the original "type" specimen of that species—the one to which the scientific name refers, and the one that is used as a standard for comparison with new finds. In the years immediately following the discovery, further fossilized remains of the same creature were found, including some complete skeletons that clearly exhibited features intermediate between reptiles and birds. One fine example, known as the Thermopolis *Archaeopteryx* (see p.376) is displayed at the Wyoming Dinosaur Center.

The fine-grained limestone beds in which the fossils were found were formed by a process of accumulation in the late Jurassic period, around 150 million years ago, most probably in a shallow tropical marine environment. When dead animals settled on the seabed, they were gradually covered by more and more limestone-rich sediment. Many other creatures, including ammonites, crinoids, crustaceans, fish, and insects, were beautifully preserved by the thin beds of fine limestone and shale material.

Although *Archaeopteryx* had feathers, which probably enabled gliding at least, it also had reptilian features. The skull (see image 4) had a long, tapering snout and small teeth, and the animal had a small pelvis, a long tail, and fingers with claws—this last characteristic it shares with the modern hoatzin (see p.378). In all these respects, *Archaeopteryx* was remarkably similar to many small meat-eating dinosaurs of the time. Similarities in anatomy between reptiles and birds had already been noticed, and *Archaeopteryx* lent tangible confirmation to close evolutionary relationships between the two groups. The animal had clearly reptilian features, but was also feathered over much of its body and its fore limbs were quite obviously wings. In addition to feathers, the creature also possessed several more birdlike features: a stout breastbone with flanges for muscle attachment; shoulder blades (scapulae) on the back, near the spinal column; a shoulder (pectoral) girdle with fused wishbone (furcula); somewhat fused metatarsal bones of the lower leg; and toes reduced to three pointing forward and a small opposing hind toe. When it became increasingly clear to biologists that *Archaeopteryx* was a bird, albeit a reptilelike bird, the schemes of classifying dinosaurs and birds were gradually changed. The process was aided by the use of cladistics (see p.18) and underscored by many other dinosaur and bird fossil finds in the following years. Deciding whether *Archaeopteryx* and other birdlike fossils are true birds is complicated by the difficulty of assigning labels to what is a gradual evolutionary process. *Archaeopteryx* and its more

recent cousins demonstrate beautifully the evolution of increasingly birdlike creatures toward the living birds of today. They validate the theory of Charles Darwin (1809–82) on evolution through the process of natural selection, while illustrating the difficulty of pinpointing where birds began.

A number of modern techniques have been used to extract increasing amounts of valuable information from *Archaeopteryx* fossils. Researchers using a CT (x-ray computed tomography) scanner have studied the skull of the Natural History Museum's specimen. The work revealed a three-dimensional brain structure and shape very similar to those of modern birds. *Archaeopteryx* fossils have also been investigated by a new technique using a giant x-ray machine at the European Synchrotron Radiation Facility in Grenoble, France. A narrow but very powerful x-ray beam was passed through each specimen to produce detailed three-dimensional images. These showed the anatomy much more clearly than ever before, including much sharper traces of feathers and even details of blood vessels within the bones.

Current consensus considers *Archaeopteryx* to be a bird, largely because of its skeletal features. Feathers are no longer the sole preserve of birds, since many dinosaurs that definitely were not birds possessed them (see p.310). Conversely, *Archaeopteryx* had many dinosaur features that are absent from modern birds. For that reason it is generally regarded as belonging to a transitional group—closely related to the dinosaurs that led eventually to modern birds, but probably not on the direct ancestral line to birds today.

The precise point, or points, in the postulated evolutionary tree leading back to the origins of birds are not universally agreed; some paleontologists believe that their true ancestry is yet to be uncovered. Nevertheless, the prevailing view is that birds evolved from a group of terrestrial dinosaurs known as the Theropoda (beast-feet). The nonbird theropods (meat-eating dinosaurs) lived from the late Triassic period, about 230 million years ago, through to the end of the Cretaceous, about 66 million years ago. The group contained such well-known and impressively large dinosaurs as *Tyrannosaurus* (see p.302) and *Spinosaurus* (see p.300), as well as many much smaller types, some of which have decidedly birdlike features, such as *Troodon* (see image 5). The great extinction event that marked the close of the Cretaceous (see p.364) saw the demise of about three-quarters of all plant and animal species, including the nonbird dinosaurs. However, the specific dinosaur lineage that led to modern birds survived the event to become a dominant feature of the fauna of much of the Earth today.

Within the theropod dinosaurs, it is probably the coelurosaurs (hollow-tailed lizard/reptiles) that led to modern birds. This fairly widespread assemblage includes the small *Compsognathus* and its kin, the great tyrannosaurs, the ostrich dinosaurs (ornithomimosaurs; see image 6) and the generally smaller, faster, more agile "hand-graspers" (eumaniraptorans). The last had long arms, a distinct breastbone, and a backward-pointing hip: features also seen in birds. Some eumaniraptorans showed birdlike adaptations of their lungs, as well as hollow, air-filled bones. In certain fossil finds there is even evidence of birdlike nesting behavior. The bones of their shoulder, wrist, and hand had shapes and joints that, with a few small evolutionary changes, would realign and become suited to a wing that moved up and down, but with a twisting motion to provide powered flight forward. This view of bird ancestry has been reinforced by systematic anatomical analysis to create family trees of relationships, regarded as more faithfully indicating true evolutionary pathways, compared to traditional classification based on arguably more subjective comparisons. More and more evidence from recent fossil finds, cladistically analyzed, tends to support this now-orthodox view of bird origins among the eumaniraptoran dinosaurs. **MW**

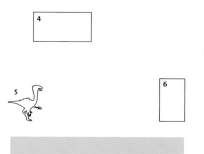

4 The skull of *Archaeopteryx*, reconstructed with the aid of computerized x-ray scans, demonstrates small sharp teeth, large eyes, and a rear casing that housed a brain similar to that of modern birds.

5 The small eumaniraptoran *Troodon* is often cited as similar to the small dinosaurs that evolved into birds.

6 Despite their general resemblance to birds, ornithomimosaurs, while in the coelurosaur group, are not especially closely related.

Archaeopteryx
LATE JURASSIC PERIOD

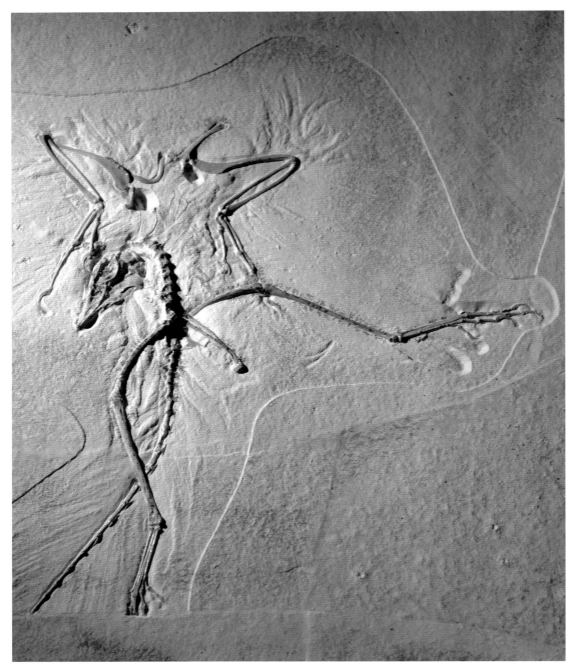

SPECIES *Archaeopteryx siemensii*
GROUP Archaeopterygidae
SIZE 20 inches (50 cm)
LOCATION Solnhofen, Germany

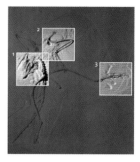

Of the dozen or so fossils of *Archaeopteryx* so far discovered, the one known to archeologists as the Thermopolis Archaeopteryx was discovered, like most of the others, in Bavaria, Germany. Officially described in 2005, it was owned privately in Switzerland before an anonymous buyer donated it to the Wyoming Dinosaur Center in Thermopolis. It is one of the finest and most complete of all the specimens of this early bird, and it is celebrated for its well-preserved head and feet. It shows that the second toe was hyperextensible and had a much greater range of movement than the other two forward-facing toes, a feature also seen in raptor-type dinosaurs, or dromaeosaurs, such as *Velociraptor* (see p.308) and *Deinonychus* (see p.306). The breastbone appears stout with flanges for the attachment of muscles. A coracoid bone, linking the shoulder blade to the breastbone, braces against the stresses of movement; this evolutionary feature persists in modern birds.

Two fossils that complement the Thermopolis find are the London Archaeopteryx and the Berlin Archaeopteryx. The London specimen is well-known as the first near-complete fossil of the bird and is also the type specimen of *A. siemensii*. The Berlin specimen, which was purchased by the Humboldt Natural History Museum in that city, is spectacularly well preserved and shows many details, notably of the head. **MW**

👁 FOCAL POINTS

1 SKULL AND TEETH
X-ray analysis has revealed that the skull contained a brain comparable in size, shape, and structure to that of modern birds. *Archaeopteryx* had the teeth of a reptile, small and sharp, indicating a carnivorous diet, chiefly insectivorous but possibly including small creatures such as worms.

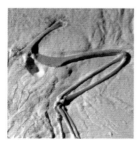

2 WINGS AND FEATHERS
The fore limbs are modified into wings and are well covered by feathers that include types very similar to those of living birds. It is unlikely that *Archaeopteryx* was capable of flying in a powerful, sustained way; it is more likely that it simply fluttered and glided over short distances.

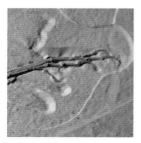

3 FEET AND TOES
The toes are intermediate between reptile and bird. There are three toes pointing forward; the single, albeit small, opposing hind toe is not fully reversed to enable perching, as in modern birds. Of the three toes pointing forward, the second toe is hyperextensible, as in raptor-type dinosaurs.

SLEEPING DRAGON

That nonbird dinosaurs sometimes behaved like birds is demonstrated dramatically by a feathered fossil found preserved almost perfectly in a bird-like pose: its head tucked under one wing or fore limb, and its legs folded under its body, as if asleep (above). This fossil, discovered in 2004 in Liaoning province, China, has been given the full name *Mei long*, meaning "sleeping dragon" in Mandarin. In fact, the rocks in which it was found indicate that far from sleeping peacefully, it may have been engulfed rapidly in volcanic ash, so was probably trying to protect itself. It may also have been poisoned by volcanic gases. *M. long* dates from 124½ million years ago and had a length of about 1½ feet (45 cm) and a wingspan of about 1 foot (30 cm). It is classified in the dinosaur family Troodontidae, very close to the lineage of the true birds.

Hoatzin
PLEISTOCNE EPOCH—PRESENT

SPECIES *Opisthocomus hoazin*
GROUP Opisthocomidae
SIZE 28 inches (70 cm)
LOCATION Northern South America
IUCN Least Concern

The hoatzin, *Opisthocomus hoazin*, is surely one of the strangest living birds, and its place in avian evolution has long puzzled scientists. Classified in its own family, the Opisthocomidae, it inhabits tropical swamps, mainly in the Amazon and Orinoco river basins of northern South America. The female lays two or three eggs in an untidy stick nest, usually above water, and the chicks hatch after about 4 weeks. All of these features stand out as unusual when compared to the other birdlife of its habitat.

Even stranger, and like *Archaeopteryx* and many ancient birds, the hoatzin chick has claws on its wings, which it uses to scramble about in branches (see panel). The adults continue to spend much of their time clambering and perching, even though they are capable of somewhat clumsy flight. Because of these claws, it was at first concluded that the hoatzin was a so-called living fossil with a direct link to a dinosaur ancestry. However, it is now thought that evolution has gone backward, and that the unusual feature is a relatively recent adaptation to the hoatzin's highly specialized lifestyle. Fossil hoatzins have come to light recently. *Hoazinavus lacustris* from Brazil, found in 2011, has been dated at 24 million to 22 million years ago; it is very similar to the living form, although rather smaller. At various times the hoatzin has been classified with pheasants and other game birds in the Galliformes, and with the cuckoos and relatives in the Cuculiformes. **MW**

👁 FOCAL POINTS

1 HEAD AND NECK
The adult hoatzin has a rather small head; the face is mainly bare of feathers and the bird has a large, spiky crest. The small head, long neck and long tail create a general shape that is unusual among birds, especially among species that live mainly in trees and require agility. The bird attains a length of 24 to 28 inches (60–70 cm). The oddity of the hoatzin is reinforced by the way they communicate: a mixture of hissing, grunting sounds and high-pitched barks.

2 CROP AND DIGESTION
The leaves and fruits of the hoatzin's diet are digested partly by fermentation; the smell of this accounts for its common name, stinkbird. Fermentation is common in the digestion of ruminant mammals but is unusual in a bird. The fermentation is slow, so the hoatzin must take in a lot of food, which makes it bulky and heavy. The food storage organ, the crop, is adjacent to the bird's main flight muscles, adversely affecting their efficiency.

CLAWED WINGS

If the adult hoatzin is one of the world's oddest birds, the chicks are even stranger. Naked when they first hatch, they very soon grow a covering of down. They do not stay in the nest for long. Instead, they clamber about on twigs and branches using a remarkable feature: claws on their wings. Each wing has two claws, at the wrist where the wing first bends (above). Clawed wings are highly reminiscent of the early ancestors of birds, including *Archaeopteryx*, which retained its wing claws into adulthood. The hoatzin chick loses its wing claws after about 2 months, when it can fly, if rather weakly. If alarmed, the chick climbs away or drops down into water below the nest before swimming back to grasp overhanging branches.

THE DEVELOPMENT OF FLIGHT

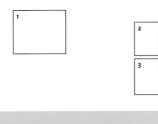

1 *Sinosauropteryx*, named in 1996, was the first nonbird dinosaur discovered with feathers. A relative of *Compsognathus* (see p.262), it was not on the evolutionary line to true birds.

2 Today's southern flying squirrel, *Glaucomys volans*, although a mammal, shows how true sustained flight may have arisen through a gliding stage.

3 This isolated fossil feather, a 50-million-year-old specimen from the Green River Formation, Wyoming, has a fluffy plumaceous or downy structure for insulation.

The thought of birds immediately conjures up images of flight because the majority of bird species alive today do fly, although there are examples of species that have since lost this ability, such as the ostrich, kiwi, and many birds of isolated oceanic islands, notably rails. Faced with the numerous fossils of bird ancestors, including flightless forms and those that definitely flew, a key question arises: how did flight first evolve? Several main theories attempt to answer this question.

Some early birdlike dinosaurs had sizeable feathers that gave them extra bounce and speed as they ran. The "ground-up" or cursorial (running) theory suggests that over time this enabled them to occasionally take to the air to glide for short distances. Such gliding flight could have been gradually interspersed with weak flapping to extend the time in the air. From this, it is not hard to imagine creatures becoming ever more adept at the flapping phase, resulting in an evolutionary path of increasing flying efficiency that led eventually to full, powered flight. This could perhaps have evolved in the ancestors of today's birds as a mechanism of escaping from terrestrial predators. Any adaptation that developed to reduce the pressure of being chased and caught is likely to have prospered, and even flapping or gliding for a short distance might well have given the protobird a better chance of survival.

KEY EVENTS

215 Mya	165 Mya	160 Mya	150 Mya	125 Mya	125 Mya
The first single-filament "protofeathers" possibly evolve. They have since been found in ornithischian (bird-hipped) dinosaurs of the late Jurassic period.	*Pedopenna* (foot feather), a small and early eumaniraptoran dinosaur with puzzling, highly feathered feet, inhabits China.	The small, feathered dino-bird *Anchiornis* has flight feathers on its fore and hind limbs, and feathered feet, though it is probably not a good flyer.	*Archaeopteryx* has well-developed flight feathers. It could certainly glide well, if not fly in a more sustained way.	Multiple-filament feathers, joined at the base to a central filament, are found on some dinosaurs related to *Tyrannosaurus* (see p.302).	Multiple downlike filament feathers are found on dinosaurs such as *Sinosauropteryx*.

There are many examples today of living bird species that have lost the ability to fly in situations where there are no native predators, such as on oceanic islands. Therefore the predation pressure theory is certainly plausible. Against this theory is the application of aerodynamic criteria to the wing area and likely body weight of a species such as *Archaeopteryx* (see p.376). It seems unlikely that *Archaeopteryx* at least would have been able to achieve sufficient speed to take off from level ground—although ground is seldom open and level.

Another theory is "pouncing" (proavis), in which the creature's leap onto its prey progressively became more efficient with the use of feathered fore limbs. The pounce gradually became a swoop and enabled the incipient flyer to change direction easily and even follow airborne prey such as flying insects. The "tree-down" (arboreal) theory assumes that the first flying birds used gravity by launching themselves from a height after clambering up to a sufficiently lofty vantage point—similar to how a flying squirrel takes off into a glide (see image 2). Although the feet of *Archaeopteryx* are not well suited to climbing, having a poorly developed hind toe, the claws on its wings could have been for this purpose. Again the evolutionary process would favor individuals with bigger, stronger feathered fore limbs, leading to better aerial abilities.

Many feathered creatures have been found, including *Confuciusornis* (see p.382) and nonbird dinosaurs. Feathered wings may well have evolved in more than one possible evolutionary line—notably among small meat-eating dinosaurs. One of the most remarkable of these was *Microraptor* from China, named in 2003 from specimens found in Liaoning province. *Microraptor* dates from about 120 million years ago in the early Cretaceous period and is regarded as a member of the eumaniraptoran group (see p.375). Not only were its front limbs feathered and winglike, but it also had small winglike groups of feathers on its hind legs, giving it an almost butterflylike appearance. Mainly tree-living, it certainly had enough wing area (on its legs as well as its arms) to allow for gliding, if not for a weak form of powered flight. Such dinosaurs give many insights into how birds evolved their form of true, sustained flight, and whether they did so only once, or several times in different early groups.

The origin of flight is linked closely to the origin of feathers (see image 3). These lightweight structures seem to have evolved several times in reptiles, especially in nonbird dinosaurs. Some simple plumelike feathers may have been for insulation (see image 1). Aerodynamic flight feathers—one of the prerequisites for active, powered flight—are very different. Flight also required other adaptations. Bones became partly hollow and thus retained rigidity and strength while being much lighter; the breastbone developed a distinct keel to serve as a firm anchor point for the strong muscles that powered the wings; and the dinosaurian snout, with its true teeth, evolved into the strong yet lightweight, horny toothless bill of true birds. These developments reduced the overall weight of the body, thus increasing the ability to become airborne. **MW**

125 Mya	124 Mya	124–120 Mya	120 Mya	100 Mya	70 Mya
In China, *Yutyrannus*, an early cousin of *Tyrannosaurus*, is one of the largest feathered creatures known, at 30 feet (9 m) in length.	Vaned feathers become more widespread, notably in oviraptor dinosaurs and among many kinds of birds.	The fish-eating bird *Yanornis* from China has powerful flight muscles and is probably one of the best flyers of its time.	Asymmetrical vaned feathers, well designed for flight, are by now found in many fossil birds from different groups.	Multiple feather types occur, including flight feathers and more conservative body feathers, typical of those on living birds.	The biggest-ever flying animals live in North America and Europe. They are not birds, but azhdarchid pterosaurs with wingspans of more than 33 feet (10 m).

Confuciusornis
EARLY CRETACEOUS PERIOD

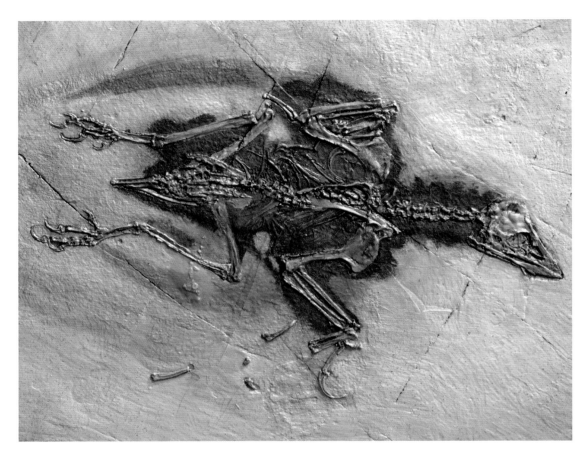

SPECIES *Confuciusornis sanctus*
GROUP Confusciusornithidae
SIZE 15 inches (37 cm)
LOCATION China

◆ NAVIGATOR

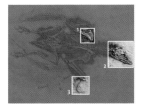

First described in 1995, *Confuciusornis* (Confucius bird), named after the famed Chinese philosopher, is yet another important fossil genus from Liaoning province, northeast China. Since its discovery, several hundred well-preserved specimens have been unearthed. This crow-size early reptile-bird lived about 125 million years ago and is represented by a number of species, notably *C. sanctus*, *C. dui*, and *C. jianchangensis*. *Confuciusornis* had a head-body length of 15 inches (37 cm) and a wingspan of 24 to 28 inches (60–70 cm). Its bill resembled that of many modern birds, being sharply pointed and lacking teeth. Several specimens show well-preserved feathers, with long primary (wing-tip) and rather short secondary (inner-wing) feathers, comparable with those of living birds. Although the body plan and feathered wings suggest flight, the structure of the shoulder blades implies that it could not have raised its wings in an upstroke, as in modern flying birds. It was probably capable of only weak powered flight, interspersed with short periods of gliding. Perhaps *Confuciusornis* used its wing claws to clamber up trees, then launched itself into the air, to flap and glide between trunks and branches. The tail was very short, with the tail bones much reduced in number and compressed into a blunt stump: the pygostyle—again very like those of modern birds. Some fossil specimens have preserved signs of two, long, streamerlike tail feathers, which trailed behind the birds in flight. Other specimens lack these ribbonlike streamers. They could have been present in one sex only, probably the male, as a display of readiness to mate. **MW**

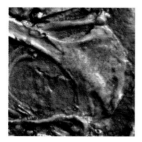

1 WEAK FLIGHT
The structure of the shoulder blades suggests that *C. sanctus* was not capable of the upstroke necessary for sustained active flight, even though its wings were well feathered. It also had a rather small breastbone, implying that its wing muscles were not very powerful.

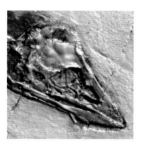

2 MODERN BILL
C. sanctus had a bill very like those of living birds, lacking teeth and sharply pointed. It seems suited to a generalized diet of plant matter and also small animal prey. Some specimens have been found with the remains of fish in their stomachs, giving an indication of their habits and habitat.

3 WINGS WITH CLAWS
In addition to well-developed long, narrow wings, *C. sanctus* had finger claws. Similar to those of *Archaeopteryx*, they consisted of one large thumb claw and two smaller ones on fingers two and three of the hand. It probably used these claws to help it clamber up into trees.

⊥ CONFUCIUSORNIS RELATIONSHIPS

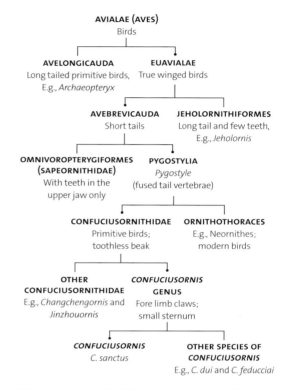

AVIALAE (AVES)
Birds

AVELONGICAUDA
Long tailed primitive birds,
E.g., *Archaeopteryx*

EUAVIALAE
True winged birds

AVEBREVICAUDA
Short tails

JEHOLORNITHIFORMES
Long tail and few teeth,
E.g., *Jeholornis*

OMNIVOROPTERYGIFORMES (SAPEORNITHIDAE)
With teeth in the upper jaw only

PYGOSTYLIA
Pygostyle
(fused tail vertebrae)

CONFUCIUSORNITHIDAE
Primitive birds;
toothless beak

ORNITHOTHORACES
E.g., Neornithes;
modern birds

OTHER CONFUCIUSORNITHIDAE
E.g., *Changchengornis* and *Jinzhouornis*

CONFUCIUSORNIS GENUS
Fore limb claws;
small sternum

CONFUCIUSORNIS
C. sanctus

OTHER SPECIES OF CONFUCIUSORNIS
E.g., *C. dui* and *C. feducciai*

▲ Birds can be grouped in different ways; here, the teeth and tails are used. Another system focuses on the keel, carina (on the breastbone). Most systems include Neornithes, all living birds.

FORGED ARCHAEORAPTOR

The study of evolution has occasionally been misled by forged fossils, an occurrence that is particularly common for birds, especially with recent finds from China. In some cases, forged specimens were created for profit by combining parts of separate fossils into composite forms. The most well-known, possibly inadvertent, forgery was "Archaeoraptor" (right), which was initially considered to be a missing link, rather like *Archaeopteryx*. However, it was later found to have been created from parts of *Microraptor zhaoianus* and *Yanornis martini*. Such specimens have the unfortunate effect of discrediting serious scientific study in the eyes of skeptics, but they should not detract from the immense value of genuine fossil finds.

BIRDS DIVERSIFY

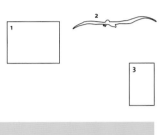

1 *Hesperornis*, which grew to 6 feet (1.8 m) tall, is generally reconstructed in an almost upright, penguinlike posture, with feet set well back on the body.

2 Remains of the huge pseudotooth *Pelagornis*, the largest flying bird known, were discovered while constructing a new airport at Charleston, South Carolina.

3 The great auk, *Pinguinus impennis*, of the North Atlantic, highly adapted to marine life, was extinct by the 1850s due to human hunting and specimen collection.

Bird diversity began to accelerate during the early Cretaceous period, 145 million to 100 million years ago, and it has led to there being 10,000 plus species around the world today. Early birds probably evolved from the small meat-eating dinosaurs known as maniraptorans (see p.304), and fossils show that some of these, as well as certain other dinosaurs, possessed feathers. Early feathers had a simple filamentous structure, but certain birdlike fossils from the middle Jurassic period had quill-like feathers on the arm, which is suggestive of an early winglike limb. Some maniraptorans had vaned feathers similar to those of flying birds and also an alula (bastard wing), a feature that is important in preventing stalling during flight. Further transitions for bird evolution involved the loss of the long lizardlike series of tail bones, which was reduced to little more than a stump carrying lightweight tail feathers, and changes in shoulders and fore limbs, which allowed vertical movements of the wings during flapping flight. The breastbone developed a distinct keel as a firm anchor point for the powerful muscles required for flight. Whereas the early birdlike reptiles had jaws with embedded teeth, later fossil forms replaced these with a much lighter and more birdlike horny beak.

Of the many important recent fossil reptile/bird finds from China, one of the more significant is *Aurornis* (dawn bird). This fossil is provisionally dated at 160 million years old, toward the end of the Jurassic, and may therefore predate

KEY EVENTS

145–100 Mya	125 Mya	84–78 Mya	80 Mya	70 Mya	66 Mya
Bird diversity begins: some birds take to flight, others live on the ground, and some adopt a life on, or even in, water.	*Iberomesornis* is a sparrow-size member of the Enantiornithes, which includes many birds of the Cretaceous.	*Hesperornis* is a flightless marine bird of North America and northern Asia, similar to the large diver or loon of today.	*Ichthyornis* has strongly developed wings. It is thought to be very close to the evolutionary branch leading to living birds.	*Gobipteryx* (now extinct) leaves fossils that show that its young are well developed and able to walk and run when they hatch.	The mass extinction at the end of the Cretaceous signals the demise of toothed birds. Modern groups of toothless birds spread.

Archaeopteryx (see p.376). *Aurornis* was rather chickenlike, with a length of around 20 inches (50 cm) and a covering of feathers, including on its long, winglike arms. Its teeth were small and sharp, its tail relatively long. Although it may not have used powered flight, it is possible that *Aurornis* glided from tree to tree. It may represent an evolutionary stage similar to *Archaeopteryx*, but in a parallel group that was more central to the origins of the bird family tree.

Several kinds of Cretaceous birds took to life by or in the sea, preying on fish and similar food, although they did not belong to modern seabird groups such as the Procellariiformes (albatrosses and petrels). Some evolved to swim and hunt in the water rather than fly above and dive down for prey. *Hesperornis* (see image 1) was clearly adapted for swimming underwater, in the manner of modern divers, and like divers, it was probably very ungainly on land, able only to shuffle along. It had a streamlined body with strong legs set far back, a long neck and a long, tapering bill. The wings were very small so *Hesperornis* was probably flightless, a secondary adaptation to an extreme aquatic lifestyle. The extinct great auk (see image 3) was also flightless (although the living auk species can fly), as is today's emperor penguin (see p.388). In contrast, *Ichthyornis* (fish bird; see p.386) resembled living seabirds, such as terns and gulls. Early birds such as *Hesperornis* and *Ichthyornis* are thought to lie close to the evolutionary line leading to modern birds (Neornithes), the only branch of the dinosaur-bird evolutionary tree that has persisted. The Neornithes include all the birds alive today and a few that went extinct recently, some of which were quite remarkable creatures—notably the terror birds (see p.404), the elephant birds (see p.412), and the moas (see p.414).

One of the most spectacular of the relatively recent fossil finds is *Pelagornis* (see image 2). Like the giant birds of prey, it had a very large wingspan, estimated at more than 20 feet (6 m). It was the largest known flying bird, probably close to the maximum size possible for flight. The several known species of *Pelagornis* were in a group known as pseudotooth birds, which refers to the toothlike sharp bony pegs that lined their long jaws. They probably soared over long distances in the manner of albatrosses, using rising wind deflected from the waves of the sea. *Pelagornis* may have fed by trailing its lower jaw into the surface water to catch shoaling fish or squid, as skimmers do today. The earliest birds had true teeth and later birds lost these. The pseudoteeth of these specialist fish-eaters seem to have served a similar function to true teeth, and are comparable to the horny projections in the bills of living mergansers, known as sawbills. **MW**

62–58 Mya	55 Mya	35 Mya	35 Mya	25 Mya	3 Mya
An early type of penguin, *Waimanu*, evolves in New Zealand. It is up to 3 feet (1 m) tall and well-suited to life in water.	Bird groups diversify with the appearance of the earliest song birds, parrots, loons, swifts, and woodpeckers.	By this time, the ancestors of most of the bird orders recognizable today are in existence.	*Colymboides* is an early member of the bird group called the true Gaviiformes, which contains the divers and loons.	The seabird *Pelagornis* has an enormous wingspan, more than twice the width of today's record holder: the wandering albatross.	*Aptenodytes ridgeni* (Ridgen's penguin) is about 3 feet (1 m) tall. It is a close relative of the living emperor and king penguins.

Ichthyornis
LATE CRETACEOUS PERIOD

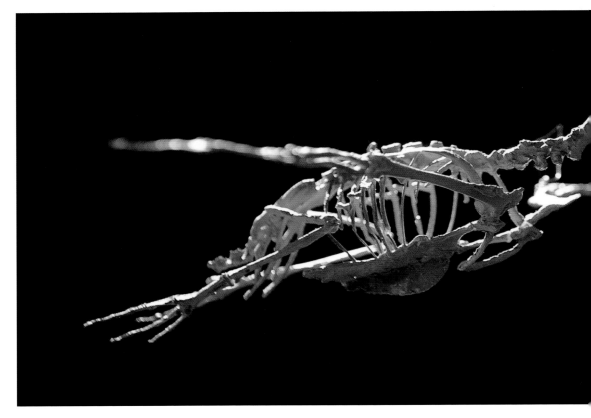

SPECIES *Ichthyornis dispar*
GROUP Ichthyornithidae
SIZE 1 foot (30 cm)
LOCATION North America

The first *Ichthyornis* fossils were discovered by American geologist Benjamin Franklin Mudge (1817–79) in the 1870s in Kansas, and date from approximately 80 million years ago, during the late Cretaceous. Further fossils have been found in Alabama, California, and Texas. Current thinking places these fossils firmly within the defining group Aves (birds) and, within that, very close to the branch Neornithes, which leads to modern birds.

On average about the size of a pigeon, *Ichthyornis* is immediately recognizable as a bird, and indeed more resembles a waterbird or seabird. Its most prominent features are its long, straight jaws, which contain a set of true teeth embedded in the middle of the upper and lower jawbones. These teeth were not serrated but projected backward and would have helped *Ichthyornis* to catch its prey—probably fish and other aquatic animals—by enabling it to keep hold of slippery victims. *Ichthyornis* had a wingspan of about 2 feet (60 cm). Its legs were very short, and its feet may well have been webbed. It probably looked and behaved similar to a modern small gull or skimmer, plucking its prey from the water or perhaps plunging into the water to catch fish swimming close to the surface. Its wings and breastbone are very similar to those of living birds, and the deep keel would have provided a firm anchor for quite powerful flight muscles. An interesting point is that although *Ichthyornis* was discovered approximately a decade after *Archaeopteryx*, the latter was not fully described until 1884, whereas *Ichthyornis* appeared in scientific literature during the early 1870s. Consequently, in around 1880, it was *Ichthyornis* and the related *Hesperornis* that Charles Darwin (1809–82) considered gave substantial support to his theory of evolution. **MW**

◆ NAVIGATOR

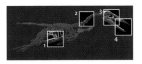

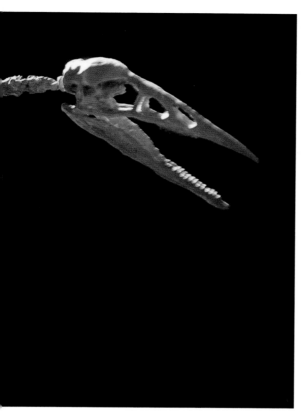

▲ The first skeletal restorations of *Ichthyornis* were made from the original specimen, or holotype, found by Mudge, along with other fragments assumed to be from the same creature, including the lower jaw, rear skull, and much of the fore limb or wing.

👁 FOCAL POINTS

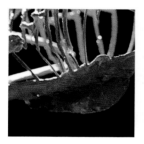

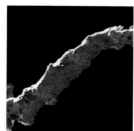

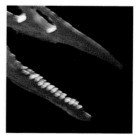

1 BREASTBONE
The breastbone was large and had a very deep keel, which is the down-pointing, flangelike area that served as the attachment and anchorage site for the main muscles that enabled flight. The keel is also characteristic of modern birds, an indication that *Ichthyornis* was capable of sustained, powered flight.

2 NECK BONES
The cervical (neck) vertebrae of *Ichthyornis*'s long, flexible neck are unusual in shape. The vertebrae are concave at the front and at the back, similar to those found in fish. This explains the name "fish bird" given by renowned US fossil collector and paleontologist Othniel Charles Marsh (1831–99) in 1872.

3 JAWS
The long, straight jaws of *Ichthyornis* contain true teeth embedded in the bone. This feature would be more expected in a reptile than in what is obviously a bird. The teeth slant backward, giving *Ichthyornis* a powerful grip on its aquatic prey and helping to move it down into the gullet with each successive bite.

4 BILL
The tip of the bill (beak) was similar to that seen in some living seabirds, such as albatrosses and petrels. It consisted of a number of quite distinctive plates fused together. However, this may be an example of parallel evolution, because *Ichthyornis* is not on the lineage leading to modern birds.

Emperor Penguin
PLEISTOCENE EPOCH—PRESENT

E volution has produced a wide range of body types and lifestyles among flightless birds, both extinct and living. One of the most specialized of all living bird groups is the penguin family (Spheniscidae), whose wings are reduced to small flippers, without flight feathers. Penguins use these flippers to power through the water in what is almost underwater flight. Largest of the penguins is the majestic emperor penguin, *Aptenodytes forsteri*, found only on Antarctica and nearby islands. Remarkably, emperor penguins breed during the Antarctic winter, when conditions are among the harshest on the planet. The penguins huddle together in breeding colonies numbering hundreds or thousands of birds. When the single egg is laid, it is the male who broods it, resting it on his feet to keep it clear of the ice and covering it in a fold of skin and feathers. He persists, without feeding, for about 2 months, while the female goes out to sea to forage. It is difficult to understand how evolution led to breeding in such an extreme environment. Perhaps, long ago, when conditions were more equable, penguins had to contend with predators and therefore sought out more remote places for their rookeries. As ice ages came and went, the penguins were pushed to more severe places. It is an example of an evolutionary race: penguins developed cold-coping adaptations that enabled them to survive, while most predators adapted to other prey or succumbed to extinction. **MW**

SPECIES *Aptenodytes forsteri*
GROUP Spheniscidae
SIZE 48 inches (122 cm)
LOCATION Antarctica
IUCN Near Threatened

👁 FOCAL POINTS

1 PACKED FEATHERS
Emperor penguin insulation is provided by three layers of short, densely packed feathers covering the entire body surface, supplemented by a layer of fatty blubber up to 1¼ inches (3 cm) thick beneath the skin. Similar adaptations are seen in many cold-water species, from whales to seals.

2 CONTRASTING PLUMAGE
The plumage colors of the back and belly are highly contrasting: the back is very dark, while the underside is white, with a pale yellow breast. This provides camouflage from predators such as leopard seals, especially when at sea: from below, the birds are hard to spot against the water's bright surface.

3 SPECIALIZED WINGS
The fore limbs of the emperor penguin have evolved to become quite unlike flight wings: they have been reduced to stiff, leathery paddles. Using these, emperor penguins can reach maximum speeds of 14 miles per hour (22 km/h), with a typical sustained speed of about 7 miles per hour (11 km/h).

▲ Penguins have a rich fossil record compared to other birds. Their bones are robust and preserve well as they do not need to be light for flight. This skeleton of the banded penguin, *Spheniscus urbinai*, from South America, dates to 5 million years ago.

PERCHING BIRDS

1 The vegetarian finch, *Camarhynchus crassirostris,* is one of the well-known Galapagos (Darwin's) finches, members of the perching bird family Thraupidae.

2 The typical passerine has its first toe facing backward, with the next three directed forward; tendons and ligaments give a tight grip on a perch with little muscle effort.

3 *Resoviaornis jamrozi,* named after its fossil site and discoverer, Polish paleontologist Albin Jamróz, gives clues to passerine evolution.

O f the roughly 10,500 species of bird alive today, by far the majority— about 5,700 species—belong to the group called Passeriformes, the passerines (perching birds). Many birds outside the group have feet that enable them to perch, but in the passerines the anatomy of the feet is particularly well adapted for gripping tightly when perching on a branch or twig, for example. The key feature of the passerine's foot is that the hind toe is very strong and flexible (see image 2). Working with the three forward-pointing toes, it grasps tightly and secures a firm grip, even when the bird is asleep.

The diversity of living perching birds is apparent from their classification into no fewer than 80 different families. A number of features account for their great success. They include, in addition to the specially adapted foot, the fact that most are small and agile; their brains are large and many exhibit intelligent

KEY EVENTS

60–55 Mya	29 Mya	7¼–5¾ Mya	5¼–2½ Mya	2–3 Mya	1 Mya
Estimating from typical species origin rates, the first perching birds probably appear at this time, but clear fossil evidence is yet to appear.	*Resoviaornis jamrozi* hunts for insects and other food on the ground in Poland.	The ancestor of the Hawaiian honeycreepers group, a member of the Eurasian rosefinches, reaches the Hawaiian archipelago.	Perching birds diversify rapidly and spread to most habitats on most continents.	The ancestors of the Galapagos finches, also known as Darwin's finches, reach the Galapagos Islands from mainland South America.	The long-legged bunting, *Emberiza alcoveri,* is heading for extinction at this time.

behavior; they possess a sophisticated vocal system that allows for efficient communication by means of complex calls and songs; and they learn quickly, which enables them to adapt swiftly to changed conditions.

Perching birds are divided into two main branches: the suboscines and the oscines. A third branch, containing only the New Zealand wrens, such as the rifleman (*Acanthisitta chloris*, see p.394), is thought to be an early offshoot of the passerine evolutionary tree, and these tiny birds are probably the most primitive of all living perching birds. The majority of passerine families and species belong to the oscines, and these are commonly referred to as songbirds. The oscines have a distinctively complex vocal organ or "voice box" (equivalent to the human larynx) called the syrinx, while the vocal organ of the suboscines is less complex. The syrinx contains a membrane, the tension of which can be altered by muscles, thus changing the pitch of the sounds made by the bird. Air is forced past this membrane, and, by manipulating this special mechanism, many songbirds produce amazingly complicated calls and songs with apparent ease and efficiency. Slowed-down recordings of bird songs often reveal highly complex sounds, some of which are inaudible to the human ear when heard naturally. The songbird families are very diverse, ranging from swallows, larks, and shrikes to thrushes, warblers, finches (see image 1), and buntings. All possess the same elaborate vocal system, although some birds, such as the crows and their relatives, utter simple calls, many of which sound harsh to the human ear.

How did this huge diversity of fascinating and complicated birds evolve? Until recently, not much was known about their origins and lineages. However, detailed studies of comparative anatomy, combined with some revealing fossil finds and techniques such as DNA hybridization, have helped to shed more light on the subject. The fossil record is quite scant for the usual reasons related to birds. First, their very thin, light, and mostly hollow bones soon decay or are shattered by scavengers. Second, many perchers of the past were small, just as they are now, and it is likely that smaller species are less well represented in the fossil record because they decay at a faster rate. Third, being small, they are more likely subjected to predation before preservation becomes possible. That said, several studies have shown that extinction rates for nonpasserine, mostly larger, birds are generally much higher than those for passerines.

The earliest known perching birds date from around 55 million years ago, at the start of the Eocene epoch, although they may have appeared as long ago as 80 million years, in the late Cretaceous period. One early passerine fossil is that of *Resoviaornis jamrozi* (see image 3). The species was discovered in Poland in a shale outcrop near the city of Rzeszów (Resovia in Latin) in 2013, and is dated at some 29 million years old, from the early

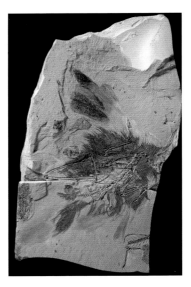

Oligocene epoch. The second, specific part of the scientific name honors the finder's surname, Jamróz. About the size of a blue tit (*Cyanistes caeruleus*), *Resoviaornis* had a thin bill, suggesting that it was mainly insectivorous. The fact that its legs were quite long indicates that it may have spent much of its time feeding on the ground. Interestingly, *R. jamrozi* has characteristics of both branches of the passerines: the suboscines and the oscines. Another Polish fossil find, also from the early Oligocene, is *Jamna szybiaki* (see image 4), named after the village of Jamna in Poland and the surname of its finder, Szybiak. With feathers on the wing and a clearly visible tail, this fossil is more complete than that of *R. jamrozi*. The bird's anatomy suggests that it lived in scrub or forest, feeding on insects and possibly fruits.

Such fossil finds imply that the passerines first evolved in the late Paleocene epoch, between 60 million and 56 million years ago, in the huge southern continent known as Gondwana, from which Australasia eventually separated. It is therefore significant that Australia now contains many of what are considered the most primitive of the living passerines: birds that are perhaps closest to the original ancestral stock. These include the previously mentioned New Zealand wrens, and songbirds such as bowerbirds, fairy wrens, honeyeaters, and lyrebirds (see image 5). The two living species of lyrebird are well-known for their marvelously varied songs and calls, and their amazing ability to mimic not only the calls and songs of other bird species, but also many other sounds, including the vehicles and chainsaws of foresters.

The passerines flourished and diversified, especially during the Pliocene epoch, between 5½ million and 2½ million years ago, and ever since they have remained prominent in a diverse array of Earth's habitats. Comparisons conducted among the living families indicate that the ancestors of the whole group may have been related to the modern order of Coraciiformes (kingfishers, bee-eaters, and their relatives) or to the Piciformes (woodpeckers, toucans, and their relatives), although the picture is far from clear. Fossils of long-extinct birds give clues as to how species became extinct through natural processes that predate human interference. The factors involved are varied and include environmental changes and pressure of competition from other animals.

Since the arrival of the human species, however, the pressures on birds have increased exponentially. Direct human exploitation, perhaps for their meat, eggs, or plumage, is one cause of their decline, but probably more important is the indirect pressure exerted by human destruction of their habitats. Many birds, including perching birds, have been driven to extinction in recent history, and such extinctions continue today. For example, fossils of the long-legged bunting, *Emberiza alcoveri*, were found in a volcanic cave on Tenerife in the Canary Islands in 1994. This bunting was flightless, like the Stephens Island wren of New Zealand, and it probably fell victim to a combination of habitat loss and introduced predators. The fossils date from the middle Pleistocene epoch to the early Holocene epoch, from 1 million years ago to possibly as recent as 11,000 years ago.

Among the most endangered perching birds are those surviving on isolated oceanic islands, where very often they have evolved in the absence of major predators, but where predators have since been introduced. Islands that have seen recent passerine extinctions include New Zealand's Chatham Island, the Hawaiian archipelago, and Australia's Norfolk Island. For example, the Chatham bellbird, *Anthornis melanocephala*, was last recorded in 1906 and is now presumed extinct. It belonged to the honeyeater family, Meliphagidae, and lived in forests on Chatham, Mangere, and Little Mangere islands. Again, the honeycreepers of Hawaii, which belong to the finch family, Fringillidae, are small perching birds endemic to their islands and many have gone extinct in historical times. For example, the Kona grosbeak, *Chloridops kona*, was last seen in 1894, and the Hawaii mamo (*Drepanis pacifica*, see image 6) and black mamo, *D. funerea*, were last seen in 1898 and 1907 respectively. The Hawaiian honeyeaters, of the family Mohoidae, comprise five species that went extinct between 1859 and 1987. These fine songbirds mainly inhabited forest and succumbed to a combination of habitat loss, predation by introduced species, notably black rats, and disease from introduced mosquitoes.

The Hawaiian honeycreepers are of major evolutionary interest because they demonstrate a well-known example of rapid adaptive radiation—similar to the case of Darwin's finches on the Galapagos Islands, but even more diverse. Adaptive radiation is said to occur when one kind of living thing, or a small group, evolves rapidly and adapts quickly to many different environments and surroundings, becoming a number of separate species in the process. (Another example is that of the blue chaffinch, *Fringilla teydea*, of the Canary Islands, see p.396.) The remarkable honeycreepers probably evolved from a common ancestor and adapted to a range of different habitats and behavior patterns on the Hawaiian Islands, in the absence of serious competition from other birds. Originally there were at least 56 species, but only 18 survive to give a hint of their diversification; of these, several are endangered, and some are on the brink of extinction. In addition to differences in plumage, the various species show a wide range of bill sizes and shapes, adapted to suit different diets.

Recent analysis of the DNA of Hawaiian honeycreepers, which make up the finch subfamily Drepanidinae, has revealed something about the origins of the various species alive today. Perhaps surprisingly, this recent research shows that an ancestor of the rosefinches of Europe and Asia was the closest relative to all the living Hawaiian honeycreepers. This rosefinch ancestor may have reached Hawaii during an irruption, when large numbers of birds fly long distances after suffering food shortages. Only a small proportion of such birds would have needed to reach Hawaii to found the dynasty of the endemic honeycreepers. The researchers estimated that the ancestral rosefinch colonists reached Hawaii between 7¼ million and 5¾ million years ago, indicating just how rapidly speciation can proceed, given favorable conditions. **MW**

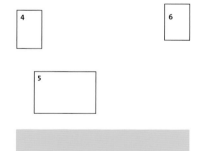

4 Approximately 30 million years ago *Jamna szybiaki*, discovered by Polish fossil expert Robert Szybiak, was an efficient flyer with toes adapted to scraping in earth for grubs and worms.

5 The superb lyrebird, *Menura novaehollandiae*, is a member of an ancient passerine group dating back at least 16 million years.

6 The Hawaii mamo, *Drepanis pacifica*, succumbed not only to habitat loss and predation by and competition from introduced species, but also to collecting of its startlingly bright feathers, both by local people and European settlers.

New Zealand Wrens
PLEISTOCENE EPOCH—PRESENT

SPECIES *Acanthisitta chloris*
GROUP Acanthisittidae
SIZE 3 inches (8 cm)
LOCATION New Zealand

A mong the plethora of living passerines, one family with few species offers special insight into passerine evolution. These are the New Zealand wrens, so distinctive that they are classified in their own family: the Acanthisittidae. The New Zealand wrens have proved so difficult to classify within the two main branches of the passerines—the ocines and subocines—that some taxonomical authorities suggest they deserve their own group of equal status, termed the suborder Acanthisitti. It is thought that the wrens have evolved on the isolated islands of New Zealand for many millions of years, and like the flightless kiwis, kakapo (night parrot), and other species there, they gradually reduced their dependence on flight, without entirely losing the power to fly.

Probably splitting off very early in the passerine evolutionary timeline, the New Zealand wrens are today represented by only two, or possibly three, species: the rifleman, *Acanthisitta chloris*; the South Island wren or rock wren, *Xenicus gilviventris*; and the bush wren, *Xenicus longipes*. The rifleman has a large range across both islands of New Zealand and is common locally, although it is declining mainly due to habitat destruction. The South Island wren, however, has now disappeared from North Island and is also decreasing, partly through direct predation and losses of eggs to introduced stoats and mice. The bush wren is thought to have gone extinct, probably also due to predation by introduced carnivores. It was last recorded in 1972. The rifleman is New Zealand's smallest bird, and one of the most charming. It weighs a mere one-fifth of an ounce (6 g) and is only about 3 inches long (8 cm). Found mainly in mature forests, especially at high altitudes, these wrens are active birds, foraging on tree trunks and in the canopy, seeking out insects and other small invertebrates. **MW**

 NAVIGATOR

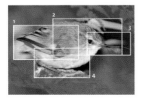

👁 FOCAL POINTS

1 WINGS AND FLIGHT
The rifleman has short, stubby wings proportional to its tiny body. It is far from an expert flyer, having a small breastbone and weak muscles for flapping its wings. Rather than making long, sustained flights, it tends to whirr or hum in short bursts through the canopy or undergrowth using fast wingbeats, or hop and jump rapidly among twigs, stems, and stones.

2 WRENLIKE HEAD
The rifleman and its New Zealand cousins are called wrens because they are very wrenlike in their overall size and appearance, especially in the proportions of the head and bill. However, this is a result of convergent evolution (see p.19). They are not closely related to the true wrens of the family Troglodytidae, which are almost all American apart from the common wren of Eurasia and Africa.

3 SLENDER BILL
Like their true wren namesakes, New Zealand wrens have a relatively long, slender bill with a slightly upturned tip. The bill is adapted for poking into tree bark and rock crevices, and picking out small insects and similar items with maximum efficiency. The slender bill is also well suited for probing into leaf litter on the forest floor and rapidly flicking leaves aside to expose food beneath.

4 PLUMAGE COLORS
The bird's common name of "rifleman" is a reference to the similarity of the bird's plumage to the colors of the uniform of a rifleman in one of New Zealand's regiments in colonial times. The male has bright green feathers on its back, while the green of the female ranges from speckled olive to olive-brown. The underside plumage of both sexes is a mixture of gray and white.

THE LAST FLIGHTLESS WREN

The Stephens Island wren, *Traversia lyalli* (right) may have been the only flightless passerine of recent history. Having lived throughout New Zealand, it eventually survived only on Stephens Island, at the northern end of South Island. This outpost was untouched by humans or their introduced animals until workmen came to build a lighthouse in the 1890s. In 1894, the assistant lighthouse keeper's cat caught at least one of the remaining birds, and specimens were sent to naturalist Walter Rothschild (1868–1937) in England; he described and named the species, choosing *lyalli* for the assistant lighthouse keeper, David Lyall. By 1895, cats had caught all remaining birds; it was a graphic demonstration of how specialized species are very vulnerable to introduced predators.

Blue Chaffinch
HOLOCENE EPOCH—PRESENT

SPECIES *Fringilla teydea*
GROUP Fringillinae
SIZE 6¾ inches (17 cm)
LOCATION Canary Islands
IUCN Near Threatened

Evolution has produced some remarkable birds on isolated islands, and many of these are closely adapted to a particular specialist habitat. This is true of the blue chaffinch, *Fringilla teydea*, of the Canary Islands. Classed as Near Threatened, this finch is restricted to forests of Canary pine (*Pinus canariensis*; see panel) on the islands of Tenerife and Gran Canaria. Here, it breeds at altitudes of between 3,000 and 6,000 feet (1,000–2,000 m). This chaffinch usually builds its nest toward the end of a branch, hidden by bunches of long pine needles. The female is a dull gray and olive-brown color, but the male lives up to its name, with a bright, smoky-blue head, back, and underparts, and a black tail.

There are two distinct subspecies of blue chaffinch: *Fringilla teydea teydea* on Tenerife, and *Fringilla teydea polatzeki* on Gran Canaria. Although the islands are separated by only a short sea crossing, reproductive isolation and adaptive radiation have led to the evolution of two distinct races. The Gran Canaria subspecies is 10 percent smaller and has a slightly smaller bill and duller plumage. It is classed as Highly Threatened and there are only around 250 individuals left. In common with many other island birds, the blue chaffinch relies upon a habitat that is itself under threat—from felling for timber and from the fires that periodically devastate areas of forest, especially in the summer months on the mid-altitude slopes of Tenerife's volcanic mountains. An outbreak of forest fires could seriously reduce the population of *F. teydea*, which is currently thought to stand at 2,000 pairs, mostly on Tenerife. Various measures have been undertaken to protect this finch, including a ban on hunting and trapping (which is mainly for the illegal cage-bird trade), captive breeding programs, and the restoration of fire-damaged forest. **MW**

✧ NAVIGATOR

👁 FOCAL POINTS

1 BILL TIP
Although strong, the bill of the blue chaffinch narrows steeply to a fine tip, allowing the bird to forage for insects in the cracks and fissures of the pine bark. The adults feed mainly on pine seeds, but they also consume beetles, moths, and other invertebrates, and the nestlings are fed a range of items, including caterpillars.

2 BILL
The bill of this finch is strong and sturdy, and markedly heavier than that of its very common close relative: the Eurasian chaffinch (*Fringilla coelebs*). This is an adaptation that enables the blue chaffinch to extract seeds from the tough cones of the Canary pine, a coniferous tree restricted to the Canary Islands.

3 COLOR
The unusual color of the male blue chaffinch can appear quite gaudy. Perhaps the islands' lack of native mammalian predators has allowed the evolution of this conspicuous plumage. In fact, the blue-gray hue lends a subtle camouflage when the bird is seen against the sky, or among the blue-green foliage of the Canary pine.

CANARY PINE

The Canary pine, *Pinus canariensis* (below), grows only in the Canary Islands, and the remaining large stands are restricted to the islands of Tenerife, Gran Canaria, and La Palma. The ecology of this magnificent tree is inextricably linked to the blue chaffinch, and is the result of a long period of isolated co-evolution. On Tenerife, fine forests dominated by this pine encircle the central peak, Mount Teide. The tallest tree on the islands, it can reach up to 195 feet (60 m), and its straight trunks are part of the reason why so much forest has been destroyed for timber. Its unusual features include blue-green foliage, very long needles, and sturdy cones that often remain closed for several years. While the volcanic soils are porous and dry, and the pine very drought-tolerant, the clusters of long, fine needles trap moisture from the damp prevailing winds and channel water down to the soil below.

◀ A male blue chaffinch perches on a branch of Canary pine. The tight clusters of long needles are typical of this endemic conifer, and the finch is highly dependent on the tree as its main habitat.

BIRDS OF PREY

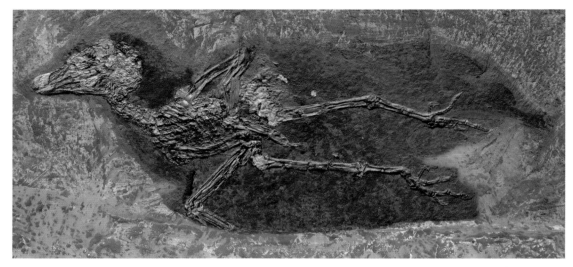

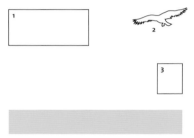

1 *Masillaraptor parvunguis* represents one of the earliest birds of prey, with an unusually long bill and relatively small talons—"parva" means small or little, and "unguis" means a claw, nail, or hoof.

2 *Teratornis merriami* resembled the two living species of condor, though its bill was significantly larger.

3 The American harpy eagle, *Harpia harpyja*, is adapted to swoop onto the rain forest canopy and seize monkeys and sloths from the branches.

Birds of prey are not common in the fossil record. This is somewhat surprising, given their success and diversity in the modern age—about 290 species survive in varied forms, including vultures, eagles, hawks, kites, harriers, and falcons. Many of the birds of prey known only from fossils were very large species, having eaglelike and condorlike shapes, and they include some of the biggest of all flying birds. The fossil finds are from the relatively recent past, in paleontological terms, mostly from the Miocene epoch (23 million to 5 million years ago).

The ancestry of the whole group of raptors, or birds of prey as they are commonly called, remains unclear. One of the oldest fossils so far discovered is a skull and part of the spinal column, found in Germany, of *Masillaraptor parvunguis* (see image 1), which lived in the middle Eocene epoch 45 million to 40 million years ago. The species, a small bird of prey with a hooked beak, large eyes and long legs, shows similarities to living falcons. Its existence indicates that the lineage leading to modern falcons may already have been established. In contrast, some extinct birds of prey were huge, especially those belonging to the Teratornithidae (teratorn) family. These monster raptors lived in both North and South America from the Miocene to the Pleistocene epochs, from 20 million to only a few thousand years ago, and they were the largest birds of prey ever known. Of these, the best-known species is *Teratornis merriami* (see image 2). This vulturelike raptor had a wingspan of about 12 feet (3.6 m) and was notably larger than today's California condor (*Gymnogyps californianus*, see p.402); it went extinct about 10,000 years ago. The related species *T. woodburnensis* was

KEY EVENTS

60 Mya	45 Mya	36 Mya	20–1 Mya	12 Mya	10 Mya
Fossils of owls such as *Berruornis* and *Ogygoptynx* show that this group had already split from day-flying birds of prey and from the nightjars.	*Masillaraptor parvunguis*, a small falconlike species, is one of the earliest known raptorial birds of prey, known from fossils in Germany.	Fossils suggest that the first members of the eagle group split from the kites, becoming sea predators of fish and squid.	Teratorns thrive through this long period of prehistory. The huge birds of prey are vulturelike and live in North and South America.	A preserved part of a wing bone is suspected to be oldest known remnant of an early true eagle, the Bullock eagle, named after an Australian creek.	The falcon genus *Falco* is present by this time in North America. It later diversifies to include more than 35 species.

even larger. Its fossil bones suggest that it had a wingspan of 14 feet (4.5 m) or more; fossils date this species to between 12,000 and 11,000 years ago. However, the largest-ever bird of prey, and one of the biggest-ever flying birds, was *Argentavis* (see p.400), another vulturelike teratorn from the Miocene.

Another species was *Aiolornis incredibilis*, the biggest flying bird known to have lived in North America. *A. incredibilis* had a wingspan of about 16 feet (5 m) and, with its very powerful bill, was almost certainly an active hunter. It survived longer than *Argentavis*, and the fossil finds, mostly from the west and southwest of the United States, date from the early Pliocene to the late Pleistocene epochs, 4 million to 12,000 years ago. Some paleontologists have wondered whether the legendary "thunderbird" mentioned in Native American traditions is a reference to this giant bird of prey. It is not impossible that some of the teratorns could have been contemporary with early human societies, and because of this they may have entered folk memory. Thunderbirds featured in the culture and art of certain Native American tribes; carved effigies of them, for example, were often set at the top of totem poles.

An even more recent extinction was New Zealand's Haast's eagle (*Harpagornis moorei*). The largest eagle ever known, this mighty predator probably lived in forested habitats. In common with living forest eagles, its wings were broad but relatively short, giving it good maneuverability around trees. Its anatomy chimes well with that of the largest forest eagles of today: the harpy eagle (*Harpia harpyja*, see image 3) of South and Central America and the Philippine eagle, *Pithecophaga jeffreyi*, found only on a few islands in the Philippines. Haast's eagle may have been driven to extinction partly through the loss of its chief prey, the moas, and also by deforestation brought about by the actions of early human settlers. The enormous bird is thought to have evolved from smaller eagles, becoming larger as it gradually adapted to feed on big prey. In fact, some moa skeletons bear puncture marks that correspond exactly to the Haast's eagle's talons. Humans arrived in New Zealand around 1280, and Haast's eagle probably went extinct around 1400, possibly as late as 1600 (see p.401).

Did some of the recent giant birds of prey, which overlapped with early people, actually include humans in their diet? Studies of bones collected from beneath the nests of the African crowned eagle, *Stephanoaetus coronatus*—one of the continent's most powerful eagles—show this to be quite possible. The bones, including those of mangabey monkeys, show signs of eagle attack. Mangabeys are too heavy for the eagles to carry, so the eagles rip them apart before transporting them to the nest. Skulls of mangabeys caught by the eagles' talons show a characteristic pattern of damage. Similar damage was found in the skull of a juvenile human cousin, *Australopithecus africanus* (see p.530), the so-called Taung Child, from about 2¾ million years ago, strongly suggesting that it could well have been killed by a large bird of prey. **MW**

8–6 Mya	4 Mya–2,000 Ya	2¾ Mya	12–11,000 Ya	10,000 Ya	600 Ya
The giant teratorn *Argentavis* is the largest flying bird of prey so far discovered, with a wingspan perhaps exceeding 23 feet (7 m).	The giant condor, *Aiolornis incredibilis*, is the biggest known flying bird in North America, with a wingspan of about 16 feet (5 m).	The Taung Child, a juvenile early humanoid, has skull puncture marks consistent with attack by a large bird of prey.	The North American *Teratornis woodburnensis* has a wingspan of 14 feet (4.5 m). Mastodons share its environment.	The best-known teratorn, *Teratornis merriami*, is well known from La Brea Tar Pits of Los Angeles; it soon goes extinct.	Haast's eagle, *Harpagornis moorei*, is thought to go extinct, although some records suggest it may have survived until about 400 years ago.

Argentavis
MIOCENE EPOCH

One of the largest flying birds so far discovered, *Argentavis magnificens* (magnificent Argentine bird) was a true giant. This relative of modern condors, sometimes referred to as the giant teratorn, belonged to the teratorn group of the Miocene, between 12 million and 5 million years ago. Fossils of this extinct species were first found in Argentina in 1980. The skeletal remains reveal a splendid raptor with a wingspan of perhaps 23 feet (7 m) or more. The huge wings would have enabled the immense bird to fly, even though its body must have weighed as much as 220 pounds (100 kg). While it is possible that even larger flying birds of prey may have existed, calculations and computer models show that *Argentavis* was close to the physical limits of flight, so their increase in size would not have been great. The giant teratorn would have behaved like today's condors, relying mainly on gliding and soaring on updrafts of air. Its breastbone was relatively small, implying that its flight muscles were not powerful enough for sustained flapping. Take-off would have been a mid-air launch from a cliff or other high perch, using headwinds where possible. Like the condors of today, the bird probably fed largely on carrion rather than hunted live prey. In South America the top carnivores of the Miocene were large flightless terror birds such as *Phorusrhacos* (see p.410), and this species may well have descended to scavenge on the prey of those grounded avian killers. **MW**

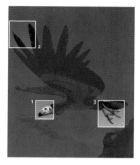

SPECIES *Argentavis magnificens*
GROUP Teratornithidae
SIZE 4 feet (120 cm)
LOCATION Argentina

👁 FOCAL POINTS

1 SKULL
The skull was more than 22 inches (55 cm) long and 6 inches (15 cm) wide. The large beak was sharply hooked at the tip, an indication of the bird's carnivorous diet. The predator may have eaten small rodents, common on the plains of Argentina in the Miocene, in addition to carrion.

2 HUGE WINGS
Some specialists calculate that *Argentavis* had a wingspan of as much as 27 feet (8 m). That makes it one of the largest known flying birds, and the upper arm bone alone was about the length of an entire human arm. Significantly larger birds could exist but it would be impossible for them to fly.

3 LEGS AND FEET
The legs of *Argentavis* were long and powerful, suggesting that the bird was adapted for walking as well as flying. The feet were also large and strong. Perhaps, when there were no upcurrents for flight, the bird strode around on the ground searching for carrion and small mammals and reptiles.

THE LARGEST EAGLE

Remains of the extinct Haast's eagle, *Harpagornis moorei*, were first discovered in New Zealand in 1871 by German geologist Julius Haast (1822–87); dozens more skeletons have been found since. It is the largest eagle known to have lived, with a wingspan of 8 feet (2.5 m), possibly more, and weighing as much as 33 pounds (15 kg). Like today's eagles, it had a sharply hooked beak, short and powerful legs, and sharp talons up to 3½ inches (9 cm) in length. Haast's eagle was an active predator, feeding on relatively large prey, including flightless birds up to the size of moas (above). The species only went extinct quite recently, possibly surviving to the 1600s; it almost certainly existed when the first people arrived in New Zealand 800 years ago.

California Condor

PLEISTOCENE EPOCH—PRESENT

Two living species of condor exist: the Andean, *Vultur gryphus*, and the Californian, *Gymnogyps californianus*. The Andean condor is still relatively common in parts of the Andes, but the California condor is critically endangered. The latter bird was rescued from extinction through a controversial and expensive conservation program involving captive breeding. By 1987, the remaining 27 wild birds had been caught. This was followed by carefully monitored release into suitable habitats, notably along the coast of California. By the middle 2010s the numbers had risen to more than 430, of which about half were living back in the wild, in California, Baja (Mexico), and Arizona.

Both of the surviving condor species are eerily reminiscent of their gigantic prehistoric forebears and cousins, such as the teratorns. California condors have evolved the largest wing area of any living bird. They feed mainly on carrion, gliding and soaring for long periods on their long, broad wings, making full use of updrafts and thermals to conserve energy as they scan the ground below for food. Condors are closely related to American vultures but not to the Old World vultures of Europe and Asia, despite superficial similarities (see panel). They live for more than 50 years and feed mainly on animal carcasses, ranging from small mammals, such as rabbits, to larger prey, such as deer and in the case of the California condor, stranded marine mammals. **MW**

◉ NAVIGATOR

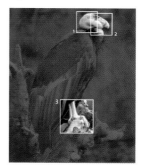

SPECIES *Gymnogyps californianus*
GROUP Cathartidae
SIZE 4 feet (1.2 m)
LOCATION California, Mexico, and Arizona
IUCN Critically Endangered

◉ FOCAL POINTS

1 BARE HEAD

The California condor's head is almost entirely featherless, and this is an adaptation to feeding on carrion. Blood or flesh from rotting prey might otherwise adhere to the bird's feathers, attract flies, and possibly cause infection or disease. Without feathers, the head and neck are kept clean more easily.

2 HOOKED BILL

The sharp, hook-ended bill is a highly effective tool for tearing at large prey items and cutting them into smaller chunks for swallowing. This shape is seen in many similarly adapted birds whose prey is naturally large and too big to swallow whole. The powerful beak completes the reduction process.

3 LEGS AND FEET

The California condor's feet are equipped with short claws, a relatively poorly developed hind toe, and an elongated middle toe. Unlike most birds of prey, condors do not use their claws to catch prey. In fact, they are not even particularly adept at perching and supporting their weight.

RESEARCH INTO RELATIVES

Research, partly using DNA, has shed new light on the relationships of condors and their living relatives. Condors are classified in the family Cathartidae, along with the five species of New World vultures, including the small turkey vulture (above). However, in an example of classification evolving as new information arrives, recent findings have suggested that the condors share a closer ancestor with storks and other birds of the order Ciconiiformes. This is partly because New World vultures hunt using a keen sense of smell, and the condors do not. Old World vultures, on the other hand, are thought to comprise two separate lineages: one including bearded and Egyptian vultures; the other, griffon, black vultures, and relatives.

GIANT AND TERROR BIRDS

1 A reconstruction of North American *Titanis*, 8 feet (2.5 m) tall, twice the weight of a human adult, with a huge beak and tiny wings, portrays the heavyweight size and power of the phorusrhacids or terror birds.

2 Even for phorusrhacids, *Kelenken* had an enormous bill, rivaling that of today's pelicans, but far more deadly.

3 The bill of *Andalgalornis* suggests it was adapted as a hatchet or ax to stab and chop victims.

Fossils of several types of giant birds have been discovered around the world, and some of these birds persisted until only a few hundred years ago. The extinction of nonavian dinosaurs 66 million years ago opened up opportunities for other land animals, and some birds took on the role of massive flightless animals within their local fauna. Some, with their huge, hooked beaks and clawed feet, were definitely predators; others used their heavy, blunt beaks to browse on vegetation and crush large seeds and fruit before eating them. Some of these giants have been dubbed "terror birds." This dramatic name refers to a remarkable group, mainly in the family Phorusrhacidae, of mostly very large, flightless, carnivorous birds that lived mainly in South America between about 62 million and 2 million years ago. Reconstructions of these impressive birds, with their stocky, powerful builds and huge, deeply hooked beaks, suggest that an encounter with them would indeed have been terrifying. In build and size they resembled an ostrich of today, but their thick necks, massive heads and oversize beaks seem disproportionately huge.

Most fossil finds of terror birds have been in Argentina, Brazil, and Uruguay, and it seems that the creatures probably flourished across much of South

America. *Phorusrhacos* (see p.410) is the genus that gives its name to the family. The species *P. longissimus* was described in 1887 from a fossil of the bird's remarkably large jawbone, unearthed in Patagonia, Argentina, by Florentino Ameghino (1854–1911), a largely self-taught naturalist and paleontologist who was much influenced by Charles Darwin (1809–82).

Although most of the fossils are from South America, one of the larger forms, *Titanis* (see image 1), is known from finds in Texas and Florida. Fossils of *Titanis* date from around 5 million to 1¾ million years ago, from the Pliocene epoch to the early Pleistocene epoch. It seems that *Titanis* reached North America from South America during the Great American Interchange 3 million years ago, when volcanic and tectonic activity created a land bridge between the two continents. This formed dry corridors along which fauna of the two formerly separate landmasses were able to travel and mix. The oldest fossils of *Titanis*, however, predate this time, which leads to speculation that the animal possibly migrated north via a series of islands before the continents made full terrestrial contact. Standing 8 feet (2.5 m) tall, *Titanis* is thought to have been an agile, fast-running carnivore. Its prey is likely to have been mainly small mammals such as rodents, plus reptiles and possibly also carrion. Another terror bird genus, *Gastornis* (see p.408), previously known as *Diatryma*, belongs to a different, but possibly related, family and is known from finds in North America, Europe, and China.

Some terror birds were even more fearsome creatures than these: larger and probably capable of running at high speed on their long, powerful legs. Many species had a huge, hooked bill and sharp claws, which they would have used to catch and rip up their prey. One example from Argentina, named *Kelenken* (see image 2) in 2007, had the largest skull and bill of any known bird. This terror bird stood around 10 feet (3 m) tall and had a skull more than 28 inches (70 cm) long, including its 18-inch (45 cm) bill. Although only this skull and a lower leg bone have been found, comparison with other terror bird skeletons has enabled an approximate reconstruction, revealing a giant, flightless avian predator. About 15 million years ago it roamed the pampas (grasslands) of Argentina, pursuing a diet of mainly reptiles and small mammals.

Another phorusrhacid, *Brontornis*, is known mainly from leg bones and skulls discovered in Santa Cruz province, Argentina. This mighty creature is the heaviest known terror bird, weighing up to 1,100 pounds (500 kg), and one of the tallest at more than 9 feet (2.8 m). It would have been a top predator and capable of catching and killing large prey. It probably hunted using a technique of surprise ambush, pouncing from the cover of forest or scrub before subduing and killing its victim with its huge, hooked bill and heavy, sharp talons. Another terror bird, *Andalgalornis steulleti* (see image 3), known from fossils in deposits from the late Miocene and early Pliocene epochs in

5 Mya	5–1¾ Mya	2 Mya–500 Ya	40,000 Ya	400 Ya	300 Ya
The terror bird *Andalgalornis steulleti* is yet another fearsome carnivorous bird. It inhabited Argentina at this time.	The terror bird *Titanis*, up to 8 feet (2.5 m) tall, stalks Florida.	Moas evolve and spread around New Zealand. These giant herbivores forage in forests but are extinct before the arrival of European settlers.	The large flightless bird *Genyornis*, a close relative of *Dromornis*, is thought to have persisted in Australia until around this time.	This date is one of the later estimates of the approximate time of the last moa in New Zealand.	The great elephant bird, *Aepyornis*, may have become extinct in Madagascar at this time.

Argentina, has provided much useful information about how these fierce predators may have killed their prey. Although the bird was relatively small, a mere 4½ feet (1.4 m) tall, its bill was massive, and the upper jawbone was deeply hooked at its tip. The skull was heavy and inflexible, though rather narrow. It is thought that this species, and many other terror birds, used its hooked bill in a vertical motion, bringing it sharply down in the manner of an ax to stab deeply into its prey.

Another terror bird known from fossils discovered in Argentina is *Mesembriornis*. Although relatively modest in size compared to the true giants of its family, it is one of the last known to have gone extinct. It probably lived from approximately 10 million to 2 million years ago. About 5 feet (1.5 m) tall, *Mesembriornis* seems to have been a swift runner, hunting down its prey in open country rather than using the ambush method of its slower, bulkier relatives. Terror birds seem to have been particularly prominent in the savanna habitats of South America between about 25 million and 2½ million years ago, when they were among the top carnivores of their day. What brought about their final demise was probably competition from large incoming carnivorous mammals, notably species of cats and dogs crossing south from North America during the Great American Interchange. Until their arrival, the terror birds had reigned supreme and unchallenged as the dominant carnivores of the South American mainland. Climate change may also have played a role in their disappearance. The Andes mountains were thrusting upward, and the young mountain ranges were creating rain shadows, resulting in drier conditions in those areas.

In the relatively recent past there were hunters with even bigger beaks. Stirton's thunderbird (*Dromornis stirtoni*, see image 5) was a huge bird known from late Miocene fossils found in Australia's Northern Territory. It stood at more than 10 feet (3 m) tall and weighed up to 1,100 pounds (500 kg). As in other flightless birds, the breastbone was not keeled and the wings were very short. The legs were extremely sturdy and the toes strong and hooflike. However, the most notable feature was the beak, which resembled that of a giant parrot—very deep and strong, and a giant nutcracker in effect. In the absence of further evidence it is not known whether it used its beak to slice and rip through plant stems and crack open tough seeds, or to tear into animal flesh.

Closely related to *Dromornis*, and also classified in the Dromornithidae family, was *Genyornis*. This large, flightless bird was about 6½ feet (2 m) tall and also lived in Australia until 40,000 years ago. Through the dating of its egg fragments, researchers concluded that it went extinct over a short period of time, probably as a result of human activity, through hunting and habitat loss. Aboriginal Australian rock paintings 40,000 years old depict what was probably this species. Surprisingly, some of the most impressive and frightening of giant birds, and some of the biggest known kinds, survived until relatively recently. Among the latest to evolve, and also the last to go extinct, were the moas (see p.412) of New Zealand and the great elephant birds (see p.414) of Madagascar. Both forms took advantage of a lack of large herbivorous mammals on their isolated island homes to assume the role of major herbivore in their environment, and both eventually succumbed to human hunting. Some idea of what these giants may have looked like, and how they may have behaved, may be gained by comparing them with the large flightless birds still living today: the ostrich (see image 4), rhea, emu, and cassowary (see image 6).

Although superficially similar to emus and cassowaries, the moas are now thought be more closely related to the tinamous of South America, though they are usually placed in their own separate group, the order Dinornithiformes. The giant moas, two species in the genus *Dinornis*, were the tallest of all birds known to have existed; at up to 12 feet (3.6 m), they were taller even than the living ostrich, which attains about 9 feet (2.8 m). Moas laid only one or two eggs per clutch, and these were often collected for food by humans—one probable reason for their demise. Interestingly, most, if not all, of the giant moa bones collected from human camps have been shown, by genetic analysis, to have been females. This may have been because the moas were mostly caught when they were foraging in the forests, while the smaller males, probably incubating the eggs, were less easily found. Today's cassowaries and kiwis show a similar discrepancy in size between the sexes, and in these birds it is the smaller male that does most of the egg incubation, as the male moas very likely did. **MW**

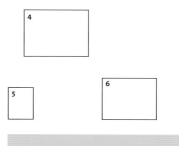

4 The largest living bird, the common ostrich, *Struthio camelus*, reaches a maximum height of 9 feet (2.8 m), but it is half the weight of the largest moas and one-third the weight of elephant birds.

5 Stirton's thunderbird, *Dromornis stirtoni*, was an herbivore with a nonhooked "nutcracker" bill.

6 Three species of wattled, helmeted cassowaries, *Casuarius*, have evolved in New Guinea and Australia. Their probable closest relative is another tall flightless bird from the region: the emu, *Dromaius novaehollandiae*.

Gastornis
PALEOCENE EPOCH—EOCENE EPOCH

SPECIES *Gastornis giganteus*
GROUP Gastornithidae
SIZE 6½ feet (2 m)
LOCATION France, North America, and China

One of the first and largest birds described from fossils, *Gastornis* is known from remains discovered near Paris (see panel). The name commemorates Gaston Planté (1834–89), an eager young French fossil hunter who became a well-known physicist and invented the lead-acid battery. As often happens with difficult-to-assign fossils, *Gastornis* was placed in its own family, Gastornithidae, which is thought to be derived from the waterfowl line of bird evolution, order Anseriformes, rather than the cranes, Gruiformes. There were at least four *Gastornis* species including Planté's original *G. parisiensis* in Europe and *G. giganteus* (formerly *Diatryma gigantea*) from North America.

Like *Phorusrhacos*, *G. giganteus* was a tall and rather fearsome bird, 6½ feet (2 m) high, with short wings and a heavy, large head and bill. But was it a carnivore or herbivore? Its beak was barely hooked at the tip, unlike those of the phorusrhacids, and was used more like a nutcracker. Measurement of the calcium composition of the bones showed similarities with modern herbivores rather than carnivores. It seems that *Gastornis* used its massive bill to feed on tough plant material, including nuts and seeds. Fossil footprints preserved in sandstone, uncovered in 2009 in Washington State and thought to be those of *Gastornis*, are consistent with a vegetarian habit because they lack signs of sharp talons that would have marked out the bird as a carnivore. **MW**

◈ NAVIGATOR

◉ FOCAL POINTS

1 HEAVY BILL

The bill of *G. giganteus* provides the most convincing evidence that the bird was an herbivore. The bill was very deep top to bottom, but laterally compressed (narrow), side to side. It was powerful enough to break up tough plant materials, but quite unsuited to piercing and gripping live prey.

2 STURDY NECK

The vertebrae of the neck were short and heavy, so the neck itself would have been relatively short and sturdy. This may have been an adaptation for picking out heavy fruits, nuts, and similar plant foods. Likewise, the body was relatively short and inflexible, with heavily muscled hips for powerful running.

3 SHORT WINGS

The wings of *G. giganteus* were vestigial, having been lost as it evolved along the path to flightlessness. They may have served a display function, perhaps to potential mates at breeding time. If so, it is conceivable that the feathers were brightly colored, as is the case in many living birds.

IDENTIFICATION

In 1855 Planté discovered the remains of a giant, flightless bird in deposits not far from Paris. The naming of *Gastornis* (right) demonstrates how complicated it can be to designate a fossil discovery to a particular species. The fossils found by Planté were very incomplete, and an early reconstruction of the entire skeleton suggested that *Gastornis* had a long neck like a modern stork or crane, and stood about twice as high as a person. This reconstruction was later discovered to have used some bones from other animals, thus giving a misleading indication of anatomy. Later finds, notably by US paleontologist Edward Cope (1840–97) in 1874 in New Mexico, and in 1916 in Wyoming, helped resolve the matter. The American fossils were first named *Diatryma*, but were later realized to be so similar to *Gastornis* that they were reclassified as belonging to the latter genus.

Phorusrhacos
MIOCENE EPOCH

SPECIES *Phorusrhacos longissimus*
GROUP Phorusrhacidae
SIZE 8 feet (2.5 m)
LOCATION Argentina

 **NAVIGATOR**

The terror bird that gave its name to the group Phorusrhacidae, the South American *Phorusrhacos* stood 8 feet (2.5 m) tall and had one of the largest beaks ever known. The first discovered fossils, found in Argentina, date from the early to middle Miocene, about 17 million years ago. First to be found, in Santa Cruz province, was a piece of a lower jawbone. When the fossil was first described, in 1887, it was assumed to belong to a mammal, but as more specimens came to light it became clear that this was indeed a bird, albeit one on a massive scale. The wings were very short, so it would certainly have been flightless. In general build it was comparable to an emu, except that it had a very large head, equipped with a fearsome, deeply hooked, heavy bill. It also had hooklike claws on its wings, like some of the early bird ancestors.

Details of the bill's structure indicate that it was inflexible and rigid and would not have enabled a powerful bite. Instead, this terror bird probably used its hammerlike bill in a vertical mode, bringing it sharply down in a stabbing and pecking motion while using its strong clawed feet, and possibly also the claws on its wings, to hold its prey fast. With its long, powerful legs, it would have chased prey animals across the open savannas of Patagonia, perhaps ambushing them from thickets. Some estimates put the top running speed of *Phorusrhacos* and its relatives at 30 miles per hour (50 km/h) or more, fast enough to catch small mammal prey, especially in a surprise attack. **MW**

👁 FOCAL POINTS

1 TINY WINGS
Relative to the size of the bird, the wings were very small and useless for flight, which would have been impossible for a bird of this weight—about 280 pounds (130 kg). The hooklike claws attached to the wings suggest that the wings were used as weapons during hunting, and possibly in fights with other birds, too.

2 LONG LEGS
Standing 8 feet (2.5 m) tall on its long, powerful legs, *Phorusrhacos* was not unlike an ostrich in build. The limb anatomy suggests it was capable of running flat out, and also of trotting or jogging for long distances. Its height would have helped in sighting prey in the grasslands, scrub, and woodlands of its habitat.

3 LARGE CLAWS
The large, strong foot of *Phorusrhacos* bore powerful, heavily muscled toes each of which had a large, sharp claw at the tip. These would have been fearsome weapons, used to kick and slash at prey animals, and to rip open their bodies for the meat. They also gripped the ground for traction and control when running.

MODERN RELATIVES

The terror birds are now thought to be related to modern cranes and their relatives, and are sometimes classified with them as Gruiformes. Within this order they may be linked to the Cariamidae, the family represented today by the seriemas, also of South America—although some bird experts regard seriemas to have their own distinct order, Cariamiformes. The two species of seriema alive today—the black-legged seriema, *Chunga burmeisteri*, and the red-legged seriema, *Cariama cristata* (above)—show some similarities to the extinct terror birds. The overall body plans are similar, if smaller, and the seriemas' hunting method—smashing prey against the ground—may also have been one of the techniques used by their much larger and stronger prehistoric relatives.

◄ *Phorusrhacos* is best known for its massive beak. The skull was up to 2 feet (60 cm) long and the upper jawbone was very deep and sharply hooked at its tip, like that of a giant eagle. Consequently, there can be little doubt that this terror bird was carnivorous.

Giant Moa
PLEISTOCENE EPOCH—HOLOCENE EPOCH

T he moas were among the strangest of all birds. As giant land herbivores found only in New Zealand, they roamed supreme for more than a million years. Around a dozen species have been named, and the best known are the two giant species, *Dinornis robustus* and *D. novaezealandiae* (*Dinornis* means "terrible bird"). These giant moas were about 12 feet (3.6 m) tall and would probably have weighed about 510 pounds (230 kg). Moas lacked wings entirely and plodded about on stout legs. Preserved stomach contents show that they were herbivores, plucking berries, leaves, and twigs from the forest trees and shrubs. Resembling the flightless emus and cassowaries of Australasia, the moas are thought to be closer in evolutionary terms to the tinamous of South America. Remnants of the feathers show that they were brownish red in color, and filamentous, so that they superficially resembled mammalian fur. Similar feathers are known from kiwis, some early birds, and dinobirds. It is thought that they functioned mainly as a form of insulation.

The moas were able to dominate the scrub and forests of New Zealand's two main islands because there were no terrestrial mammals to compete with them. However, after the arrival of the first humans, the Maori, 800 years ago, the birds' demise was very rapid. They were hunted for their eggs, meat, bones, and feathers, and being defenceless, were easy to catch and kill. By about 1600 the moas were probably all gone. **MW**

◆ NAVIGATOR

SPECIES *Dinornis elephantopus*
GROUP Dinornithidae
SIZE 5¾ feet (1.8 m)
LOCATION New Zealand

◉ FOCAL POINTS

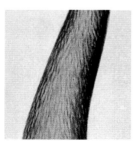

1 SMALL HEAD
In comparison to the rest of its body, the head of a giant moa was small, with a short, flattened beak that curved downward slightly at its tip. As a food-gathering device the beak had a small capacity, and consequently the giant moa had to graze constantly for many hours each day.

2 LONG NECK
With the advantage of its long neck, a giant moa could stretch high into the scrub and trees to pluck and eat leaves and other plant materials that were well out of the reach of other ground herbivores, just as a giraffe does. The neck also allowed ground feeding because it was as long as the legs.

3 POWERFUL FEET
Supporting the legs, which were very long and accounted for about half of the birds' height, the feet were stable, sturdy, and equipped with short claws. The feet were well adapted to enable the heavy birds to roam over large territories as they foraged for food in the woods and scrubs of their habitat.

MAORI LEGEND
There is evidence from moa bones found in Maori camps that these settlers hunted moas (above). Habitat loss and collection of moa eggs probably also played a big part in the birds' extinction. When exactly the moas went extinct is uncertain. A diary extract from missionary James Watkin (1805–86), dated September 27, 1841, reads: "The New Zealanders have many fables . . . of immense birds which were formerly said to exist, and the bones of which are said to be often met with, but the oldest man never saw one of these gigantic birds, neither his father nor grandfather." Watkin concluded that the date of extinction was around 1620, which compares well with evidence found later.

Great Elephant Bird
PLEISTOCENE EPOCH—HOLOCENE EPOCH

W hole skeletons, parts of skeletons, subfossils, eggs, and suspected fragments of feathers and skin survive to prove that the flightless great elephant bird, *Aepyornis,* was the biggest known bird. It would not have been the tallest— that title belongs to New Zealand's giant moa (*Dinornis*), which exceeded 11½ feet (3.5 m), or Stirton's thunderbird, *Dromornis stirtoni,* at 10½ feet (3.2 m); the latter was heavy, too, weighing up to 1,000 pounds (450 kg). However, with an estimated bulk approaching 1,100-pounds (500 kg), the great elephant bird is likely to have surpassed them both. The *Aepyornis* genus, belonging to the Ratites group and native to Madagascar, contained four species. The largest, *A. maximus,* resembled a powerful version of Africa's tallest and largest living bird, the ostrich, but it was not an especially close relative of that or other members of its group, such as the emu and rhea. More than 70 million years ago, the bird's native island of Madagascar was isolated from the African mainland and so it evolved along its own path from that time. Covered by hairlike, bristly feathers, and with no competition from large, herbivorous mammals, it became the island's biggest plant-eater. Ecological considerations suggest that Madagascar could have supported only a limited, slow-breeding population, but the elephant bird flourished until humans appeared, about 1,500 years ago. Habitat clearance for agriculture drove the birds into remote areas, and the arrival of European sailors and settlers with firearms ensured their extinction. **SP**

◉ FOCAL POINTS

1 CONICAL BEAK
Set on a relatively small skull, the long, cone-shaped beak would have served various purposes, and was adapted well for plucking leaves and fruits from the branches of trees and for pecking and grubbing in the soil. The long, flexible neck gave immense reach both vertically and horizontally.

2 BREASTBONE AND WINGS
The breastbone of *A. maximus* was unkeeled, lacking the down-pointing flange that anchors the powerful pectoral wing-beating muscles of flying birds. The bird did have wings but they were vestigial—not used for flying, they gradually diminished in size as the bird evolved.

3 LEGS AND FEET
The legs and feet were both long and powerful. Sturdy legs were needed to support the bird's immense weight, but the proportions and thickness of the bones indicate that the bird was probably not a fast runner. Three long toes with stout, sturdy claws served to scratch in soil for edible plant matter.

COLLECTORS' ITEMS

Even huge dinosaurs would have had trouble laying eggs as large as those of the great elephant bird, which could be more than 14 inches (35 cm) on the longer axis. Preserved parts of shell are often found in Madagascar, and entire eggs are held in several museums and institutions worldwide. During the collecting mania of the 19th century, two eggs were taken to Australia where one was acquired by the Melbourne Museum in 1862 for $160. In April 2013, a very rare example of a whole egg, measuring 12 by 8¼ inches (30 x 21 cm), fetched more than twice the expected price, selling for almost $101,000. In the photograph (above), the egg sizes of three species, the great elephant bird, ostrich, and hummingbird, are compared.

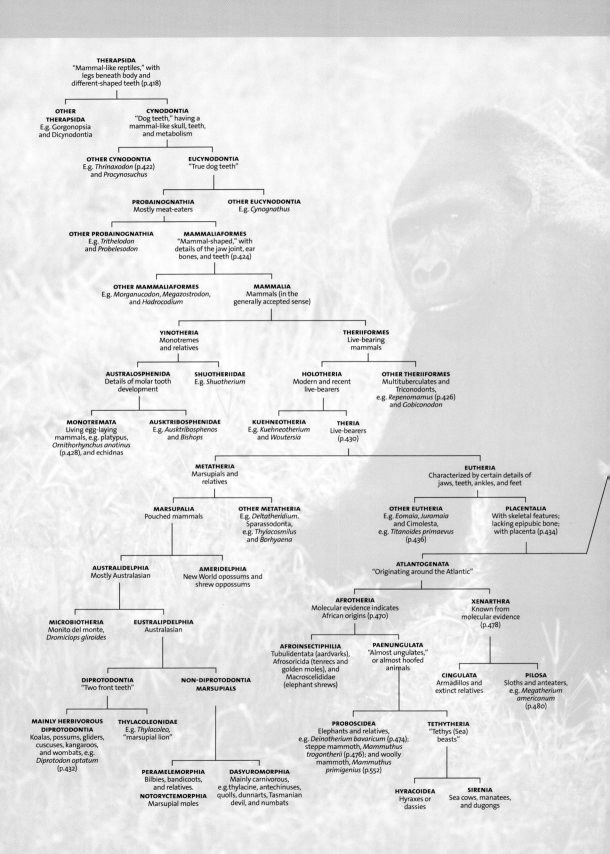

THERAPSIDA
"Mammal-like reptiles," with legs beneath body and different-shaped teeth (p.418)

OTHER THERAPSIDA
E.g. Gorgonopsia and Dicynodontia

CYNODONTIA
"Dog teeth," having a mammal-like skull, teeth, and metabolism

OTHER CYNODONTIA
E.g. *Thrinaxodon* (p.422) and *Procynosuchus*

EUCYNODONTIA
"True dog teeth"

PROBAINOGNATHIA
Mostly meat-eaters

OTHER EUCYNODONTIA
E.g. *Cynognathus*

OTHER PROBAINOGNATHIA
E.g. *Trithelodon* and *Probelesodon*

MAMMALIAFORMES
"Mammal-shaped," with details of the jaw joint, ear bones, and teeth (p.424)

OTHER MAMMALIAFORMES
E.g. *Morganucodon, Megazostrodon,* and *Hadrocodium*

MAMMALIA
Mammals (in the generally accepted sense)

YINOTHERIA
Monotremes and relatives

THERIIFORMES
Live-bearing mammals

AUSTRALOSPHENIDA
Details of molar tooth development

SHUOTHERIIDAE
E.g. *Shuotherium*

HOLOTHERIA
Modern and recent live-bearers

OTHER THERIIFORMES
Multituberculates and Triconodonts, e.g. *Repenomamus* (p.426) and *Gobiconodon*

MONOTREMATA
Living egg-laying mammals, e.g. platypus, *Ornithorhynchus anatinus* (p.428), and echidnas

AUSKTRIBOSPHENIDAE
E.g. *Ausktribosphenos* and *Bishops*

KUEHNEOTHERIA
E.g. *Kuehneotherium* and *Woutersia*

THERIA
Live-bearers (p.430)

METATHERIA
Marsupials and relatives

EUTHERIA
Characterized by certain details of jaws, teeth, ankles, and feet

MARSUPALIA
Pouched mammals

OTHER METATHERIA
E.g. *Deltatheridium.* Sparassodonta, e.g. *Thylacosmilus* and *Borhyaena*

OTHER EUTHERIA
E.g. *Eomaia, Juramaia* and Cimolesta, e.g. *Titanoides primaevus* (p.436)

PLACENTALIA
With skeletal features; lacking epipubic bone; with placenta (p.434)

AUSTRALIDELPHIA
Mostly Australasian

AMERIDELPHIA
New World opossums and shrew oppossums

ATLANTOGENATA
"Originating around the Atlantic"

MICROBIOTHERIA
Monito del monte, *Dromiciops gliroides*

EUSTRALIPDELPHIA
Australasian

AFROTHERIA
Molecular evidence indicates African origins (p.470)

XENARTHRA
Known from molecular evidence (p.478)

AFROINSECTIPHILIA
Tubulidentata (aardvarks), Afrosoricida (tenrecs and golden moles), and Macroscelididae (elephant shrews)

PAENUNGULATA
"Almost ungulates," or almost hoofed animals

CINGULATA
Armadillos and extinct relatives

PILOSA
Sloths and anteaters, e.g. *Megatherium americanum* (p.480)

DIPROTODONTIA
"Two front teeth"

NON-DIPROTODONTIA MARSUPIALS

MAINLY HERBIVOROUS DIPROTODONTIA
Koalas, possums, gliders, cuscuses, kangaroos, and wombats, e.g. *Diprotodon optatum* (p.432)

THYLACOLEONIDAE
E.g. *Thylacoleo,* "marsupial lion"

PROBOSCIDEA
Elephants and relatives, e.g. *Deinotherium bavaricum* (p.474); steppe mammoth, *Mammuthus trogontherii* (p.476); and woolly mammoth, *Mammuthus primigenius* (p.552)

TETHYTHERIA
"Tethys (Sea) beasts"

PERAMELEMORPHIA
Bilbies, bandicoots, and relatives.
NOTORYCTEMORPHIA
Marsupial moles

DASYUROMORPHIA
Mainly carnivorous, e.g.thylacine, antechinuses, quolls, dunnarts, Tasmanian devil, and numbats

HYRACOIDEA
Hyraxes or dassies

SIRENIA
Sea cows, manatees, and dugongs

7 | MAMMALS

Since the 1990s, established mammal groups, such as egg-laying monotremes, pouched marsupials, and placentals (those nourishing unborn young via the womb's placenta), have been compared to new clade-based groupings derived from DNA analysis and other evidence. As a result, traditional accounts of mammal evolution are being revised. When it comes to humans, recently found fossils of now-extinct relations have helped to clarify certain areas of our ancestry—while also presenting scientists with new puzzles.

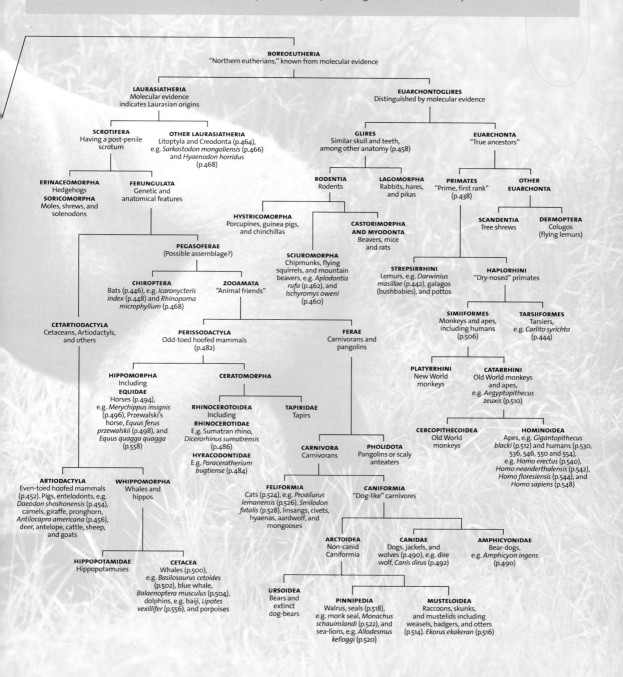

BOREOEUTHERIA
"Northern eutherians," known from molecular evidence

LAURASIATHERIA
Molecular evidence indicates Laurasian origins

EUARCHONTOGLIRES
Distinguished by molecular evidence

SCROTIFERA
Having a post-penile scrotum

OTHER LAURASIATHERIA
Litoptyla and Creodonta (p.464), e.g. *Sarkastodon mongoliensis* (p.466) and *Hyaenodon horridus* (p.468)

GLIRES
Similar skull and teeth, among other anatomy (p.458)

EUARCHONTA
"True ancestors"

ERINACEOMORPHA
Hedgehogs
SORICOMORPHA
Moles, shrews, and solenodons

FERUNGULATA
Genetic and anatomical features

RODENTIA
Rodents

LAGOMORPHA
Rabbits, hares, and pikas

PRIMATES
"Prime, first rank" (p.438)

OTHER EUARCHONTA

HYSTRICOMORPHA
Porcupines, guinea pigs, and chinchillas

CASTORIMORPHA AND MYODONTA
Beavers, mice and rats

SCANDENTIA
Tree shrews

DERMOPTERA
Colugos (flying lemurs)

PEGASOFERAE
(Possible assemblage?)

SCIUROMORPHA
Chipmunks, flying squirrels, and mountain beavers, e.g. *Aplodontia rufa* (p.462), and *Ischyromys oweni* (p.460)

STREPSIRRHINI
Lemurs, e.g. *Darwinius masillae* (p.442), galagos (bushbabies), and pottos

HAPLORHINI
"Dry-nosed" primates

CHIROPTERA
Bats (p.446), e.g. *Icaronycteris index* (p.448) and *Rhinopoma microphyllum* (p.468)

ZOOAMATA
"Animal friends"

SIMIIFORMES
Monkeys and apes, including humans (p.506)

TARSIIFORMES
Tarsiers, e.g. *Carlito syrichta* (p.444)

CETARTIODACTYLA
Cetaceans, Artiodactyls, and others

PERISSODACTYLA
Odd-toed hoofed mammals (p.482)

FERAE
Carnivorans and pangolins

PLATYRRHINI
New World monkeys

CATARRHINI
Old World monkeys and apes, e.g. *Aegyptopithecus zeuxis* (p.510)

HIPPOMORPHA
Including
EQUIDAE
Horses (p.494), e.g. *Merychippus insignis* (p.496), Przewalski's horse, *Equus ferus przewalskii* (p.498), and *Equus quagga quagga* (p.558)

CERATOMORPHA

RHINOCEROTOIDEA
Including
RHINOCEROTIDAE
E.g. Sumatran rhino, *Dicerorhinus sumatrensis* (p.486)

TAPIRIDAE
Tapirs

CERCOPITHECOIDEA
Old World monkeys

HOMINOIDEA
Apes, e.g. *Gigantopithecus blacki* (p.512) and humans (p.530, 536, 546, 550 and 554), e.g. *Homo erectus* (p.540), *Homo neanderthalensis* (p.542), *Homo floresiensis* (p.544), and *Homo sapiens* (p.548)

HYRACODONTIDAE
E.g. *Paraceratherium bugtiense* (p.484)

CARNIVORA
Carnivorans

PHOLIDOTA
Pangolins or scaly anteaters

ARTIODACTYLA
Even-toed hoofed mammals (p.452). Pigs, enteledonts, e.g. *Daeodon shoshonensis* (p.454), camels, giraffe, pronghorn, *Antilocapra americana* (p.456), deer, antelope, cattle, sheep, and goats

WHIPPOMORPHA
Whales and hippos

FELIFORMIA
Cats (p.524), e.g. *Proailurus lemanensis* (p.526), *Smilodon fatalis* (p.528), linsangs, civets, hyaenas, aardwolf, and mongooses

CANIFORMIA
"Dog-like" carnivores

HIPPOPOTAMIDAE
Hippopotamuses

CETACEA
Whales (p.500), e.g. *Basilosaurus cetoides* (p.502), blue whale, *Balaenoptera musculus* (p.504), dolphins, e.g. baiji, *Lipotes vexillifer* (p.556), and porpoises

ARCTOIDEA
Non-canid Caniformia

CANIDAE
Dogs, jackels, and wolves (p.490), e.g. dire wolf, *Canis dirus* (p.492)

AMPHICYONIDAE
Bear-dogs, e.g. *Amphicyon ingens* (p.490)

URSOIDEA
Bears and extinct dog-bears

PINNIPEDIA
Walrus, seals (p.518), e.g. monk seal, *Monachus schauinslandi* (p.522), and sea-lions, e.g. *Allodesmus kelloggi* (p.520)

MUSTELOIDEA
Raccoons, skunks, and mustelids including weasels, badgers, and otters (p.514). *Ekorus ekakeran* (p.516)

THERAPSIDS

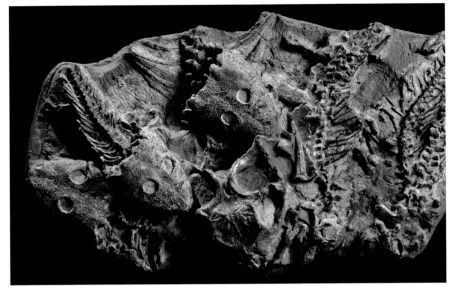

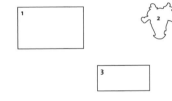

1

2

3

1 This fossil skull of *Lystrosaurus*, staring out from bottom center, is accompanied by three amphibian skulls to the left and above.

2 An *Estemmenosuchus* fossil from 267 million years ago has the elaborate bony horns of dinocephalians.

3 About 262 million years ago, rival *Moschops* prepare to fight in what is now South Africa.

bout 315 million years ago, the early amniotes—vertebrate animals that protect their developing young with shelled eggs—split into two evolutionary lines. One led to reptiles; the other led to mammals, via synapsids. Named after the single opening in the skull behind the eye, early synapsids were superficially reptilelike. Around 275 million years ago one of these groups, the pelycosaur synapsids (see p.236), began to evolve distinctive features and move into a new grouping, Therapsida (mammal-like reptiles), of which the widespread *Lystrosaurus* (see image 1) was one. Ultimately, certain therapsids gave rise to the mammals that dominate terrestrial life today.

Key therapsid features preserved in the fossil record include more diverse and specialized teeth, and in particular a change of posture: while their pelycosaur cousins retained the sprawling gait typical of modern lizards, therapsids had legs positioned vertically beneath their shoulders and hips, and feet parallel to their bodies rather than splayed outward. These features combined to give the early therapsids an increasingly mammal-like appearance. Although the evolution from pelycosaurs was a gradual one, the earliest probable therapsid is *Tetraceratops insignis*—a dog-size animal that had six horns (rather than the four that its name might suggest). The creature is known only from a single fossil skull discovered in Texas in 1908, and consequently many details of its anatomy and relationships are uncertain. *Tetraceratops* dates to 275 million years ago in the middle Permian period, and within a few

KEY EVENTS

275 Mya	272–260 Mya	260 Mya	251 Mya	247 Mya	245 Mya
The small synapsid *Tetraceratops*, known from a single fossil skull found in Texas, is thought to be the earliest therapsid.	The enormous dinocephalians briefly flourish across a large swathe of the Pangaean supercontinent before becoming extinct.	The most successful therapsid group, Cynodontia, appears at this time.	The mass extinction at the end of the Permian wipes out several major therapsid groups and thins out many others.	The small cynodont *Lumkuia* is generally acknowledged as the earliest member of the Probainognathia group that gives rise to true mammals.	A group of surviving dicynodonts, the lystrosaurs, evolves into dominant land herbivores, giving rise to the Kanneymeyeriidae.

million years the therapsids had displaced the pelycosaurs as the dominant big land animals.

The earliest and most primitive major group of therapsids was the dinocephalians (awesome/terrible heads). They retained many similarities to their pelycosaur ancestors, but combined them with unique new features. Although their posture was still distinctly sprawled compared to later therapsids, they held themselves higher above the ground, particularly at the front of the body. They included both carnivores and herbivores, with the largest growing up to 15 feet (4.5 m) long and reaching body masses of 2¼ tons (2 tonnes). Herbivores such as *Moschops* (see image 3) retained a small head with a downturned face. They frequently had thickened skulls, thought to be used in headbutting contests with rivals during mating, while some, such as *Estemmenosuchus* (see image 2), also sported elaborate bony horns that may have been used for defence. Dinocephalian carnivores evolved much longer heads and more powerful jaws. They rivaled the largest herbivores in length, although they did not develop such elephantine bodies. One common feature of all dinocephalians was an interlocking set of front incisor teeth.

The dinocephalians' reign was short-lived. Around 260 million years ago the entire group became extinct in a very short period of time, leaving the way open for much smaller therapsids to take its place. These later groups can be broadly divided into the anamodonts, biarmosuchians, and theriodonts. The anamodonts were mostly toothless herbivores that had already diversified into several distinct groupings, and all but one of these groups went extinct around the same time as the dinocephalians. The survivors were the dicynodonts (two dog teeth), named for their distinctive two-tusked appearance. They rapidly

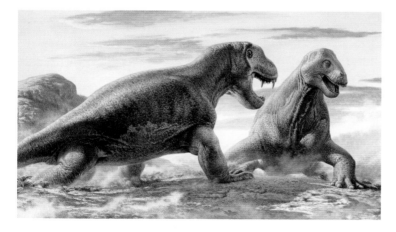

245 Mya	235 Mya	205 Mya	201 Mya	193 Mya	105 Mya
Andescynodon is the most primitive of the herbivorous traversodonts, a cynodont group that will survive into the Jurassic.	The therocephalians, last of the major theriodont groups, become extinct.	Found in both Wales and China, *Morganucodon* is a small insectivorous mammaliaform—the closest known cousin of the true mammals.	The traversodonts become extinct, leaving the probainognathans as the major surviving therapsid group.	*Sinoconodon* is a surviving member of a group of close mammal relatives that retain reptilian features.	The possible extinction of the last surviving dicynodonts occurs in a southern refuge of Australia.

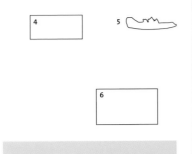

4 Lending its name to its group, *Dicynodon* from 252 million years ago was about 4 feet (1.2 m) long.

5 *Morganucodon*'s mammalian molar teeth were shaped to chew small prey such as insects.

6 *Sinoconodon* retained the reptilian feature of continuously losing old teeth and growing new ones, yet its jaw joint evolved to be almost typically mammalian.

evolved and diversified in size and habitat, while developing larger skull openings to accommodate more powerful jaw muscles (see image 4).

The carnivorous biarmosuchians, meanwhile, were the most primitive therapsids. Lightly built compared to their herbivorous prey, they retained many primitive pelycosaur features, and they flourished once the dinocephalian competition had been removed. The theriodonts, in contrast, were a new, more mammal-like group of therapsids that appeared late in the dinocephalian reign. They were distinguished by a heavier lower jaw and other associated changes that gave them much better hearing. Theriodonts evolved rapidly into two more groups, Gorgonopsia and Therocephalia, which were joined late in the Permian by Cynodontia (see image 6). The gorgonopsians were relatively primitive carnivores, and the largest grew to the size of a polar bear. The somewhat smaller therocephalians are also thought to have been carnivores, and displayed more specialized teeth and changes to the skull that would have allowed a significant increase in brain size and intelligence. They shared these features with the even more mammal-like cynodonts (dog teeth).

The driving forces behind the evolution of all these therapsids are uncertain. They rose to prominence on the brink of the coming together of all the major landmasses into the single supercontinent of Pangaea, which rendered large parts of the continental interior inhospitable deserts. The ability to graze on tougher plants, or to move faster and hunt prey more efficiently, might have been crucial factors to their survival and rise to dominance over the other synapsids. However, 252 million years ago, at the end of the Permian, life on planet Earth faced the biggest crisis in its history. The extinction event known as the "Great Dying" (see p.224) laid waste to the vast majority of species. Only a few lineages survived, and those that did most probably endured as a result of luck rather than particular evolutionary suitability. The dicynodont line was reduced to a couple of families. One, the lystrosaurs, flourished in the aftermath and became the most successful herbivores around the globe. These mid-size, piglike animals had tusks and beaks to assist their foraging: within a few million years they evolved into, and were supplanted by, the Kanneymeyeriidae.

Of the other groups, the biarmosuchians died out completely in the cataclysm. The theriodonts fared a little better: their gorgonopsian branch went extinct, but a number of therocephalian families made it through—only to go extinct by the middle Triassic period (around 235 million years ago). The cynodonts, however, thrived and became increasingly similar to modern mammals. Carnivorous species grew smaller, with more specialized and relatively larger brains, while a distinctive group of mid-size herbivores seems to have branched off the main cynodont line in the middle Triassic.

One significant challenge in tracing the development of early mammals is the fact that the fossil record cannot capture many typical mammalian traits,

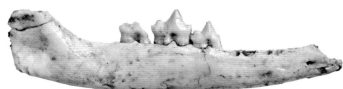

such as live birth of young and insulating body hair. Fortunately, however, modern mammals also share some unique bone structures that do fossilize. As well as their variety of teeth, all mammals have a secondary palate, a structure in the roof of the mouth to separate the nasal and oral cavities. This allows them to breathe while eating without the risk of choking and is thought to have played an important role in the evolution of thermoregulation. Maintaining a constant body temperature not only allows an animal to be more active, but also demands a higher, more regular intake of food. Fossils confirm that the secondary palate was present in both the cynodonts and therocephalians. *Thrinaxodon* (see p.422), which lived in the early Triassic around 245 million years ago, is thought to show a key step in the evolution from cynodonts to true mammals.

In the wake of the end-of-Permian extinction, the surviving therapsids faced increasing competition from the rapidly evolving dinosaurs and their reptilian relatives. By the end of the Triassic, around 201 million years ago, the dicynodonts had mostly been driven to extinction. Many of the cynodonts also faced extinction, but their smaller size and evolution of thermoregulation allowed some to find unique ways of surviving in the Mesozoic era. The most successful group was the eucynodonts, divided into the Cynognathia and the Probainognathia. Fossil evidence suggests that, under pressure from the rise of dinosaurs and others during the Triassic, two lifestyles emerged. The Cynognathia became small- to medium-size herbivores that survived until the early Jurassic period around 201 million years ago. The Probainognathia, meanwhile, became smaller and developed into a number of families, including herbivores and carnivores. Reduction in size and a resulting loss of ability to retain body heat drove the evolution of a faster metabolism and better insulation, which led to increasingly furry creatures that only just fail to be true mammals because they retained certain primitive reptilian features in the jaw.

Around 225 million years ago, one group of cynodonts gave rise to the Mammaliaformes. This group includes the common ancestor of all living mammals, their nearest extinct relatives—exemplified by the shrewlike *Morganucodon* (see image 5)—and the monotremes represented today by the platypus (see p.428). For the next 160 million years, these creatures evolved in the shadows of the dinosaurs, until the extinction event at the end of the Cretaceous period allowed their re-emergence to dominate the world as true mammals. **GS**

Thrinaxodon
EARLY TRIASSIC PERIOD

SPECIES *Thrinaxodon liorhinus*
GROUP Cynodontia
SIZE 20 inches (50 cm)
LOCATION South Africa and Antarctica

The therapsid *Thrinaxodon* is often considered a key species in the evolution from reptiles toward true mammals. Its fossils have been found in both South Africa and Antarctica, countries that were side by side within the great supercontinent of Pangaea 245 million years ago in the early Triassic. *Thrinaxodon* was a member of the most successful therapsid group, Cynodontia, which first appeared around 260 million years ago in the late Permian. Cynodonts survived the end-of-Permian mass extinction and rapidly diversified throughout the Triassic. On the evolutionary family tree, it is thought that *Thrinaxodon* is situated just outside the main cynodont subgroup, Eucynodontia, which includes most of the later cynodonts, including mammals. It is believed to be a primitive member of a broader clade (group with one common ancestor), the Epicynodontia.

Growing to 20 inches (50 cm) long, the tetrapod (four-limbed vertebrate) *Thrinaxodon* would have held its badger-like body low above the ground on short legs with broad feet. Its general body shape, coupled with the fact that its fossils have been found in the remnants of underground tunnels, suggests that it was a burrowing animal. Harsh conditions in southern Pangaea may have forced it to estivate—spend the harsh summer months sleeping underground (see panel). Specimens of *Thrinaxodon* teeth suggest that it was a carnivorous animal and it is frequently depicted with hair. There is no conclusive evidence for this, but the dimples on its pitted snout could show that it had whiskers, and so perhaps hairs on its body. However, similar dimples feature on the skulls of modern lizards, which are hairless. Overall, *Thrinaxodon* shows a mix of reptilian and mammalian characteristics. For example, it has the bony secondary palate shared by all modern mammals, but it would have laid eggs, which is a reptile trait. An important innovation that separates *Thrinaxodon* from its reptilian ancestors is the ribcage, which divides the torso into a distinct thorax and abdomen. In life, these two torso compartments are likely to have been separated by a diaphragm that would have increased breathing efficiency and would have raised the animal's metabolism, which is a feature of later mammals, including humans. **GS**

✧ NAVIGATOR

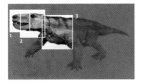

◉ FOCAL POINTS

1 JAW AND SNOUT

Thrinaxodon had an enlarged lower jawbone, which improved bite strength, and reduced secondary jawbones, linking the jaw to the skull. Dimples in the fossil skull bone are similar to those at the roots of whiskers in mammals. These sensitive hairs would have been useful for a burrowing animal that spent much of its life in darkness.

2 BRAINCASE

Scans of *Thrinaxodon* fossils suggest that, while the back of its cranium was larger (for its size) than those of its ancestors, its brain occupied only a small region, so it had a lower brain-to-body size ratio than any modern mammal. Therapsid skulls included several bones that evolved to produce the braincase in true mammals.

3 RIBCAGE

Thrinaxodon is the earliest known animal to display distinct chest and abdominal regions. The thoracic (upper back) vertebrae led to ribs that protected the organs at the front of the body, but these ribs were absent in the abdomen. A diaphragm compressed the digestive organs and allowed lung capacity to expand.

⊥ THRINAXODON RELATIONSHIPS

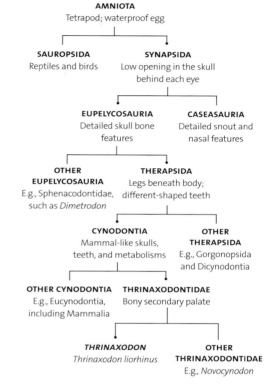

▲ *Thrinaxodon* was a member of the Cynodontia (dog-tooths), a varied and widespread group throughout southern continents. After the end-of-Permian mass extinction they took advantage of more new lifestyles and eventually give rise to mammals.

SUMMER HIBERNATION

A discovery from South Africa's Karoo Basin supports the theory that *Thrinaxodon* spent summers in a deep sleep in its burrow (right). Using a synchrotron scanner, researchers in France peered inside a rare fossilized cast of an underground burrow from early Triassic rocks. They discovered a fossil *Thrinaxodon* in a sleeping posture, and also amphibian *Broomistega putterilli*. The two animals were buried alive. *B. putterilli*, which seems to have been injured weeks before its death, could not have dug such a tunnel. It probably sought shelter in the burrow. The fact that the predatory *Thrinaxodon* remained asleep instead of attacking the intruder suggests it was in a dormant state, perhaps due to excessive heat and lack of food.

THE FIRST MAMMALS

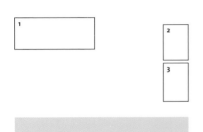

1 *Castorocauda*, up to 20 inches (50 cm) long, showed adaptations for a semi-aquatic lifestyle, similar to a mix of the modern beaver and otter.

2 *Volaticotherium* showed a remarkable resemblance to the modern "flying" squirrel—an example of convergent evolution in very different groups toward a similar way of life.

3 The single known fossil of *Eomaia* is remarkably complete, and even shows its fur. It represents a mammal 4 to 5 inches (10–12 cm) long.

The first true mammals appeared during the middle Triassic period, around 225 million years ago, just after the earliest dinosaurs. The mammals that coexisted with dinosaurs were originally depicted as timid, shrewlike insectivores cowering beneath the feet of the giant reptiles. This view suggests they owe their survival to small size and a high metabolism that gave them a chance of running to safety if they were noticed. However, in the past few years, several mid-size mammal-like animals have been discovered that show distinct specializations, and that clearly coexisted with dinosaurs as an important part of at least some ecosystems. The most significant of these finds are from the Chinese regions of Inner Mongolia and Liaoning province. They include the badgerlike nest-raider *Repenomamus* (see p.426) from the early Cretaceous period, 125 million to 123 million years ago, the beaverlike *Castorocauda* (see image 1), and the gliding *Volaticotherium* (see image 2), both from 164 million years ago, in the middle Jurassic period. These finds reveal unexpected diversity among early mammals, but it is unclear where they fit in the evolutionary picture.

A combination of genetic and fossil evidence suggests that the first "crown group" mammals—the direct ancestors of all mammals living today, including placentals, marsupials, and monotremes—appeared in the early Jurassic, around 190 million years ago. However, mammals from other groups lived earlier. Fossil evidence, mainly bones and teeth, points to the existence of several other closely related groups that have not survived to the present time, unlike the monotremes, marsupials, and placentals. In the absence of genetic evidence and fossilized soft parts, opinions differ as to whether some of these are true mammals (extinct descendants of the crown group) or "cousins" of the crown group, called mammaliaforms.

KEY EVENTS

205 Mya	195 Mya	193 Mya	170 Mya	164 Mya	160 Mya
The shrewlike *Morganucodon* is one of the earliest true mammals. It has a tiny brain and weighs less than 3 ounces (85 g).	*Hadrocodium*, found in China's Lufeng Basin, shows developments to the brain and inner ear and is considered a close relative of the true mammals.	*Sinoconodon*, another Chinese mammal relative, is considered to be the most basal of mammals and their closest relations, mammaliaforms.	Extant mammal groups —monotremes, marsupials, and placentals—may have separated at this time, as mitochondrial DNA studies have suggested.	*Castorocauda* and *Volaticotherium* live in Inner Mongolia. The Daohugou fossil beds later reveal diversity among early mammals of this time.	The approximate date that *Juramaia*, a tree-climbing animal, lived in China. Some researchers claim this is a very early placental mammal.

The most significant of these extinct groups is the multituberculates, named because of lumps or "tubercules" on their molar teeth. These rodentlike animals have a fossil record stretching across 120 million years, and finally became extinct when they were outcompeted by true rodents (placental mammals) during the Oligocene epoch, 30 million years ago. They may have been either an independent line of mammal evolution that separated from the therians (placentals and marsupials) some time after the monotremes or deceptive-looking mammaliaforms that lay outside the original crown group. At the present time, exactly where the newly discovered mammals of the Mesozoic era fit into the overall picture of mammalian evolution is open to debate.

Identifying the earliest known representatives of these major mammal groups is challenging. The earliest known monotreme is the 123-million-year-old *Teinolophos*, which appears to be some way along the evolutionary path toward the modern platypus (see p.428). It must have descended from a much earlier and more generalized basal (original) monotreme that separated from the other living mammal orders well before its time. The oldest known marsupial is *Sinodelphys*, a possumlike animal recovered in immaculate condition from Liaoning province. It dates to 125 million years ago in the early Cretaceous. Unlike the specialized *Teinolophos*, *Sinodelphys* appears to have been a fairly generalized or basal tree-dweller, closer to the evolutionary junction point where marsupials and placentals split from one another. The discovery of near-contemporary mammals such as *Eomaia* (see image 3) and *Acristatherium*—both also from China and described as placentals—tends to support a date in the early Cretaceous for this split. However, the evidence is conflicting. Some researchers claim that *Eomaia* and *Acristatherium* are not as advanced as originally believed and that the placental group did not arise until shortly after the end-of-Cretaceous mass extinction, 66 million years ago. Conversely, others claim that the earlier *Juramaia* is a placental mammal, potentially pushing back the marsupial–placental split another 35 million years into the late Jurassic.

Contradictory genetic evidence adds further confusion. Molecular clock measurements based on the mitochondrial DNA of living mammals (see p.17) suggests that today's three major mammal groups separated from one another during the middle Jurassic, 170 million years ago. The same genetic evidence dates the origins of several modern mammal subgroups to the Cretaceous, despite little evidence to support this. In order to reconcile these dates, one theory is that the ancestors of today's varied mammals remained small and indistinguishable from one another in the fossil record until the sudden disappearance of the dinosaurs cleared the way for them to evolve, diversify, and spread. Another idea is that the molecular clock "ran fast" as mammal evolution exploded in the postdinosaur period from 66 million years ago, thereby producing deceptive results from a method that assumes a slow, steady rate of genetic change. **GS**

130 Mya	125 Mya	125 Mya	123 Mya	65½ Mya	35 Mya
Dog-size triconodont carnivores of the genus *Repenomamus* apparently thrive alongside dinosaurs around this time.	The opossumlike *Sinodelphys*, found in China's Liaoning province and dated to this period, is the earliest identified marsupial mammal.	Roughly contemporary with *Sinodelphys*, the Chinese fossil *Eomaia* was possibly an early placental mammal, although this is doubted by some.	The earliest known monotreme, a platypuslike mammal called *Teinolophus*, lived in Australia.	The sudden extinction of the nonbird dinosaurs leaves the field open for an explosion in size and diversity among surviving mammals.	Multituberculates become extinct. They are a major group of rodentlike mammals with a separate line of descent from the placental true rodents.

Repenomamus
EARLY CRETACEOUS PERIOD

SPECIES *Repenomamus robustus*
GROUP Gobiconodontidae
SIZE 20 inches (50 cm)
LOCATION China

The largest and most impressive dinosaur-age mammal so far discovered, *Repenomamus* was a badger-size carnivore that is known to have eaten dinosaur hatchlings. Two species have so far been found, both known solely from the Yixian Formation, a unique fossil bed in China's Liaoning province that is well known for its preservation of several feathered dinosaurs. The two species are the smaller *R. robustus*, which was around 20 inches (50 cm) long and weighed up to 13 pounds (6 kg), and *R. giganticus*, which grew up to 3 feet (1 m) long and weighed up to 31 pounds (14 kg). Fossils of these animals are frequently found in a curled up posture. Paleontologists believe that they may have died in their sleep as a nearby volcano released toxic fumes, subsequently burying them in volcanic ash that preserved their remains in immaculate detail. Both species are dated to the early Cretaceous, around 125 million to 123 million years ago.

Analysis of teeth shows that *Repenomamus* was a triconodont mammal, characterized by a row of three raised peaks along the molar teeth. Triconodonts have long been considered as close relatives of the living mammals, but they are not part of the same group because they do not share a common ancestor with living mammals. Indeed, there seem to be several distinct triconodont groups related to modern mammals. According to this classification scheme, *Repenomamus* belongs to a small group called the Gobiconodontidae, along with another fairly large, opossumlike extinct mammal called *Gobiconodon*. Features of the jaws and teeth of *Repenomamus* point to a generally carnivorous lifestyle—and this suggestion was confirmed graphically by the discovery of juvenile *Psittacosaurus* (see p.340) dinosaur bones in the stomach area of one *Repenomamus* fossil. This is the strongest evidence so far that active, predatory mammals lived alongside dinosaurs, preying on them and perhaps even competing with them for food sources. **GS**

✪ NAVIGATOR

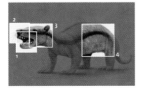

1 BITING JAW

R. robustus had a jaw roughly the same length as that of a fox, but more heavily built, with large surfaces for muscle attachment. This would have given it a powerful bite that, combined with elongated, sharp incisor and canine teeth, enabled it to keep hold of prey and tear it apart before eating.

2 NOSE

Fossil braincases show that the brain was small compared to later marsupials and placental mammals. The most developed area was the olfactory bulb (smell region), suggesting that *R. robustus* hunted by scent. Some claim that growth of this area triggered the expansion of mammalian brains.

3 JAW AND EAR

The jaw had reached the full mammalian state, with the postdentary bones reduced to small ear bones (ossicles), leaving the dentary as the main lower jawbone. However, the ear was not fully modern: cartilage linked the ossicles to the jaw. This would have limited the range of frequencies heard.

4 STOMACH PLACEMENT

The discovery of *Psittacosaurus* bones within a *R. robustus* specimen offers insights into the animal's internal organs. This last meal lies on the lower-left side, in the same position as the stomach of a modern mammal, confirming the basic body plan developed early in the history of mammal evolution.

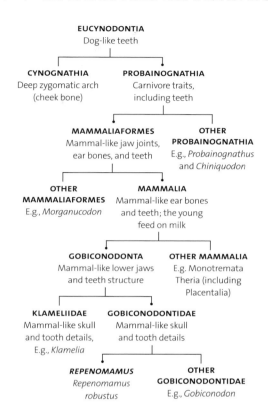

EUCYNODONTIA
Dog-like teeth

CYNOGNATHIA
Deep zygomatic arch (cheek bone)

PROBAINOGNATHIA
Carnivore traits, including teeth

MAMMALIAFORMES
Mammal-like jaw joints, ear bones, and teeth

OTHER PROBAINOGNATHIA
E.g., *Probainognathus* and *Chiniquodon*

OTHER MAMMALIAFORMES
E.g., *Morganucodon*

MAMMALIA
Mammal-like ear bones and teeth; the young feed on milk

GOBICONODONTA
Mammal-like lower jaws and teeth structure

OTHER MAMMALIA
E.g. Monotremata Theria (including Placentalia)

KLAMELIIDAE
Mammal-like skull and tooth details, E.g., *Klamelia*

GOBICONODONTIDAE
Mammal-like skull and tooth details

REPENOMAMUS
Repenomamus robustus

OTHER GOBICONODONTIDAE
E.g., *Gobiconodon*

▲ The features in *Repenomamus* do not allow it to be placed with a degree of certainty in any of the major mammal categories. So until further evidence is known, it is usually assigned to a high-level group, Gobiconodonta, which is mostly east Asian.

▼ In 2005, small dinosaur bones were discovered in the stomach of a fossil *Repenomamus*, which questioned the relationships between dinosaurs and mammals. The remains were of a hatchling *Psittacosaurus*, nests of which show that this dinosaur formed social groups. It is unknown whether *Repenomamus* caught unprotected hatchlings or circumvented parents. The partially articulated *Psittacosaurus* remains confirm that it was the result of deliberate predation.

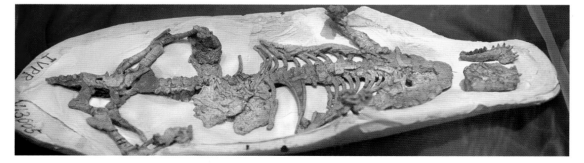

Duck-billed Platypus

MIOCENE EPOCH—PRESENT

SPECIES *Ornithorhynchus anatinus*
GROUP Monotremata
SIZE 20 inches (50 cm)
LOCATION Australia
IUCN Least Concern

W hen the first remains of the platypus, *Ornithorhynchus anatinus*, came to the attention of European anatomists in the late 18th century, most of them suspected a hoax. With its ducklike beak, webbed feet, and beaverlike body, this curious animal looks bizarre in comparison to the more familiar placental mammals found elsewhere. However, the platypus is adapted perfectly to its Australian environment and is neither primitive nor an evolutionary throwback. Nevertheless, its family, Ornithorhynchidae, has followed its own evolutionary path for 60 million years and perhaps more, and it also preserves some unique features that echo the earliest mammals, including the therapsids.

The three existing subgroups of mammals are the monotremes—including the platypus—the marsupials, and the placentals. These groups are thought to represent different stages in mammal evolution. The therians were the common ancestors of marsupials and placentals, and the monotremes are thought to have separated from the therians at an early stage. Thereafter, different evolutionary pressures meant that monotremes preserved a mode of reproduction that disappeared in their mammal cousins. They do not bear live young, but instead lay leathery eggs through a single body opening called a cloaca. The spiny anteaters (echidnas) also reproduce this way, and it offers a rare glimpse of a primitive, reptilelike means of reproduction that was common in early mammals and their direct ancestors more than 200 million years ago. Despite its egg-laying habit, the platypus is undoubtedly a mammal; most of its body is covered in thick fur and it produces milk to nourish its offspring after they have hatched (although it lacks well-defined nipples). Monotremes have a high metabolic rate and are homeothermic, meaning that they regulate their own body temperature. However, they are not as warm-blooded as placental mammals or the pouched marsupials. Due to the fragmentation of the Gondwana supercontinent 180 million to 150 million years ago, Australia was geographically isolated. This meant that the platypus was protected from competition against placental mammals, which allowed it to find and evolve into its own ways of life. Only a few early monotremes are known (see panel): the giant platypus, *Obdurodon tharalkooschild*, grew up to 4 feet (120 cm) long during the much more recent Miocene epoch, around 10 million years ago. **GS**

✪ NAVIGATOR

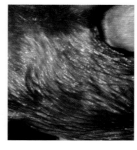

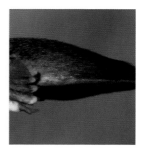

1 DUCK BILL
The platypus hunts underwater with its eyes closed, finding prey with its leathery snout. The bill is lined with sensitive nerve endings for detecting pressure changes in the water caused by movement. It also has rows of electroreceptors that detect the electric currents generated by the muscles of other animals.

2 JAW STRUCTURE
The lower jaw is formed entirely of the dentary bone, and the three additional jawbones have reduced to form part of the ear apparatus (a feature of all mammals), but have not moved to the side of the head. As a result, the platypus's external ears are slitlike openings at the rear of the jaw.

3 DENSE FUR
The platypus is covered in short, dense, brown, waterproof fur that traps a layer of insulating air close to the body. This prevents it from cooling down during dives that typically last 30 seconds. A platypus needs to eat 20 percent of its body weight each day and spends half of each day hunting.

4 FAT TAIL
The platypus's broad tail bears a resemblance to that of a beaver. It is a powerful swimming aid, but has evolved a secondary function. The platypus's bulky tail stores fat reserves. This feature, along with low body temperature, is a likely adaptation for survival in a harsh climate.

FOSSIL MONOTREMES

There is a huge gulf in the fossil record between the estimated divergence of monotremes from other mammals and their first fossil remains. A few monotremes are known from teeth or jawbones. Based on this, it is likely that early monotremes were similar to the platypus. The earliest known monotreme fossils are of *Steropodon* (below), a 110-million-year-old species that was similar in both size and appearance to the platypus and the smaller *Teinolophus*, about 123 million years old and only 4 inches (10 cm) long.

▲ The platypus is a rare example of a venomous mammal. The hind foot of the male is equipped with an "ankle spur" that can inject toxic venom for defence, or perhaps during fights with rival males. The presence of spurs in echidnas suggests they were inherited from monotreme ancestors.

MARSUPIALS

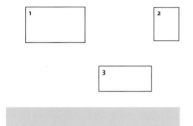

Marsupials are, in a sense, an intermediate stage between the basal (original) egg-laying mammals and the advanced placental mammals known as eutherians. Marsupials bear live young after a short gestation period, producing partially developed infants that are highly dependent on the mother, and grow in a pouch (marsupium). Modern marsupials are members of a single evolutionary clade (group with one common ancestor) known as the Metatheria. Marsupials and placental mammals form a broader group called the Theria. Molecular clock studies (see p.17) based on the degree of genetic difference between the groups suggest that the therian line split from other ancient mammals in the Jurassic period and soon diverged into the placental Eutheria and the marsupalian Metatheria. The earliest fossil evidence to support this are the 160-million-year-old *Juramaia* and the 125-million-year-old *Eomaia*, both of which have features marking them out as the oldest eutherians. The earliest confirmed metatherian, *Sinodelphys szalayi* (see image 1), is 125 million years old, which indicates that the earliest stages of marsupial evolution were older than this and remain undiscovered.

Modern marsupials are divided into seven groups containing around 330 species. Four of today's seven orders are confined to Australasia and share common features, while three are confined to the Americas. Two of the American orders—the opossums and the shrew opossums—are close relations. Intriguingly, the third, represented by a single dormouselike species called the monito del monte (little monkey of the mountains), *Dromiciops gliroides* (see image 2), is more closely related to the Australian groups and is the most primitive member of those groups.

KEY EVENTS

160 Mya	125 Mya	70 Mya	55 Mya	25 Mya	20 Mya
Divergence occurs between marsupial-containing metatherians and placental-containing eutherians.	*Sinodelphys* is the earliest known fossil marsupial, and is later discovered in rocks dating from this time in China's Liaoning province.	North American *Didelphodon*, a predatory opossum with an otterlike body and powerful jaws, is a large Mesozoic era mammal.	The Tingamarra Fauna from southeastern Queensland includes the earliest Australian marsupials and fossil placentals such as bats and condylarths.	Fossils are laid down in Riversleigh, Australia, which show the evolution of the marsupial megafauna over more than 15 million years.	Large sparassodont mammals (marsupial cousins) such as bearlike *Borhyaena*, flourish alongside placental mammals in South America.

The marsupial-containing metatherians evolved in Asia during the Age of Dinosaurs and spread across the northern continents of Laurasia into North America, and into Europe in lesser numbers. Fossils suggest that they arrived in South America shortly after the mass extinction at the end of the Cretaceous period where they flourished as their Laurasian relatives died out. They then reached Australia, most likely via the joined intervening land mass of Antarctica, in the early Eocene, 55 million years ago. Perhaps these groups had begun to diversify before they arrived in Australia; if so, they left no fossil trace in South America, and the monito del monte's ancestors returned to South America before going extinct in Australia. Alternatively, one group of monitolike animals arrived from the Americas in Australia by chance, aboard floating vegetation, suggesting all Australian marsupials evolved from a monitolike animal.

Since the colonization of Australasia, marsupials have developed along two distinct pathways. South American species retained an opossumlike lifestyle and appearance, while the ancestors of the Virginia opossum made their way back into North America during the Great American Interchange 3 million years ago. Australian species, in contrast, diversified in an environment in which they faced little competition from placental mammals. The fossil site Riversleigh in Queensland offers insights into dozens of marsupial species up to 25 million years old. They gave rise to relatives of Tasmanian devils, koalas, wombats, and kangaroos. Convergent evolution saw them develop similarities to placental species elsewhere. Through ice ages over the past few million years, evolution gave rise to an entire now-extinct megafauna of marsupial cats, moles, rats, wolves, and tigers—such as the thylacine (Tasmanian wolf/tiger) that survived until the 1930s (see image 3)—as well as species with no obvious placental counterparts, such as the giant wombat, *Diprotodon* (see p.432). Most of these giants began to go extinct around 50,000 years ago, shortly after the arrival of humans and other placentals, such as dingos. **GS**

4 Mya	2½ Mya	2 Mya	2 Mya	1½ Mya	50,000 Ya
The thylacine (Tasmanian wolf) evolves in Australia—it survives until the 1930s before succumbing to human persecution.	The last of the sparassodonts, including the South American marsupial big cats, die out at the end of the Pliocene epoch.	Australia's marsupial lion, *Thylacoleo*, evolves and flourishes until around 46,000 years ago.	The giant short-faced kangaroo, *Procoptodon*, is another member of a marsupial megafauna that evolves in Australia.	*Diprotodon*, a rhinoceros-size relative of the modern wombat, is the largest marsupial ever to have lived.	The Australian megafauna enters a sudden decline coinciding with the arrival of humans.

Giant Wombat
PLEISTOCENE EPOCH

SPECIES *Diprotodon optatum*
GROUP Diprotodontidae
SIZE 12½ feet (3.8 m)
LOCATION Australia

The giant (or hippopotamus) wombat was a very large marsupial mammal that lived in Australia through much of the Pleistocene epoch, from about 1½ million years ago until about 50,000 years ago, although it may have survived until as recently as 25,000 years ago. The largest marsupial known, it measured up to 12½ feet (3.8 m) long and stood 6½ feet (2 m) at the shoulder. This giant herbivore browsed on leaves, grasses, and other plant material, most probably in open woodland, and on semi-arid plains and savannas. The genus name *Diprotodon* means "two forward-facing teeth," which is a major feature of the skull. The closest living relatives of this marsupial are the three species of wombat—the common wombat, *Vombatus ursinus* and the two species of hairy-nosed wombats, genus *Lasiorhinus*—plus the koala, although these familiar marsupials are much smaller. *Diprotodon* was first described in 1838 by British biologist Richard Owen (1804–92) from fossils discovered in New South Wales. Fossils have also been found in South Australia, Queensland and Victoria, mainly in caves or in marshland. Lake Callabonna in South Australia has yielded fossils of hundreds of individuals that might have sought refuge there during a period of drought. Humans are thought to have arrived in Australia 50,000 years ago, so it is likely that they would have encountered *Diprotodon*. Some of the bones discovered have marks that could have been made by a spear, and it is plausible that hunting by these first indigenous Australians may have been partly responsible for the extinction of this remarkable mammal. As a relatively slow-moving herbivore, it would have been easy to hunt. **MW**

 NAVIGATOR

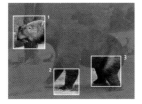

⊙ FOCAL POINTS

1 LARGE SKULL
The skull was massive yet lightweight due to its many air spaces. The lower jaw had two forward-directed incisors and the molars were large and simple, suitable for crushing vegetation. Fossils show that two sizes of *Diproodon optatum* existed: potentially males and females, rather than two species.

2 SMALL FEET
The feet of *D. optatum* were relatively small for such a heavy beast and the front feet in particular carried long claws. The gait was plantigrade (walking on the soles of its feet). Fossilized prints from various habitats indicate that the feet had a covering of hair.

3 STURDY LIMBS
The legs were sturdy and straight; the upper bones (humerus and femur) were longer than the lower bones (ulna/radius and tibia/fibula). Like living wombats, *D. optatum* walked with a pigeon-toed gait, its feet turned inward. It is likely to have moved slowly as it browsed on vegetation.

THE REAR-FACING POUCH

The marsupium, where young develop and which gives marsupials their name, has shown considerable evolutionary plasticity. In common with wombats (below), the koala *Phascolarctos cinereus,* and the mole genus *Notoryctes,* the pouch of *Diprotodon* opened toward the rear. This would have given developing young protection as the mother searched for food among vegetation. For burrowing wombats and moles, this prevents the pouch filling with soil. Marsupials living in open areas, such as the familiar kangaroos and wallabies, have forward- or upward-opening pouches.

◄ These tracks are likely to have been made by *Diprotodon.* A common scenario is for the animal to walk on a soft substrate, such as a muddy riverbank or silty seashore, and leave impressions, which dry and bake hard. Then erupted volcanic ash and cinders cover them and fossilization begins. The smaller curved tracks are the front feet, the larger ones, the rear feet; in some places the two overlap.

MODERN MAMMALS

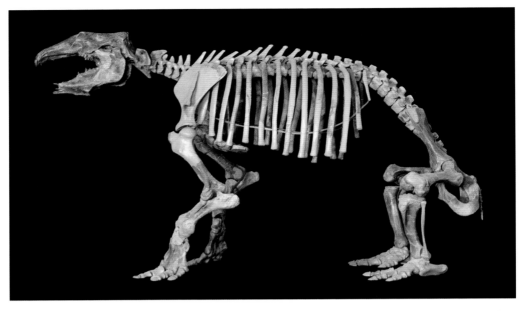

The mass extinction at the end of the Cretaceous period that wiped out an estimated 75 percent of Earth's species, including the nonavian dinosaurs and the marine and flying reptiles of the Mesozoic era, left behind a devastated environment. However, many new opportunities in food and lifestyle soon appeared for the survivors to exploit. Birds and mammals rapidly evolved to take advantage, with mammals in particular "exploding" in both size and diversity, exemplified by *Titanoides* (see p.436). Evidence from molecular clock studies (see p.17) suggests that the placental mammals that dominate the modern world (outside of Australasia) had already split into most of the major living groups during the Cretaceous, but this seems to contradict the fossil record and many paleontologists currently put greater faith in the physical evidence.

Molecular and genetic studies do, however, illuminate the close relationships between living mammal groups and the order in which they split from one another during their evolution. Although these studies have given rise to some groupings that currently have only poor support in the fossil record—controversially resulting in a redrawing of the mammalian family tree—there seems little doubt that the basic patterns they reveal are likely to be accurate. Much of the discussion of modern mammal groups is therefore based on a synthesis of evidence from genetics, fossils, and comparative anatomy of living mammals.

KEY EVENTS

120 Mya	66 Mya	60 Mya	56 Mya	35 Mya	34 Mya
Tectonic forces separate Australia (with Antarctica) from the southern continents, later allowing marsupial mammals to evolve in isolation.	Placental mammals appear, although some fossils and genetic studies point to much earlier dates of origin.	By this time the herbivorous pantodonts have spread to North America and started to diversify.	Global temperatures soar in the Paleocene –Eocene Thermal Maximum (PETM). Mammals, such as artiodactyls and primates, appear.	Multituberculates, a group of rodentlike mammals with a separate descent from the placental true rodents, become extinct.	Mammals around the world, but particularly in Europe, are affected by an extinction event known as the Grande Coupure (Great Break).

According to genes, living mammals are split into a few deep-rooted groups. The Xenarthra are South American sloths, anteaters, and their kin. The Afrotheria include a variety of African mammals, including elephants, aardvarks, and hyraxes. These two groups have few physical similarities, but one theory is that they share a more recent common ancestor with one another than with any other mammals. This is partly backed up by geological evidence from plate tectonics and continental drift, which suggests that Africa and South America were the last parts of the ancient southern supercontinent of Gondwana to separate from one another. Further evidence points to other relationships with the group Boreoeutheria. This name means "northern eutherians," which alludes to its origins on what was then the northern equivalent of Gondwana—the great supercontinent of Laurasia (North America and Eurasia). Within this group is the Laurasiatheria, which includes hoofed mammals, whales, bats, and carnivorans, and Euarchontoglires, which includes tree shrews, primates, rabbits, and rodents. It is unknown exactly how these groups are truly related.

These four major groups of living placental mammals included significant extinct groups. The Afrotheria, for example, once encompassed the carnivorous Ptolemaiida, the anteaterlike Afredentata, the hippopotamuslike desmostylians (see image 1), and the rhinoceroslike embrithopods (see image 2). The Laurasiatheria, meanwhile, included the early hoofed mammals known as condylarths, the wolflike mesonychids, and the horned, herbivorous dinoceratans. The Euarchontoglires also lost a major group known as the Plesiadapiformes, which were close relatives of the primates. Of the four, only the Xenarthra avoided the loss of a complete branch of its family tree.

Over the years since dinosaurs disappeared, mammals have adapted to a huge variety of situations that have shaped their evolution. Large-scale climate change saw temperatures soar to a peak around the Paleocene and Eocene epochs, roughly 56 million years ago, then cool gradually through the Eocene and Oligocene epochs, with ice ages over the past few million years. Climate change and insect evolution, coupled with the rise of flowering plants (see p.86), led to the dominance of mammals. The retreat of once-abundant forests and the spread of new plants, such as grasses, also affected their evolution. On the longest timescales of all, however, the slow but inexorable rearrangement of the continents resulted in the separation and isolation of populations in some places, leading to the development of new species; elsewhere intercontinental collisions brought once-isolated species into competition with one another. This is seen clearly in the Great American Interchange of North and South American mammals from 10 million years ago, as today's Central American connection gradually formed and separated the Pacific and Atlantic oceans. Such factors have produced great ebbs and flows in the pattern of evolution during the Age of Mammals, with some families rising to prominence and others dwindling to extinction. **GS**

1 The 7-foot-long (2 m) desmostylian *Paleoparadoxia*, from around 15 million years ago, had a mouth and teeth adapted to gulp mouthfuls of marine plants, such as seagrasses.

2 The afrotherian *Arsinoitherium* from Egypt was 10 feet (3 m) in length, lived 30 million years ago, and evolved massive rhinoceroslike horns.

30 Mya	23–5¼ Mya	9–3 Mya	2¼ Mya	50,000 Ya	11,700 Ya
In the aftermath of the Grand Coupure, the original mammals of Europe are largely replaced by those from Asia.	The Miocene epoch is marked by a general cooling of global climate, the spread of grasslands and the rise of grazing mammals.	The approach of North and South America, and the formation of the Isthmus of Panama, lead to the Great American Interchange of once-isolated species.	The onset of the Pleistocene epoch sees repeated ice ages. Mammals grow larger to survive, producing "ice age megafauna" (see p.550).	The arrival of humans in Australia coincides with the start of rapid extinctions among the large marsupials or megafauna.	At the end of the most recent ice age, most of the megafauna decline toward extinction, marking the beginning of the current Holocene epoch.

Titanoides
PALEOCENE EPOCH

SPECIES *Titanoides primaevus*
GROUP Pantodonta
SIZE 10 feet (3 m)
LOCATION North America

After the great end-of-Cretaceous extinction 66 million years ago, the pantodonts were among the first mammals to take advantage of the evolutionary vacancies left by the dinosaurs. This group is usually considered part of a large and now-extinct group of early mammals called cimolestids. Pantodonts had an overall similarity to modern wolverines and weasels but they followed herbivorous lifestyles. Their earliest known fossils have been found in China and date to the early Paleocene, only several million years after the mass extinction. By 60 million years ago pantodonts had spread to North America and had started to diversify. While most were roughly the size of modern dogs, the smallest were opossumlike tree-dwellers and the largest grew to the size of bears. Size may have been an advantageous trait for protection from the smaller mammalian predators of the time.

The first widespread genus of these large herbivorous pantodonts seems to have been the huge, slow-moving *Barylambda*, of which three North American species are known, all from around 60 million years ago. It is likely that they gave rise to the smaller and highly successful *Coryphodon*, the dominant herbivore across much of the northern hemisphere between 57 million and 46 million years ago. However, perhaps the most impressive pantodont of all was *Titanoides*. Although somewhat smaller and lighter than *Barylambda*, the genus carried a fearsome set of characteristics, including huge, overdeveloped canine teeth and sharp claws on all four feet. Spectacular though it was, *Titanoides* thrived for only 3 million years before becoming extinct 57 million years ago. However, the pantodont group as a whole continued to flourish for much longer, with its last members becoming extinct in the late Eocene, around 34 million years ago, in Mongolia. **GS**

⊕ NAVIGATOR

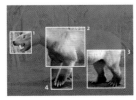

FOCAL POINTS

1 CANINE TEETH
Specialized canine teeth could have fulfilled a dual role. The hooklike lower pair may have dug up buried food, which the saberlike upper canines then sliced for processing by the crushing molars. If the upper teeth were used as weapons, it is likely they were only for self-defence.

2 LIMBS AND BULK
Powerfully built, stocky limbs supported *T. primaevus*'s large bulk, and made it a powerful digger. Size alone would have protected it from the largest mammalian predators of the time. Contemporary pantodonts from Asia are smaller and may have been capable of climbing trees.

3 FEET
T. primaevus's feet ended with five thick claws, showing that it walked on its soles (plantigrade) instead of its toes (digitigrade). Fossil footprints linked to *T. primaevus* have been found from the Svalbard Archipelago in Norway, confirming its bearlike gait and that it migrated from its US origins.

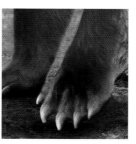

4 CLAWS
Elongated claws were likely for digging up tough roots and tubers, but could also have been for defence. *Titanoides* is the only known pantodont with claws rather than hooflike toes, and its habitat—the Dakotas—was a tropical swamp where crocodiles lay in wait to ambush herbivores.

ADAPTATION OF THE SKELETON

Most specimens of *Titanoides* have been identified only from fossil teeth and jaws (below), but rare body skeleton remains suggest that the animal grew up to 10 feet (3 m) long and reached a weight of up to 330 pounds (150 kg). Intriguingly, a succession of known species suggests that *Titanoides* evolved smaller body size over time, perhaps adapting to changing climate conditions. For example, a drying trend would make plant material less common and tougher, meaning that *Titanoides* would have more trouble gaining enough nutrition to fuel its large body. Over several generations, this could encourage the survival of smaller *Titanoides* individuals, causing evolution by natural selection.

▼ Bryan Patterson (1909–79), curator of mammals at the Field Museum, Chicago, from 1942 to 1955, demonstrates the size of the front feet of *Titanoides* and how it walked. The claws would have been at the tips of the toe bones.

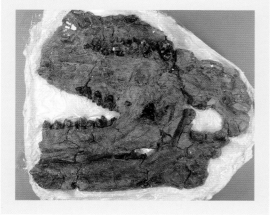

PRIMATES

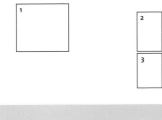

1 Cheirogaleidae, the mouse lemurs, are the smallest primates. Like all members of the lemur group, they evolved in isolation on the island of Madagascar.

2 *Teilhardina magnoliana* is one of the oldest known primates in North America but its relationship to the Asian species within the same genus is uncertain. If clarified, this could shed light on the split between the Old and New World primates.

3 The sole *Archicebus* specimen, found in China, has focused attention on Asia for early primate evolution.

In traditional animal classification, the primates—order Primates—originated with 18th-century Swedish botanist and taxonomist Carl Linnaeus (1707–78), who produced the first classification of animals and plants to be recognized by modern science. Linnaeus adopted the name "Primates" from the Latin *primus*, meaning "first," which reflected the prevailing culture of science in the Western world at the time. Its dominant religious orthodoxy placed humans above all other forms of life, and humanlike creatures as first among animals. In terms of biological reality, the Primates is a small group of predominantly tropical mammals, which evolved more than 55 million years ago and, according to some experts, perhaps as early as 90 million to 80 million years ago. Today, the group includes more than 200 species, and possibly as many as 370 if the closely related kinds that are regarded as subspecies are classed as separate species. By comparison, there are more than 1,200 living bat species and over 2,200 rodent species.

The living primates include lemurs, lorises, galagos (bushbabies), tarsiers (see p.444), marmosets, monkeys, gibbons, apes, and humans. Typically they are small- to medium-size omnivores, ranging from 1-ounce (30 g) mouse

(see p.444)

KEY EVENTS

90–80 Mya	75 Mya	66 Mya	57 Mya	56 Mya	55 Mya
The primates and their relatives, the colugos and tree shrews, may have originated at this time—the Age of Dinosaurs.	The ancestors of lemurs may have reached Madagascar by crossing the sea from Africa on vegetation rafts.	The end-of-Cretaceous extinction, when several groups of large reptiles die out, allows many kinds of mammal groups to replace them.	*Altiatlasius koulchii* from Morocco may be the oldest of the now-extinct omomyids, which were tarsierlike primates.	*Teilhardina*, a possible omomyid, occupies Mississippi and is one of North America's earliest primates.	Lemurlike adapiform primates, such as *Notharctus* from Europe and North America, evolve at this time.

lemurs (see image 1), with a body length of 4 inches (10 cm), to the 550-pound (250 kg) gorilla, with a standing body height of 5¾ feet (1.75 m). Additionally, there is a much larger diversity of extinct relatives and ancestors, especially from the Miocene epoch, more than 10 million years ago, when their geographical distribution extended into North America and Eurasia. The primate body, with its four limbs and tail, enlarged brain, and acute stereoscopic vision for distance judgment, is adapted for an arboreal (tree-dwelling) way of life. Significant evolutionary advances from the primates' small insectivore ancestors include the acquisition of unique features such as an opposable thumb, flat nails instead of claws, a shortened snout and closely spaced eyes pointing forward. Primates also show an enlarged brain for body size, delayed sexual maturity, reduced numbers of offspring, and increased maternal care and social behavior. Most are relatively defenceless animals, depending for survival on heightened senses of danger and cooperative awareness of danger, along with the ability to escape into the tree canopy to avoid predators.

Apart from human primates, the group consists primarily of arboreal woodland-dwellers in the tropics and subtropics around the world, with the exception of Australasia and the islands of the Pacific. However, the Chinese and Japanese macaque monkeys are adapted to survive cold, snowy winters. Fossilization in most woodland and forest regions is rare because the remains are usually scavenged and decompose rapidly; consequently, the overall primate fossil record is poor. It largely consists of teeth and other tough bony skeletal remnants left by predators and scavengers. These fragmentary fossils show a deep-seated division between primates of the Old World, mainly Africa and Asia, and the New World of the Americas, as exemplified by *Teilhardina* (see image 2); the long-term continental separation of these landmasses by ocean waters started during the Cretaceous.

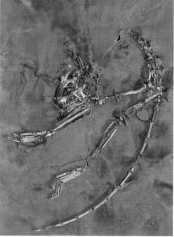

Traditionally, primates have been subdivided into the prosimians—lemurs, lorises, bushbabies, and tarsiers—and the simians or anthropoid primates, which include the monkeys, marmosets, gibbons, apes and humans. However, recent molecular analysis produces a different division: the strepsirrhines, lemurs, and lorises; and the haplorhines (simple noses), tarsiers, and anthropoids. According to fossil evidence, the evolutionary divergence of these two groups dates back 55 million years to the Eocene epoch. However, molecular clock estimates (see p.17) push the divergence further back, again into the late Cretaceous. There is also conflicting evidence over the site of primate origins and the fundamental divergence of the major groups. The fossil primate *Archicebus* (see image 3), from 55-million-year-old Eocene rocks in China, has moved back the evidence for the haplorhine/strepsirrhine divergence. This strengthens the argument that Asia, rather than Africa, was the center of origination for these early primates.

55 Mya	50 Mya	50 Mya	47 Mya	40 Mya	15–9 Mya
Archicebus from China lays down fossils that are later used in the analysis of the haplorhine/ strepsirrhine divergence.	The earliest fossil tarsiers are the Eocene *Shoshonius* from North America and *Afrotarsius* from Egypt.	In Madagascar, lemurs are rapidly diversifying into numerous species, with particular diets and behaviors.	*Darwinius*, from Messel, Germany, is the best preserved of the adapiforms from the Eocene.	*Adapis*, which gave its name to the Adapiformes group of primates, is found at several locations in Europe.	Adapiform genus *Indraloris* inhabits the Indian subcontinent.

A near complete fossil skeleton of *Archicebus* shows that it was a tiny, 1-ounce (20–30 g), slender-limbed, long-tailed primate with a mixture of tarsierlike features in its skull, teeth, and limb bones, yet more anthropoidlike features in its heel and foot bones. *Archicebus*'s long legs and grasping feet suggest that it was a tree-dweller and used its long tail as a balance when leaping from branch to branch. And, since its eyes were not enlarged, it was most likely active in daylight, when it fed on insects. Detailed analyses of some 1,200 anatomical features suggests that the evolutionary position of *Archicebus* is near to the origin of the tarsier group, and it shows that the split between the haplorhines and the strepsirrhines had occurred prior to 55 million years ago.

Among the many extinct primate groups were the omomyids. With more than 40 genera of somewhat tarsierlike early primates, they evolved and died out across North America, Eurasia, and North Africa during the Eocene, 56 million to 34 million years ago. Most of the omomyids were large-eyed animals with grasping hands and feet, nails instead of claws, and relatively small body weights of less than 18 ounces (500 g). Their body form was like that of the living dwarf lemurs and they were almost certainly tree-dwelling nocturnal primates. It is generally accepted that the omomyids gave rise to the living tarsiers. The fossil species *Altiatlasius koulchii*, discovered in 1990 in 57-million-year-old rocks of the late Paleocene epoch in Morocco, may be the oldest omomyid. However, it is known only from a few teeth and bony jaw fragments, and some experts suspect that it could be a strepsirrhine, therefore more closely related to the lemurs.

Another extinct group, Adapiformes, encompassed 35 extinct genera. The earliest species, and the namesake of the group, is *Adapis* (see image 4). It evolved in the Eocene around 55 million years ago, and some *Indraloris* species survived through into the late Miocene, around 9 million years ago. They had a similar geographic distribution to the omomyids, but in contrast, many of the adapiforms were more lemurlike and somewhat larger, with body weights of around 2¼ pounds (1 kg); they also had small eyes, more elongate snouts, and cheek teeth (molars and premolars) adapted for a plant diet. They were mostly day-active tree-dwellers with a wrist and ankle anatomy more similar to strepsirrhines than haplorhines.

However, the adapiforms lacked some of the special adaptations of the living strepsirrhines, such as the distinctive closely spaced "toothcomb" structure of the front teeth, which living lemurs and lorises use to groom their fur. One adapiform is known worldwide, namely *Darwinius masillae* (see image 5; see p.442). Discovered in recent decades from 47-million-year-old Eocene rocks in Messel, Germany, *Darwinius* attracted a great deal of interest and debate as to whether it was an early haplorhine foreshadowing later features, or an early strepsirrhine and more closely related to the lemurs. When found in 1983, the fossil slab was split in two and the separate pieces sold to different owners. The two parts were not reunited until 2007; only then could a complete study of the specimen be made. Already informally named "Ida," examination revealed no trace of a penis bone (os penis or baculum), a bone present in many mammals—but not humans—that helps to support and stiffen the male organ during sex. This evidence pointed to the name "Ida" being suitable because the specimen was female.

The evolutionary origins of the primates lie within a wider mammalian group called the Euarchonta. This is based largely on molecular evidence that shows a close relationship between the Scandentia (tree shrews), the

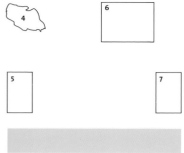

4

6

5

7

4 Fossil skulls of *Adapis* show the large eye sockets and, toward the rear, the relatively large braincase, similar to most primates.

5 A reconstruction of *Darwinius* sports the long slim limbs and very long tail typical of tree-climbing primates.

6 Colugos, "living parachutes," use their incredible aerial skills for many things, including to escape predators.

7 Tree shrews, or tupaiids, such as this northern tree shrew, *Tupaia belangeri*, form a sister group to the primates.

Dermoptera (colugos) and Primates. The colugos are another Southeast Asian group of small, tree-dwelling, plant-eating mammals notable for their ability to glide considerable distances from higher to lower parts of the forest canopy using a skin membrane that stretches between their limbs. The colugos' poor fossil record suggests that they were more diverse in the past: they originated in the Paleocene, with three extinct families and a much wider geographical distribution in North America and Eurasia. Today the group is reduced to two living species: the Philippine colugo, *Cynocephalus volans*, and the Sunda or Malayan colugo, *Galeopterus variegatus* (see image 6). Considering their gliding skills, they are surprisingly large animals, and grow to 16 inches (40 cm) in length and weigh up to 4½ pounds (2 kg). The slender limbs are of moderate length, as is the tail. The skull is relatively small with an extended snout and large, forward-facing eyes with stereoscopic vision and excellent distance judgment, necessary for gliding and landing on vertical tree trunks. However, they lack the opposable thumbs of the primates, so cannot grasp branches and instead use their sharp claws for climbing. These fascinating creatures, like the tree shrews, are a living storehouse of evolutionary features, and new findings about them continue to shed light on primate evolution.

The tree shrews (see image 7) include 19 living species of small, arboreal, omnivorous mammals, which live in the forests of Southeast Asia. They have a body form somewhat like that of squirrels, but their skull anatomy and relatively large brains have similarities to primitive primates—a physical feature supporting the molecular evidence. Long before genetic and molecular studies, 19th-century evolutionary biologists considered that tree shrews had the size, shape, and habits of the earliest mammals, even if they were not actually these earliest kinds. The early fossil record of these close primate cousins—tree shrews and colugos—is even poorer than that of the early primates. It does not extend back any further than the late Paleocene or early Eocene. The molecular clock method, however, places their origin further back still, 88 million years ago in the late Cretaceous, when the great dinosaurs still roamed the planet. Widening out from the Euarchonta—primates, tree shrews, and colugos—the next most closely related mammal groups are the rodents and the lagomorphs (rabbits, hares, and pikas). It seems that primates had their deep origins in mainly herbivorous ancestors before developing omnivorous species. **DP**

Darwinius
EOCENE EPOCH

SPECIES *Darwinius masillae*
GROUP Caenopithecinae
SIZE 23 inches (58 cm)
LOCATION Germany

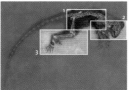

In 1983, a small fossil primate called *Darwinius masillae* was discovered. When later studied in 2007, the exceptionally preserved remains caused controversy. With its long legs and tail, the 23-inch-long (58 cm) primate was lemurlike in its body form and was included in the subfamily Caenopithecinae, in the Adapiformes, all of which are extinct. *Darwinius*—nicknamed "Ida" after the daughter of one of the fossil's leading researchers—is the third fossil primate from Messel (see panel) to be recognized as a member of this family.

The evolutionary relationships of these primitive primates to the living groups have been debated. The experts who first described and analyzed *Darwinius* claimed that it showed anatomical features unique to the haplorhine primates: tarsiers, monkeys, and apes. These include a deep jawbone (mandibular ramus), a short snout (cranial rostrum), and the loss of grooming claws. Together, these features indicate the early diversification of haplorhines, from which later primates evolved, including apes. Because it was from early haplorhines that later monkeys and apes—including humans—evolved, it was further claimed that *Darwinius* belonged to an early common branch or stem group that formed an important transition between primitive primates and, eventually, humans. Previously, the extinct adapiform primates were considered more closely related to the strepsirrhines: the other group of living primates, which includes the lemurs, galagos (bushbabies), pottos, and lorises. Other experts maintained that the analysis of *Darwinius* was flawed. They instead claimed that *Darwinius* does indeed indicate a relationship between the adapiforms and the strepsirrhines and so is not a distant direct cousin of humans. This latter interpretation has been supported by subsequent, more detailed, analysis. **DP**

FOCAL POINTS

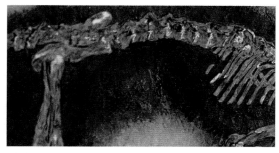

1 STOMACH CONTENTS
The last meals of plant food, leaves, and fruit are preserved in the fossil, along with soft tissues, such as traces of body hair. *Darwinius* is typical of Messel's best preserved fossils and has a near complete skeleton, with only the left hind leg missing. The animal, an immature or sub-adult, had reached about 80 percent of its final body length, with relatively long legs and also a long tail.

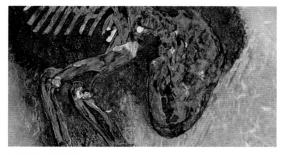

2 JAWS AND TEETH
The sharp, pointy ridges on the molars, also seen in other herbivorous primates and certain mammals, are typical of a plant diet that includes leaves (folivorous) and seeds (granivorous). The jaws also contain molars that are not yet erupted, or grown out of the gum, indicating that "Ida" was probably 8 to 10 months of age at the time of her death.

3 HANDS AND FEET
Both the hands and feet have grasping fingers and toes with nails rather than claws, along with opposable thumbs and big toes. This arrangement is ideal for climbing trees and for grasping, holding, and manipulating plant food. There is no grooming claw, which is the elongated clawlike nail on the second digit of the rear foot present in modern lemurs, bushbabies, and lorises.

DARWINIUS RELATIONSHIPS

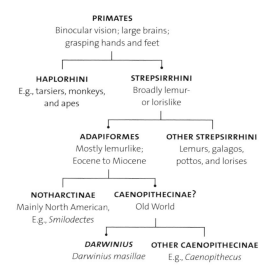

PRIMATES
Binocular vision; large brains; grasping hands and feet

HAPLORHINI
E.g., tarsiers, monkeys, and apes

STREPSIRRHINI
Broadly lemur- or lorislike

ADAPIFORMES
Mostly lemurlike; Eocene to Miocene

OTHER STREPSIRRHINI
Lemurs, galagos, pottos, and lorises

NOTHARCTINAE
Mainly North American, E.g., *Smilodectes*

CAENOPITHECINAE?
Old World

DARWINIUS
Darwinius masillae

OTHER CAENOPITHECINAE
E.g., *Caenopithecus*

▲ *Darwinius* is certainly a remarkable specimen. Much discussion has centered on whether it was on the lemur-based, strepsirrhine branch of the primate evolutionary tree, or the haplorhine lineage leading to monkeys and apes, including humans.

DIVERSITY AT MESSEL PIT

The once quarried early Eocene oil shales at Messel in Germany (below) have been a World Heritage Site since 1995 due to the outstanding richness and preservational quality of the fossils found there. About 47 million years ago, a maar lake (flooded volcanic crater) offered a range of habitats, from open water, swamp, bankside, and damp forest, to drier banks and higher ground wooded with pines, beech, chestnut, and oaks. The abundant vegetation and warm climate encouraged a diversity of mammal life, from dog-size primitive horses, anteaters, and rodents, to bats and early primates, as well as varied birds, fish, and insect life.

Tarsier
PLIOCENE EPOCH—PRESENT

The 16 living species of tarsier are small, tree-dwelling, carnivorous primates. Within the order Primates, tarsiers have been classified as haplorhines, as supported by genetic evidence. Haplorhine features include a proper nose instead of a naked nose pad (rhinarium) and a unique structure to the external ear canal, which is shared with later evolved haplorhines, such as baboons and apes. Tarsiers, such as *Carlito syrichta*, are characterized by a large, round head, big eyes, large ears, and the ability, like owls, to turn their heads 180 degrees—adaptations for life at low light levels in tropical rain forests and scrub forests. Excluding the tail, body length ranges from 3½ to 6 inches (9–15 cm) and weight from 3½ to 4½ ounces (100–135 g). The hind legs are longer than the arms and tarsiers are proficient jumpers with excellent distance judgment, thanks to their flat faces, forward-facing eyes, and stereoscopic vision. Most of the fingers and toes carry flat nails, but the second and third toes have grooming claws. Tarsiers on Southeast Asian islands are the sole survivors of this group of small primitive primates, whose evolutionary history extends back to the Eocene. Their fossil distribution is much more widespread than the living species. In the early Eocene, more than 50 million years ago, they spread from North Africa, with the genus *Afrotarsius* from Egypt, across Eurasia to North America, with genus *Shoshonius* from Montana. **DP**

✦ NAVIGATOR

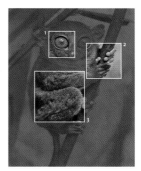

SPECIES *Carlito syrichta*
GROUP Tarsiidae
SIZE 6 inches (15 cm)
LOCATION Philippines
IUCN Not Yet Assessed

👁 FOCAL POINTS

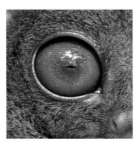

1 HUGE EYES
The eyes of *C. syrichta* are about ½ inch (16 mm) wide—enormous relative to the size of the skull and brain. Big eyes are of vital importance to this nocturnal creature, which hunts insects and other small animals. As a result, the skull and brain anatomy have unique features to accommodate them.

2 HANDS AND FEET
The finger bones are lengthy, and the third finger is almost the same length as the upper arm. The legs are usually folded against a branch, but when extended they are longer than the body. The ankle bones are unusually long, and the tibiofibular joint between the two lower leg bones is fused.

3 LIMBS
C. syrichta's limb bones are extremely elongate and "spring-loaded," adapted for clinging to tree trunks and vertical branches. Even in near darkness, the tarsier can make sudden jumps from standstill to catch its prey in mid-air and then land and grip the tree safely.

NOCTURNAL POTTO

A secretive member of the loris family in the strepsirrhine primate group, *Perodicticus potto* (above) is a nocturnal inhabitant of tropical forests in West and Central Africa. Weighing 2¼ pounds (1 kg), it has a body length of 14 inches (35 cm) and a short tail. A long tail is not needed because the potto is adapted for creeping through the branches slowly, unlike primates such as tarsiers, which leap. It has evolved unusual hands: the second digit (index finger) has almost disappeared, but the thumb is opposable so the potto is able to grasp branches. It is also one of the few primates with a specialized anatomy for defence. Its neck bones have lumpy points, and when threatened the potto "neck-butts" its enemy, occasionally raising its head to deliver a bite.

BATS

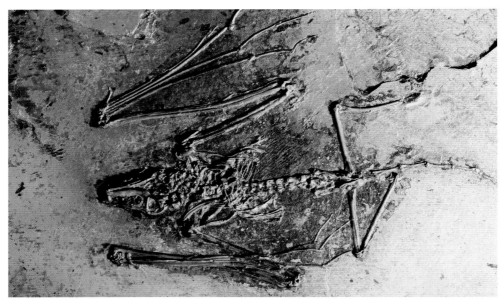

1 *Palaeochiropteryx*, from 48 million years ago, had a wingspan of 12 inches (30 cm).

2 The gray-headed flying fox, *Pteropus poliocephalus*, with its doglike face, exemplifies the megabat group.

3 Rousette megabats, such as Geoffroy's rousette, *Rousettus amplexicaudatus*, produce clicks for echolocation that can be heard by humans.

In most parts of the world bats are elusive, seen only briefly at dawn or dusk, yet they are the second largest group of living mammals after rodents, with more than 1,200 known species. This is evidence of a long evolutionary history, which may have begun as early as 80 million years ago, with plenty of opportunity to diversify. Bats, from the order Chiroptera, are broadly divided into two distinct groups. The microbats, or microchiropterans, are mostly insectivorous with remarkable echolocating abilities that allow them to hunt at night, and the megabats, or megachiropterans, are fruit-eating bats that lack echolocation abilities. Despite these names, megabats are not always larger than microbats—the smallest megabats are smaller than the largest microbats. Microbats account for about 80 percent of all bat species and have a long but somewhat sparse fossil record stretching back to *Icaronycteris* (see p.448) and *Palaeochiropteryx* (see image 1), one of the earliest known bats, from 48 million years ago. Megabats, such as those of the genus *Pteropus* (see image 2), appear to have evolved much later—their earliest known fossil is *Archaeopterus*, dating from the Oligocene epoch, 30 million years ago. However, some argue that *Archaeopterus* shows distinctive microbat features.

For several years, the lack of echolocation in megabats, and differences in claw arrangement on their wings, dentition, and the structure of the skull, raised doubts about whether the two bat groups were closely related. However, recent

KEY EVENTS

80 Mya	56 Mya	54½ Mya	52½ Mya	52½ Mya	47 Mya
A mammal tooth identified from the late Cretaceous of South America may be a sign of early bat evolution.	The Paleocene–Eocene Thermal Maximum (PETM) encourages evolution of mammals; several kinds of bats begin to appear soon after.	The Australian bat *Australonycteris* may have hunted fish as well as insects and appears to display echolocating adaptations.	*Icaronycteris* inhabits Wyoming's Green River Formation and shows signs of echolocation abilities. Later, it produces the first fully preserved bat fossil.	*Onychonycteris*, also from the Green River Formation, is a well-developed bat that appears to lack echolocating apparatus.	*Palaeochiropteryx*, from Germany's Messel Pit, has short broad wings for flying through dense forests, and is believed to have been insectivorous.

DNA analysis confirmed that megabats and microbats are more closely related to one another than they are to any other mammals. This study also suggested that bats should be divided into two new groups: the Yinpterochiroptera, including the megabats and three distinctive microbat families, and the Yangochiroptera, accounting for all remaining microbats. This grouping is based on the theory that megabats evolved from echolocating insectivorous ancestors and subsequently lost the ability as their diet changed. There are still many questions surrounding the timing of bat evolution and their closest fossil relations, and where genetic evidence places them in the mammal family tree. Anatomical similarities previously placed them as close relatives of the tree shrews, primates, and gliding colugos, yet the genes suggest—rather surprisingly—that they are most closely related to hoofed mammals, carnivores, and whales, in a large group or superorder known as the Laurasiatheria.

The delicate anatomy of bats means that they are fossilized only in exceptional circumstances, so there are currently no fossils showing the intermediate stages through which bats presumably evolved from earlier gliding mammals. However, these early stages in evolution can be deduced from the anatomy of both fossil and modern bats. A bat's wing is a modified arm, with the individual finger bones of its hand greatly extended and far more flexible than normal bone in order to support the membrane that stretches between them. Bat bones are also much lighter and thinner than those of other mammals, to reduce their flying weight. Unfortunately, this adaptation diminished their preservation as fossils.

A number of fossil teeth from various small mammals of the late Cretaceous appear similar to those of bats, and this has prompted several contrasting theories about which mammal group the bat ancestors belonged to. However, even if a single origin could be agreed upon, the teeth alone would reveal very little about the animals that preceded bats. Without further fossil discoveries, it will not be known whether ancestral bats flew (or glided) over the last big dinosaurs. However, looking at circumstantial evidence, one favored theory on the factors that drove the evolution of bats connects them with the explosion in flowering plant types during the late Cretaceous and the accompanying diversification among insects. This created a natural opportunity for small mammalian insectivores to evolve, which, in turn, allowed agile mammal-hunting dinosaurs to thrive, including some suspected arboreal (tree-dwelling) species. In this habitat, the ability to glide from tree to tree in order to escape predation would have given bat ancestors an obvious evolutionary advantage, and as their flying became more skilled they would have used it for hunting. Intriguingly, the megabat genus *Rousettus* (see image 3) seems to have re-evolved a new form of echolocation based on tongue clicks, but since they are frugivorous (fruit-eaters), they use their skill primarily for navigation rather than for hunting. **GS**

46 Mya	41–34 Mya	37 Mya	34 Mya	30 Mya	3 Mya
Tanzanycteris is the earliest known placental mammal from sub-Saharan Africa. An enlarged inner ear suggests advanced echolocation.	Several species of the widespread and successful genus *Archaeonycteris*, from the early and middle Eocene epoch, live mainly in Europe.	*Phasmatonycteris butleri*, from the Egyptian Sahara, is the earliest fossil sucker-footed bat, and part of a family now limited to Madagascar.	Most of the 25 or more families of bats, living and extinct, have evolved by this time, the end of the Eocene.	The long-tailed *Archaeopterus* inhabits northern Italy.	*Desmodus archaeodaptes* inhabits Florida during the late Pliocene epoch. It is the earliest known vampire bat.

Icaronycteris

EOCENE EPOCH

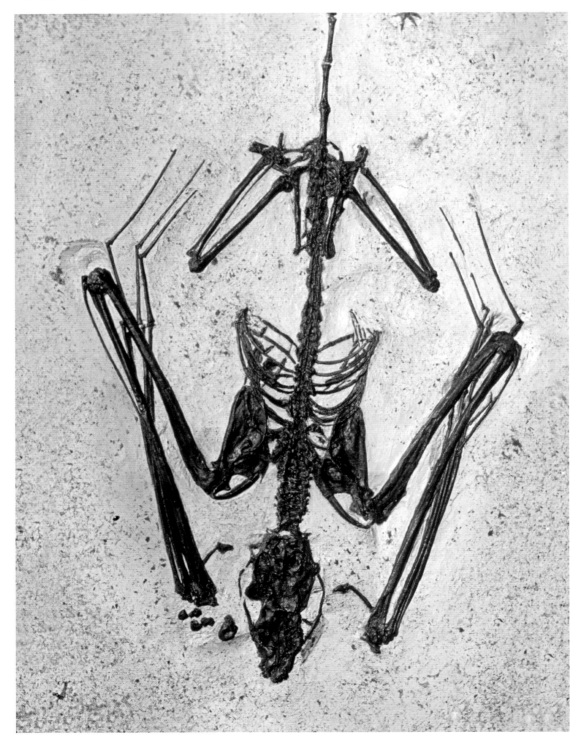

The fragility of bat bones means that bats are among the rarest vertebrates in the fossil record. The oldest known fossil bat, *Icaronycteris*, was found in 52½-million-year-old rocks from the early Eocene, within the Green River Formation—a layer of rocks that formed at the bottom of a chain of lakes in present-day Colorado, Wyoming, and Utah. Animals that died and fell into parts of the lakes were buried rapidly in lime muds, which preserved them in exquisite detail. Not only bones survive, but also traces of soft parts such as skin. In *Icaronycteris*'s case, this means that the outline of its wing membranes is preserved. Its body was only 6 inches (15 cm) long and its wings spanned 14 inches (35 cm). Four main fossil specimens have been found at Green River, all apparently a single species named *I. index*. (Fossil fragments from contemporary French rocks suggest the existence of a second species.) Preserved moth scales have been found in what would have been the stomach of one Green River specimen, suggesting that *I. index* had already evolved the echolocation necessary to hunt flying insects at night. The general structure of *I. index*'s body was fairly flexible compared to later bats. While modern chiropterans have a fused or one-piece breastbone with a keel for attachment of flight muscles (similar to birds), *I. index*'s breastbone is unfused. However, this could also be explained if the fossil was a juvenile in which the sternum had not yet fused. **GS**

SPECIES *Icaronycteris index*
GROUP Icaronycteridae
SIZE 6 inches (15 cm)
LOCATION North America

👁 FOCAL POINTS

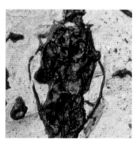

1 INNER EAR
I. index's cochlea is within the size range of modern microchiropterans, suggesting that it already had well-developed echolocation. It is conjectured that echolocation first evolved as an aid to navigation in the dark, and then was additionally adapted to move in on prey by sound.

2 ELONGATED TAIL
I. index's tail was long compared to its body and almost rodentlike, in contrast to most living bats, where the tail is greatly reduced to a stump. It also lacked a uropatagium—the aerodynamic membrane that stretches between the hind legs and tail of living bats. This probably limited its flying ability.

3 LEGS, WINGS AND ROOSTING
The leg structure suggests the typical chiropteran attribute of hanging upside down to rest and sleep. A bat's long finger bones and wing membrane make resting on the ground difficult and walking slow. Hanging upside down was almost certainly an evolutionary adaptation to avoid predators.

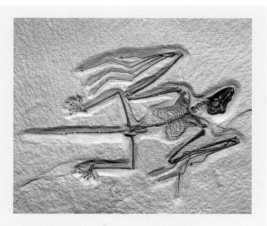

ANCESTORS' ANATOMY

The discovery in 2008 of an early bat species contemporary to *Icaronycteris* has shed further light on the evolutionary story. *Onychonycteris* (above), also from the Eocene Green River Formation, has the unique primitive feature of claws on all five of its distended fingers, hence its name, meaning "clawed bat." This gives clues to the anatomy of the bats' ancestors, because all other living or fossil bats have either two or three claws. Also, the structure of *Onychonycteris*'s ear shows none of the specialized features associated with echolocation, in contrast to the circumstantial evidence that *Icaronycteris* and even earlier bats could hunt by sound. Therefore it seems likely that bats evolved flight before they evolved echolocation and lost their claws.

Greater Mouse-tailed Bat
PLEISTOCENE EPOCH—PRESENT

SPECIES *Rhinopoma microphyllum*
GROUP Rhinopomatidae
SIZE 3 inches (7.5 cm)
LOCATION North Africa, Middle and Near East
IUCN Least Concern

The modern mouse-tailed bats that make up the Rhinopomatidae family are limited in diversity, with a maximum of six recognized species in the single genus *Rhinopoma*, but they are widespread in terms of numbers and distribution. They can be found from North Africa across the Arabian Peninsula, and as far as Southeast Asia. They favor arid and semi-arid regions and typically roost in caves and empty houses—and also take shelter in ancient ruins across the Middle East, including in the Egyptian pyramids. Colonies range in size from only a few individuals to several thousand, and they can be an impressive sight when they emerge from their roosts at sunset. Their "flutter and swoop" pattern of flight means that individual bats in search of insect prey, such as moths and beetles, can easily be mistaken for birds.

The most prominent feature of these bats is their distinctive long tail and reduced uropatagium—the thin flight membrane between the hind legs. Both of these features are reminiscent of the very early fossil species *Icaronycteris*. Such apparently primitive physical traits can be misleading in species that are in fact relatively modern. However, in the case of the mouse-tailed bat, genetic analysis suggests that these features are genuinely primitive: they have been inherited with little alteration from earlier species, instead of being secondary characteristics resulting from later evolution that changed the bats' appearance to closely resemble an ancestral form. Despite this, other aspects of mouse-tailed bat anatomy are advanced and are clearly specialized for adaptations to a desert lifestyle. These include a slitlike nostril shape—analogous to that of camels, which is designed to keep sand and dust away from breathing passages—and kidneys that can produce highly concentrated urine in order to retain as much water as possible in the body. Mouse-tailed bats have also evolved a system whereby they avoid complete hibernation; they instead enter a state of torpor (sluggish, reduced activity) during cold conditions. **GS**

 NAVIGATOR

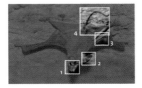

FOCAL POINTS

1 EYESIGHT
Despite their nocturnal habits and reliance on echolocation and sense of touch, none of the known bat species are blind. But sight in *R. microphyllum* is reduced in importance, especially compared to day-active megabats, which commonly have large eyes.

2 INNER AND OUTER EAR
R. microphyllum's inner ear has evolved great size, with adaptations to its shape and the number of twists allowing it to hear very specific frequencies. The outer ear is large and mobile, and channels sound from different directions toward the inner ear.

3 WING MEMBRANE (PATAGIUM)
The skin between the finger bones is thin and prone to tearing, but heals quickly. Tiny bumps on the wing contain touch-sensitive microreceptors (Merkel cells). A hair on each bump allows the bat to gather information about the air, including flight currents of prey.

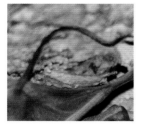

4 TAIL
R. microphyllum's tail extends beyond its uropatagium and is covered in fine fur. A bulge at the base of the tail stores fat reserves that allow these tiny animals to survive cold periods by entering a state of torpor, when body temperature and physiology are reduced.

RHINOPOMA RELATIONSHIPS

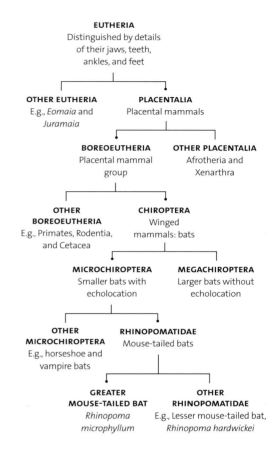

EUTHERIA
Distinguished by details of their jaws, teeth, ankles, and feet

OTHER EUTHERIA
E.g., *Eomaia* and *Juramaia*

PLACENTALIA
Placental mammals

BOREOEUTHERIA
Placental mammal group

OTHER PLACENTALIA
Afrotheria and Xenarthra

OTHER BOREOEUTHERIA
E.g., Primates, Rodentia, and Cetacea

CHIROPTERA
Winged mammals: bats

MICROCHIROPTERA
Smaller bats with echolocation

MEGACHIROPTERA
Larger bats without echolocation

OTHER MICROCHIROPTERA
E.g., horseshoe and vampire bats

RHINOPOMATIDAE
Mouse-tailed bats

GREATER MOUSE-TAILED BAT
Rhinopoma microphyllum

OTHER RHINOPOMATIDAE
E.g., Lesser mouse-tailed bat, *Rhinopoma hardwickei*

▲ Eutherian mammals are mainly nonmarsupial, and the placental group includes all living nonmarsupials. Placentalia has three or four major groupings, according to genetic and other molecular analysis, with bats (Chiroptera) in the Boreoeutheria.

SOPHISTICATED ECHOLOCATION

Bats are among several mammal groups that have independently evolved echolocation, or sonar (others include dolphins, toothed whales, and some shrews). Bat echolocation relies on calculating the distance to an object based on the time it takes its sounds to return (right). This sophisticated system has evolved from simple beginnings—the equivalent to humans shouting at a cliff and timing the echoes. When alone, a mouse-tailed bat emits sound pulses of less than 1/20th of a second, at a frequency of 32½ kilohertz (too high for human ears). When flying in a group, bats alter their calls to frequencies of 30 to 35 kilohertz in order to avoid confusion.

HOOFED MAMMALS: CETARTIODACTYLS

One of the largest mammal groups today is the hoofed mammals. These are herbivores that walk on their toes (digitigrade), which have evolved thickened tips of keratin— the same hard, waterproof substance that forms the fingernails and claws in other mammals. The hard toe tips are called hooves (see image 2) and this method of walking is accompanied by further modifications in the number or size of individual toes. Hoofed mammals are often referred to as ungulates, and they were formerly considered to be one group, Ungulata, with a shared ancestry. However, recent genetic studies and new fossil evidence hint at a more complex evolutionary story. Indeed, ungulates are one of the earliest placental mammal groups for which there is fossil evidence. From fossils found in Montana's Hell Creek Formation, we know that *Protungulatum* (see image 1) dates back to the late Cretaceous period around 70 million to 60 million years ago. If it was an ungulate, it shows that the first hoofed mammals coexisted with the last big dinosaurs, such as *Tyrannosaurus* (see p.302). However, the nonavian dinosaurs needed to die out in order to clear the way for ungulates to develop and diversify. From fossil finds from the early Paleocene epoch at Bug Creek Anthills, it is known that *Protungulatum* was one of the genera that survived the mass extinction event 66 million years ago.

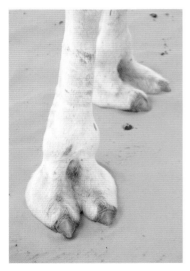

Modern ungulates are traditionally divided into two major groups. In the artiodactyls (even-toed hoofed mammals), the third and fourth toes have evolved to carry the weight, while in the perissodactyls (odd-toed hoofed mammals; see p.487) the third toe alone carries most of the weight, as in *Macrauchenia*. However, genetic studies have shown that whales evolved from artiodactyl ancestors, which suggests two surviving evolutionary lineages: the Cetartiodactyla (artiodactyls and whales) and the Perissodactyla (all other ungulates). These two groups are included in the broader Euungulata and share an ancestry with many other animals in the group known as Laurasiatheria.

KEY EVENTS

66 Mya	65–60 Mya	60 Mya	55–46 Mya	54 Mya	46 Mya
Protungulatum, the earliest known hoofed mammal ancestor, is contemporary with some of the last dinosaurs.	Arctocyonidae, the bearlike condylarths that will eventually give rise to the artiodactyls and the carnivorous mesonychids, flourish.	The ancestors of the artiodactyl groups most probably split at this time, with Whippomorpha splitting from the ruminants soon after.	The small deerlike *Diacodexis* lives at this time. Although it is an artiodactyl, it retains five toes on each foot.	The evolutionary split between ancestors of cetaceans and hippopotamuses most probably occurs.	The Hypertragulidae, a family of deerlike animals with basal (original) ruminant stomachs, evolves and flourishes until around 13½ million years ago.

Several groups of mammals previously thought of as close cousins of ungulates, including elephants, hyraxes, sirenians, and aardvarks, are now understood to have developed their ungulatelike characteristics through convergent evolution.

The Paleocene and Eocene epochs, 66 million to 34 million years ago, gave rise to a number of now-extinct ungulate groups whose evolutionary placement is uncertain. Dinoceratans were probable ungulates that sported huge paired horns and tusks, while litopterns were camel-like, three-toed South American creatures that flourished until the recent ice ages, dying out only a few thousand years ago. Notoungulates were also unique to South America, with a range of sizes and a similarly recent extinction date for their last survivors. The ancestry of all of these groups is thought to be traced back to the condylarths: a broad and diverse group of primitive ungulates that do not necessarily share a unique ancestor and are therefore considered an evolutionary "grade."

The first artiodactyls appeared during the early Eocene around 55 million years ago, evolving from a group of bearlike condylarths known as the Arctocyonidae. Their "first cousins" were the fearsome wolflike mesonychids, also descended from arctocyonids. This was the first major group of predatory mammals; until this time, the major hunters of evolving mid-size mammals came from crocodylians (see p.254) and giant flightless terror birds (see p.404). The earliest fossil artiodactyl, *Diacodexis* (see image 3), had roughly the size and build of a small deer, such as a chevrotain, and was widely distributed across the northern hemisphere. It retained five toes on each foot, but the third and fourth toes show signs of elongation and hoof development.

By the Paleocene to early Eocene epochs, artiodactyls had diversified and split into the three major subgroups that survive today: Cetruminantia, including the ancestors of whales, cattle, sheep, and goats; Suina, which are pigs and other swine; and Tylopoda: camels and their relatives. Throughout this period, artiodactyls were greatly outnumbered by various perissodactyl species—a reversal of the present-day situation, in which horses are the only widespread perissodactyls. Due to their specific evolutionary roles, artiodactyls developed highly modified stomachs with multiple chambers capable of fermenting vegetation in order to extract greater nutrition. Variations on this distinctive digestive system can be found in ruminants, camels, and to a lesser extent swine, which have remained largely omnivorous. Their digestive system meant that artiodactyls were pre-adapted to take advantage of the sudden explosion in grasslands during the later Eocene and Miocene epochs. The modern dominance of artiodactyls owes much to an explosion of species among the bovid family of ruminants—cattle, bison, sheep, goats, antelopes, and gazelles—that accompanied the spread of grasslands. **GS**

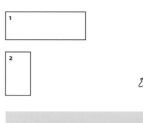

1 *Protungulatum*, only 16 inches (40 cm) long, was possibly an early ungulate. However, some experts place it with an earlier, less specialized mammal group, the arctocyonids.

2 The two-hoofed, wide feet of this dromedary camel, *Camelus dromedarius*, are adapted to crossing the dry sands of the Sahara Desert.

3 *Diacodexis* demonstrates the evolving form of hooves, which are in effect toe tips comparable to the nails and claws of other mammals.

46–45 Mya	40 Mya	42 Mya	35 Mya	20 Mya	17 Mya
The Protoceratid family of horned artiodactyls flourishes. Despite appearances, it is not closely related to deer.	Giant artiodactyl *Andrewsarchus* (see p.455), the largest known carnivorous land mammal, roams the region of Mongolia.	The dog-size *Protylopus* from North America is the oldest known ancestor of the modern Tylopoda (camels).	The piglike oreodonts —grazing animals that are closely related to camels—appear. They flourish for 30 million years before extinction.	Grasslands rapidly expand due to climate change. Bovid ruminants undergo a burst of evolution to take advantage of new food sources.	The Antilocapridae family evolves in the grasslands of North America, but only one species survives today: the pronghorn (see p.456).

Terminator Pig
LATE PALEOGENE PERIOD—EARLY NEOGENE PERIOD

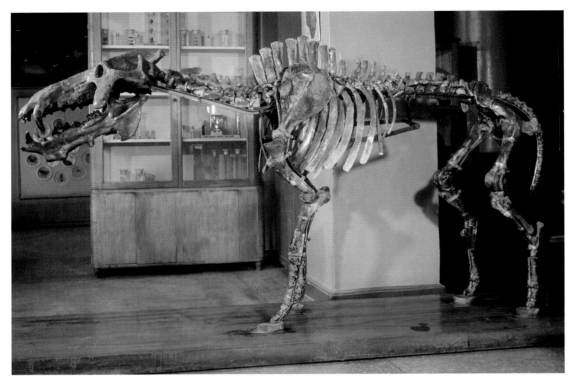

SPECIES *Daeodon shoshonensis*
GROUP Entelodontidae
SIZE 10 feet (3 m)
LOCATION North America

The entelodonts were piglike artiodactyls that flourished from around 37 million to 16 million years ago across Europe, Asia, and North America. The largest species reached a height of 7 feet (2 m) at the shoulder, and they were equipped with a fearsome set of teeth and pronounced bony lumps on its face, so its intimidating appearance has earned the nicknames "terminator pig" or "hell pig." They were mostly omnivorous scavengers and opportunistic predators. Despite the superficial similarities to living warthogs, recent studies of anatomical features have shown that entelodonts were not as closely related to living swine as once thought. Instead, their nearest living relatives seem to be the hippopotamuses and cetaceans (whales and dolphins). This places them in the Cetancodontomorpha group, alongside the gigantic, wolflike carnivore *Andrewsarchus* (see panel), which is known only from an enormous 33-inch (83 cm) skull fossil and other fragmentary remains. It is not known exactly how predatory these enormous terminator pigs were, but distinctive bite marks found on the bones of fossil herbivores show that they certainly had a taste for flesh. Their surprisingly slender—though comparatively short—legs had adaptations that would have allowed them to run at high speeds over short distances.

The largest known entelodonts were the North American *Daeodon* (dreadful tooth) and the Eurasian *Paraentelodon* (near perfect tooth). The genus *Daeodon* (an amalgam of several other genera, including *Ammodon* and *Boochoerus*) flourished between about 29 million and 19 million years ago, from the late Oligocene to the early Miocene, with *Daeodon shoshonensis* being the largest species. *Paraentelodon* is found slightly earlier in the fossil record. This, and various anatomical similarities, has led to suggestions that *Daeodon* could have evolved from *Paraentelodon,* which crossed the land bridge from Asia to North America some time during the late Oligocene. **GS**

✦ NAVIGATOR

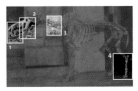

1 TEETH

The jaws of *D. shoshonensis* were lined with a full set of mammalian teeth: heavily built incisors at the front, large canines, and distinct premolars and molars. This suggests they had a varied omnivorous diet similar to that of modern pigs. The frontmost teeth are so robust that they may well have been used to dig out tough roots and tubers as well as to tear at carrion, or even live prey.

2 CHEEK BONES

The skull of *D. shoshonensis* had broad cheek bones forming wide flanges on either side of the face, which, combined with other bony protrusions, looked terrifying. The cheek bones formed attachment surfaces for powerful jaw muscles, but they also varied in size between males and females, which suggests that they played a role in either attracting mates or in duels between males.

3 NECK

The bones of *D. shoshonensis*'s neck were surprisingly small and lightly built considering the enormous size of the skull. However, as is the case in modern bison, these neck bones supported a powerfully muscled neck. These muscles were anchored to the tall flanges (neural spines) projecting from the backbones of the chest region, termed the thoracic vertebrae.

4 LEGS

The short, slender legs were strong and would have been capable of reasonably high speeds. The fibula and tibia (lower leg bones) were fused together for additional strength. Artiodactyl legs were initially shorter and better suited for picking through the dense vegetation of the tropical early Eocene; longer legs evolved in response to the spread of open grasslands.

▲ The warthog looks like a modern, smaller version of an entelodont and it follows a similar omnivorous lifestyle. However, it is not an especially close relation and is instead included in the pig family, Suidae.

CONTROVERSIAL BEAST

Andrewsarchus (below) is acknowledged as the largest known carnivorous land mammal. Estimates suggest a shoulder height of 7 feet (2 m) and a length of 11 feet (3.5 m) excluding the tail. However, a powerful jaw coupled with blunted teeth suggest *Andrewsarchus* may have been more of a scavenger than a hunter. Its position in the mammalian family tree has been controversial since its discovery in Mongolia during an expedition in 1923 led by American paleontologist Roy Chapman Andrews (1884–1960), whose name it honors. Original classification attempts placed *Andrewsarchus* in the Mesonychia, a group of extinct five-clawed carnivores. However, experts now favor a position within the ungulates, but closer to the entelodonts, in the group Cetancodontomorpha.

Pronghorn
PLEISTOCENE EPOCH—PRESENT

With a reputation as the second fastest land animal after the cheetah, North America's pronghorn (*Antelocapra americana*) is also known as the prairie ghost, on account of its ability to appear and disappear silently by moving at great speed. It is the only surviving member of the antilocaprid family and its closest living relatives are not antelopes, despite it often being referred to as the "pronghorn antelope," but the giraffe and okapi. It is an intriguing example of an animal that evolved to cope with evolutionary pressures that have now disappeared. The pronghorn runs at up to 55 miles per hour (88 km/h)—three-quarters of the speed of an African cheetah, but it can sustain high speeds over longer distances. It presumably developed this to avoid predators, but no current North American land carnivores move this fast. The pronghorn's one-time threat is in the fossil record: the American cheetah, *Miracinonyx*. DNA evidence has shown that this feline was a highly specialized cougar relative. It developed many of the advanced features of the African cheetah due to convergent evolution, and was a top prairie predator. Its extinction 12,000 years ago left wolves, coyotes, and cougars as pronghorn predators. This allowed it to flourish until the 19th century, when it came under threat from human hunting and fragmentation of its habitat. However, it now thrives in many hostile territories, thanks to a wide-ranging diet including plants that other herbivores find inedible. **GS**

SPECIES *Antilocapra americana*
GROUP Antilocapridae
SIZE 5 feet (1.5 m)
LOCATION North America
IUCN Least Concern

FOCAL POINTS

1 HORNS
Males have longer and more complex horns than females. The core of each horn is a flattened blade of bone directly from the skull—in males this creates a forward point. A layer of skin grows over the core, hardening into a sheath of protective keratin that is shed and regrown each year.

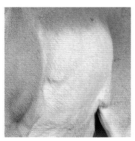

2 INTERNAL ANATOMY
The internal anatomy facilitates sudden bursts of speed: unusually large windpipe, lungs, and heart allow great amounts of oxygen to be absorbed and passed into the bloodstream to fuel muscles. The bone structure is extremely light, and even the hairs are hollow to minimize weight.

3 CUSHIONED TOES
Each long, slender leg ends in two pointed toes, cushioned by soft tissue. This reduces the shocks passing into the animal's body as it pounds the ground. Pronghorns display more than a dozen different gaits (walking or running styles), but unlike true antelopes, are not built for leaping, so remain fairly level.

THE FIRST SIGHTING

The pronghorn first came to the attention of Western scientists during Meriwether Lewis and William Clark's expedition in 1804 to 1806. While searching for practical routes from eastern to western America, they collected information about wildlife and plants, describing more than 200 species. They came across grazing pronghorn in September 1804, in what is now South Dakota. Recognizing what would later be known as convergent evolution, Clark recorded shooting a buck that was "more like the Antilope or Gazella of Africa than any other species of goat." The pronghorn first appeared to the public 40 years later when John James Audubon published *The Viviparous Quadrupeds of North America* illustrating them for the first time (above).

THE RISE OF RODENTS

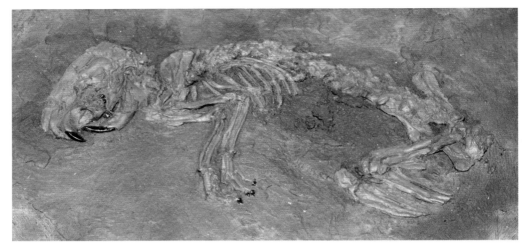

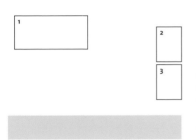

1 This fossil skeleton of the early rodent *Masillamys*, from 50 million years ago, clearly displays its forward, ever-growing incisor teeth (black).

2 The North American porcupine, *Erethizon dorsatum*, a member of the caviomorph group, has evolved similar features to its Old World namesake, despite not being especially closely related.

3 The capybara, *Hydrochoerus hydrochaeris*, is a skilled swimmer and evades land predators by diving into water.

The vast group of gnawing mammals known as rodents contains more than 2,280 living species. They occur in all terrestrial environments and represent more than 40 percent of all living mammal species. These include squirrels, beavers, dormice, mole-rats, rats, mice, voles, hamsters, gerbils, jerboas, porcupines, guinea pigs, capybaras, and chinchillas. Several characteristics combine to make rodents a successful mammal group. Primarily, virtually all of their evolution has occurred at small body size, and almost all breed at a young age, producing large numbers of fast-growing babies. These traits mean that rodents can rapidly fill up ecosystems with new individuals and evolve quickly in the face of changing conditions. Secondly, rodents possess a generalized anatomy, which means they can adapt to virtually any habitat: most rodents are quick-moving, scurrying omnivores.

The key anatomical feature that distinguishes rodents from other placental mammals is the presence of two pairs of enlarged, constantly growing, incisor front teeth: one pair in the upper jaw and one in the lower. Because hard enamel covers only the face of the teeth, the tips develop chisel-shaped surfaces, with the front edge forming a sharp blade. These teeth have big curved roots that take up huge amounts of space within the skull and jaws. Canine teeth and some of the premolars are absent, so a large space is present between the incisors and the chewing teeth farther back. Many features of rodent teeth and jaws are also present in lagomorphs, the group that includes rabbits and hares, and fossil and molecular data show that they are closely related. Together they form the group called Glires. Several early members of Glires appear in fossils up to 60 million years old.

KEY EVENTS

65 Mya	60 Mya	54 Mya	52 Mya	50 Mya	32 Mya
Rodents diverge from a common ancestor shared with lagomorphs, the group that includes rabbits and hares.	The first rodents appear. *Acritoparamys* from North America is the oldest known member of the rodent group.	Myodonts appear. This is a huge group that includes mice, rats, and many more, as well as most major living rodent lineages.	The evolutionary lines that lead to dormice and squirrels diverge from their ancestral line, which has existed for some time.	Ischyromyids—a loose group of ancestral rodents—diversify in North America. They include ground-dwellers and possible burrowers and hoppers.	African rodents possibly cross the Atlantic and colonize South America, leading to the caviomorphs: guinea pigs, capybaras, and kin.

The fossil record shows that early rodents—the classic example is *Ischyromys* from North America (see p.460)—were ground-dwelling generalists that combined climbing, scurrying, and burrowing abilities. From such ancestors tree-climbers evolved, as did dedicated burrowers, swimmers, and grassland- and desert-dwellers. Climbing rodents gave rise to gliders on at least three occasions, and amphibious lineages evolved several times among mice and voles. Strongly aquatic groups such as beavers and capybaras also evolved, mostly within the past 25 million years. There are many different views on how the rodent groups are related. Classically, rodents were grouped together on the basis of their jaw muscle anatomy. However, molecular and genetic evidence contradicts some of these classic relationships. According to the most recent studies, since 2000, the oldest major divergence event in rodent history occurred between the group comprising dormice and squirrels, Sciuromorpha—including *Masillamys* (see image 1) and, perhaps oddly, the mountain beaver (see p.462)—and the group that includes all remaining lineages, Castorimorpha-Myodonta. Within this latter group, Old World porcupines (see image 2), African mole-rats, and caviomorphs, including guinea pigs, capybaras, and kin, belong to one branch of the rodent evolutionary tree; beavers and pocket gophers to another; and mice, voles, gerbils, hamsters, and their relatives to another still. The last group, called Myodonta, has the largest number of species and is the most widely distributed.

Some living rodent groups, including dormice and beavers, have fossil records that go back 40 million years. However, most data indicates that modern rodent lineages originated in a huge explosion of diversity about 25 million years ago, an event probably driven by the spread of grasslands and the drier conditions that then prevailed. Today's biggest rodent, the South American capybara (see image 3), can exceed 200 pounds (90 kg). However, it is dwarfed by several extinct caviomorphs, some of which were the size of bears or small rhinoceroses. *Phoberomys*, from Venezuela, may have weighed more than 1,500 pounds (700 kg) while some estimates put *Josephoartigasia*, from Uruguay, at 5,500 pounds (2,500 kg). How these gigantic caviomorphs lived is uncertain, but it has been suggested by some evolutionary scientists that they were amphibious creatures that lived on wetlands and riverbanks.

How and when caviomorphs arrived in South America has been a source of controversy. One idea is that they moved south from a North American ancestor, but no such ancestors have yet been identified. The rodents most similar to caviomorphs are African mole-rats, and a more popular theory is that their ancestors crossed the Atlantic on floating vegetation—massive tangles washed out of great rivers and pushed across the oceans by currents and wind. (Live animals traveling across large stretches of sea in this way have also been observed in modern times.) Another possibility is that early members of the caviomorph group used ancient land connections between Asia, Australia, and Antarctica to get to South America. **DN**

25 Mya	24 Mya	15 Mya	10 Mya	8 Mya	3 Mya
Myomorphs (mice, voles, hamsters, gerbils, and cousins) rapidly diversify and evolve into species that live in habitats all over the world.	The sigmodontines, a new myomorph group, invade South America from the north. They become one of the most successful mammal groups.	Amphibious beavers evolve in Europe or Asia from small, burrowing, terrestrial beavers. Fossils later indicate that some attain huge sizes.	The mylagaulids group of burrowing rodents inhabit the plains of North America. Several species have short, paired horns on their snouts.	Caviomorphs as big as bears and rhinoceroses inhabit wetland and grassland habitats in South America. They are the largest rodents of all time.	South American caviomorphs move north on new islands and land bridges. New World porcupines and capybaras now occur in North America.

Ischyromys
EOCENE EPOCH

SPECIES *Ischyromys oweni*
GROUP Ischyromyidae
SIZE 2 feet (60 cm)
LOCATION North America

Rodents are enormously well represented in the fossil record due to their abundance and their distinctive, durable teeth, as well as the fact that their often tiny remains are helpfully transported and regurgitated in neat packages—pellets—by predatory birds. However, good complete skeletons are rare, especially of early kinds. Therefore, the discovery of *Ischyromys* from 35-million-year-old rocks in the United States during the 1850s crucially shaped ideas about early rodent evolution and diversification.

Long-tailed and robust, *Ischyromys* (strong mouse) was an early plains-dwelling rodent, similar in size to a large prairie dog or marmot. The proportions of its skeleton suggest that as a ground-dweller it may have used (but not necessarily dug) burrows and tunnels. Not only is *Ischyromys* important for being a "classic" early rodent, but also it is familiar from a large number of remains, the majority of which are lower jaws: more than 4,000 are known. These fossils come from sedimentary rocks preserved across a span of several million years and reveal numerous small differences. As a result of this variation, paleontologists have named several *Ischyromys* species and have also suggested that various evolutionary changes can be observed; for example, some species seem to have increased in size over time. A large number of additional fossil rodents found in North America, Europe, and Asia have been regarded as close relatives of *Ischyromys* and grouped within the Ischyromyidae family. These are mainly generalized early rodents lacking the detailed anatomical features of modern rodent lineages. This suggests they are not all related to one another, but represent an assortment of archaic ancestral forms. **DN**

✦ NAVIGATOR

1 TOOTH FORMULA
The reduced tooth count of *Ischyromys oweni* is typical of rodents. In the upper jaw there is a single pair of incisors, no canines, two premolar pairs, and three molar pairs. The formula is similar in the lower jaw, the only difference being that only one pair of premolars is usually present.

2 SKULL
I. oweni and its relatives have a more archaic skull anatomy than living rodents. The snout and braincase are comparatively long and the skull is shallow and approximately rectangular, which is a shape seen widely in rodents that scamper across the ground, climb and inhabit shallow burrows.

3 FEET
Early rodents, such as *I. oweni*, had five fingers, five toes, hind limbs that were somewhat longer than the fore limbs, and a long tail. Later groups evolved a lifestyle that included running or hopping; consequently the numbers of fingers and toes were reduced accordingly.

◄ *Ischyromys* and other rodents can be defined by their ever-growing incisor teeth. Few were larger than this fossil specimen, 6 inches (15 cm) long, from *Castoroides ohioensis*, an almost pony-sized giant beaver that had become extinct in North America by 10,000 years ago.

AN ADAPTABLE GROUP

Over time, ischryomyids evolved more complex cheek teeth (molars and premolars) and an array of body sizes and proportions. One of the biggest was *Manitsha* from North America; it was larger than modern beavers, which are among the largest living rodents. The smallest, such as *Microparamys* from North America and Europe, were similar in size to living mice. This diversification illustrates the evolutionary potential of animals that breed often and rapidly and that have many offspring. Lemmings, *Lemmus sibiricus* (right), are a classic example of this because their features are typical of the rodent group. They have an increased chance of producing new adaptations, and at a faster rate than other animals, so they are able to move into new habitats and lifestyles as necessary.

Mountain Beaver
PLEISTOCENE EPOCH—PRESENT

SPECIES *Aplodontia rufa*
GROUP Aplodontiidae
SIZE 14 inches (35 cm)
LOCATION North America
IUCN Least Concern

 **NAVIGATOR**

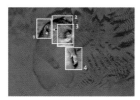

One of the world's most unusual and archaic rodents is the mountain beaver, also known as the boomer or sewellel: a chunky-bodied, short-limbed rodent of northwest North America. Despite its name, it is not solely associated with mountains and can be found at most elevations, from near the sea to uplands, although more often in cool, damp forest environments. Nor is it a true beaver, as it is the only member of the Aplodontiidae family, while beavers are in the Castoridae family.

As is common with modern animals, the mountain beaver, *Aplodontia rufa*, is the only living member of a once large and diverse group of rodents, the fossil record of which extends back more than 40 million years. Termed aplodonts, the group was present in Europe and Asia before moving across the Bering land bridge to evolve and diversify in North America. The relationship of aplodonts to other rodents has been controversial, partly because aplodonts lack the complex jaw-closing muscles present in other living rodent groups. For this reason, it has been presumed that aplodonts belong to an unusual side branch that evolved directly from *Ischyromys*-like creatures early in rodent history. However, aplodonts resemble squirrels in the shape of their tooth cusps, so it has also been suggested that they evolved from the same common ancestor that gave rise to squirrels. This theory receives support from molecular and genetic evidence, as it appears that dormice, aplodonts, and squirrels are close relatives, forming a group that diverged from other rodents more than 50 million years ago. The mountain beaver is tailless, and it uses its stout limbs and broad, prominently clawed hands and feet to dig tunnels. The small eyes, external ears, and general body shape are all consistent with this burrowing (fossorial) lifestyle. However, this rodent, which typically reaches 14 inches (35 cm) in length, can swim well and climb trees when needed. **DN**

👁 FOCAL POINTS

1 EAR
The internal ear anatomy of the mountain beaver is unusual compared to other rodents, and the part of the brain associated with hearing is unusually large. It seems that the mountain beaver is especially good at detecting air pressure changes within its burrows. However, the evolutionary advantage of this is not entirely clear—it could possibly serve to detect invaders or predators.

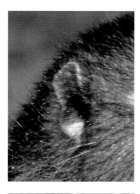

2 CHEEK TEETH
The tall, constantly growing cheek teeth have a simple look compared to those of many other rodents; in fact, the group name *Aplodontia* means "simple tooth." Teeth of this sort are suited for a diet that includes bark and twigs as well as fern fronds, shoots and leaves. The molars have unique spurlike projections that enable fossils to be identified, although their function is unclear.

3 JAW
The lack of the specialized jaw-closing muscle arrangement typical of modern rodents means the mountain beaver has been regarded as a living fossil—a late surviving member of a group that was in existence long ago. However, this terminology is misleading; the mountain beaver is not especially ancient and its fossil record goes back only a few million years.

4 CLAWS
The claws on the mountain beaver's front toes are long, narrow from side to side, curved and quite sharp, considering the amounts of scrabbling and digging they endure. The thumb claw (pollex) is shorter and blunter, more like a nail. This digit is opposable, meaning it is movable enough to press against the other digits (like the human thumb), which helps when handling food materials.

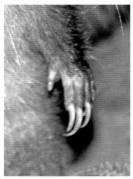

▲ The skeleton has adaptations for regular digging. The muscle attachment sites on the limb bones are large, and the prong that forms the elbow region—the ulnar olecranon—is huge, meaning the muscles that pull the hand downward during digging are especially big. Similar features also occur in fossil relatives.

HORNED GOPHERS

The mountain beaver and other aplodonts share several anatomical features with an unusual group of extinct rodents that were specialized for a digging lifestyle. These were the mylagaulids, known as "horned gophers," which were especially remarkable for the short paired horns certain species possessed on their snouts (below). Features of the skull and neck indicate that mylagaulids used their short, tough snouts during digging, and a suggested explanation for the use of horns is that they played an important role in helping to dislodge and move soil during burrowing.

CREODONTS

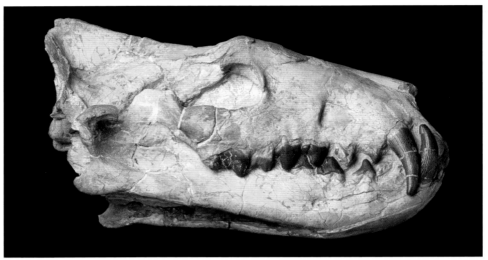

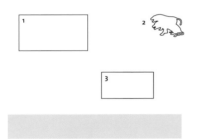

1. A *Hyaenodon horridus* skull (see p.468 for the whole animal) reveals the slicing and shearing teeth (carnassials) in the cheek tooth (molar and premolar) row, with the robust canines at the front.

2. Enormous *Megistotherium* had typical creodont proportions, whereby the head appears oversized when compared to living carnivorans.

3. About 55 million years ago the 7-foot-long (2 m) *Oxyaena*, which lent its name to one of the main creodont groups, hunted in the woodlands of North America.

The early history of a major group of placental mammals termed Carnivora, typically called carnivorans, overlaps with that of the group of predatory mammals called creodonts (early mammalian meat-eaters), which have often been regarded as less sophisticated carnivorans. In some respects, creodonts were primitive compared to carnivorans, but were successful and innovative in the variation among their anatomies. Some species resembled dogs, foxes, or civets, while others were saber-toothed or otterlike. Specialized teeth suited for shearing (carnassials) along with large, curved, canine teeth and sharply curved claws confirm that creodonts were predators. Their carnassials were modified molars, so were farther back in the mouth than in modern carnivorans, where they are formed from the last premolar in front of the molars and the first molar. The creodont jaw joint seems to have allowed more flexibility than that of today's carnivorans. This is advantageous for an omnivorous diet, but disadvantageous in predators and scavengers because it increases the risk of jaw dislocation as force is transferred to the bones and joints. Perhaps creodonts ate fruits and other plant products more regularly than carnivorans, such as cats and dogs, did.

Many creodonts had proportionally short limbs with feet mostly in contact with the ground, which suggests they were ambush predators suited to forested habitats. Longer, slender limbs evolved in the superficially doglike hyaenodontid creodonts, such as *Hyaenodon*, (see image 1; see p.468), which most likely pursued prey over short distances. Compared to modern carnivorans, creodonts had smaller brains and shallower skulls. Large crests for the attachment of

KEY EVENTS

65 Mya	60 Mya	55 Mya	55 Mya	55 Mya	53 Mya
Creodonts evolve from a small, long-tailed, insectivorous ancestor that also gives rise to carnivorans, the group that will include cats, dogs, and bears.	The shorter-headed oxyaenids and longer-skulled hyaenodontids are already distinct. Early hyaenodontids are foxlike in form and lifestyle.	Early types of hyaenodontid—such as *Koholia*—inhabit northern Africa, suggesting that they originated there. They later spread elsewhere.	A group of oxyaenids with heavy jaws and blunt-tipped teeth, the palaeonictines, inhabits North America. They may have lived a hyenalike lifestyle.	New hyaenodontid species inhabit Europe, Asia and North America. Their history is complex, involving migrations along the edges of the Tethys Sea.	Machaeroidines—a group of saber-toothed creodonts—evolve in North America, most likely from an ancestor shared with oxyaenids.

powerful jaw muscles were present along the top midline of the skull. Typical creodonts ranged in size from 3 to 10 feet (1–3 m) in total length, although diminutive species less than 16 inches (40 cm) long also evolved. A few were especially large; in the African hyaenodontid *Megistotherium* (see image 2), the fossil skull was 25 inches (65 cm) long. Scaling up, this could have been a monstrous hyperpredator more than 20 feet (6 m) long. However, the skulls of hyaenodontids were large for their bodies, and the biggest species were unlikely to have exceeded 13 feet (4 m) in total length.

The relationships between creodonts and their ancestors, between different creodont groups, and between creodonts and modern carnivorans are difficult to ascertain because many kinds of early placental mammals possessed a generalized anatomy, making it hard to identify close relationships with the groups that evolved later. Despite this, recent studies have confirmed that creodonts and carnivorans are close relatives. The armor-plated pangolin mammals are part of the same group, which originated 65 million years ago. Presumably, the ancestor was a small, long-tailed mammal adept at climbing and at eating insects and small animals. Genetic evidence shows that pangolins and carnivorans are part of the superorder Laurasiatheria: the major mammal group that includes shrews, moles, bats (see p.446), and hoofed mammals.

Within the creodont group, it was previously thought that an early divergence led to the evolution of the doglike hyaenodontids and the shorter-limbed, shorter-faced oxyaenids. Only the teeth of both groups look similar, so the two are now thought to have separate origins. Oxyaenids (see image 3) are often depicted as big cats because they possess short, deep faces. There may have been a superficial similarity but they were significantly different from cats overall. Their limbs were shorter, their wrist and ankle joints were held far closer to the ground and their bodies were longer and closer to the ground. **DN**

45 Mya	45 Mya	35 Mya	22 Mya	15 Mya	13 Mya
Apterodontines inhabit Africa and Europe, with features linked to swimming and digging. They evolve from foxlike hyaenodontid ancestors.	Hyaenodontid-like creodonts, the limnocyonids, inhabit the Rocky Mountain corridor. Skeletons suggest they had a digging lifestyle.	Oxyaenid *Sarkastodon* (see p.466) evolves in Mongolia, alongside big hoofed mammals and plant-eaters. It is likely to have been a hyena-style predator.	Huge predatory gigantihyaenodontids *Hyainailourus* and *Megistotherium* live in Africa, Asia, and Europe.	*Hyaenodon* becomes extinct in North America and creodonts are absent from the continent. Other creodont species persist elsewhere.	The last creodonts disappear from Asia after 50 million years. Their roles in ecosystems are occupied by members of the Carnivora.

Sarkastodon
LATE EOCENE EPOCH

SPECIES *Sarkastodon mongoliensis*
GROUP Oxyaenidae
SIZE 10 feet (3 m)
LOCATION Mongolia

NAVIGATOR

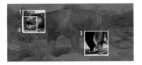

Between approximately 60 million and 35 million years ago, oxyaenid creodonts inhabited wooded, forested, and scrub-covered landscapes across North America, Europe, and Asia. Early kinds were 3 feet (1 m) long or less, and their proportions and body shapes indicate that they were ambush predators of smaller mammals, birds, and reptiles. However, members of at least one oxyaenid lineage became larger, culminating in the youngest member of the group: an especially big, heavily built predator possibly up to 10 feet (3 m) long. This was *Sarkastodon* (flesh-tearing tooth) from 35-million-year-old rocks in Mongolia, discovered in 1930 and named in 1938.

Sarkastodon mongoliensis is known only from its skull and jaws, and theories about it are based on what is known of smaller related species. Similar to other oxyaenids, it was presumably long-tailed with short limbs and the overall look may have been that of a heavy-bodied, bearlike predator with bulbous, hyenalike teeth. Oxyaenids may have embodied a strange combination of catlike and bearlike features. The fossil record shows that *Sarkastodon* and its oxyaenid relatives became extinct 35 million years ago, with only a few species persisting beyond this. It is not entirely clear why creodonts and other archaic placental mammals declined at the end of the Eocene epoch, but a slow period of climatic cooling, followed by a rapid warming, may have disrupted ecosystems and initiated extinctions. If *Sarkastodon* relied on the abundance of big herbivores as prey, such as the rhinoceroslike brontotheres, it may have been forced into extinction by their decline. **DN**

1 HEAD

The snout of *Sarkastodon* was deep and robust and the palate extremely broad. The massive lower jaw was worked by powerful muscles at the cheeks and skull sides. The jaw and face arrangement, with stout, pointed front teeth and broad, rounded cheek teeth, is reminiscent of modern hyenas.

2 TEETH

Unusually, there were only two incisors per side in the upper jaw and one per side in the lower. In the upper jaw, the inner pair was tiny, behind the large outer pair. The array of large canines and caninelike incisors suggests that it inflicted wounds in the tissues of prey, and perhaps carried them.

3 LIMBS

With short bones in the fore and hind limbs, the wrists and ankles would have been close to the ground. This carnivore's short limbs, large size, and powerful proportions implies that it could exploit the carcasses of big mammals, and perhaps take over the kills of other predatory mammals.

HYPERPREDATORS

Giant predators frequently evolved when giant prey species appeared in an ecosystem, the predators lagging behind their prey by a few million years. Approximately 45 million years ago, a burst of evolution led to the appearance of a diversity of new, big-bodied herbivorous mammals. They included many kinds of brontotheres (odd-toed hoofed mammals), which had startling external similarities to the rhinoceroses of today and grew even larger, such as *Megacerops* (right). Despite this similarity, they would have been more closely related to horses than to the rhinoceros family. In turn, it appears that these animals drove the evolution of the hyperpredators that were capable of attacking and killing them, including creodonts such as *Sarkastodon*.

Hyaenodon
LATE EOCENE EPOCH—EARLY MIOCENE EPOCH

SPECIES *Hyaenodon horridus*
GROUP Hyaenodontidae
SIZE 6 feet (1.8 m)
LOCATION North America

Among the best known of the creodonts is *Hyaenodon*, originally named in 1838 from fossils in France and later discovered in Africa, Asia, and North America. Numerous *Hyaenodon* species have been identified, varying in proportions and size. The smallest were fox-sized while the biggest were 10 feet (3 m) long, similar to the largest living lions or tigers. Due to its long, doglike snout and doglike proportions, *Hyaenodon* has typically been reconstructed as resembling a robustly built wolf with large teeth. It is often given a patchy coat, which is reasonable among the species that inhabited wooded environments. Both the upper and lower canine teeth of *Hyaenodon* are proportionally larger than those of dogs, suggesting that *Hyaenodon* latched onto big prey with the front of its mouth, an idea consistent with the depth and robust look of the skull (see p.464). The other teeth were suited for slicing—flesh and bones could be sheared off the body of prey in a manner similar to that of modern hyenas and dogs.

The comparatively stiff limbs present in *Hyaenodon* were good for running but little else. The lack of rotational ability at the elbow and wrist shows that the fore limbs could not be used to manipulate prey. These predators may have been good diggers, so could have excavated underground dens for shelter and family, or dug up rodents and other burrowing prey. The overall trend seen in the evolution of the hyaenodont lineage leading to *Hyaenodon* reveals how a group of small, generalist predators that lived fox- or civetlike lives moved out of the forests into more open habitats, becoming large, fast-running predators capable of pursuing big herbivorous mammals over short distances. **DN**

⚽ NAVIGATOR

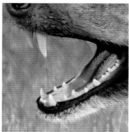

1 TEETH

The teeth of *H. horridus* show sexual dimorphism. In one set of fossils, the canine teeth are proportionally bigger than they are in another set. Presumably the bigger-toothed animals were males, equipped with bigger teeth than females for use in fights associated with territorial or mating rights.

2 FLAT HEAD

The top surface of the braincase and muzzle were flat. This made the shape of the head very different from that of dogs and other modern carnivorans, most of which have a bulging forehead. As this skull shape suggests, *H. horridus*'s brain was small compared to that of its modern equivalents.

3 REAR END

Carnivore droppings in places where *H. horridus* was the most abundant large predator are preserved close to, or on top of, skeletons of herbivorous mammals. This suggests that *H. horridus* marked kills with its dung, a territorial action that modern predators use to deter others from eating their catch.

VARIED CARNIVORES

Compared to many mammals, the genus *Hyaenodon* was extremely long-lived and varied. The first described species, *Hyaenodon leptorhynchus* (right), was named in 1838 and many more followed at regular intervals. Estimates show that these carnivores lasted for more than 25 million years, with species evolving and fading. Clearly the body plan of *Hyaenodon* suited the time it lived in and could adapt to a range of climates, conditions, and prey. Some were as small as a cat at 9 pounds (4 kg), while others became tiger-size, at 660 pounds (300 kg). The amount of different kinds varies according to the most recent classification scheme, but regularly amounts to 30 or even 40 species. Their distribution was also huge, with fossils ranging across North America to Europe, North Africa, and China.

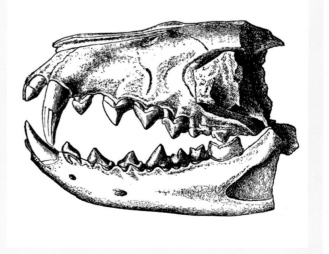

ELEPHANTS AND KIN: AFROTHERIA

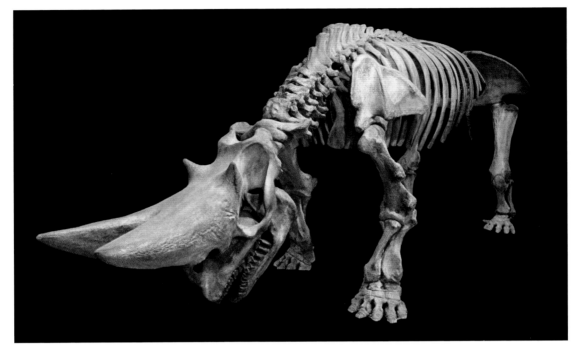

1 *Arsinoitherium* is a 10-foot-long (3 m) member of the group Embrithopoda.

2 The West Indian manatee (*Trichechus manatus*) is the largest living sirenian.

3 Hyraxes are remnants of a widespread group that dates to 35 million years ago.

One of the most recent breakthroughs in understanding mammal evolution has been the recognition of a group consisting of a variety of African mammals. Furthermore, this clade (a group with one common ancestor) known as the Afrotheria (African beasts) may represent the earliest and deepest-surviving split in the great family tree of placental mammals. The existence of the Afrotheria was first proposed in 1998 as a way of reconciling information about genes in a variety of species. It has since been backed up by the recognition of previously overlooked anatomical similarities among living and fossil species. The Afrotheria includes some important extinct groups, such as Desmostylia, Embrithopoda, and Ptolemaiida. Living members of the group include small mammals, such as golden moles, elephant shrews, tenrecs, and hyraxes (see image 3), as well as the larger aardvarks—previously assumed to be a relative of the South American anteaters—the enormous sirenians (sea cows; see image 2), and the proboscideans (elephant group).

Some mammals among the Afrotheria had been recognized previously as significant groupings. For example, the unlikely similarity between elephants and hyraxes has long been noted, and biologists group them together with sea cows as the Paenungulata, meaning "almost ungulates (hoofed mammals)." Even so, the realization that these diverse mammals evolved from a shared and unique common ancestor surprised the scientific community. With hindsight, perhaps this should have been predicted because Africa was an island continent for a very long period, between its separation from South America about 100 million years ago and the beginning of its current and ongoing collision with Eurasia. Furthermore, this period of isolation continued throughout the traumatic events of the mass extinction at the end of the Cretaceous period (see p.364), 66 million years ago. The disappearance of the great dinosaurs among many other species at the time would have left huge opportunities for surviving mammals to develop and prosper.

Exactly when the afrotherians separated from other placental mammals, however, is disputed. Molecular clock estimates (see p.17), which assume a constant rate of change due to random gene copying errors, date the origin of the Afrotheria to around 105 million years ago, but the fossil record suggests it was far more recent. Undisputed afrotherian remains appear only shortly after the disappearance of the large dinosaurs. So is this evidence for a molecular clock that occasionally "runs fast"? Fossils, genetics, and the plate tectonics of drifting continents suggest to some that that the South American xenarthrans—sloths and kin (see p.478)—may be close living relatives of afrotherians. Yet some researchers have raised doubts about whether the afrotherians are truly African animals. Contentious fossils suggest that they may also have been present in the northern supercontinent of Laurasia early in their history, before dying out everywhere except Africa.

The earliest well-described afrotherian is *Ocepeia*, a small mammal from Morocco, dated around 61 million years old. Two species are known from skulls, jaws and teeth. Their remains suggest an animal that shared features with the subgroups Paenungulata and Afroinsectiphilia; it is therefore likely to be located near the point on the family tree at which the two groups diverged.

Afrotheria produced extinct orders alongside those that survive to the present day. The precise relationships are often clouded by lack of evidence; for example, the wolf-size, predatory Ptolemaiida is assumed to be a subgroup due to the similarities between its skulls and those of aardvarks, and the fact that it evolved in Africa during its period of isolation. The desmostylians and embrithopods are two extinct groups that are better understood. Embrithopods are mostly known from a number of fragmentary fossils and teeth. They thrived between approximately 40 million and 27 million years ago and culminated in the enormous, superficially rhinoceroslike *Arsinoitherium* (see image 1), which is unmistakable because of its pair of enormous horns that emerged side by side from above its nose. Desmostylians were large amphibious animals with a somewhat hippopotamuslike appearance, although in contrast to the living hippopotamus, they were marine

23 Mya	20 Mya	11½ Mya	7¼ Mya	5¼ Mya	12,000–4,000 Ya
The huge proboscidean *Deinotherium* roams Africa and Eurasia alongside its distant elephantiform cousins.	Four-tusked *Stegolophodon* is the earliest member of the stegodonts, the closest relatives of the true elephants.	*Primelephas*, the earliest member of the elephantid family, still retains four tusks, although the lower pair is much reduced.	The African elephants *Loxodonta* separate from the common ancestors of mammoths and Asian elephants.	The Asian elephant genus *Elephas* first appears, most likely evolving from ancestors shared with the mammoths.	The end of the ice age sees mammoths and other elephant-type families decline to extinction, with only two genera surviving: African and Asian.

amphibians, living along beaches and swimming in the ancient Tethys Sea. In fact, they are grouped with their closest relatives: sea cows and the proboscidean elephants.

The best-known afrotherian order is the Proboscidea—modern elephants and their many extinct relatives; living cousins are the small, agile, rodentlike hyraxes or dassies. Proboscideans first evolved during the Paleocene epoch, with the earliest recognized genera, *Eritherium* and *Phosphatherium*, dating to 58 million and 56 million years ago respectively. Small and squat, they most likely bore a resemblance to the modern pygmy hippopotamus, but had a much narrower snout, hinting at evolutionary developments to come. The next pivotal stage in proboscidean development was marked by *Moeritherium* (see image 5) from the late Eocene epoch around 38 million years ago. Based on fossils found in North Africa, it was about the size of a modern pig or tapir and fairly similar in appearance, with short legs and broad, flat-toed hoofed feet to support its substantial bulk. *Moeritherium* was partially aquatic: its skull was notably elongated, with the eyes set close to the front, while the nostrils had moved during evolution from the tip of the nose toward the top of the skull. These collective features allowed it to wade through swampy environments while keeping its head above water. There were also significant changes to *Moeritherium*'s teeth, with the second incisors greatly enlarged in both the upper and lower jaws and several other teeth lost. This can be seen as an early step toward the evolution of tusks that would have allowed it to dig out submerged roots and other vegetation.

Moeritherium is thought to be only distantly related to modern elephants: a cousin rather than a direct ancestor. It survived until fairly recently and was one of many cousins of the Elephantiformes (the immediate ancestors and close relatives of modern elephants), which developed similar traits due to convergent evolutionary pressures. Another well-known cousin was *Deinotherium* (see p.474), an elephant-size proboscidean with a short trunk and backward-curving tusks that emerged from the lower jaw. It first appeared in the early Miocene epoch, around 23 million years ago, and survived until 1 million years ago. The Elephantiformes are thought to have developed during the same burst of evolution as the first deinotheres, around 37 million years ago. They most likely descended from a genus called *Eritreum*, located on the horn of Africa, and then gave rise to four major families: the mammutids, gomphotheres, stegodonts, and the Elephantidae themselves (listed in the order they branched from the main line of descent). By the time the mammutids had appeared, the trunk had evolved into a remarkably versatile "fifth limb," familiar in modern species.

Despite their deceptive scientific name, the Mammutidae were in fact not the mammoths, but the mastodons, a somewhat different group. Evolving around 34 million years ago in East Africa, they were distinguished by adaptations of their molar teeth for a shearing action to cut through tough vegetation. Other features included enlarged upper tusks that curved upward, and lower tusks that reduced in size over time. Certain species reached North America, and *Mammut americanum* survived until around 10,000 years ago, long after the group had become extinct in Europe.

The gomphotheres, despite being more closely related to living elephants, often showed greater physical differences from them than the mastodons. This was largely due to the arrangement of their tusks. In animals such as *Gomphotherium* and *Platybelodon*, the upper pair curved down and outward, while the lower pair formed a flattened "shovel." This bladelike surface was once assumed to be useful for scooping up aquatic vegetation in swamps, but it is now thought to have been a multipurpose tool for stripping bark from trees and even hacking through small branches. Gomphotheres first appeared around 34 million years ago and spread successfully across Eurasia and into the Americas, including South and Central America, where the last species, which were superficially elephantlike in appearance, survived until 6,000 years ago.

Stegodonts, the third group and closest relatives of the modern elephant family, appeared around 20 million years ago, evolving from gomphothere ancestors. They were distinguished by the possession of molars with low crowns and ridged upper surfaces—features that experts argue are adaptations for browsing a variety of vegetation in ancient forests. Two genera are known: the earlier four-tusked *Stegolophodon* was the probable ancestor of the two-tusked *Stegodon* (see image 4). The latter proliferated into numerous species, remains of which are mostly found across Asia and various Southeast Asian islands. Their mainland representatives included some of the largest elephants, growing up to 13 feet (4 m) at the shoulder and rivaling even the steppe mammoth (see p.476) and the woolly mammoth (see p.552) in size. In contrast, other species succumbed to the phenomenon of "island dwarfism." One buffalo-size species survived alongside the well-known "hobbit" humans (see p.554) on the Indonesian island of Flores until at least 12,000 years ago.

The living group, elephantids, evolved from the gomphotheres 12 million years ago in Africa, with early genera including *Stegodibelodon* and the four-tusked *Primelephas*. The latter may have been the direct ancestor of three later elephant genera: *Loxodonta*, the African elephants (see image 6), *Elephas*, the Asian elephants and *Mammuthus*, the extinct mammoths. All three originated in Africa, where *Elephas* and *Mammuthus* dispersed and evolved into Asia, Europe and—in the case of *Mammuthus*—the Americas. *Elephas* and *Mammuthus* eventually went extinct in their African homeland, and the latter became completely extinct during the last ice age (see p.550). The exact relationship between these three groups is still uncertain, but genetic evidence supports the idea that living African elephants split from the others first, leaving the living Asian elephant and the extinct mammoths as closely related "sister" groups. **GS**

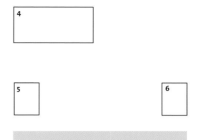

4 Species of *Stegodon* were part of an adaptable, long-lived group, ranging from enormous, long-tusked types to pony-size versions.

5 *Moeritherium*'s enlarged canines indicate the trend toward big tusks, used to uproot plants and vegetation.

6 Proboscideans have long specialized in large size. The largest land animal living today is the African bush elephant, *Loxodonta africana*.

Deinotherium
EARLY MIOCENE EPOCH—PLEISTOCENE EPOCH

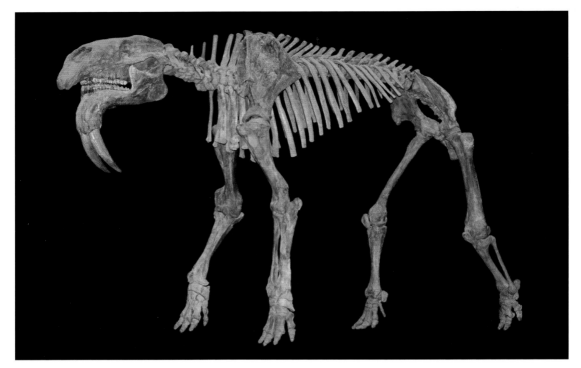

SPECIES *Deinotherium (Prodeinotherium) bavaricum*
GROUP Deinotheriidae
SIZE 15 feet (4½ m)
LOCATION Europe

A round 37 million years ago, a burst of evolution among the early proboscideans produced a variety of new groups that roamed Africa, Asia, and Europe alongside the direct ancestors of modern elephants. Deinotheres (terrible beasts) were one such group and they eventually gave rise to *Deinotherium*, the largest land animal of its day. Three genera of deinotheres are known and the earliest is the tapir-size *Chilgatherium* of Ethiopia from the late Oligocene epoch, 27 million years ago. This seems to be the direct ancestor of the much larger *Prodeinotherium*, which grew up to 9 feet (2.75 m) at the shoulder and was found in Southern and East Africa. As its name suggests, this genus gave rise to the enormous *Deinotherium* itself, around 23 million years ago in the early Miocene.

A full-grown bull *Deinotherium* reached at least 15 feet (4.5 m) at the shoulder and stood as tall as 16½ feet (5 m), with a weight of 9 to 11 tons (8–10 tonnes). At this size, it may have outgrown even the largest mammoths and stegodonts, making *Deinotherium* the second largest land mammal ever recorded after the enormous rhinoceros relative *Paraceratherium* (see p.484). Alongside the general increase in size, the fore limbs of *Deinotherium* are notably longer than those of its ancestors, indicating an adaptation for a longer, more efficient stride on the rapidly spreading grasslands. The most striking features of this giant proboscidean, however, were two large backward-pointing tusks that developed from the lower, rather than the upper, incisor teeth. Several species are recognized. *D. giganteum*, the first named, was identified in Europe in 1845; *D. bosazi* was widespread in Africa; and *D. indicum* was from India and Pakistan. *D. bosazi* appears earliest in the fossil record, indicating an African origin for these enormous creatures, and persisted for the longest. The last deinotheres became extinct in India around 7 million years ago and in Europe around 3 million years ago, yet persisted in Africa until 1 million years ago. **GS**

⬡ NAVIGATOR

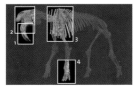

FOCAL POINTS

1 BACKWARD TUSKS
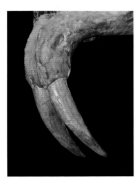
The tusks were located in the lower jaw and curved backward. They were most probably used for stripping leaves and bark from branches and they indicate a very different diet and lifestyle from that of *Deinotherium*'s elephant cousins. Browsing on softer forest vegetation would wear down molar cheek teeth far more slowly than chewing on gritty grasses.

2 TRUNK AND SKULL
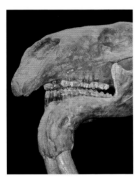
Although the soft tissues of *Deinotherium*'s trunk did not fossilize directly, a large nasal opening set high on the skull suggests that the trunk would have been strong and well developed, and was most probably used along with the tusks for collecting vegetation. Overall, *Deinotherium*'s skull was surprisingly small and flat compared to those of most proboscideans.

3 SHOULDERS

The extremely heavy skull, tusks, teeth, and trunk of *Deinotherium* required very powerful neck muscles to support and move them. These were attached to the vertebrae of the neck and chest region, especially the tall, upward-pointing, straplike projections called neural spines. Similar adaptations are seen in other heavy-headed animals, from dinosaurs to buffalo.

4 FEET
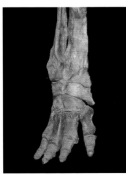
Advanced proboscideans, such as *Deinotherium*, developed the hoofed-mammal style of walking on toes (digitigrade). All five toes retained full contact with the ground to distribute the animal's weight, and an enlarged "fat pad" developed around and between the toes to spread it further. Earlier proboscideans had walked in a plantigrade style (on the soles of their feet).

▲ This *Deinotherium* specimen was excavated by paleontologist and anthropologist Mary Leakey (1913–96; see p.733). She is visible upper right, which gives an idea of the scale of this enormous, ancient proboscidean.

ORIGIN OF THE TRUNK

The trunk of a modern elephant (below) is an incredibly sensitive and dexterous organ, containing more than 150,000 muscles and countless nerve endings that allow it to function as a "fifth limb." It developed from adaptations of the nose and upper lip that may have been linked to the early semi-amphibious stage of proboscidean evolution, when a prehensile nose allowed animals such as *Moeritherium* to breathe easily in water. However, judging by fossil evidence, the trunk evolved to its current state among proboscideans with land-based lifestyles. One theory is that as the animals grew, their jaws and neck shortened to support the head, and the trunk extended to compensate for the head's immobility, while the tusks enlarged to assist with feeding.

Steppe Mammoth
PLEISTOCENE EPOCH

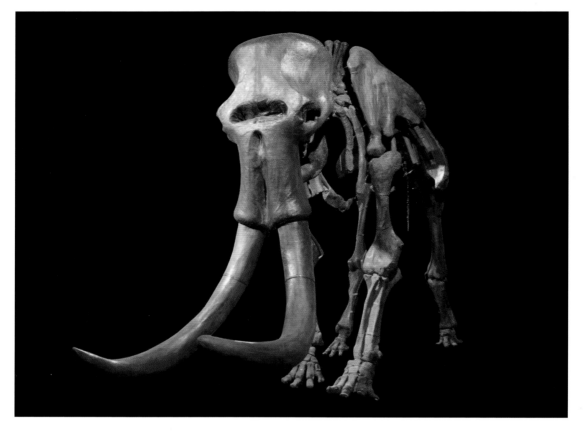

SPECIES *Mammuthus trogontherii*
GROUP Elephantidae
SIZE 26 feet (8 m)
LOCATION Northern Eurasia

NAVIGATOR

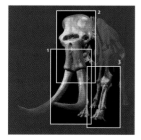

The mammoth genus *Mammuthus* is thought to have originated in Africa around 5 million years ago when it diverged from its nearest relatives in the genus *Elephas*, represented today by the Asian elephant. The earliest known mammoths were the relatively compact *M. subplanifrons* and its descendant *M. africanavus*. A population of these moved north into Eurasia, where it gave rise to the significantly larger southern mammoth, *M. meridionalis*, before becoming extinct 4 million to 3 million years ago. *M. meridionalis* led to the Eurasian steppe mammoth, which is likely to have appeared in Siberia, and the New World Columbian mammoth, *M. columbi*, which is most probably the descendant of a southern mammoth population that crossed the ice age land bridge around 1¾ million years ago. An important and previously widespread cousin of modern elephants, the steppe mammoth (*Mammuthus trogontherii*) thrived during the middle to late Pleistocene epoch from 700,000 to 125,000 years ago. It roamed the steppe and tundra grasslands that spread around the northern hemisphere with the onset of the most recent ice age. Both steppe and Columbian mammoths grew significantly larger than their ancestors, and were among the largest proboscideans of all time, second only to the earlier *Deinotherium*. As the climate declined into the penultimate glacial period of the ice age, known as the Riss glaciation, beginning around 200,000 years ago, one population of steppe mammoths gave rise to the woolly mammoths. This somewhat smaller variety, the size of a modern African elephant, adapted to the extreme conditions of ice age life. The steppe mammoths finally died out at the end of the Riss glaciation, but their Columbian cousins and woolly descendants survived into the Holocene epoch. **GS**

1 TUSKS

The steppe mammoth's tusks were among the longest of any proboscidean, reaching up to an impressive 17 feet (5 m) from root to tip in some adult males. They then curved back on themselves at the tips— particularly in males. They grew in an annual cycle, similar to tree rings, and careful study reveals not only the mammoth's age, but also the prevailing climate conditions at the time of tusk growth. The tusks fulfilled various functions, including display to rivals and mates, defence against predators, and feeding duties.

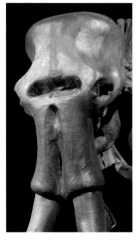

2 SKULL

Mammoth skulls changed significantly over time, seemingly in order to accommodate high-crowned teeth that had ridges that were used to grind tough grasses. Although the skulls of other elephants are long and relatively shallow, those of mammoths grew increasingly short and deep, giving the animals a more forward-looking face. The steppe mammoth had less of a high-domed skull compared to other mammoths, and muscle attachment areas indicate that its ears were probably smaller.

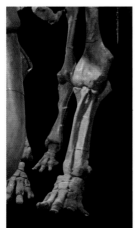

3 FORE LEGS

The front limbs of the steppe mammoth were longer and more robust than the rear ones, to support the heavy head and shoulders, and the back sloped downward. All limbs were relatively long, allowing a considerable stride for such a bulky animal—the largest steppe mammoths reached a height of at least 13 feet (4 m) at the shoulder and some may have been as tall as 14¾ feet (4.5 m). Their style of walking covered a lot of ground efficiently as the mammoths regularly ate the vegetation in one area and had to move on.

▲ The North American Columbian mammoth, *M. columbi*, approached the steppe mammoth in size. Its tusks were more spiral-shaped, curving down, then up, outward and then inward, often crossing at the tips. It went extinct only around 12,000 to 10,000 years ago.

BERGMANN'S RULE

In the 1840s, biologist and anatomist Carl Bergmann (1814–65) noticed that in a widely distributed genus or species of warm-blooded animal (mammals and birds), the size of individuals increases in colder conditions. This can be seen in deer, with moose; cats, with Siberian tigers; bears, with polar bears (below); and elephants, with mammoths. This may be due to the relationship between body volume and surface area: larger bodies have less area compared to volume, so lose heat slowly and keep warmer in the core. Therefore, evolution would favor a bigger bulk. Known as Bergmann's Rule, this has been used to explain many situations, such as why the body size of certain extinct genera increased during ice ages and fell in mild climates.

SLOTHS AND ARMADILLOS

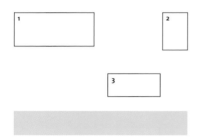

1 The herbivorous pampathere *Holmesina*, five times larger than modern armadillos, had partly flexible protection.

2 The pale-throated sloth, *Bradypus tridactylus*, from the three-toed group Bradypodidae, follows the typical sluggish sloth lifestyle.

3 The giant anteater, *Myrmecophaga tridactyla*, is highly adapted to consuming termites and ants, eating as many as 25,000 daily.

A mong living mammals, the smallest and most geographically limited major group is the Xenarthra: the sloths, anteaters, armadillos, and their extinct relatives. The name comes from the Greek for "strange joints," due to the extra articulation found in their vertebral joints. They are native only to the Americas, and fossil evidence suggests that the xenarthrans first emerged in South America 59 million years ago, during the Paleocene epoch. Left in isolation on this continent while the supercontinent of Gondwana broke up, they diversified to become the dominant land mammals. A land bridge then formed and allowed species to move into Central and North America, beginning several million years ago. Until recently, these mammals were grouped with some unusual Old World animals—the pangolins and aardvarks—in a group known as the Edentata (toothless ones). However, it is now recognized that the New World xenarthrans and the Old World species are from different lineages.

Living and extinct xenarthrans can be divided into two distinct groups: the Cingulata and Pilosa. Cingulata are represented today only by the armadillos: small animals covered with flexible armor of osteoderms (bony plates) embedded within their skin and covered by thin scutes (waterproof scales made of keratin—the same substance as fur, nails, and claws). Two other major cingulate families survived until relatively recently: the glyptodonts and pampatheres (see image 1). Both looked like larger, more heavily built armadillos, but had distinctive differences in anatomy. Pampathere armor was only partially flexible, due to three bands of movable scutes, whereas that of glyptodonts was entirely inflexible. Pampatheres grew to 5 feet (1.5 m) long and reached 440 pounds (200 kg), while the largest glyptodonts, such as *Glyptodon* itself, grew to

KEY EVENTS

59 Mya	47 Mya	37 Mya	25 Mya	20 Mya	6 Mya
The new mammal group Xenarthra most probably emerges at this time.	The heavily built armored pampatheres separate from the glyptodonts and living armadillos.	The two-toed and three-toed sloth families split from one another, although they will later produce remarkably similar modern species.	The anteater family, Myrmecophagidae, emerges, although their development almost certainly dates further back.	Several xenarthran species reach the islands of the Antilles (West Indies) after being set adrift on rafts of vegetation.	A land bridge forms between North and South America; the Great American Interchange of formerly separated species begins.

more than 10 feet (3 m) in length and could weigh up to 2¼ tons (2 tonnes). True armadillos are insectivorous or omnivorous, but the teeth and jaws of their car-size relatives suggest that they grazed on pampas grasslands.

Pilosa contains both the sloths, suborder Folivora, and the anteaters, Vermilingua. The tree-dwelling sloths (see image 2), with a total of six living species, are remnants of a formerly more diverse and ancient group. The Folivora once consisted of five families, most of which were ground-dwellers and larger than today's small arboreal (tree-dwelling) sloths. The bear-size *Nothrotheriops shastensis* (Shasta ground sloth), became widespread in North America after the land bridge formed across Central America, while *Megatherium* (see p.480), though not quite so widely distributed, reached elephantine proportions. Many sloths evolved protective osteoderms similar to those seen in armadillos, and one genus, *Thalassocnus*, developed an aquatic way of life, swimming out to sea to graze on kelp and seagrass. A range of evolutionary pressures, such as changing climate and vegetation, and the introduction of competing or predatory species from North America saw the majority of sloths decline and disappear during the Pleistocene epoch, leaving only a handful of species.

Anteaters have a toothless, tubelike snout and a long flexible tongue for catching and consuming small insects, such as ants and termites—adaptations also seen in Old World pangolins and aardvarks due to convergent evolution. There are four living anteater species: the pampas-dwelling giant anteater, which uses its powerful claws to rip open termite mounds (see image 3), the two mid-size, forest-living tamanduas, and the small, short-faced silky anteater.

Genetic and anatomical evidence shows that xenarthrans are the most basal (original) placental mammals alive today, meaning that their ancestors were the first to diverge from the main line of evolution and progress along a separate line. As a result, they lack many of the later, more advanced features found in other placental mammals. Xenarthrans have a slow metabolic (body-chemistry) rate, internal testes in the male and a relatively small brain. **GS**

5–2 Mya	2 Mya	2 Mya	¾ Mya	12,000 Ya	12,000–10,000 Ya
Aquatic sloths, *Thalassocnus*, live in western South America (Peru and Chile), and successive species gradually adapt to marine grazing.	*Glyptotherium*, up to 7 feet (2 m) long, lives in areas from southern North America to northern South America.	The appearance of the enormous *Megatherium* marks the apex of size evolution among the many and varied giant ground sloths.	The Pleistocene ice age produces the largest armored glyptodont, the 13-foot (4 m) *Doedicurus*, with a heavy defensive "mace" on its tail.	The Pleistocene ends. Pampatheres become extinct after more than 40 million years.	Natural climate change and human arrival in the Americas sees a catastrophic decline in xenarthran variety, with larger species suffering the most.

Megatherium
PLEISTOCENE EPOCH—HOLOCENE EPOCH

SPECIES *Megatherium americanum*
GROUP Pilosa
SIZE 20 feet (6 m)
LOCATION South America

The huge ground sloth *Megatherium* flourished across the Americas until just 10,000 years ago. Anatomical and genetic evidence suggests that modern sloths are the surviving members of two of five sloth families known from the fossil record. The vast majority of prehistoric sloths were ground-dwellers, and they used their elongated claws not to maneuver in the jungle canopy, but to pull tree branches to the ground to eat the best leaves. The largest ground sloths, the family Megatheriidae, appeared 23 million years ago, but took some time to reach maximum size—*Megatherium americanum* only appeared around 2 million years ago. This monstrous mammal weighed up to 4½ tons (4 tonnes), comparable to a large African elephant, and could grow up to 20 feet (6 m) long. An inhabitant of grassland and less dense woodland, it was also capable of standing, and even walking for short distances, on its hind legs. The megatheriids survived the Great American Interchange and the invasion of North American mammals—in fact, they appear to have thrived at that time. *Eremotherium*, closely related to, if not the same species as, *Megatherium* moved northward into Central and North America. The eventual disappearance of these two animals around 10,000 years ago coincides with the end of the Pleistocene ice age and the arrival of humans in South America. Hunting probably played a role in their extinction, as did climate change and the demise of their preferred habitat. **GS**

◈ NAVIGATOR

◉ FOCAL POINTS

1 MOUTH

M. americanum is often depicted with a long prehensile tongue, although this is disputed. Studies of the fossil jaw anatomy and points of muscle attachment suggest that it had flexible, prehensile lips for grasping vegetation, and powerful jaws for grinding its food prior to swallowing.

2 LEGS

A pillarlike tail and powerful hind legs allowed the creature to stand upright, supporting itself in the manner of a tripod to reach the highest branches, or sit back on its haunches. Fossil footprints also show that *M. americanum* was capable of walking bipedally, carrying its weight on its hind legs alone.

3 FEET

M. americanum's five long claws prevented both plantigrade (walking on the soles of its feet) and digitigrade (walking on its toes) gait. Ground sloths evolved a form of movement in which their weight was supported on the sides of the feet—also displayed by their living distant cousin, the giant anteater.

THE TOES OF SLOTHS

Living sloths fall into two families: the three-toed Bradypodidae (above) and the two-toed Megalonychidae. Species in the former are all living arboreal sloths, but the latter family once included large ground-dwelling species. Living sloths are all similar in appearance and lifestyle, with long hook-shaped claws that allow them to hang securely from high branches without muscular effort—an adaptation that helps these slow mammals survive on a diet mostly of leaves with poor nutritional value. Surprisingly, the two sloth families are only distantly related. Genetic analysis suggests that they last shared a common ancestor in the late Eocene epoch, more than 35 million years ago.

HOOFED MAMMALS: PERISSODACTYLS

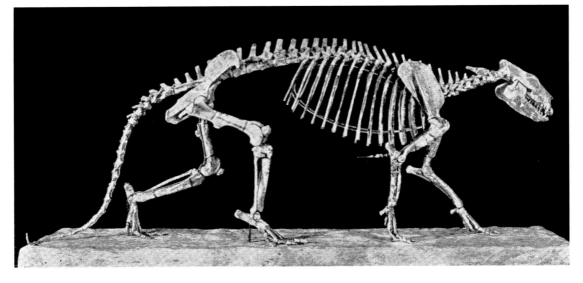

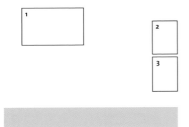

M any artiodactyl groups (even-toed hoofed mammals; see p.452), have survived to the present day, but their odd-toed cousins (see image 2), the perissodactyls, are not so well represented. Only three families are living: the tapirs, rhinoceroses, and equids, comprising horses, asses, and zebras. Only this last group is widespread. However, all of these families were much more diverse in earlier times, and one gave rise to the largest land mammal ever to exist: a giant cousin of the rhinoceros known as *Paraceratherium* (see p.484).

Perissodactyls, like artiodactyls, are thought to have their evolutionary roots among the ancient hoofed mammals known as condylarths, and specifically among the Phenacodontidae family, named after *Phenacodus* (see image 1). These dog-size herbivores evolved 57 million years ago and survived for about 15 million years into the middle Eocene epoch. They had slender legs suited to running and five hoofed toes on each foot, although they supported most of their weight on the large central digit and its immediate neighbors. The earliest true perissodactyl may have been *Radinskya*, known only from the partial remains of a 55-million-year-old skull found in China. The structure of the few surviving teeth shows that it had none of the adaptations of the later perissodactyl families, so is likely to have been a very early type in the group. Other experts disagree and suggest that it was simply a close cousin. Odd-toed hoofed mammals evolved and diversified rapidly during the Eocene, driven by factors such as climate change, evolving food sources (especially trees), and the changes of various predators, such as the more primitive

1 *Phenacodus*, from 55 million years ago in Europe and North America, was one of the earliest hoofed mammals.

2 The toe bones of *Macrauchenia* show their robust weight-bearing nature. This ungulate was a litoptern, a now-extinct group that evolved in isolation in South America.

3 Four of the five living species of tapir inhabit South and Central America, including this Baird's tapir *Tapirus bairdii*. One species lives in Southeast Asia.

KEY EVENTS

57 Mya	56–33 Mya	56 Mya	55 Mya	55–49 Mya	55–45 Mya
The Phenacodontidae, a group of small hoofed mammals with five toes and a body suited to running, emerges from the larger group Condylarthra.	Brontotheres, or Titanotheriidae, cousins of the horses, flourish in North America and Asia, eventually growing to enormous size.	The Ancylopoda branches off from other perissodactyls, and their front hooves develop into curved claws.	The small Chinese mammal *Radinskya* may be a member of an early perissodactyl group but has none of the advanced features seen in later families.	*Heptodon*, an early member of the Tapiroidea, is very similar to modern tapirs.	*Hyracotherium* is a highly successful palaeothere, a dog-size member of an early perissodactyl group that would lead to horses.

creodont mammals (see p.464) and the later carnivoran mammals. They became the largest and most widespread land mammal group on Earth. At least 15 distinct families developed.

Among the extinct groups, the Titanotheriomorpha contained only one family, the brontotheres. These enormous and superficially rhinoceroslike animals had four toes on their front feet and three on their hind feet, and survived through the Eocene until its end 34 million years ago. Despite their appearance, they are more closely related to horses than rhinoceroses, as the small Eocene herbivores share a common ancestor known as palaeotheres. They steadily grew, beginning at a relatively small size and developing increasingly elaborate bony growths on their skulls that looked similar to rhinoceros horns. However, a rhinoceros horn has no bony core and is made only from the protein keratin. Another extinct group, Ancylopoda, included the chalicotheres, such as *Moropus*. These herbivorous mammals survived until only a couple of million years ago. Growing to the size of horses, they developed elongated fore limbs and shortened hind limbs, walking with their bodies in a diagonal posture. Furthermore, their front hooves developed into long claws, similar to those found on modern anteaters, which they probably used for stripping vegetation from trees. The impracticality of walking on clawed feet forced them to adopt a "knuckle-walking" posture similar to that of living gorillas.

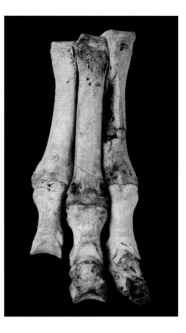

Two other major groups of perissodactyls are represented by living families. One is the Hippomorpha, with the horses (Equidae; see p.494), and the other is the Ceratomorpha, with both rhinoceroses and tapirs (see image 3). Each of these living groups has several families of extinct cousins, which in the case of the rhinoceroses were particularly diverse. The Amynodontidae family, which flourished from 46 million to 7 million years ago, lived in semi-aquatic habitats and increasingly resembled hippopotamuses. The Hyracodontidae, from 56 million to 20 million years ago, developed longer limbs and produced the largest ever land mammals, culminating in *Paraceratherium*. True rhinoceroses, family Rhinocerotidae, have a history stretching back to the late Eocene, 35 million years ago. They went through several major bursts of evolution before the ancestors of the modern species developed in Asia during the Miocene epoch 20 million years ago. This produced not only the five species that are endangered today (see p.486), but also the *Coelodonta*, the woolly rhinoceros of the last ice age. *Elasmotherium*, another giant ice-age rhinoceros with bladelike horns, was the last survivor of a separate subfamily until extinction 50,000 years ago.

While rhinoceroses have been through many evolutionary experiments, the tapirs have been more conservative. They first appeared in North America during the early Eocene, around 56 million years ago. Early examples, such as *Heptodon*, were very similar to the modern species. Their main innovations over time seem to have been a doubling in size and the development of a short prehensile trunk formed by modifications of the upper lip. **GS**

52 Mya	46–7 Mya	40 Mya	30–22 Mya	26 Mya	350,000 Ya
Eohippus is the earliest known member of the equid family that gives rise to modern horses.	The Amynodontidae group of "aquatic rhinos" evolves, most likely from within the hyracodontids, and flourishes until 7 million years ago.	The first chalicotheres, horselike animals with long clawed fore limbs, evolve from earlier Ancylopoda.	*Paraceratherium*, a giant member of the hyracodontid family roams the rain forests of central Asia. It is the largest-ever land mammal.	The three major tribes of modern rhinoceroses most probably split from one another at around this time.	The woolly rhinoceros, *Coelodonta*, becomes widespread across northern Eurasia during the last ice age, going extinct around 10,000 years ago.

Paraceratherium
OLIGOCENE EPOCH—EARLY MIOCENE EPOCH

SPECIES *Paraceratherium bugtiense*
GROUP Hyracodontidae
SIZE 26 feet (8 m)
LOCATION Pakistan

The largest land mammal ever to exist, *Paraceratherium* (near horned beast) flourished during the Oligocene epoch, between 30 million and 22 million years ago. A giant hornless rhinoceros relative in the Hyracodontidae family, it grew up to 16 feet (4.5 m) tall at the shoulders, had a head to body length of 26 feet (8 m) and long, pillarlike legs carrying a body weight of 16 tons (15 tonnes) or more. Deceptively named, because they are only distantly related to the hyraxes, the hyracodontids first evolved during the early Eocene, 56 million years ago. They emerged from the Ceratomorpha group that includes the modern rhinoceroses and tapirs. They were initially small or medium-size herbivores with long faces and slender legs for fast running, and flourished in central Asia. While some ceratomorphs evolved the squat forms of the rhinoceros and tapir, the subgroup Indricotheriinae evolved in a different direction, to later produce *Paraceratherium*. As their bulk increased, their legs and necks grew longer and sturdier, enabling them to reach vegetation on high branches in the rain forests. The pattern of *Paraceratherium*'s teeth indicates that it fed by stripping away vegetation with modified incisor teeth. Along with other hyracodontids, *Paraceratherium* became extinct in the early Miocene due to large-scale climate change: the Indian subcontinent collided with mainland Asia, pushing up mountain ranges, which increased aridity and caused a retreat of the rain forests. **GS**

◆ NAVIGATOR

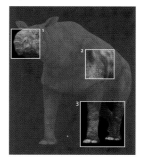

◉ FOCAL POINTS

1 TRUNK OR LIP?
The elongated nasal area of *P. bugtiense*'s skull, along with a lengthy slit down the middle, indicates that this giant animal had at least a prehensile upper lip, as seen in today's rhinoceroses and tapirs. It possibly had a fully developed trunk to increase its reach into the highest trees.

2 HAIRLESS SKIN
It is presumed that *P. bugtiense* was hairless to alleviate the problem of overheating. All large mammals lose heat slowly, therefore tend to lack insulating body hair. Some researchers suggest that this creature also had elephantlike flapping ears to increase the rate of heat loss.

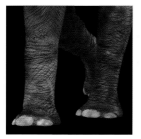

3 LEGS
P. bugtiense's huge weight required thick, pillarlike legs to support it. These grew longer at the front, giving it a deep chest. Its enormous stride would have reduced the energy expended in moving its bulk, and the legs could have delivered a powerful defensive kick toward any predators.

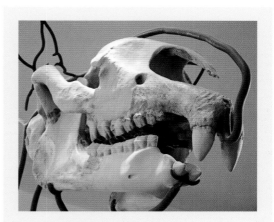

NAMING THE GENUS

The remains of *Paraceratherium* (above) have been known by a number of names: *Baluchitherium*, *Thaumastotherium*, *Aralotherium*, *Dzungariotherium*, and *Indricotherium*. This is because the fragmentary nature of the fossils unearthed across Asia made it difficult to identify these animals as members of the same genus. Early finds in 1908 from Balochistan in Pakistan were at first placed in another genus: *Aceratherium*. Based on other fossils, *Paraceratherium* was described in 1911, *Baluchitherium* in 1913 and *Indricotherium* in 1915. A review of the fossil evidence in 1989 suggested that all of these were not sufficiently different to be genera in their own right, and the earlier name—*Paraceratherium*—took priority.

Sumatran Rhinoceros
MIOCENE EPOCH—PRESENT

SPECIES *Dicerorhinus sumatrensis*
GROUP Rhinocerotidae
SIZE 8 feet (2.5 m)
LOCATION Sumatra, Borneo, and the Malay Peninsula
IUCN Critically Endangered

✦ NAVIGATOR

The Sumatran rhinoceros is one of the most endangered animals on Earth, with around only 200 living in the wild. It is the smallest of the five living rhinoceros species and has an average weight of 1,540 to 1,870 pounds (700–850 kg) and a head–body length of 8 feet (2.5 m). It is also the sole survivor of the ancient subgroup Dicerorhini, one of numerous divisions within the rhinoceros family. The Rhinocerotidae family is divided into two major subfamilies—the Rhinocerotinae and the extinct Elasmotheriinae. The Rhinocerotinae group includes several extant subgroups; the Sumatran rhinoceros's Dicerorhini; the one-horned Rhinocerotini with Indian and Javan rhinoceroses; and the two-horned Dicerotini, which includes the African black and white rhinoceroses. Geography implies that the Sumatran species, *Dicerorhinus sumatrensis*, is closely linked to its neighbors in Asia, but its two horns suggest a relationship with the African rhinoceroses. Recent studies indicate that the ancestors of all three split from one another at the same time—26 million years ago during the Oligocene—then evolved separately. The Sumatran rhinoceros's closest relative is likely to be the extinct woolly rhinoceros, *Coeldonta antiquitatis* (see panel), and it is generally agreed to be the least derived of the living rhinoceros species, which means it bears the most similarity to the common ancestor of all living rhinoceroses. The Sumatran rhinoceros is found in rain forests, cloud forests, and swamps on the Malay Peninsula, Sumatra, and Borneo, having probably migrated to these islands when sea levels were lower during the last ice age. Its range once encompassed Myanmar and much of the Indian subcontinent. Three subspecies are recognized—western, northern, and eastern, or Bornean—and all are endangered by poaching and habitat loss. The northern subspecies may already be extinct. **GS**

⊙ FOCAL POINTS

1 HORNS
The Sumatran rhinoceros has two horns, but the rear one has been reduced to little more than a bump. The nasal horn is 6 to 10 inches (15–25 cm) long, but can grow longer. The horn consists entirely of keratin (the same protein as hair); there is no bony core.

2 SENSES
Rhinoceroses are short-sighted but have extremely good hearing, with movable radar-dish ears. They also have an impressive sense of smell, with a large nasal chamber. These senses are useful in dense forest habitat, where sight can be obscured by undergrowth.

3 SKIN
The Sumatran rhinoceros's skin is thin and relatively loose-fitting, but toughened by large amounts of collagen that form a weblike matrix within it. Folds of skin encircle the body around the leg joints and on the neck. The loose skin eases the rhinoceros's movements.

4 HAIR
This rhinoceros has a noticeable coat of hair, which is unique among living rhinoceroses. The hair is usually reddish brown and often becomes matted with mud as the animal wallows to stay cool. It is longest around the ears and tail.

⊥ RHINOCEROS RELATIONSHIPS

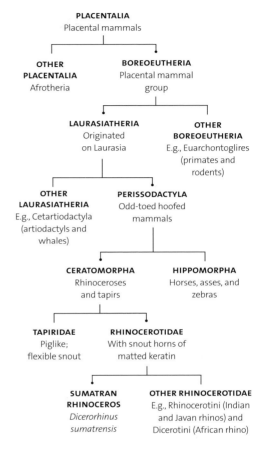

PLACENTALIA
Placental mammals

OTHER PLACENTALIA
Afrotheria

BOREOEUTHERIA
Placental mammal group

LAURASIATHERIA
Originated on Laurasia

OTHER BOREOEUTHERIA
E.g., Euarchontoglires (primates and rodents)

OTHER LAURASIATHERIA
E.g., Cetartiodactyla (artiodactyls and whales)

PERISSODACTYLA
Odd-toed hoofed mammals

CERATOMORPHA
Rhinoceroses and tapirs

HIPPOMORPHA
Horses, asses, and zebras

TAPIRIDAE
Piglike; flexible snout

RHINOCEROTIDAE
With snout horns of matted keratin

SUMATRAN RHINOCEROS
Dicerorhinus sumatrensis

OTHER RHINOCEROTIDAE
E.g., Rhinocerotini (Indian and Javan rhinos) and Dicerotini (African rhino)

▲ As perissodactyls, rhinos, horses, and tapirs are included in the Laurasiatheria—a huge grouping of mammals that had their early evolution on the northern supercontinent of Laurasia (now North America, Europe, and Asia).

WOOLLY RHINOCEROS

One of the iconic species of the last ice age, the woolly rhinoceros, *Coelodonta antiquitatis* (right), was widespread across Europe and Asia from 350,000 to 10,000 years ago. It is one of the best-known prehistoric animals thanks to fossils, mummified remains in the permafrost of Siberia, cave paintings, and the preservation of its DNA. These have been combined to show a creature well adapted to its cold environment. *C. antiquitatis* was larger than the Sumatran rhinoceros, up to 6½ feet (2 m) tall, with stocky limbs and thick, shaggy hair. Both of its horns were longer than those of the Sumatran rhinoceros, perhaps to clear snow for feeding, to intimidate rivals, or to deter predators.

WOLVES, DOGS, AND FOXES

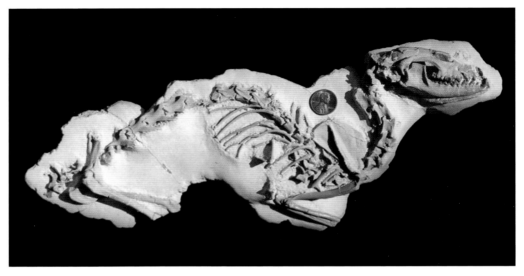

1 The almost recognizably doglike *Hesperocyon* was around 2½ feet (75 cm) long and lived in North America between 42 million and 31 million years ago.

2 The raccoon dog, *Nyctereutes procyonoides*, indigenous to East Asia, is the only extant species of its genus.

3 Also known as the warrah, the Falkland Islands wolf was the only native land mammal of those islands.

The canids, family Canidae, are one of the main groups of the modern mammal order Carnivora, and the two principal surviving branches of the family—wolflike canines and foxlike vulpines—probably diverged approximately 10 million to 7 million years ago. There are more than 30 species—some extinct—including wild and domestic dogs, wolves, jackals, and foxes. Two of the living species have similarities with earlier, more primitive members of the group, especially with regard to the skull and teeth, making it difficult to term them either canine or vulpine. These are the bat-eared fox, *Otocyon megalotis*, and the raccoon dog, *Nyctereutes procyonoides* (see image 2), which is unrelated to the raccoon but looks similar. The latter, in particular, gives a glimpse of what the earlier, more generalized members of the canid group might have looked like.

The Carnivora first evolved around 58 million to 53 million years ago, during the Paleocene epoch. In the succeeding Eocene epoch, the doglike canids and catlike felids became established, and by 40 million years ago, the first proper canid—*Prohesperocyon*—had appeared. Known from a set of fossils found in southwest Texas, *Prohesperocyon* was at first thought to be a miacid and therefore an ancestor of the whole Carnivora group, but details at the base of its skull suggest that it was an early canid; in appearance it resembled a small, slim fox. *Hesperocyon* (western dog; see image 1), also from North America, is much better known than *Prohesperocyon* and is one of the earliest members of the group, dating to 40 million years ago. It was a combination of accepted notions

KEY EVENTS

58–53 Mya	44–40 Mya	40–30 Mya	30 Mya	30 Mya	30–20 Mya
Early members of the Carnivora order (primarily meat-eating mammals) appear.	The bear dogs or amphicyonids branch out along their own evolutionary path (see p.490).	The first recognizable canids—*Prohesperocyon* and *Hesperocyon*—appear.	*Daphoenodon* is an early digitigrade amphicyonid with long limbs and long feet.	*Rhizocyon* appears; it is a member of the borophagins (bone-crushing dogs).	*Mesocyon*, a carnivorous descendant of *Hesperocyon*, inhabits North America.

of a cat and a dog, perhaps resembling the modern animal known as the civet, with a small head, long, slim body and short legs.

Hesperocyon represents one of the three main groups of canids—hesperocyonins, borophagins, and canins—that established around this time. Only the canins survive today. Another hesperocyonin was *Cynodesmus*, whose time span centered on 30 million years ago. This canid was an altogether larger, heftier animal. Although similar overall to a wolf, it looked more snub-nosed and had shorter legs. It also had five proper toes per paw, rather than the four and a dewclaw of modern dogs. Some experts consider the hesperocyonins to be ancestors of the other two canid groups: the canins and borophagins.

The borophagins (gluttonous eaters)—nicknamed "bone-crushing dogs"—were a large and successful group with more than 60 known fossil species. Early forms, such as *Rhizocyon* (root dog), weighed only several pounds (a few kilograms), but the group rapidly evolved into big, powerful hunters and scavengers, with crushing jaws and teeth reminiscent of today's hyenas (which are not in the canid group). *Borophagus* from North America was one of these later members and had the appearance of a heavy-set, muscle-faced wolf. During this time, North America had extensive grasslands, and herbivores were evolving to graze them; animals such as *Borophagus* doubtless preyed on these herbivores, although perhaps as scavengers because their shortish legs could not have sprinted after such fleet victims. Borophagins persisted until less than 5 million years ago, but all have since disappeared.

Around 10 million years ago, as climates changed, the canins, a modern subgroup of canids, began to spread out of North America during a rapid period of evolutionary diversification known as a radiation. By around 8 million years ago, the canins were dividing into the canines (modern dogs and wolves) and the vulpines (foxes). One of the earliest of the latter was Riffaut's fox, *Vulpes riffautae*. Its fossils come from Chad in Africa, and at 7 million years old, they are some of the earliest of any canid to appear in the Old World—the group had been known previously only from the Americas. *V. riffautae* was small and light-bodied, resembling today's tiny fennec fox and slightly larger Rüppell's fox. One representative of the way modern dogs and wolves were evolving about 1 million years ago was *Xenocyon*, which was probably similar to the living African wild dog and may be its ancestor. By this time, several kinds of wolves had become common, especially in North America, including Edward's wolf, *Canis edwardii*; the dire wolf, *C. dirus* (see p.492); and, largest of all, Armbruster's wolf, *C. armbrusteri*, which went extinct 250,000 years ago. Several other canids have become extinct much more recently. The Sardinian dhole, *Cynotherium sardous*, was probably exterminated when humans settled on the Mediterranean island of Sardinia around 20,000 years ago. The Falkland Islands wolf, *Dusicyon australis* (see image 3), was recorded by Charles Darwin (1809–82) when HMS *Beagle* stopped there in 1833, but the last live specimen was seen in the 1870s. **SP**

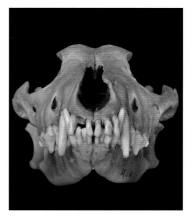

25–20 Mya	10 Mya	7 Mya	300,000–10,000 Ya	100,000 Ya	30,000–20,000 Ya
Wolves and dogs approximately resembling today's species begin to evolve.	New canids evolve rapidly following the evolution of carnassials (shearing teeth).	*Vulpes riffautae* appears: its fossils are the earliest evidence of canids in the Old World.	*Canis dirus* is similar to the gray wolf but built more heavily.	The gray wolf, *Canis lupus*, first inhabits North America, Eurasia and North Africa.	The domestic dog, *Canis lupus familiaris*, related to the grey wolf, is widely adopted by humankind.

Amphicyon
MIDDLE–LATE MIOCENE EPOCH

SPECIES *Amphicyon ingens*
GROUP Amphicyonidae
SIZE 8 feet (2.5 m)
LOCATION Europe, North America, Asia, and Africa

The Amphicyonidae was a large and diverse family of carnivorous mammals that encompassed 34 different genera. They roamed the northern hemisphere between around 40 million and 9 million years ago, from the middle Eocene to the Miocene epochs. Often known as "bear dogs" on account of their similarities to both bears and dogs, the Amphicyonidae family was in fact not closely related to these animals. Moreover, whether they were solitary hunters like bears or pack hunters like dogs is a matter of speculation, with little evidence for either theory. Amphicyonids were geographically widespread: most fossils have been discovered in Europe and North America, but finds have also been made across Eurasia and Africa. Not all amphicyonids were particularly large. Some of the earliest species probably weighed only around 4½ pounds (2 kg), which is the size of a small domestic cat. It was not until the early Miocene that larger forms evolved and diversified into formidable wolf- and bearlike species that weighed 440 pounds (200 kg) on average.

Amphicyon ingens was one of the largest of the amphicyonids and a member of one of the better-known bear dog genera. It was a powerfully built animal with a thick neck and strong limbs that could measure 8 feet (2.5 m) in length. The largest known specimen has been estimated to weigh in the region of 1,300 pounds (600 kg), although most are likely to have been smaller. It was most probably an omnivore with a mixed diet similar to that of the living brown bear. The name *Amphicyon* means "ambiguous dog," and it appears in the fossil record of western Europe in the Oligocene epoch, around 25 million years ago. A few million years later, during the early Miocene, the genus migrated to Africa and North America via the Bering land bridge. **RS**

✦ NAVIGATOR

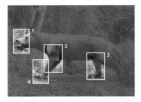

FOCAL POINTS

1 TEETH
The dentition of *A. ingens* combined fearsome sharp-edged, flesh-tearing canines with bone-crushing carnassials and grinding molars. These extremely strong teeth were supported by powerful jaw musculature, which gave a very forceful bite compared to that of most other mammals. The overall appearance of the teeth was similar to that of living bears and would have been suited to an omnivorous diet.

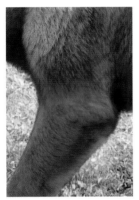

2 FORE LIMBS
The well-developed fore limbs may have been used to wrestle prey to the ground. It is probable that the evolution of *A. ingens* favored strength over speed, meaning that *A. ingens* was likely to prey on large, slow-moving animals, which it caught from ambush, rather than to chase prey animals for long distances. After the ambush, *A. ingens* would hold down the prey under the fore limbs and tear with its powerful teeth.

3 FEET
A. ingens was a plantigrade animal. This means that it walked on the sole of its foot with the whole of the foot touching the ground: a trait that it shared with both bears and humans. Dogs, by contrast, are digitigrade animals, walking with only their toes in contact with the ground. Preserved footprints show that *A. ingens* moved its two left legs and two right legs alternately in the "rocking" or "pacing" gait.

4 CLAWS
The strong front claws of some species of *A. ingens* indicate that they were probably burrowing animals that could dig out prey animals from their burrows. The smaller species may also have been light enough to climb trees in pursuit of prey or potentially to find the honey of wild bees. The teeth of *A. ingens* confirm that its diet probably consisted of roots and buds, small animals, larger prey, and scavenged carcasses.

MIOCENE MIGRATIONS

The Miocene, 23 million to 5 million years ago, saw great changes in the global climate. Episodes of mountain building in the Americas and Asia, as the Andes, Cascades, and Himalayas were formed, altered weather patterns, which resulted in cooler, drier conditions overall. This suited the growth of grasslands and the animals that grazed them, such as deer and antelope, and these habitats began to dominate the Americas and eastern Eurasia. Predators such as *Amphicyon* (right) quickly took advantage. Land bridges, such as the Bering land bridge between present-day Siberia and Alaska, formed and allowed waves of animal migration between Africa, Eurasia, and North America. Evolution was spurred by the movement into new habitats and by increasing competition between species. Fossils in Namibia show that *Amphicyon* had migrated south through the African continent by the middle Miocene.

Dire Wolf
PLEISTOCENE EPOCH—HOLOCENE EPOCH

SPECIES *Canis dirus*
GROUP Canidae
SIZE 5 feet (1.5 m)
LOCATION North and South America

NAVIGATOR

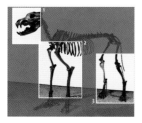

The dire wolf, *Canis dirus*, is likely to have first appeared in the Americas around 300,000 years ago during the Pleistocene epoch. It evolved earlier than the gray wolf, although the two were contemporaneous in North America for most of the dire wolf's existence. It became extinct during the end of the last major ice age (see p.550), 10,000 years ago, which the gray wolf survived. The dire wolf is a well-known extinct canid, partly due to the huge numbers of specimens preserved in the La Brea Tar Pits in Los Angeles. The first specimen was found on the banks of the Ohio River in Indiana, by Joseph Norwood (1807–95), the first state geologist of the time. He sent the fossil to paleontologist Joseph Leidy (1823–91) at the Academy of Natural Sciences in Philadelphia. Leidy determined that the fragment belonged to a then unknown species of wolf and named it *C. primaevus* in 1855. He later discovered that the name *primaevus* had already been claimed for another discovery, so he renamed it *C. dirus* (dire wolf) in 1858. The dire wolf is the most common of the North American fossil canids. Similar to the gray wolf, which it superficially resembled, the dire wolf's origins can be traced back to Eurasia; it is possibly a descendant of the great wolf *C. armbrusteri*. Its appearance in the North American fossil record as a fully evolved animal suggests it had traveled to that continent and had already evolved elsewhere. It expanded its range down to the north and west coasts of South America, and during that time it is likely to have been one of the most formidable of the New World predators. It may have lived a hyenalike lifestyle, because its powerful teeth and jaws were well adapted to scavenging the bones discarded by other predators of the time, such as the sabre-toothed cat *Smilodon* (see p.528). **RS**

◉ FOCAL POINTS

1 TEETH

The canines of dire wolves had greater bending strength than those of modern wolves. The flesh-shearing, upper jaw, fourth premolars were larger than those of modern gray wolves, although the first molars on the lower jaw were similar to those of typical canids. Wear patterns on teeth show blunting associated with bone chewing, but no evidence of bone-crushing.

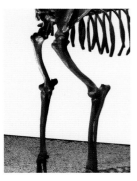

2 STOCKY LIMBS

Based on the dire wolf's stocky legs and powerful build, some paleontologists speculate that it was an ambush predator rather than a hunter that pursued prey across distances with great stamina, like the gray wolf. It is likely that dire and gray wolves had different prey preferences, which meant that they were not competing with one another and could coexist.

3 FEET AND CLAWS

The dire wolf's feet and claws show slight differences to those of the gray wolf. The claws are more robust, and probably better at scrabbling in soil rather than scratching sharply. The vestigial fifth toe, also known as the dewclaw, is loosely attached and raised above the ground. It evolved in the wolf line during the Miocene, around 20 million to 10 million years ago.

◷ SCIENTIST PROFILE

1823–46

Born in Philadelphia, Joseph Leidy graduated from medical school there in 1844. In 1846 he became the first scientist to use a microscope in forensic examination when he identified human blood stains, despite a murder suspect claiming that the blood belonged to chickens.

1847–58

Leidy left his medical practice to study life sciences, and became one of the United States's authorities on microscopy and the founder of US vertebrate paleontology. In 1858 he reconstructed and named *Hadrosaurus*, the world's first near-complete large dinosaur skeleton, from bones collected in New Jersey.

1859–60

In his copious writings, Leidy described variation within species and how new species appear while others become extinct. When Charles Darwin (1809–82) published his theory of natural selection in 1859, Leidy became an enthusiastic supporter.

1861–91

As professor of anatomy at the University of Pennsylvania, Leidy spent years building on his knowledge of extinct mammals. His fossil studies showed that horses had existed and become extinct in the Americas prior to their reintroduction by the Spanish conquistadors. He also demonstrated that species of lion, tiger, camel, and rhinoceros were all present in prehistoric western North America.

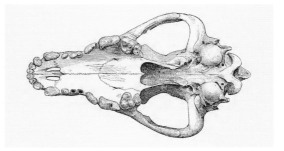

▲ A view from below of the upper jaw and palate shows, on each side, three incisor teeth, the large pointed canine, four premolars, and two molar teeth. The fourth premolar is the upper carnassial.

VARIATIONS IN SIZE

Dire wolves (right) could vary considerably in size, ranging from animals that were similar in proportion to a large gray wolf, to populations that were the largest specimens of the *Canis* genus ever to have existed. An average-sized dire wolf is likely to have been around 5 feet (1.5 m) in length from head to tail and weighed in the region of 140 pounds (65 kg). The living gray wolf is likewise variable according to geographic range, the heaviest individuals in the far north also scaling 130 pounds (60 kg). In comparison, the modern German Shepherd dog weighs around half of this.

HORSES

1 Once classed as a true equid, *Hyracotherium* is now regarded as a palaeothere, a group of herbivorous mammals that preceded equids.

2 The skull of *Mesohippus* displays six grinding teeth in each side of each jaw, and a huge, deep jawbone, which would have anchored powerful chewing muscles.

3 *Miohippus* was well on the way to evolving the long slim legs, strong neck and long face typical of the horse family.

The family Equidae, which includes horses, zebras, and donkeys, is the most successful group of living perissodactyls (odd-toed hoofed mammals). Bulky rhinoceroses and tapirs bear their weight on the middle three hoofed toes, with the central one enlarged, while modern horses stand on a single enlarged middle toe, with the second and fourth digits much reduced. Ancient equid fossils trace the loss of the outer toes and are often discussed as if they were a single chain of descent, but in reality, very few species can be identified by direct ancestor to descendant relationships. The equid family tree is wide-ranging, with dozens of close cousins evolving, flourishing, and going extinct on the path toward the modern species.

The first equids, like many other perissodactyl groups, separated into a distinct lineage during the early Eocene epoch 56 million years ago. The earliest horse ancestor was *Hyracotherium* (hyrax beast; see image 1), a small herbivore about 2 feet (60 cm) long, which placed its weight on four hoofed toes on its front feet and three on the hind feet. It thrived from 56 million to 45 million years ago. Its molar teeth displayed distinctive ridges seen in later horses, and for more than 140 years it was believed to be an early member of the Equidae. However, recent studies have shown that *Hyracotherium* was a member of the extinct Palaeotheriidae family, and as a result of this reclassification, *Eohippus* (dawn horse) regained its place as the earliest true equid. It looked like *Hyracotherium* but may have grown larger. Its slender legs were well suited for running, although they lacked the specialized adaptations seen in later equids. The remains of *Eohippus* species are found across North America, where most of its evolution took place. Diversification through the Eocene gave rise to a variety of other genera, such as the tapirlike *Propalaeotherium*, an equid found in Europe. By 50 million years ago, *Eohippus* had evolved into the more slender *Orohippus* (mountain horse). It had the same number of walking digits as *Eohippus* but had lost the vestigial toes.

Various middle Eocene horses developed innovations that are still seen in their modern descendants, but the next major step is marked by *Mesohippus* (middle horse; see image 2), that lived between 40 million and 30 million years ago, from the late Eocene to the early Oligocene epochs. Only slightly bigger than its predecessors, it was more slender and walked with three toes on both fore and hind feet, with the fourth front toe reduced to a vestigial bump on the side of the leg. *Mesohippus*'s face was longer and more equine, its brain larger and its teeth developed the modern arrangement of three premolars and three molars, usually grouped as six molars in each side of each jaw, upper and lower.

About 36 million years ago grasslands began to supplant forests. One equid took advantage of the new habitat. Evolving from within a population of *Mesohippus*, new species *Miohippus* (small horse; see image 3) lived alongside its cousins for some time before replacing them 32 million years ago. Significantly larger than previous equids, it was suited to running and grazing on the prairies of North America. However, once the smaller competing species had disappeared, evolution effectively went into reverse, as some *Miohippus* moved back into woodlands, diminished in size, and reverted to a four-toed front foot that offered stability on softer ground. Some researchers classify forest-dwelling *Miohippus* as a separate species, *M. intermedius*, while others consider it a different genus, *Kalobatippus* (stilt horse). It eventually produced a *Mesohippus*-like descendant, *Anchitherium* (near beast), which crossed the land bridge into Eurasia 20 million years ago and led to the Chinese *Sinohippus* (Chinese horse) and the European *Hypohippus* (low horse), which became extinct 5 million years ago. Meanwhile, the main line of horse evolution continued in North America, as the plains-roaming *Miohippus* gave rise to *Parahippus* (near horse) and, 20½ million years ago, *Merychippus* (ruminant horse; see p.496). The latter evolved into many species and several descendant genera of increasing size. From this emerged *Dinohippus* (powerful horse), which is recognized as the direct ancestor of the modern horse, *Equus ferus*, around 10¼ million years ago. **GS**

24 Mya	20 Mya	15 Mya	10¼ Mya	3½ Mya	43,000 Ya
A population of *Mesohippus* returns to a forest-dwelling life, evolving "backwards" into the primitive-looking *Kalobatippus* and its descendants.	The hugely successful *Merychippus*, the earliest known grazing horse, appears.	A burst of evolution gives rise to new species and genera of *Merychippus*, such as hipparions, protohippines and true horses.	*Dinohippus* appears in North America. It is believed to be the closest relative to *Equus*.	*Equus simplicidens*, the earliest member of the modern horse genus, evolves in North America.	Przewalski's horse, *Equus ferus przewalskii* (see p.498), splits from the main line of evolution that will lead to the modern horse.

Merychippus
MIOCENE EPOCH

SPECIES *Merychippus insignis*
GROUP Equidae
SIZE 5 feet (1.5 m)
LOCATION North America

The most successful equid genus of the Miocene epoch, *Merychippus* flourished in North America between 20 million and 10 million years ago. Growing up to 3 feet (1 m) tall at the shoulder, it was the largest horse to exist until that time, and the first to resemble the modern horse in appearance, with a long face, widely set eyes, and long legs. Molars with characteristic high crowns indicate that it was adapted for a life grazing on prairie grasslands—an environment in which its improved all-around eyesight and legs longer than other animals made it easier to flee from predators. *Merychippus* means "ruminant horse," because initially it was thought to belong to the ruminants group.

Merychippus evolved from the plains-dwelling species of *Miohippus*—the genus that first moved from forests to plains in response to the environmental changes of the late Oligocene and early Miocene, around 30 million years ago. Approximately 15 million years later, *Merychippus* underwent a burst of evolution that produced at least 19 different species in several different genera. The major groups of merychippine descendants were small three-toed animals, such as *Protohippus* and *Callippus*; larger three-toed equines, typified by the hugely successful *Hipparion*; and the "true horses," which led to the modern *Equus*, in which only a single toe touched the ground and the other toes were increasingly reduced. The *Merychippus* burst of diversification led to various evolutionary lines, encompassing between 15 and 20 species, which has demonstrated that the evolution of horses is more complicated than previously was believed. From *Merychippus* came the small, three-toed *Callippus*, which appears to have given rise to the much larger, single-toed *Pliohippus*—this was long thought to have been the direct ancestor of modern *Equus*. However, this theory has now been dismissed due to differences in tooth structure and facial shape. Instead, it represents another line that died out, and the true ancestry of modern horses lies with another *Merychippus* offshoot, *Dinohippus*. **GS**

◆ NAVIGATOR

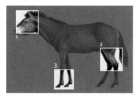

👁 FOCAL POINTS

1 FACIAL FOSSAE
The head of *Merychippus* shows pronounced facial fossae—indentations in front of the eyes with the form of an elongated concave dish. These are key diagnostic features for understanding the relationships between ancient horses. They are greatly reduced, or absent, in most modern horses.

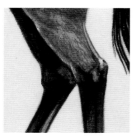

2 CHANGING LEGS
Changes to the length of leg bones and joints in horses took place over tens of millions of years. As the ability to run fast became more important, the trend was for the ankle joint to move higher up the leg, forming the hock of modern horses, while the feet and toe bones became increasingly elongated.

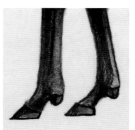

3 ENLARGED CENTRAL TOE
The feet of *Merychippus* had an enlarged central toe and greatly reduced second and fourth toes. The first and fifth toes had long since disappeared. While earlier horses had support pads behind their hooves, in *Merychippus,* the pad had disappeared, making it the first horse to walk entirely on the tips of its toes.

⊔ HORSE RELATIONSHIPS

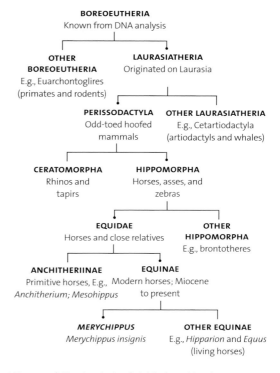

BOREOEUTHERIA
Known from DNA analysis

OTHER BOREOEUTHERIA
E.g., Euarchontoglires (primates and rodents)

LAURASIATHERIA
Originated on Laurasia

PERISSODACTYLA
Odd-toed hoofed mammals

OTHER LAURASIATHERIA
E.g., Cetartiodactyla (artiodactyls and whales)

CERATOMORPHA
Rhinos and tapirs

HIPPOMORPHA
Horses, asses, and zebras

EQUIDAE
Horses and close relatives

OTHER HIPPOMORPHA
E.g., brontotheres

ANCHITHERIINAE
Primitive horses, E.g., *Anchitherium; Mesohippus*

EQUINAE
Modern horses; Miocene to present

MERYCHIPPUS
Merychippus insignis

OTHER EQUINAE
E.g., *Hipparion* and *Equus* (living horses)

▲ Horse evolution is notoriously intricate and involves many subgroups and sub-subgroups. *Merychippus* is classed in the subfamily Equinae, which is relatively modern and includes all living horses, asses, and zebras (genus *Equus*).

RUMINATION

The name *Merychippus* was inspired by the apparent similarity between this equid's teeth and those of ruminants, such as cows (right). However, there is little evidence that early horses were true ruminants. Rumination is the ability to extract nutrition from vegetable matter by fermenting it in the foregut (digestive stomach chambers) then re-chewing, known as chewing the cud. This system appears to be unique to the artiodactyls (even-toed hoofed mammals), such as cows, sheep and goats. *Merychippus*, like modern horses, was probably a "hind-gut fermenter" with an additional digestive chamber in the intestine. Here, large quantities of stored bacteria would break down cellulose to release energy.

Przewalski's Horse
PLEISTOCENE EPOCH—PRESENT

SPECIES *Equus ferus przewalskii*
GROUP Equidae
SIZE 7 feet (2 m)
LOCATION Mongolia
IUCN Endangered

◆ NAVIGATOR

The only truly wild horse surviving today, Przewalski's horse is a short-legged, stocky equine that roams the steppe grasslands of Mongolia in Central Asia. Genetic analysis shows that it separated from other modern horses, *Equus ferus*, at least 43,000 years ago and has never been domesticated, unlike other so-called "wild horses," which are in fact feral descendants of domesticated ancestors. Other wild equines, such as zebras, onagers, and African wild ass, are more distant relations. Przewalski's horse is named after Russian explorer and army colonel Nikolai Przewalski (pronounced "zhevalski"), who brought the animal to the attention of the scientific community in 1881. However, it went into drastic decline in the middle of the 20th century due to harsh winters and hunting by humans, and the last truly free-roaming individuals were seen in the 1960s. At that time only a handful of animals survived in zoos and parks. However, a carefully managed program of breeding at various centers around the world has seen the total captive population rise to more than 1,500, with an estimated 300 animals released back into the wild.

After much discussion, Przewalski's horse is now regarded as a subspecies, *E. ferus przewalskii*, of the living true horse species, *E. ferus*. Modern horses of the genus *Equus* first evolved in North America 3½ million years ago. The earliest species is known by a variety of names, including *E. simplicidens*, the "Hagerman horse" (after the location of its original discovery in Idaho) and the "American zebra." As happened previously in equine history, it went through a period of rapid evolution as it spread around the world (see panel). **GS**

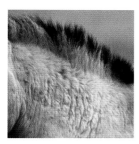 FOCAL POINTS

1 PRIMITIVE MARKINGS

Przewalski's horse has a number of distinctive coat markings: a dark mane, a "dorsal stripe" along the back, and faint stripes on the legs. These are said to be primitive markings because they frequently occur in several breeds of domestic horse, despite not being chosen during the selective breeding process.

2 STAY APPARATUS

The passive stay apparatus is a system of muscles and tendons that locks the upper leg in place and prevents the joints buckling when the horse relaxes. This enables Przewalski's horse to stand for long periods and it can see, hear, and smell potential predators more easily in this upright position.

3 HOOVES

Przewalski's horse has slightly longer, narrower hooves than the domestic or feral horse. These hooves have a thicker horny layer and overall are harder and sharper, which helps the horse dig through sand and loose soil in its arid habitat to reach water or succulent underground plants.

▲ The genome (total gene collection) of Przewalski's horse has been studied in detail thanks to conservation efforts. It has the largest number of chromosomes (gene packages) of any known horse species: 66, which is one pair more than the domestic horse. The uniformity of this high count indicates that *E. ferus przewalskii* did not interbreed significantly with its domestic cousins. Nevertheless, the two subspecies, *E. f. przewalskii* and *E. f. caballus*, are able to mate and produce viable offspring.

EQUUS WORLDWIDE

Genetic and archeological evidence suggests that once the genus *Equus* appeared in North America, it spread rapidly. Some wild horses crossed the ice age land bridge into Asia where they diversified to produce species such as the mountain zebra, *E. zebra* (right), the plains zebra, including *E. quagga* (see p.558), and Grévy's zebra, *E. grevyi*, as well as the onager, *E. hemionus*, the kiang, *E. kiang*, and the African wild ass, *E. africanus*. Others moved into South America. In North America, the modern horse *E. ferus* appeared 2 million to 1 million years ago and spread across the northern continents before becoming extinct in its native land. It has since produced three subspecies— Przewalski's *E. f. przewalskii*; the tarpan, *E. f. ferus*, (extinct since 1909); and the successful domesticated horse, *E. f. caballus*. Archeological evidence of the relationship between humans and horses from Ukraine and Kazakhstan dates to 6,000 years ago.

WHALES

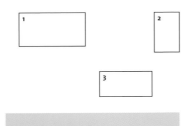

1 *Pakicetus*, which was up to 7 feet (2 m) long, probably lived a part-amphibious lifestyle, wading often, but resting on land.

2 *Ambulocetus* showed transition to greater aquatic adaptation, habitually swimming and diving.

3 Baleen plates (the white teethlike combs in the mouth of the whale) allow "whalebone" whales to filter water for food, as seen here in the gray whale, *Eschrichtius robustus*.

A s the oceans recovered from the mass extinction at the end of the Cretaceous period, 66 million years ago, the disappearance of giant marine reptiles such as mosasaurs (see p.352), plesiosaurs (see p.282), and ichthyosaurs (see p.244) left ecological vacancies for new, large aquatic creatures. However, it was 20 million years before animals fully exploited this. Mesozoic era mammals of the dinosaur age had experimented with an aquatic lifestyle as early as the Jurassic period, but it would take a group of placental mammals (see p.424) to truly conquer the oceans: the Cetacea—whales, dolphins, and porpoises.

For more than a century, fossil triangular teeth appeared to link the earliest whales to a group of carnivorous hoofed mammals: the extinct mesonychids. However, new studies suggested that all cetaceans evolved from the artiodactyls (see p.452) and that hippopotamuses may be among their closest living cousins. An ancient whale relative is the small, deerlike *Indohyus*, remains of which were found in Pakistan from 48 million years ago when the region was isolated in the Tethys Sea. Several species important in whale evolution have been found here. *Indohyus* showed adaptations for life foraging in an aquatic environment, such as dense bones that would have counteracted its natural buoyancy in water. It also displayed a specific mode of bone growth found only in whales today. However, it is included in the family Raoellidae, a sister group to the cetaceans. Other artiodactyls are the first cousins of *Indohyus*. Among the earliest true cetaceans was the Pakicetidae family (see image 1), a group of four-legged "walking whales" that flourished 50 million years ago, in the early Eocene epoch, around the banks of seasonal rivers in an arid desert in north Pakistan. The Ambulocetidae family soon followed. These hunters had limbs that were better

KEY EVENTS

50 Mya	49 Mya	48 Mya	45 Mya	44 Mya	44 Mya
The wolf-size, long-legged *Pakicetus* from Pakistan is regarded as the earliest true cetacean, and would have spent time wading in water.	The 10-foot-long (3 m) *Ambulocetus* shows adaptations for the water, with legs better adapted for swimming than walking.	The deerlike *Indohyus* from the Kashmir region shows bone features that indicate a close kinship to the earliest whales.	Protocetid *Protocetus* has whalelike features, such as a streamlined body, adaptations for hearing underwater and nostrils moving back along the skull.	Cetaceans such as *Dalanistes*, with an otterlike form, show signs of adapting to a marine diet.	*Rhodocetus* from Pakistan is a protocetid that retains short fore limbs and hind limbs with elongated "hands."

adapted for an amphibious lifestyle and they would have swum in both fresh and salt water (see image 2). The skull indicated the start of a characteristic fat pad that transmits sound from the lower jaw to the ear, and also the movement of the eye sockets toward the sides of the skull.

The middle Eocene, 49 million to 43 million years ago, saw more evolutionary experiments in an aquatic lifestyle for whales, with a drift away from fresh water toward the sea. This is apparent in the Remingtonocetidae group. Their limbs reduced in length but had not yet adapted into fins—they would have propelled themselves through the water by undulations of their elongated bodies. By the late Eocene, 41 million years ago, the Remingtonocetidae had given rise to the huge Basilosauridae (see p.502) and the smaller Dorudontidae. Members of these families had small fins in place of fore limbs and the beginnings of tail flukes. Basilosauridae were the first whales to be entirely aquatic, with eyes on the sides of the head and nostrils nearer the position of the blowhole in modern whales. They roamed the open ocean rather than coasts and shallows, so their fossils are distributed more widely than those of their ancestors. As their brains grew in relation to their size, they developed the intelligence and social abilities seen in their descendants. Adaptations of the dorudontids, particularly their way of swimming, suggest that the ancestors of modern whales lie within this family.

During the early Oligocene epoch, 30 million years ago, whales split into two lineages. One group, the odontocetes (toothed whales) developed echolocation abilities to hunt fish—as is evident from the growth of a large, sound-emitting organ, the melon, on the forehead. The other group, mysticetes, which includes baleen whale (see image 3) such as the blue whale (see p.504), developed comblike plates in their mouths. These allowed the animal to gather large quantities of small animals, such as plankton, in a single gulp, while draining seawater. This is known as filter feeding and it enabled baleen whales—particularly rorquals—to grow into the largest animals seen on Earth. **GS**

37 Mya	35 Mya	33–14 Mya	25 Mya	25 Mya	20 Mya
Basilosaurus is among the first obligate aquatic whales, confined entirely to the water by its anatomy and lifestyle.	Close relatives of *Dorudon*, resembling a smaller *Basilosaurus*, are likely to be the direct ancestors of modern whale groups.	Squalodontid whales show the first evidence of echolocation and are thought to be the cousins of modern odontocetes, rather than their ancestors.	During the late Oligocene, the cetotheriid family evolves baleen plates for use in various types of filter feeding.	The small early baleen whale *Mammalodon* shows a unique combination of teeth and baleen plates.	Some toothed whales develop highly sophisticated echolocation and ultimately give rise to modern dolphins.

Basilosaurus

EOCENE EPOCH

SPECIES *Basilosaurus cetoides*
GROUP Basilosauridae
SIZE 66 feet (20 m)
LOCATION North America

One of the earliest cetacean ancestors to take on a truly whalelike form, *Basilosaurus* (emperor lizard/reptile) rivaled the largest marine reptiles, sharks, and modern whales in size, growing up to 66 feet (20 m) long and reaching a weight of 55 tons (50 tonnes). Two species are currently recognized—*B. cetoides* and *B. isis*. Both species reached enormous proportions through the development of elongated vertebrae in the back region, and would have moved through water by undulating their entire bodies. The low density of the enlarged bones made their rear ends buoyant and led to an unusual "tail-up" swimming style. In contrast, their cousins—the dorudonts—had shorter vertebrae and were therefore capable of more powerful movements. Experts believe that dorudonts swam using stronger vertical oscillations (up and down movements) of the tail fluke, similar to the movement seen in modern whales. *Basilosaurus* was a toothed predator, as were all early whales. However, remains do not show signs of the echolocating ability seen in modern kinds, such as dolphins. The inner ears of *Basilosaurus* were similar to those of modern cetaceans. The semicircular canal, important for the sense of balance in land mammals, was reduced, and a sound-transmitting fat pad filled the entire lower jaw. Its asymmetrical skull structure is likely to have improved its directional hearing, thus allowing *Basilosaurus* to determine the origin of a sound based on a time delay coming from one of three sides of a triangle. Unlike modern whales, *Basilosaurus* retained a substantial external ear, despite it serving little purpose under water. It also had vestigial hind limbs, which had reduced in size and were no longer useful. **GS**

◆ NAVIGATOR

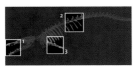

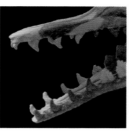

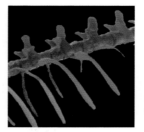

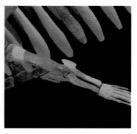

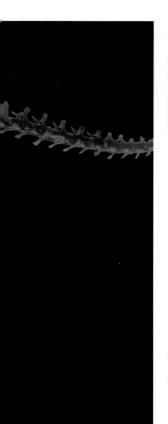

1 NOSTRILS

B. cetoides had nostrils at the tip of its snout. These, along with features that added buoyancy to the tail, suggest it swam at the surface and raised its head out of the water to breathe. The nostrils later moved along the cetacean skull to form the blowhole in modern whales

2 EXTENDED BODY

Complete fossils of *B. cetoides* are yet to be discovered, and tail bones are often fragmented or missing, raising questions about its structure. Modern estimates suggest a total of 70 bones in the spine, including 20 back and 25 tail vertebrae, giving an eel-like shape.

3 LIMBS

The front limbs resembled those of other mammals and had not yet modified into flippers. The hind limbs, at 14 inches (35 cm) long, were greatly reduced with fused ankle bones and only three toes. Joint movement was limited, suggesting they contributed little to steering.

▲ This *Basilosaurus* reconstruction was suspended from the ceiling of the Arts and Industries Building in 1896. The skeleton was based on the skull of one animal and the pelvis and vertebrae of a second.

FALSE IDENTIFICATION

Basilosaurus means "emperor lizard/reptile" and hints at the whale's misidentification. The first fossils studied, including fossil vertebrae (right), were found in Alabama and Arkansas in the 1830s and sent to the American Philosophical Society in Philadelphia, where fragments reminded physician Richard Harlan (1796–1843) of marine reptiles. He proposed the name *B. cetoides* (whalelike emperor lizard) but the mammal-like teeth gave him doubts. In 1839, specimens were inspected by biologist Richard Owen (1804–92). Supported by Harlan, Owen proposed the name *Zeuglodon cetoides* based on the teeth. But due to convention, despite the latter name being correct, *Basilosaurus* remains the official name. Since that time, at least a dozen other sets of specimens have been named as various species of *Zeuglodon* (yoke tooth) or other species of *Basilosaurus*. However, most of these were from scant remains, such as a single tooth, therefore are no longer considered valid.

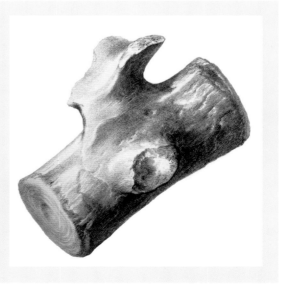

Blue Whale

PLEISTOCENE EPOCH—PRESENT

👁 FOCAL POINTS

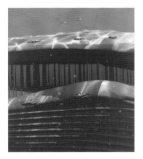

1 BALEEN PLATES

The blue whale's baleen plates form a comblike bristly fringe, descending from its upper jaw and growing up to 3 feet (1 m) in length. A wide mouth with a cavernous interior allows the whale to draw in huge amounts of food, either by lunging into a dense shoal of krill, or by swimming slowly and opening the mouth to create suction.

2 BLOWHOLE

During evolution, the "nostrils" migrated back along the upper surface of the head to form a blowhole—in fact, a pair of openings, each up to 22 inches (55 cm) across. A ridge running back from the mouth opens out into a splashguard to protect the blowhole, and fleshy plugs attached to the dividing wall close it before diving.

3 THROAT

Despite the enormous capacity of the mouth, a blue whale's throat and gullet prevent it swallowing anything more than a foot or so across. Most of its food, however, is much smaller than that— during peak feeding season in cold Antarctic waters, whales can consume 3¾ tons (3½ tonnes) of krill each day.

4 LOWER JAW

Researchers recently identified a unique sensory organ near the front of the lower jaw in rorqual whales, thought to assist with the various elements of the lunge feeding process.t detects twisting of the lower jaw and expanding of the throat pleats. . The organ uses nerves "left over" from the ancestral tooth sockets.

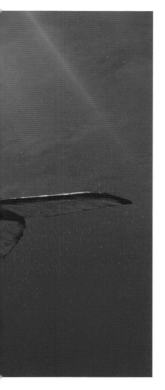

SPECIES *Balaenoptera musculus*
GROUP Balaenopteridae
SIZE 98 feet (30 m)
LOCATION Oceans worldwide
IUCN Endangered

The blue whale (*Balaenoptera musculus*) is not only the largest living animal on Earth, but by most measures, is also the largest animal ever to have lived. It can reach 98 feet (30 m) in length and weigh 187 tons (170 tonnes). It is the largest member of the baleen whale family, with at least three distinct subspecies. Its close relatives include the humpback, sei, and fin whales. These enormous baleen whales, also known as rorquals, are thought to have separated from other mysticetes in the Oligocene, 28 million years ago. Their distinguishing characteristics are a slender, streamlined shape, pleated skin folds running along the underside of the body from the lower jaw and a unique jaw anatomy. This allows the whale to open its mouth to a huge diameter, and the skin folds enable the animal to take in thousands of gallons of seawater in a single gulp. These folds also give the rorquals their name, which is from the Norwegian for "furrowed whale." Movement of the tongue and pleats forces water out through the lips via the narrow, fringed baleen plates that are in place of teeth, trapping food that is then swallowed. The substance baleen developed as a form of gum tissue, toughened by keratin. The whale's preferred food is the tiny but abundant crustacean, krill; however, fish, squid and other animals are also caught up in this "lunge feeding" method. Research has shown that blue whales evolved their large size more rapidly than comparable land-based mammals. It took around 5 million generations, starting with dog-size ancestors. Experts suspect that aquatic mammals can increase in size more easily because the water's buoyancy isolates them from the constraints of gravity, which means that fewer anatomical adaptations are required in order to grow. Whales could even continue to expand—if their food supplies remain plentiful. **GS**

▶ The blue whale skeleton highlights the huge size of the jaws, the robust front limb bones that form the flippers and the lack of hips and hind limbs.

WHALE SONG

Hearing is important for aquatic animals; visibility diminishes with distance and scent travels poorly, but the density of water allows sound to travel five times faster than it does in air. Toothed whales (left) produce clicks and whistles for both echolocation and communication, modulating sound with "phonic lips" within the main airway. Humpback and blue whales produce varied songs, with frequencies below the limits of human perception, but the mechanism that creates the sounds is unclear. Baleen whales lack phonic lips and vocal cords, yet their sounds carry across thousands of miles of ocean.

MONKEYS AND APES

1 Marmosets, such as the black and white tassel-ear (or Santarem) marmoset, *Mico humeralifera*, belong to the New World platyrrhine group of monkeys.

2 Like most baboons, the olive baboon, *Papio anubis*, has evolved away from tree-dwelling to a mainly ground-dwelling lifestyle.

3 Gibbons have exceptionally long, powerful arms for branch swinging, as shown by this white-handed or lar gibbon, *Hylobates lar*.

There are about 147 living species of monkeys and apes, ranging from the marmosets and spider monkeys, through baboons and colobus monkeys, to the apes. Apes in turn range from the Asian gibbons and orangutans, through the African chimps and gorillas, to humans—along with all their extinct fossil relatives. Collectively, the group is known as the Simiiformes or simians and has an evolutionary history reaching back to the Paleogene period, more than 30 million years ago. The two main evolutionary groups within the simians are the platyrrhines (flat noses), a small collection that includes the marmosets (see image 1), spider monkeys, and howler monkeys, all from the Americas; and the catarrhines (down noses), which includes all the other monkeys and apes.

Monkeys and apes have unique anatomical features of the skull, brain, and jaws. For example, the outer edge of the eye socket has a strengthened connection to the braincase, instead of opening rearward as seen in primate cousins such as lemurs. The cerebral cortex (wrinkled upper dome) of the brain is more highly folded than that of lemurs, and the frontal (forehead) bones of the skull roof are fused to form a single stronger, larger structure and brow ridge. Also, the dentary bones of the lower jaw are fused to form a single strong jawbone.

The platyrrhines are small- to medium-sized monkeys that weigh between 5 and 26 ounces (150–750 g). They are essentially tree-dwellers in the tropical rain

KEY EVENTS

34 Mya	25 Mya	20–15 Mya	17–14 Mya	12 Mya	10 Mya
Aegyptopithecus emerges at this time in North Africa and is the oldest catarrhine primate.	The earliest platyrrhine (New World) monkeys are the Oligocene *Branisella* from Bolivia and *Tremacebus* from Argentina.	The gibbon group of lesser apes splits from the other ape lineages and adopts a more specialized tree-swinging way of life.	Asian great apes—orangutans—separate from the African great apes some time during this period.	An early pongin, in the orangutan group, is the Miocene *Sivapithecus* from India.	A possible early gorilla, *Chororapithecus*, lives in Ethiopia. However, it is later known only from scant and inconclusive fossils.

forests of Central and South America. Many are characterized by strongly colored soft hair with distinctive tufts, manes, and moustaches. All fingernails and toenails, except on the big toe, are adapted as claws. The platyrrhines's unique features are the prehensile tail, which can wrap around and grasp objects as a "fifth limb," and the widely separate nostrils, which open sideways (hence the flat nose). The platyrrhines also have distinctive bone patterns to the skull roof and ear. Their fossil record extends back to the Oligocene epoch, 25 million years ago, with genera such as *Branisella* from Bolivia and *Tremacebus* from Argentina.

The catarrhines are a much larger group. One subgroup, the cercopithecoids, often referred to as Old World monkeys, is distributed across Africa, India, Southeast Asia, the Sunda Islands, and Japan. Most familiar are the relatively large-bodied, mostly terrestrial baboons (see image 2) and macaques, while the tree-dwelling colobus monkeys and langurs have slimmer bodies. They have tails and unique features to their molar teeth and knee joint between the thigh bone (femur) and the lower leg bones. The nostrils of the more evolved catarrhines are closely spaced and open downward (hence "down noses"). Catarrhine diversity has expanded since they diverged from the platyrrhines 34 million years ago. Their fossil record extends back to genera such as *Aegyptopithecus* (see p.510) from 34-million-year-old Oligocene rocks in Libya and Egypt. They occupy a range of habitats from tropical forest to savannas and snowy mountain peaks. Normally plant-eaters, consuming both fruit and leaves, some are occasional carnivores: for example, baboons form bands to hunt small mammals.

Another catarrhine subgroup is the hylobatids (gibbons, sometimes called lesser apes; see image 3), with about 17 species in four genera: *Hylobates* (dwarf gibbons), *Hoolock* (hoolock gibbons), *Nomascus* (crested gibbons), and the largest species, the siamang, *Symphalangus syndactylus*. Gibbons are distributed throughout monsoon and evergreen rain forests from eastern India to Southeast Asia and the Malay Peninsula. Characteristically, they are tailless with slender bodies and very long arms, uniquely evolved for a type of movement known as suspensory brachiation. They hang by their arms and swing from one hand to the other, an adaptation that required special modifications of the arm bones and musculature. Their range of head to body length is 18 to 35 inches (45–90 cm) and they weigh between 11 and 33 pounds (5–15 kg). Unusually among apes, gibbons show little difference in body size between the sexes; the color of the dense fur, facial markings, and sophisticated vocalizations are used to distinguish separate species. They live in family groups and have relatively long lifespans of 25 to 30 years. Unfortunately, as yet there is no known firm fossil record of gibbons for tracing their evolution.

A further catarrhine subgroup is the orangutans (see image 4) or pongins (*Pongo*). Although now restricted to the rain forests of the Indonesian islands of Borneo and northern Sumatra, their group was much more widespread in the

8 Mya	8 Mya	7–6 Mya	7–6 Mya	1 Mya	100,000 Ya
The gorilla-chimpanzee-human split in Africa possibly occurs at around this time.	Gibbon evolution involves the separation of the genus *Nomascus* from the main lineage.	*Sahelanthropus* inhabits Chad. It may be the earliest known creature on the chimpanzee- or human-only branch of the family tree.	The chimpanzee and human lines of evolution have probably by now fully diverged in Africa.	Common chimpanzees and bonobos separate from one another at about this time, resulting in two closely related but distinct species.	The enormous extinct *Gigantopithecus*, biggest of the whole ape group, may still roam East and Southeast Asia.

recent past. The orangutan has a long lifespan of around 35 years in the wild. Like all apes, it is tailless with a large, powerful body and a standing height, in males, of up to 4¼ feet (130 cm) and body weight of up to 165 pounds (75 kg). Despite its size and weight, the orangutan is primarily a solitary tree-dweller. It has evolved long arms and hooklike hands and feet to move around high in the tree canopy, searching for its favored fruits and leaves. Although mainly herbivorous, orangutans supplement their diet with small vertebrates, insects, and eggs. Sexual dimorphism (different male and female body forms) is also marked by the larger male's facial shape with prominent skin flaps (flanges) that widen the face and throat area; this gives a sizable double chin and throat pouch for his distinctive vocalizations. The face lacks hair in both sexes but the rest of the body has a spare covering of long, rough hair, colored orange in infants and more reddish-brown in adults. Orangutans are unusual among living apes in having a fossil record that extends back 12 million years to the Miocene in Eurasia and Africa, with genera such as *Sivapithecus* from India and *Gigantopithecus* from China (see p.512). Molecular clock studies (see p.17) suggest that the orangutan group split away from the main branch of the ape family tree, in Africa, probably more than 14 million but less than 17 million years ago.

The catarrhine subgroup known as gorillas (see image 5), with two species, includes the largest living primates, which grow to 7½ feet (2.3 m) tall and weigh up to 606 pounds (275 kg) in adult males. As in many other apes there is significant sexual dimorphism. The arms are immensely powerful and are longer than the legs, with a span of up to 9 feet (2.75 m). Apart from the face, hands, and feet, the body has a thick covering of black to gray hair, which changes to silver-gray on the back of older male silverbacks. The massive head has evolved a powerful jaw musculature, which attaches to a ridge on top of the skull known as the sagittal crest. There is also a prominent bony brow ridge above the eye sockets. Gorillas live in small groups of several females and their offspring, guarded by a dominant male, hence the marked sexual dimorphism. They are herbivores, eating mostly leaves but also fruit; much of their bulk is comprised of a large stomach, necessary for the slow digestion of tough plant tissues. Although much of the day is spent foraging in low-lying vegetation, gorillas are adept climbers and spend the night in trees. Sadly, as with many apes, their

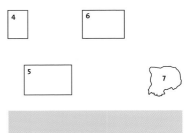

forest habitat has a very low fossilization potential, where remains are quickly scavenged or decay and there is no known fossil record for gorillas. According to molecular clock estimates, gorillas diverged from the evolutionary line leading to chimpanzees and humans between 10 million and 7 million years ago.

The catarrhine subgroup of chimpanzees (see image 6), or panins, comprises our closest living relatives, although they (and gorillas) are distinguished genetically from humans by an extra pair of chromosomes. They have 48 (to our 46), probably formed when, for each pair, two ancestral chromosomes fused into the human chromosome number two. About one third of chimpanzee body substances (proteins) are identical to those of humans, and their overall line-up of genes (or genome) is very similar to ours, with about 2.7 percent of differences. These have accumulated over the past 8 million to 6 million years or so since chimpanzees and humans went their separate evolutionary ways in the late Miocene. Today, there are only two chimpanzee species: *Pan troglodytes*, the common chimpanzee, and *P. paniscus*, the bonobo. Both are forest- and woodland-dwellers in Central Africa. Although smaller than gorillas, chimpanzees are still relatively large and powerful.

A standing male can reach a height of around 5½ feet (1.7 m) and weigh as much as 165 pounds (75 kg). As with most apes there is a distinct sexual dimorphism, especially in the common chimpanzee, but less marked in the bonobo. Like gorillas, chimps have a covering of thick hair over the body except for the face, palms of the hands, soles of the feet, and anal/genital areas. Like other African apes and many primates, chimpanzees are social. They live in small groups of 10 to 30 individuals and are as active and agile on the ground as they are in the trees. Although predominantly plant-eaters with a preference for fruit, they also eat nuts, flowers, leaves, and occasionally insects. Notoriously, chimpanzees can also combine forces actively to hunt and kill small mammals, such as monkeys and bush pigs. Like so many of the apes, the chimpanzees have a poor fossil record. Some experts claim that *Sahelanthropus* fossils (see image 7) from 7 million to 6 million years ago in Chad represent the remains of an early chimpanzee; others regard it as the oldest known member on the human branch of evolution (see p.554). **DP**

4 Both the Sumatran orangutan, *Pongo abelii*, shown here, and the Bornean orangutan, *P. pygmaeus*, are highly threatened by habitat loss, poaching, and other dangers.

5 Like most primates, gorillas live in social groups. Each small troop is dominated by a mature male silverback.

6 Chimpanzees, *Pan troglodytes*, show remarkable behaviors, such as fishing for termites.

7 *Sahelanthropus* is a much debated specimen, dubbed "Toumaï," which means "hope of life."

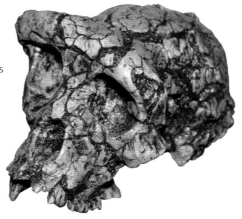

Aegyptopithecus
OLIGOCENE EPOCH

SPECIES *Aegyptopithecus zeuxis*
GROUP Propliopithecidae
SIZE 3¼ feet (1 m)
LOCATION Egypt

NAVIGATOR

Dating from the early Oligocene, around 34 million years ago, the fossil remains of *Aegyptopithecus* from Fayum in Egypt represent an extinct primate that was close to the separation of the evolutionary lines that led to the Old World monkeys and the apes. It was a small creature, less than 3¼ feet (1 m) in length, with a long, doglike snout and apelike teeth, including prominent lower canines. Weighing up to 15 pounds (7 kg), it was slightly smaller than the living howler monkeys of the Americas, but larger than contemporary fossil primates from the Fayum rocks (see panel). The fossils show strong sexual dimorphism—differences between males and females—particularly in the teeth. Males had enlarged canines, which they presumably used to threaten one another for supremacy, and in defence of their territory and mates. This behavior can be seen in monkeys and apes today; the difference is unlikely to be due to diet because males and females would have consumed the same foods. Evidence indicates that *Aegyptopithecus* would have lived in large hierarchical groups of numerous females with their offspring, which may have been controlled by a few dominant males. If so, this social behavior is surprisingly modern and indicates a major feature in ape evolution. *Aegyptopithecus* and New World monkeys may have originated from original populations in Africa. After this lived *Branisella*, the oldest known fossil primate in South America, which dates back 25 million years to the late Oligocene. Its fossil jawbones were discovered in 1969 and it is generally regarded as a platyrrhine from the beginning of the evolutionary burst, which then led to the spread and adaptation of monkeys to habitats throughout South and Central America. **DP**

FOCAL POINTS

1 SKULL
Certain older specimens of *Aegyptopithecus* show a sagittal crest: a bony ridge running along the skull from the forehead over the top of the head to the nape of the neck. This can be seen in the living gorilla, where it anchors the upper ends of the jaw muscles, giving a domed appearance to the top of the head.

2 ANKLE
The structure and joints of the tarsal bones in the ankle suggest that the ankle of *Aegyptopithecus* was capable of a wide range of movement. This would have aided climbing, allowing the foot to twist to various angles, both to grip a branch and also to exert pressure when lifting or moving the whole body.

3 LIMBS
Aegyptopithecus had long but powerful arms and legs with strong and distinctly curved fingers and also long flexible toes. The arms were adapted for grasping and hanging onto small branches to reach for fruit as the monkey moved slowly and deliberately up and down through the different levels of forest canopy.

RELATIONSHIPS

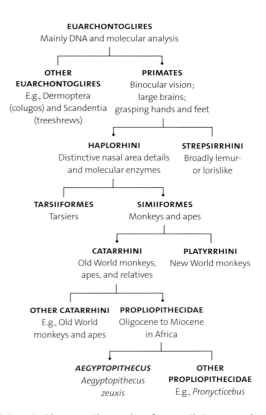

EUARCHONTOGLIRES
Mainly DNA and molecular analysis

OTHER EUARCHONTOGLIRES
E.g., Dermoptera (colugos) and Scandentia (treeshrews)

PRIMATES
Binocular vision; large brains; grasping hands and feet

HAPLORHINI
Distinctive nasal area details and molecular enzymes

STREPSIRRHINI
Broadly lemur- or lorislike

TARSIIFORMES
Tarsiers

SIMIIFORMES
Monkeys and apes

CATARRHINI
Old World monkeys, apes, and relatives

PLATYRRHINI
New World monkeys

OTHER CATARRHINI
E.g., Old World monkeys and apes

PROPLIOPITHECIDAE
Oligocene to Miocene in Africa

AEGYPTOPITHECUS
Aegyptopithecus zeuxis

OTHER PROPLIOPITHECIDAE
E.g., *Pronycticebus*

▲ *Aegyptopithecus* was the member of a group that was very close to the catarrhine split between the Cercopithecoidea (Old World monkeys) and the Hominoidea (apes, including humans), which is generally dated at 35 million to 30 million years ago.

THE FAYUM DEPRESSION

The region known as the Fayum Depression (right) is in northeast Egypt and since the 1960s thousands of fossil species have been found there, dating from 40 million to 30 million years ago. Among the extinct mammal groups were large plant-eating, rhinoceroslike embrithopods and predatory hyaenodonts. Surviving mammal groups include primitive elephants, whales such as *Basilosaurus* (see p.502), rodents and early tree-dwelling primates. The Fayum fossil primates included two groups of fruit-eating monkeys: the parapithecids and the propliopithecids. The former were small, agile primates, such as *Apidium*, which leaped from branch to branch in the forest canopy. The propliopithecids were less numerous; the relatively large *Aegyptopithecus* outweighed all other Fayum primates. Interestingly, of the Fayum monkeys, perhaps only *Parapithecus* ate leaves; it lacks lower incisors and is likely to have eaten leaves within its predominantly fruit diet.

Gigantopithecus
PLEISTOCENE EPOCH

SPECIES *Gigantopithecus blacki*
GROUP Hominidae
SIZE 6 feet (1.8 m)
LOCATION China, Indonesia, and Vietnam

This extinct ape is known only from very fragmentary remains, mainly teeth and parts of jawbone. However, from the size of these it has been estimated that *Gigantopithecus* was the largest known ape. Three species have been named: *G. blacki*, *G. bilaspurensis*, and *G. giganteus*, of which *G. blacki* was the largest. Fossil remains are distributed across Asia from India (*G. bilaspurensis*) to China and Vietnam (*G. blacki*, *G. giganteus*). *G. blacki* from China is thought to range from the early to middle Pleistocene epoch, from 2 million years ago to 700,000 or possibly 100,000 years ago, while *G. bilaspurensis* from the Siwalik Hills and southern Himalayas dates to the late Miocene, between 9 and 6 million years ago. Some estimates claim that males of *G. blacki* stood nearly 10 feet (3 m) tall and weighed more than 1,100 pounds (500 kg). In reality, it was likely to have been no more than 6 feet (1.8 m) tall. Despite its name, *G. giganteus* was smaller. The first find of *Gigantopithecus* was made in 1935 by geologist and paleoanthropologist Ralph von Koenigswald (1902–82). He found a single molar in an apothecary's shop in Hong Kong, where teeth and bones were ground into powders for traditional medicine. It was from this tooth that Koenigswald named the genus *Gigantopithecus*. More than 1,000 teeth have subsequently been recovered from similar sources, and also from caves in Vietnam, Liucheng Cave in Liuzhou, China, and in the Siwalik Hills of India. **DP**

✦ NAVIGATOR

👁 FOCAL POINTS

1 TEETH AND JAWS
G. blacki's fossil teeth consist of large, low-crowned molars with thick enamel and heavily worn surfaces. The canines, although enlarged, have a blunted form rather than being sharply pointed, as is more typical of the larger apes, and the incisors are small and peglike.

2 DIET
G. blacki could have fed on fruits, seeds, leaves, and also considerable amounts of bamboo, as revealed by a microstudy of the teeth. The teeth and jaw contain features of an animal with a heavily fibrous plant diet, which requires crushing and chewing before swallowing.

3 LEGS AND GAIT
It has been suggested that *G. blacki* walked bipedally (on two legs) but there is no evidence to support this. It is more likely that it was gorillalike in its body form and behavior but, given its size, would have been primarily terrestrial. The lack of skeletal remains has led to speculation.

MULTIREGIONAL HYPOTHESIS

In the 1930s German-born, US-based anatomist and paleoanthropologist Franz Weidenreich (1873–1948) claimed that the teeth of *Gigantopithecus* showed human traits and renamed it *Giganthropus*, losing the "ape" element of the word, but his idea was not supported. Weidenreich was looking for evidence for the polycentric or multiregional hypothesis: that various geographically distributed groups, for example, of humans, had each evolved from earlier ancestors within their own region, separate from the other groups. *Homo erectus* (above, with *Gigantopithecus*) in Southeast Asia could have been ancestors for later humans in their regions. The now-accepted scenario is that all modern humans evolved from original African ancestors (see p.554).

MUSTELIDS

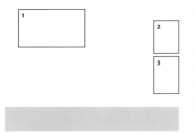

1 About 5 feet (1.5 m) in length, *Potamotherium* is generally regarded as an early mustelid, or an early member of the seal group.

2 *Enhydriodon* was a genus of large and widespread otters, with the African species *E. dikikae* named in 2011.

3 The American marten, *Martes americana*, is typical of the mustelid family: long, slim, lithe, short-legged yet athletic and a ferocious hunter.

The mustelids, family Mustelidae, are mostly small or medium-size, short-legged, slim-bodied, sinuous, short-tailed, and very fierce carnivores, represented today by approximately 60 species. This makes them the largest family in the main group Carnivora. The family includes weasels, stoats, martens, otters, and badgers, and examples are found on all continents except Antarctica and Australia. The early evolution of the mustelids is far from clear, partly because they are not particularly well represented in the fossil record until the Miocene epoch, which began 23 million years ago. This is probably due to the small size of many species, which had correspondingly small, rather fragile bones. Furthermore, many early mustelids were presumably forest animals, living in an environment where decay and recycling happen quickly and remains are less likely to be preserved. Nevertheless, a glimpse into mustelid evolution is possible from a few recently found fossils. The earliest specimens—some bearing a definite resemblance to martens—date back to the late Oligocene epoch, 30 million years ago. By the time of the Pliocene epoch, 5 million years ago, mustelids were already evolving into the various types living today.

One of the oldest fossil finds is *Megalictis*, a large weasel-like mustelid that lived in North America during the late Oligocene and early Miocene, from about 25 million to 20 million years ago, during the "cat gap" (see p.524). Now extinct, this dog-size predator would have looked similar to the wolverine and may have been the North American equivalent of the African *Ekorus* (see p.516).

KEY EVENTS

30 Mya	25–20 Mya	23–7¼ Mya	20 Mya	11 Mya	8–5½ Mya
Early mustelids inhabit parts of North America, Europe, and Asia from around this time: the early Oligocene epoch.	*Megalictis* lives in North America. This dog-size carnivore thrives during the "cat gap" with little competition from felids.	*Potamotherium* lives in Europe and North America. It is generally regarded as a mustelid, but thought by some to be an early terrestrial pinniped (seal group).	The American badgers first appear. The badgerlike *Oligobunis* also lived around this time.	By the middle to late Miocene most of the groups of mustelids known today have evolved, from marten- or badgerlike kinds.	The age of the fossils at the site of Lothagam near the southwestern shore of Lake Turkana, Kenya. Here, the large mustelid *Ekorus* is found.

Also extinct, the distinctly otter-shaped *Potamotherium* (river beast; see image 1) is relatively unknown. Dating from 23 million to 7¼ million years ago, and with finds from both Europe and North America, this unusual mustelid has been considered by some to be a terrestrial ancestor of seals and sea lions, but most experts regard it as an early mustelid. *Plesiogulo* is ancestral to the wolverine, which is the largest living mustelid on land. This early wolverine is thought to have originated in Asia and reached North America between 7 million and 6½ million years ago, during the late Miocene. It shares certain features with modern martens and may have evolved from the marten lineage. By the Pleistocene epoch, wolverine fossils were attributed to the living species.

Mustelids also evolved members that were exquisitely adapted to life in the water: otters, with 13 species surviving today. *Satherium* is a genus of extinct otter that lived in North America. Its fossils date from the middle Pliocene through to the early Pleistocene, about 3¾ million to 1½ million years ago. Fossils have been found in several sites, most notably in Idaho, and this large otter is thought to be related to the giant otter, an endangered species found in South America and the largest of the living otters. Another extinct otter was *Enhydriodon* (see image 2), one species of which, *E. dikikae*, was discovered in the Afar Valley of Ethiopia. This fossil was dated at around 3 million years old, and from its very large skull it seems to have been even bigger than the giant otter. Reflecting its bulk, *Enhydriodon* has been dubbed the "bear-otter." Its genus is thought to be ancestral to the sea otter swimming today along the Pacific coast of North America. *Enhydriodon* fossils have also been found in Europe, India, and in western North America. Like the sea otter, its teeth were adapted for crushing, rather than for shearing.

Most badgers are thought to have evolved from martenlike ancestral mustelids in forest habitats of Asia. Their lifestyle adapted to more terrestrial, muscular, omnivorous diggers, feeding on any foods available in their woodlands. Two recently extinct badger species are *Meles hollitzeri* from early Pleistocene fossils in Austria and Germany, and *M. thorali* from late Pleistocene fossils in France. These are both similar enough to living badgers to be classified in the same genus. The American badgers are thought to have appeared earlier in the evolutionary history of mustelids, possibly 20 million years ago. Certainly the single living representative, the American badger *Taxidea taxus*, differs in many ways from those of Europe and Asia. *Oligobunis* was also badgerlike and lived in the early to middle Miocene in North America, during the "cat gap." Genetic studies of living mustelids have helped to unravel their relationships, but experts are less certain about when each of the modern groups appeared. The earliest groups are likely to have resembled martens (see image 3), although some had features that were more like those of badgers. By the middle to late Miocene, about 11 million years ago, most of the groups that we see today had evolved. **MW**

7–6½ Mya	6 Mya	3¾–1½ Mya	3½–3¼ Mya	2–1 Mya	100–200 Ya
Plesiogulo, a probable ancestor of the wolverine, evolves from the marten branch of the family tree.	The badger *Chamitataxus* is similar to living badgers. Its fossils are later discovered in New Mexico.	*Satherium* swims in North America and is thought to be related to the giant otter, an endangered species of today.	*Enhydriodon dikikae* (bear-otter), which is more than 7 feet (2 m) long, is found in Ethiopia.	The badgers *Meles hollitzeri* and *M. thorali* thrive; both are similar enough to the European badger to be assigned to the same genus.	The sea mink, *Neovison macrodon*, of North American Atlantic coasts is driven to extinction by human persecution; no good specimens remain.

Ekorus
NEOGENE PERIOD

SPECIES *Ekorus ekakeran*
GROUP Mustelidae
SIZE 7 feet (2 m)
LOCATION Africa

A large member of the weasel family Mustelidae, *Ekorus* lived in Africa during the late Miocene approximately 6 million years ago. The full name of this species is *Ekorus ekakeran*, which is the local Turkana for "running badger," a very apt name for this active badger relative. In build, *Ekorus* looked more like a leopard than a mustelid; it stood around 2 feet (60 cm) tall at the shoulder, measured up to 7 feet (2 m) in length, weighed up to 110 pounds (50 kg) and had longer legs than most members of this generally short-legged family. *Ekorus* differed from the standard mustelid pattern in the structure of its limbs and feet, and was built much more like predatory running hunters, such as cats or hyenas. Like those creatures, it would have been an active hunter, using either an ambush technique or speedy pursuit, and it may have preyed upon smaller herbivores of the open plains and savannas. The nearest living African mammal to compare with *Ekorus* is the honey badger (*Mellivora capensis*; see panel), a ferocious carnivore and the largest living terrestrial African mustelid—although it is shorter legged, smaller, and considerably stockier in proportions than *Ekorus*.

Knowledge of this extinct mustelid comes from a well-preserved complete fossil discovered by Kenyan fossil hunter Kimoya Kimeu (b. 1940). It was found below an overhanging rock in a formation known as Lothagam, close to the southwestern shore of Lake Turkana. Lothagam has yielded more than 1,000 specimens of fossil animals, between about 8 million and 5½ million years old, including remains of monkeys and parts from the lower jaw of an early hominid, the human group. **MW**

✦ NAVIGATOR

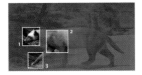

👁 FOCAL POINTS

1 HEAD
Ekorus had a shortened face and a relatively broad, blunt snout when compared to more typical modern mustelids, such as weasels, stoats, martens, and badgers. Its teeth were a combination of mustelid and felid teeth, with similarities to the cat family.

2 SHOULDERS AND PELVIS
The skeleton's pectoral and pelvic regions, as well as the limb bones, show adaptations to hold the body upright. This signifies a cursorial way of life—based on running—rather than the typical mustelid lifestyle, which involves digging and moving through burrows.

3 FLEXIBLE FEET
It is debated whether *Ekorus* was plantigrade (flat-footed) or digitigrade (walking on its toes). The fossil shows some flexibility in the foot bones, suggesting it may have been digitigrade. This would have allowed mobility, agility, and speed. Most living mustelids are plantigrade.

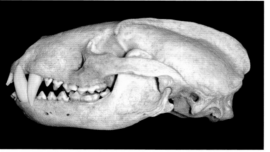

◄ The skull of the European badger, *Meles meles*, shows the typical mustelid dentition. This includes enlarged, pointed canines and strong molars (cheek teeth). As badgers are omnivorous, these teeth are flatter and more adapted for crushing, rather than shearing as in other, more predatory, mustelids.

CLOSE LIVING RELATIVE

A close living cousin to the extinct *Ekorus* is the honey badger, *Mellivora capensis* (right). It is mainly distributed throughout Africa, but its range also extends east to Nepal and India. Although *Ekorus* was considerably larger and more powerful, certain behaviors and habits of the honey badger may shed light on how *Ekorus* evolved and lived. Both mustelids are adapted as voracious predators, capable of killing and eating a range of smaller prey. The honey badger is notorious for its belligerence, strong bite, and ferocity, and it pursues a wide variety of prey, from insects and their larvae through to lizards, rodents, and snakes. It also has its own particular speciality, which was almost certainly not shared by *Ekorus*: it regularly raids the nests of wild bees to break open the combs and feed on the honey, hence its name. When prey is scarce, the honey badger turns to plant food and even digs out, crunches up, and consumes tough roots, bulbs, tubers, and similar items.

SEALS, SEA LIONS, AND WALRUSES

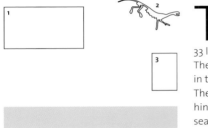

1 The Baikal seal is the only exclusively freshwater pinniped, although the group may have originated in fresh water.

2 *Puijila darwini*, around 3 feet (1 m) long, had land-adapted limbs but was evolving the seal-like skull, jaws, and teeth.

3 The walrus weighs more than 2¼ tons (2 tonnes). Its tusks are used for display, defence and chipping holes in ice.

The pinnipeds (flipper feet)—seals, sea lions, fur seals, and walrus—are cousins of dogs and cats, in the Carnivora group of mammals. All are superbly adapted for aquatic hunting but they rest and breed on land. The 33 living species, almost all strictly marine, are classified into three families. The true seals the Phocidae family, with 18 species, are the best adapted to life in the water because their hind limbs can no longer be used for moving on land. They swim with fishlike sideways movements of their bodies, using their short hind flippers for steering. The monk seals, *Monachus*, (see p.522) are the only seals from the Phocidae group that inhabit tropical warm waters. The 14 species of eared seals (sea lions and fur seals) in the Otariidae family use their hind limbs to shuffle along on land. When swimming, these eared seals use their front flippers, which are longer than those of the true seals, in addition to undulations of the body. The third family, the Odobenidae, is occupied by a single species, the walrus, *Odobenus rosmarus* (see image 3) – one of the most bizarre marine mammals. This large, tusked pinniped uses both pairs of flippers to provide propulsion under the water and, like the eared seals, can shuffle on land using its hind limbs.

The three living true seals of the genus *Pusa*, two of which are found in fresh water, are of special evolutionary interest. The Baikal seal, *P. sibirica* (see image 1), dwells exclusively in fresh water, in Lake Baikal in Russia. This huge and ancient

KEY EVENTS

29 Mya	28 Mya	25 Mya	24–21 Mya	23 Mya	22–20 Mya
The genus *Enaliarctos*, a likely ancestor of living pinnipeds, is known at this time.	The evolutionary lines leading to the eared seals (sea lions and fur seals) and the walrus group are thought to begin divergence.	One of the world's largest, oldest freshwater lakes, Baikal, forms. It will be the habitat of the living relict species, the Baikal seal.	*Puijila darwini* is known from an early Miocene lake. It may represent a stage between terrestrial carnivores and fully aquatic seals.	*Enaliarctos mealsi* exists at this time. It becomes the first of the *Enaliarctos* genus to be described and named for science.	Genetic and molecular evidence points to a split between the monk seals and other seals at about this time.

freshwater lake is thought to be 25 million years old. Certain subspecies of the ringed seal, *P. hispida*, live in freshwater lakes, for example on Baffin Island, Lake Ladoga (Russia) and Lake Saimaa (Finland), although this species is mainly marine. The third, the Caspian seal, *P. caspica*, is found in the saline waters of the Caspian Sea. The ancestor of the Baikal seal is thought to have reached Lake Baikal from the Arctic via ice sheets before becoming isolated there as the ice retreated. This transition from salt to fresh water supports the possibility that pinnipeds evolved from a terrestrial carnivore, first inhabiting fresh water, then semi-salty brackish water, and from there, the sea. There is much debate about the origin and evolution of these streamlined aquatic carnivores, specifically regarding whether they were all in one natural clade (group that shared a common ancestor). Current thinking, backed by genetic studies, suggests that despite major differences between the three living families of pinniped, they are likely to have all descended from a single bearlike or possibly mustelidlike ancestor.

One of the oldest fossil pinnipeds is *Enaliarctos*. Fossils of *E. mealsi*, uncovered in California and Oregon, were described and named in 1973 and date to around 23 million years ago, around the start of the Miocene epoch. In all, five species of *Enaliarctos* have been described, the oldest dating back to a possible 29 million years ago. Like the walrus, this ancestral pinniped probably used both sets of limbs when swimming and its spine would have been flexible, allowing it to undulate and propel itself underwater. Its hind limbs were more leglike than the flippers of living sea lions and fur seals, which suggests that *Enaliarctos* lived a more terrestrial existence, although it was certainly a proficient aquatic hunter. Its teeth were more like those of terrestrial carnivores, with sharp carnassials (shearing teeth). These are absent from living pinnipeds, whose teeth are cone-shaped and pointed, and used mostly for gripping their often slippery prey, which is frequently swallowed whole.

In 2007 another very important fossil find helped to unravel the evolution of seals and their relatives. This was *Puijila darwini* (see image 2), a small, otterlike carnivore. *Puijila* was semi-aquatic, with short limbs and large feet that were probably webbed. Its limbs were still legs, but they show a distinct reduction in size. These fossils have been found in deposits in a Canadian lake and are dated between 24 million and 21 million years ago, the early Miocene. This otterlike carnivore showed how the ancestors of seals may have evolved from a semi-aquatic to an almost entirely aquatic lifestyle, possibly via kinds such as the much more seal-like *Enaliarctos*, an early pinniped that used both its fore limbs and hind limbs when swimming. *Puijila* is likely to have inhabited fresh water rather than the sea, and, it is possible that this adaptation preceded the transition to marine life. **MW**

18 Mya	17 Mya	15 Mya	15–10 Mya	4 Mya	3 Mya
Pteroarctos lives in North America. Details of its skull and eye socket show an increasing similarity to modern seals.	*Proneotherium* inhabits North America and *Prototaria* inhabits East Asia. They are both members of the walrus group.	Many of the rich fossil deposits of the Sharktooth Hill Bone Beds of California form, preserving sharks and the ancient seal *Allodesmus*.	The seal relative *Allodesmus* (see p.520) lives in the North Pacific. This species is large—comparable in bulk with living elephant seals.	The elephant seal genus *Mirounga* separates into the two living species, the northern *M. angustirostris* and southern *M. leonina*.	A member of the walrus group *Valenictus* in North America develops two long tusks but has virtually no other teeth.

Allodesmus
MIOCENE EPOCH

SPECIES *Allodesmus kelloggi*
GROUP Desmatophocidae
SIZE 8 feet (2.5 m)
LOCATION North America and Japan

 **NAVIGATOR**

This fossil relative of seals and sea lions dates from the Miocene, and lived in the North Pacific region. It was about 8 feet (2.5 m) in length and would have weighed around 800 pounds (350 kg). Certain fossils indicate a size of more than 13 feet (4 m), making some species of *Allodesmus* almost as large as living elephant seals, which—at 16 feet (5 m) in length and with weights of more than 4½ tons (4 tonnes) —are as sturdy as elephants. Many *Allodesmus* fossils have been discovered, notably in southern California, neighboring Baja California, Mexico, and Japan. They represent between two and six species, according to features used to distinguish them. Many fossils have been found together at a single site, which indicates that *Allodesmus* lived in social groups like most of its living relatives. Differences in body sizes of the fossils suggests that *Allodesmus* had a similar breeding biology to living social pinnipeds. The larger males (bulls or "beachmasters") jousted on the shore for control over a harem of smaller females (cows), similar to living elephant seals. The bulls were aided by strong canine teeth. The eye sockets show that *Allodesmus* had large eyes and good vision. Its skull, although large, was quite narrow and the teeth, apart from the larger canines of the males, were simple, single-rooted, bulbous, and all of similar size. This early seal would have propelled itself through the warm coastal waters of the Pacific Ocean with large webbed limbs. **MW**

FOCAL POINTS

1 WEBBED LIMBS

All four limbs of *Allodesmus kelloggi* were webbed, and evolved as broad surfaces like paddles to push against the water when swimming, similar to its modern pinniped relatives. However, they retain the typical mammalian skeletal plan inside, with the usual upper and lower bones, and five sets of toe bones on each foot.

2 LARGE EYES

The fossil skulls have very large eye sockets, showing that *A. kelloggi* had good vision, presumably beneath the water as well as on land. It would have hunted mainly for fish and squid. Living elephant seals also have large eyes, adapted to the dark, deep water where they dive for prey, so *A. kelloggi* may also have been a deep diver.

3 SIMPLE TEETH

Male *A. kelloggi* used their strong canine teeth in disputes with rival males. Most of the other teeth were simple and of similar size. They came together when the jaws were closed over fish or other prey, allowing the water to drain away. This method of straining aided the efficient capture of smaller food items.

SHARKTOOTH HILL

Sharktooth Hill Bone Bed (right) is a well-known site for fossils near Bakersfield, California. Miocene rocks here, dated to around 15 million years old, have yielded large numbers of fossils of marine and other animals, including *Allodesmus*, whales, rays, bony fish, turtles, birds, and extinct species of shark—notably the giant shark, *Carcharodon megalodon*, or "Megalodon" (see p.192). More than 140 different species of fauna and flora are known from the site. The fossil layer is quite thin, only about 1 foot (30 cm), but it covers a huge area, and the density of fossils is remarkable—200 specimens per 35 cubic feet (1 cubic m) of rock. At the time, this was the site of a river delta entering the sea that covered much of central California. Silt from the river accumulated, along with the remains of sea creatures that eventually turned into fossils, many of them beautifully preserved.

Monk Seal
PLEISTOCENE EPOCH—PRESENT

👁 FOCAL POINTS

1 HEAD AND FACE
The rounded head has a large, broad muzzle; large, widely spaced eyes, and slotlike nostrils that open upward. The nostrils close when the seal dives—monk seals rarely exceed 5 minutes during regular feeding trips, but some species of seal can hold their breath for longer than 30 minutes while underwater.

2 NO EXTERNAL EARS
True seals such as the Hawaiian monk seal lack small ear flaps, or external ears, which explains their alternative name of "earless seals." The lack of flaps shows a greater adaptation to an aquatic life than the eared seals (sea lions and fur seals), because the ear flaps disrupt the streamlining of the body while swimming.

3 FLIPPERS
The Hawaiian monk seal's front flippers are relatively short and have small claws that are remnants of its ancestors' life on land. The rear flippers are more slender. Like those of other members of the true, or earless, seal family, the hind flippers cannot be rotated or twisted around for use when moving on land.

4 SLENDER BODY
The slender, sleek body of the Hawaiian monk seal makes it a very efficient diver as it searches in deep water, most often at night, for fish, squid, octopus, and even lobster. Typical sizes are 7½ feet (2.3 m) in length and nearly 600 pounds (270 kg) in weight for females, while males are slightly smaller.

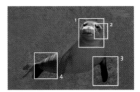
There are 18 living species of true seals—the Phocidae family—most of which belong in cold waters. But the two (formerly three) species of monk seal, genus *Monachus*, live in relatively warm waters, which indicates their interesting evolutionary position. Some experts regard them as the most primitive of the seals and sea lions, showing early features of the group; others assert that their features are the result of more recent evolution toward what looks like the primitive condition. The two extant species, the Mediterranean monk seal, *M. monachus*, and the Hawaiian monk seal, *M. schauinslandi*, are critically endangered, with very low populations. A third, the Caribbean (West Indian) monk seal, *M. tropicalis* (see panel), is now regarded as extinct; the last reliable sighting was in 1952. How did the three monk seals—the only main warm water seal species—come to live in places as far apart as the Pacific, Mediterranean, and Caribbean? One proposal is that their ancestors lived along eastern North America, and some crossed the Atlantic with the warm Gulf Stream to settle in the Mediterranean. An alternative theory is that they originated in Europe, with some heading west to the Caribbean, through the once-open gap where Central America is now located, to Hawaii and the Pacific. This latter theory is supported by genetic analysis showing that the Caribbean and Hawaiian species are most closely related.

First discovered in 1899 by German zoologist Hugo Hermann Schauinsland (1857–1937), and later named in his honor (*M. schauinslandi*), the Hawaiian monk seal can only be found around a group of islands in the Hawaiian archipelago. The population is currently estimated at a mere 1,000 individuals and is declining due to threats such as disturbance from tourism and boating, becoming caught in fishing equipment, and habitat pollution. Historically, all three species of monk seal were hunted for their meat and fur, and were perceived as a threat to commercial fishing. Because populations of the two surviving species are fearfully low, it is uncertain if their genetic variability is sufficient to allow healthy recovery. **MW**

SPECIES *Monachus schauinslandi*
GROUP Phocidae
SIZE 7½ feet (2.3 m)
LOCATION Hawaii
IUCN Critically Endangered

▲ Paralleling the monk seals in the extinction process was the Japanese sea lion, *Zalophus japonicus*. Hunting for oil, rendered from its meat, caused its numbers to crash in the early 20th century; by the 1970s it had disappeared.

EXTINCT RELATIVE

The Caribbean monk seal, *M. tropicalis* (above), was first reported by Christopher Columbus in 1494 when it was common in the Caribbean, the Gulf of Mexico, the Bahamas, the Yucatán Peninsula, the coast of Central America, and the northern Antilles. Fossils have also been found in the southeastern United States. However, it was not formally named until 1850. Thereafter, its decline was rapid, and it was regarded as rare as early as 1887. The last recorded sighting from the Texas Gulf was in 1932, and the last definite sighting of a small colony was in 1952, at a group of coral islands between Honduras and Jamaica. Centuries of persecution from the fishing industry, and hunting for its skin and blubber, finally caused its extinction.

CATS

1 About 3 feet (1 m) tall at the shoulder, *Homotherium* was a member of the saber-toothed cat subfamily Machairodontinae.

2 The upper canines of *Dinofelis* were larger than the lower ones, and were well suited to stabbing.

3 The Iberian lynx, *Lynx pardinus*, is one of the most threatened of all living cats, and is regarded as critically endangered.

The cat family, the felids, is a highly successful group of terrestrial carnivorans, represented by approximately 40 species found worldwide with the exception of Australia, New Zealand, Madagascar, and Antarctica (unless introduced to those countries by humans). Felids are the most exclusively meat-eating of all the carnivorans, and several sit at the top of their habitat food chains. The evolution of this important lineage can be traced back to the early Oligocene epoch, 30 million years ago. Possibly the oldest felid is *Proailurus* (see p.526), which lived in the late Oligocene to early Miocene epochs, from 25 million years ago. After *Proailurus*, the fossil record becomes so scarce that the period between 23 million and 17 million years ago is commonly referred to as the "cat gap." The exact reasons for this gap are unknown, but a possible explanation is that the early cat ancestors suffered competition from other emerging carnivorans, such as bears, hyenas, and also the catlike nimravids (Namravidae), also known as false saber-toothed cats.

Later in the Miocene, between about 20 million and 8 million years ago, another fossil cat appeared: *Pseudaelurus*. This is probably a closer cousin of the evolutionary lineage leading to the modern cats and is regarded as the common ancestor of both the living cat species and also the extinct saber-toothed cats, Machairodontinae, which includes the well-known *Smilodon* (see p.528). *Pseudaelurus*, inappropriately meaning "false cat," had fewer teeth than *Proailurus,* and unlike the latter, was also fully digitigrade (walking on its toes)

KEY EVENTS

25–16 Mya	23–17 Mya	20–8 Mya	18½ Mya	15 Mya	6 Mya
Proailurus is thought to be at the base of the lineage leading to living cats, or an ancestor of civets and relatives, Viverridae, as well as of cats, Felidae.	The period of the fossil record referred to as the "cat gap" occurs. The lack of cat fossils may have been caused by competition from other carnivores.	The early cat *Pseudaelurus* thrives in Europe, Asia, and North America.	*Pseudaelurus* inhabits North America, having originally appeared in Eurasia.	By this time, the established genus *Pseudaelurus* is differentiating into a number of species in a relatively short time.	The large saber-toothed *Machairodus*, about 4 feet (120 cm) tall at the shoulder, stalks North America, Europe, and Asia.

rather than plantigrade (walking on the soles of its feet). These are both features of living cats, and differentiate them from the less advanced *Proailurus*. *Pseudaelurus* first appeared in North America about 18½ million years ago, having originally appeared in Eurasia.

The essential body anatomy of cats has changed little over time, although there have been remarkable adaptations to the skull and teeth, most notably in the saber-toothed cats. By 15 million years ago there was a number of species in the genus *Pseudaelurus* and other cats appear in the fossil record thereafter. This evolutionary burst could be related to alterations in habitats through climate change—more open scrub and savanna were ideally suited to their hunting skills. *Homotherium* (see image 1) was a saber-toothed cat about the size of a modern lion that lived in the Americas, Europe, Asia, and Africa from about 5 million years ago. It may have shared the savannas with modern big cats until becoming extinct in Africa by about 1½ million years ago, although it remained in North America until about 10,000 years ago. It stood 3 feet (1 m) at the shoulder and had relatively short upper canines for a saber-tooth. *Megantereon* was another saber-toothed cat that lived on the plains of Africa until 1½ million years ago. This leopard-size predator had very long, curved upper canine teeth and would have hunted alongside many animals in the modern savannas, including the lion and leopard, preying upon the plentiful herbivores.

Dinofelis (see image 2) was a medium-size cat, about 28 inches (70 cm) tall at the shoulder. Its teeth were not as long as those of the true saber-toothed cats, but they were more prominent than those of living big cats. *Dinofelis* prowled Africa, Europe, Asia, and North America between about 5 million and 1¼ million years ago. It has achieved notoriety because its diet may have included early human relations, such as *Australopithecus africanus* (see p.530). The living big cats of Africa are not well known from fossils, although finds from Tanzania dated at 3½ million years old are recognizable as lion, leopard, and cheetah. The cats represented only in the fossil record became extinct through natural causes, such as changes in habitat, climate change, or competition from other animals. In the case of the saber-toothed cats that survived until surprisingly recently—about 10,000 years ago—their extinction may have been caused partly by the most recent ice age, which peaked between 25,000 and 15,000 years ago.

Sadly, it is now pressures from humans that are driving several of the living cat species to extinction—in some cases, rapidly. Most seriously threatened are larger cats, including the tiger, clouded leopard, snow leopard, and cheetah; several smaller species, such as the Iberian lynx, *Lynx pardinus* (see image 3), are also endangered. Some of these species risk extinction within centuries or even decades unless there are huge conservation efforts. After 25 million years of evolution, the time of big cats as top predators may be ending. **MW**

5–1¼ Mya	5 Mya–10,000 Ya	3½ Mya	2½ Mya–10,000 Ya	1½ Mya	12,000–10,000 Ya
The medium-size cat *Dinofelis* lives in Africa, Europe, Asia, and North America. Its diet may include early human relatives.	*Homotherium* lives in the Americas, Europe, Asia, and Africa. This saber-toothed cat is about the size of a lion.	Some early members of modern big cat species, including lions, leopards, and cheetahs, lay down fossils at this time.	The saber-toothed genus *Smilodon* lives at this time, with species ranging in weight from 120 pounds (55 kg) to 880 pounds (400 kg).	*Megantereon*, a leopard-size saber-tooth, lives on the plains of Africa alongside many modern animals.	Several kinds of big cat disappear due to a combination of climate change as the last ice age ends, the reduction of prey and rival humans.

Proailurus
LATE OLIGOCENE EPOCH—MIDDLE MIOCENE EPOCH

👁 FOCAL POINTS

1 JAWS AND TEETH
The jaws contained carnassial teeth—specialized upper fourth premolars that sliced onto the lower molars—used for tearing up meat, sinew, and even bone. *Proailurus* lacked typical catlike fangs, or long canines, but retained the rear molars that were eventually lost in modern cats.

2 LARGE EYES AND FLAT FACE
With its large, forward-facing eyes and flat face, *Proailurus* had good binocular vision, meaning the fields of view from the eyes overlapped to give good distance and depth perception. This is a major feature of the cat family, and it makes them well adapted for a tree-dwelling life in the forests.

3 BODY
Compared to modern cats, the body of *Proailurus* was long, slim and flexible. The backbone could curve around tightly, to twist and turn at high speeds when pursuing prey, and also for sudden adjustments to balance among the branches, as is the case in today's viverrids (civets and genets).

4 FEET AND CLAWS
The claws were only partially retractable, rather than able to retract into fleshy pockets at the end of the toes. For this reason, the claws may have resembled those of the cheetah. The foot was digitigrade and plantigrade, meaning it could take weight on both its toes and soles—another adaptation for tree life.

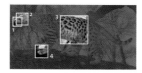
SPECIES *Proailurus lemanensis*
GROUP Felidae
SIZE 3 feet 3 inches (1 m)
LOCATION Europe, Mongolia, and North America

From around 25 million to 16 million years ago, in the late Oligocene to middle Miocene, lived *Proailurus*. It was similar in shape to modern domestic cats, although somewhat larger than most breeds, weighing up to 22 pounds (10 kg). Rather low-slung, with a long tail and sharp teeth, it had features similar to living genets and civets, of the Viverridae family, and also the fossa (see panel), which is a specialist arboreal (tree-dwelling) hunter found only in Madagascar, where its favored prey is the native lemur. The jaws of *Proailurus* contained more teeth than those of the domestic cat, although it lacked specialized long canines. Both, however, share the large carnassials (shearing teeth) typical of the cat family. An active climber, *Proailurus* would probably have hunted in the trees and on the ground, aided by its flexible backbone and the very supple bones of its wrists and ankles. The long tail would have acted as a counterbalance when negotiating the branches of its forest habitat. The head was small, with a flat face, and the eyes were large and forward-facing, giving acute binocular vision—ideal for hunting small prey in its forest habitat.

Proailurus was named in 1879 by French naturalist and comparative anatomist Henri Filhol (1843–1902). He studied many fossil mammals and drew early conclusions about their connections. Remains of *Proailurus* have been found in Europe, notably in Germany and Spain, and also in Mongolia and Nebraska. It may have been one of the first in the lineage that led directly to the true cats, Felidae; alternatively, it could have been a closely related evolutionary side branch. It has been suggested that *Proailurus* was within the evolutionary lineage not only of cats, but also of viverrids, mongooses, and hyenas. **MW**

MADAGASCAN FOSSA

The fossa, *Cryptoprocta ferox* (right), is a member of the family Eupleridae, the Madagascan mammal carnivores, and it is probably the living animal that most closely resembles *Proailurus*. The fossa is endemic to Madagascar, which is home to the lemurs—primates also endemic to the island—that are the favored prey of the fossa. In fact, lemurs comprise more than half of the fossa's diet, and other prey includes rodents, reptiles, and birds. Although the fossa is an effective predator on the ground, it does much of its hunting in trees, climbing up and down head first. It also mates on the horizontal branches of trees. In the absence of true cats on Madagascar (until introduced by humans), the fossa evolved a catlike lifestyle. Its behavior may reflect the hunting techniques employed by *Proailurus*. Unlike cats, the fossa has a mainly plantigrade method of locomotion; the sole of the foot is placed on the ground. This lends stability, but restricts speed. Cats, by contrast, employ a digitigrade method, giving a more flexible, lighter foot, and are thus capable of faster speeds.

Smilodon
PLEISTOCENE EPOCH

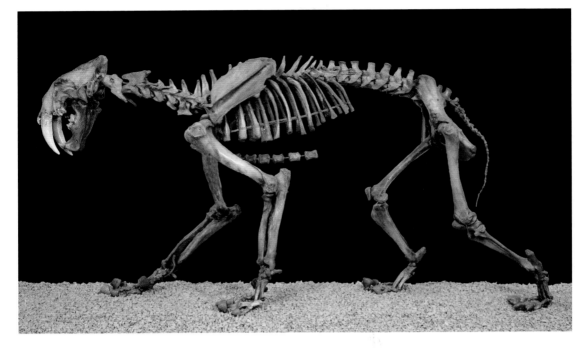

SPECIES *Smilodon fatalis*
GROUP Machairodontinae
SIZE 6½ feet (2 m)
LOCATION North and South America

Familiar mainly from fossils found in South and North America, notably in the La Brea Tar Pits, species of the well-known *Smilodon* lived in the Pleistocene epoch, from about 2½ million to 10,000 years ago. Such a recent extinction means that early humans are likely to have encountered this fearsome predator and may also have contributed to its demise. Many thousands of preserved *Smilodon* bones have been extracted from the tar pits to enable accurate reconstructions of full skeletons of this impressive carnivore. *Smilodon* was first named in 1842 by Danish paleontologist Peter Lund (1801–80), from fossils discovered in caves in Brazil, and three species have been recognized. Although classified with modern cats in the family Felidae, *Smilodon* was different enough to be placed in a separate subfamily, Machairodontinae, which is quite distinct from any living cats, having fewer molar teeth, for example. It is likely that the saber-tooths diverged early from the ancestors of living cats.

The three *Smilodon* species represent an evolutionary increase in size and power. The earliest and smallest species was *S. gracilis*, which weighed between 120 and 220 pounds (55–100 kg) and lived between 2½ million and 500,000 years ago. *S. fatalis* was larger, weighing up to 620 pounds (280 kg), and lived between 1½ million and 10,000 years ago. The largest and most recent species was *S. populator*, named by Lund. This was a very large cat, weighing up to 880 pounds (400 kg) and standing 48 inches (120 cm) at the shoulder. It is dated at 1 million to 10,000 years ago. Bearlike in shape—with a heavy build, front legs longer than the hind legs, and a short tail—it is likely to have had a lumbering gait. It would have hunted mainly by ambush, lurking under cover before charging at or jumping onto its prey in a surprise attack, rather than outrunning it in fast pursuit. Its prey is thought to have included large mammals such as bison, camels, and ground sloths. The three species of *Smilodon* show varied geographic ranges: *S. gracilis* seems to have appeared in North America and spread to South America during the Great American Interchange; *S. fatalis* followed a similar route but farther into western South America, whereas *S. populator* is known mainly from eastern South America. **MW**

 **NAVIGATOR**

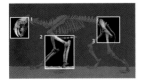

FOCAL POINTS

1 CANINE SABER-TEETH

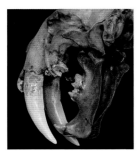

The most unusual feature of *S. fatalis* was its pair of serrated upper canine teeth, which could reach 9 inches (23 cm) long. The jaws were narrow, and the anatomy suggests that its bite was weaker than that of a tiger. The cat probably used its weight to charge and leap, then hold down prey and attack with the canines.

2 LIMBS

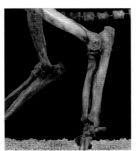

The fore limbs were longer than the hind limbs and high at the shoulder. The large shoulder blade gave ample attachment for strong fore limb muscles, indicating these were used to tackle, hold, and subdue prey. Similar proportions, with heavily muscled fore quarters, are seen in the biggest living cat, the tiger.

3 TAIL

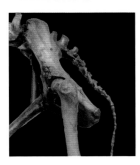

Unlike most living cats, the tail of *S. fatalis* was short, similar to that of a lynx or bobcat. A long tail is usually associated with the ability to run fast, when it acts as a counterbalance and weight-shifting rudder, and for climbing, when it serves a general balancing role. Therefore, it is doubtful that *S. fatalis* lived this type of lifestyle.

SMILODON RELATIONSHIPS

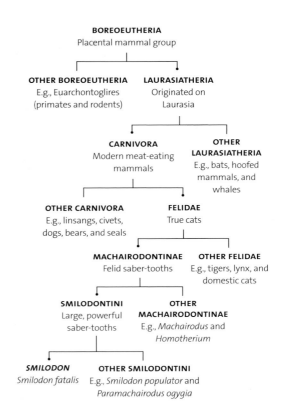

BOREOEUTHERIA
Placental mammal group

OTHER BOREOEUTHERIA
E.g., Euarchontoglires (primates and rodents)

LAURASIATHERIA
Originated on Laurasia

CARNIVORA
Modern meat-eating mammals

OTHER LAURASIATHERIA
E.g., bats, hoofed mammals, and whales

OTHER CARNIVORA
E.g., linsangs, civets, dogs, bears, and seals

FELIDAE
True cats

MACHAIRODONTINAE
Felid saber-tooths

OTHER FELIDAE
E.g., tigers, lynx, and domestic cats

SMILODONTINI
Large, powerful saber-tooths

OTHER MACHAIRODONTINAE
E.g., *Machairodus* and *Homotherium*

SMILODON
Smilodon fatalis

OTHER SMILODONTINI
E.g., *Smilodon populator* and *Paramachairodus ogygia*

▲ The term carnivore refers to an animal with a general diet of meat or flesh. Members of the Carnivora mammal group, most of which are carnivorous, are known as carnivorans. Within the true cat family, Felidae, the machairodonts are the main saber-tooths.

SABER-TOOTHED CONDITION

Saberlike teeth appeared more than once in mammalian evolution. Another saber-toothed cat, *Hoplophoneus*, lived in the Oligocene. It was smaller than *Smilodon*: the size of a lynx. Members of the extinct families Hyaenodontidae and Nimravidae also had long, daggerlike teeth, and there were even long-toothed forms among extinct marsupials and kin, such as *Thylacosmilus* (right). For this dentition to evolve more than once in unrelated forms, it must have been an efficient adaptation, but experts are unsure of its exact use. The most likely purpose was to deliver a fatal slash to the throat or belly of prey, rather than to grip with a firm bite. Broken saber-teeth are seldom found.

TOWARDS HUMANS

1 One trend in early human evolution led to heavy-jowled, massive-toothed herbivores such as *Paranthropus boisei*.

2 A speculative model of *Orrorin tugenensis* portrays an upright stance, as suggested by the fossil thigh bone.

3 The remarkable Taung Child specimen, a young *Australopithecus africanus* three years of age from South Africa, has preserved the shape of the brain within.

Charles Darwin (1809–82) predicted in 1871: "It is somewhat more probable that our early progenitors lived on the African continent than elsewhere" because that is where we found "the gorilla and chimpanzees . . . man's nearest allies." It took another 50 years before the evidence to support this evolutionary connection was found and another 20 years before it became generally accepted. According to various lines of evidence, from certain fossils to genes and molecules (see p.17), chimpanzee and human lineages diverged some 10 million to 7 million years ago. Since then, 20 or more extinct members of our wider human family are known from the fossil record, although this record is fragmentary and incomplete, especially for the early stages. However, it does show that during one phase the wider human family was dominated by an ancestral group called australopithecines (southern apes) and relatives such as *Paranthropus* (see image 1), who lived from 6 million years ago.

In 1925, Australian anatomist Raymond Dart (1893–1988) discovered and named *Australopithecus africanus* (southern ape from Africa) in South Africa (see image 3). With the help of South African paleontologist Robert Broom (1866–1951), he described the remains of a small, upright-walking, 3-foot (1 m) tall apelike species and a number of similar fossils. More discoveries and

KEY EVENTS

10–7 Mya	7–6 Mya	6 Mya	4¼ Mya	3¼ Mya	3–2 Mya
The chimpanzee and human evolutionary lines appear to diverge in Africa.	*Sahelanthropus* from Chad, later discovered from part of a skull known as "Toumaï," "hope of life," may be the oldest creature on the human lineage.	*Orrorin* (original man) from Kenya is probably on a branch that leads to humans, sidelining australopithecines.	*Ardipithecus* inhabits Ethiopia and has a puzzling mix of ape, human, and unique features.	*Australopithecus afarensis* lives in Ethiopia, including the famed "Lucy" specimen and the "First Family."	*Australopithecus africanus*, including the Taung Child and Mrs. Ples specimens, is found at various sites in Southern Africa.

studies have since revealed that some African fossils were very apelike and heavy-boned, and others less so. They are now divided into two groups: the heavier boned or robust ones are placed in the genus *Paranthropus*, while the lighter-boned forms are retained in *Australopithecus*.

In the 1950s, British anthropologist and paleontologist Louis Leakey (1903–72) and his family searched the Great Rift Valley of East Africa, specifically Olduvai Gorge in northern Tanzania. Despite finding another australopithecine, now known as *P. boisei*, in 1959, Leakey was most interested in the recent ancestry of our own genus, *Homo* (see p.536). The first fragmentary remains were uncovered in 1¾-million-year-old rocks at Olduvai and named *H. habilis* (see p.536) in 1964 by Leakey. In 1974, the 3¼-million-year-old remains of a more ancient australopithecine, *A. afarensis*, nicknamed "Lucy" (see p.534), were unearthed in Ethiopia, far to the north of Olduvai. The partial skeleton had the leg bones of an upright-walking but still very apelike being, and preserved some advanced features compared to living chimpanzees. In 1978, footprints found by Mary Leakey at Laetoli, Tanzania, dated to 3¾ million years old, thereby provided evidence that humanlike creatures of the time walked on two feet.

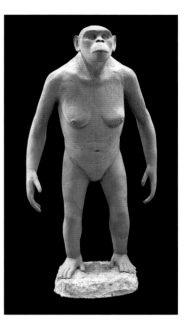

In the 1990s, 4½-million-year-old fossils from Ethiopia were named *Ardipithecus* (see p.532). Then in 2000, a French team working in the Tugen Hills of Kenya found even older but again fragmentary remains. Named *Orrorin* (see image 2) and dated to around 6 million years old, fossils included a thigh bone, the form of which is possibly indicative of upright walking. Only a year later another French team found a remarkably intact and somewhat older 6-million-year-old skull in the Djurab Desert of Chad. Named *Sahelanthropus*, it had a mixture of ape and more advanced features. Some experts suggest that it walked upright, while others argue it is no more than a fossil chimpanzee.

Taking variation into account, the overall tendencies during these stages of early human evolution seemed to be larger body size and larger brains with backbone, pelvis, legs, and feet more suited to upright bipedality (walking on two legs). The evolutionary pressures that led to an abandonment of climbing trees and the adoption of a striding gait seem to have involved phases of the gradually drying African climate, in which forests shrank and were replaced by scattered woods and grasslands. As arms were no longer used for climbing, the possibility arose for hands to become more dexterous and manipulative. Brain size also began to increase. Given the usual variation among individuals, the brain volume of Lucy and her kin was mostly around 24 to 27 cubic inches (400–450 mL), in *A. africanus* more toward 27 to 29 cubic inches (450–480 mL), while in *H. habilis* it was rising to 37 to 40 cubic inches (600–650 mL). Early evidence of tool use, with simply made, pebble-based scrapers, choppers, and hammers, appears mainly with *H. habilis* 2½ million years ago. However, more recent finds have made this general picture much less straightforward. **DP**

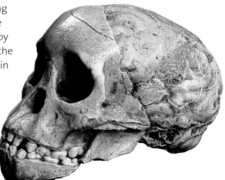

2¾ Mya	2½ Mya	2½ Mya	2½ Mya	1¾ Mya	1¼ Mya
Paranthropus is a heavier-built type with large jaws, teeth, and chewing muscles, probably descended from *Australopithecus*.	*Australopithecus garhi* wanders Ethiopia's Afar Depression; it may be the direct ancestor of the *Homo* lineage.	*Homo habilis* in East Africa is the first recognized species in the genus *Homo*.	At least three human-related genera live in Africa: *Australopithecus*, *Paranthropus*, and *Homo*.	The "Nutcracker Man" or "Zinj" *Paranthropus boisei* is present at Olduvai Gorge.	*Paranthropus robustus* or *P. boisei* may be the last of the genus *Paranthropus* to survive; this evolutionary line is near its end.

Ardipithecus

LATE MIOCENE EPOCH—EARLY PLIOCENE EPOCH

SPECIES *Ardipithecus ramidus*
GROUP Hominidae
SIZE 4 feet (120 cm)
LOCATION Ethiopia

In 2009, the 15-year-long reconstruction of a 4½-million-year-old skeleton of *Ardipithecus ramidus* (ground ape root) was complete. The spectacular result gave a new insight into early apelike ancestors and revealed a creature with a combination of features not seen in the australopithecines, nor any other human lineage relative. Like many of the first human relatives, "Ardi" stood at 4 feet (120 cm) high, had a body weight of 110 pounds (50 kg), and a brain size of 18 to 21 cubic inches (300–350 mL). The apelike arms were long relative to the legs. Additionally, the apelike toes were long and opposable, so the primate was adapted to climb trees. The foot also showed adaptations for walking upright, although in a different and much less efficient way than humans. The hands were flexible and allowed *A. ramidus* to hold onto branches. These adaptations indicate that *A. ramidus* climbed trees to obtain food and walked across open ground to move from tree to tree. Teeth remains from about 35 specimens found at the same site show that males and females were of similar size—there were no marked differences between the canine teeth of males and females. The co-occurrence of this amount of fossils suggests that the individuals belonged to the same social group and that it was overcome by a catastrophe. *A. ramidus* shows that in the human branch of the primate tree, the early evolution of our apelike relatives was more diverse and complicated than previously thought. **DP**

FOCAL POINTS

1 HEAD
Scanty skull fragments of *A. ramidus* indicate that the head was positioned on top of the backbone, like humans, rather than in front of it, shown by the position of the spinal column and foramen magnum (gap for the spinal cord). This would suit a bipedal (walking on two legs) lifestyle.

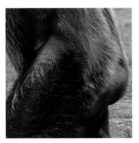

2 HIPS
The pelvis is another part of the *A. ramidus* skeleton that appears to have adaptations both for quadrupedalism when climbing trees and upright bipedality. The bone shapes and muscle attachment sites indicate that both are possibilities. But walking ability would have been very primitive.

3 FEET
The foot bones suggest a divergent (angled-out) large toe, which would have been useful for gripping branches when climbing, but awkward for walking. This was combined with a relatively rigid foot structure that indicates the opposite: a good design for walking rather than climbing.

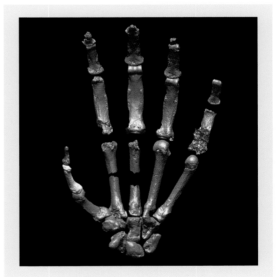

FLEXIBLE HANDS

This digital composite image (above) demonstrates that the hands of *Ardipithecus* were more flexible than those of many other apes. *Ardipithecus* probably still needed to climb trees to obtain food, despite its ability to walk upright, which it did presumably across open ground between wooded areas. Consequently, the hand structure, especially the joints between the wrist (carpal) and palm (metacarpal) bones, would have allowed this early member of the human family to hold on as it moved among branches and from tree to tree, reaching out to pick fruits.

Australopithecus (Lucy)
PLIOCENE EPOCH

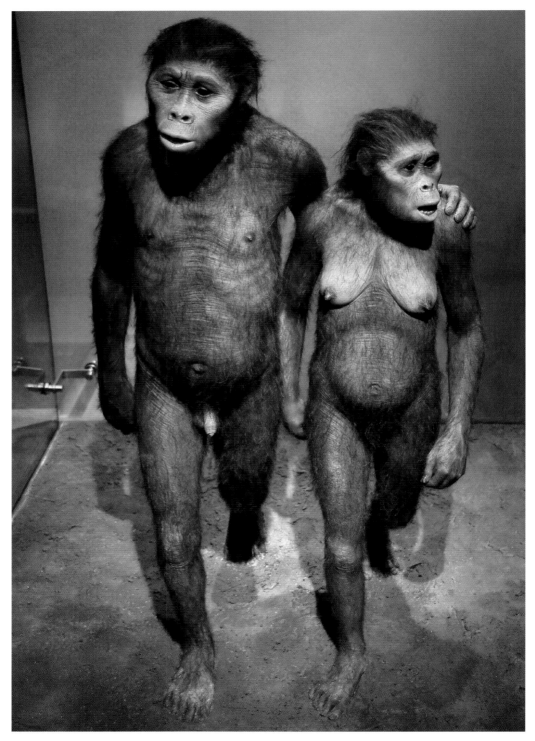

One of the best-known fossil finds of the 20th century was the most complete, early human-related fossil skeleton known at the time: specimen AL 288-1 of *Australopithecus afarensis*, nicknamed "Lucy." About 47 out of 207 bones have been preserved—23 percent of the skeleton—in addition to jawbone and skull fragments. Lucy lived an estimated 3¼ million years ago. She stood only 43 inches (110 cm) tall and weighed around 57 pounds (26 kg). Lucy was found in the Awash Valley of the Afar Depression, Ethiopia, in 1974. A year later, 200 fossil specimens were found in the same region, including a partial juvenile skull and the jaw fragments and teeth of nine adults and four juveniles. They became known as the "First Family." The discovery in 1992 of a partial adult male skull confirmed the accuracy of the reconstruction of Lucy's face. The male was distinguished by larger canine teeth and a massive jaw with indications of powerful gorillalike muscles. One of the largest australopithecine skulls known, it had a brain capacity approaching 30 cubic inches (500 mL). Heavy tooth wear shows that the front teeth were used for stripping tough outer layers from plant stems and roots, as practiced by gorillas today. The discovery in 2000 of a 3-year-old skeleton in Ethiopia with an almost complete skull confirmed the mixture of apelike and human features in *A. afarensis*. This find is known as "Lucy's baby," although it is dated to more than 100,000 years earlier than Lucy herself. **DP**

✿ NAVIGATOR

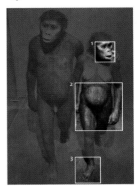

SPECIES *Australopithecus afarensis*
GROUP Hominidae
SIZE 43 inches (110 cm)
LOCATION Ethiopia

👁 FOCAL POINTS

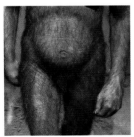

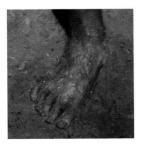

1 TEETH
The teeth were midway between the ape ancestors and later human line species. The canines were not as long and pronounced as in gorillas, but they were not as reduced as those of modern humans. Eruption of the third molar tooth shows that the skeleton is a mature female, despite its size.

2 APE AND HUMAN FEATURES
Lucy's curved finger bones suggest that she was still adept at climbing, and the flaring rib cage likely accommodated the large stomach of a mainly plant-eating ape. However, the leg bones and pelvis show that she was capable of upright walking. Overall, there is a mix of ape and human features.

3 ARCHED FEET
A foot bone of *A. afarensis* was discovered in 2011 and showed that the foot had an arch, similar to that of *Homo sapiens*. Elastic ligaments and other tissues cushioned the foot as it was placed down, then rebounded to lift it—an adaptation for comfortable, energy-efficient upright walking.

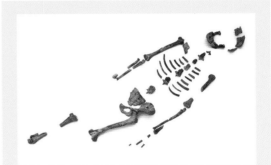

A BORROWED JAWBONE

Reconstruction of the original Lucy find (above) by paleontologists Tom Gray (b. 1952) and Donald Johanson (b. 1943) was limited by the lack of similar remains, especially skull and jaw, from previously described specimens of similar age. There was one contemporaneous fossil described by British paleontologist Mary Leakey but it came from Laetoli, Tanzania, 940 miles (1,500 km) south. Nevertheless, Johanson decided the comparison was good enough and, in 1978, "borrowed" the Laetoli jaw for his new species *A. afarensis* (southern ape from Afar). Unsurprisingly, there was outcry from many experts over the naming of a new species based on such widely separated fossils, but further finds confirmed Johanson's actions. Incidentally, the name "Lucy" came from celebrations of the discovery under a night sky studded with stars, echoing the Beatles' song "Lucy in the Sky with Diamonds."

EARLY HUMANS

1 An enlarged brain, along with evidence of stone tools, led to the inclusion of *Homo habilis* within the human genus in 1964.

2 *Homo heidelbergensis*, named from a jawbone found in 1907, may be the direct ancestor of modern humans.

3 Paleoanthropologist Louis Leakey measures a skull of the "robust" *Australopithecus (Zinjanthropus) boisei*, since renamed *Paranthropus boisei*, which lived in Africa at the same time as *Homo habilis*.

The human group, hominins, includes all the species belonging to the genus *Homo*, which numbers between nine and 12, depending on which expert opinion is taken. The disparity is partly a result of the fact that living species are defined by their capacity for reproduction—members of a living species breed to produce viable offspring, which can be observed—while fossilized species are defined primarily by their preserved anatomical features, the details of which are more open to opinion. The species of modern humans, *Homo sapiens* (wise man; see p.546), was first defined in 1753 by Swedish botanist and taxonomist Carl Linnaeus (1707–78) with the simple statement: "Know thyself." Linnaeus was the first naturalist to formally classify humans alongside monkeys and apes within the same taxonomic unit, which he named

KEY EVENTS

2½ Mya	1¾ Mya	1¾ Mya	1¼ Mya	600,000 Ya	600,000 Ya
Homo habilis in East Africa is the first recognized species in the genus *Homo*. It survives for 1 million years.	*Homo ergaster* evolves in Africa; sometimes called "African *Homo erectus*," it survives for around 600,000 years.	*Homo erectus* evolves in Eurasia of Africa, perhaps from *H. ergaster*. It survives for at least 1½ million years.	Fossils of *Homo antecessor* are found in northern Spain. The species resembles *H. ergaster* and probably persisted for 400,000 years.	*Homo heidelbergensis* evolves in Africa and spreads to Europe; it will become extinct in 400,000 years.	Modern genetic and molecular evidence indicates that the common ancestor to the Denisovans and Neanderthal people lives at this time.

the order Anthropomorpha, but later changed to the order Primates. Linnaeus's classification was primarily a hierarchical system that grouped similar creatures together. He did not intend to imply an evolutionary connection between the groups, and it was only a century later, when Charles Darwin (1809–82) published *On the Origin of Species* (1859) that the connection with evolution became explicit.

The genus *Homo* evolved from apelike predecessors and had a larger brain. Almost twice as tall as apes, fully erect and bipedal (walking on two legs), their hands were free to use and manipulate tools. The first species within the genus *Homo* was *H. habilis* (see image 1), which lived from 2½ million to 1½ million years ago. In some respects it had a body similar to that of an ape and could, from some anatomical features, be included with *Australopithecus* (see p.534). However, the size of its brain reached almost 43 cubic inches (700 mL), which, combined with the ability to use associated stone tools, qualified it for the genus *Homo*.

The first extinct member of the human family to be named was *H. neanderthalensis* (see p.542) in 1864 from fragmentary remains in Germany. By the 1890s, the debate had begun as to whether humans originated in Africa, as suggested by Darwin, or in Asia. German biologist Ernst Haeckel (1834–1919) championed an Asian origin so persuasively that young Dutch anatomist Eugene Dubois (1858–1940) traveled to Indonesia in 1887 to look for a "missing link." Fortunately, in 1891 he found humanlike fossils at Trinil in Java, and named them *Pithecanthropus erectus* (upright ape human), because the legbones clearly belonged to an individual who could stand upright. This specimen became known as "Java Man." Unfortunately, the significance of Dubois's find was not appreciated by European experts at the time, so his discovery was eclipsed by another Asian specimen of the same species. "Peking Man," *Sinanthropus pekinensis*, was found in China and described by Canadian anatomist Davidson Black (1884–1934) in 1927. These discoveries in Indonesia and China seemed to reinforce the idea that humans originated in Asia. A theory called the multiregional hypothesis stated that modern humans evolved separately from one another in different parts of the world from earlier groups of humans in their regions. By the 1940s, it was acknowledged that Java Man and Peking Man shared essentially human characteristics, and they were merged as the same species, *H. erectus,* in 1950 (see p.540). This species became recognized as an important extinct human relative, along with *H. neanderthalensis* and another species, *H. heidelbergensis*, known from a single jawbone found in Germany in 1907 (see image 2).

In 1960 the Leakey family discovered fragmentary, 1¾-million-year-old fossils at Olduvai Gorge in northern Tanzania. It was evident that these specimens had been more advanced than those previously discovered. Louis Leakey (1903–71; see image 3) saw this human species as users of tools, so named it *H. habilis*

350,000 Ya	200,000 Ya	160,000 Ya	95,000 Ya	40,000 Ya	38,000 Ya
The famed Neanderthals, *Homo neanderthalensis*, begin to appear in western Eurasia.	The modern human species *Homo sapiens* evolves, most likely from an African population of *H. heidelbergensis*.	Remains of *Homo sapiens idaltu* from Ethiopia show the beginnings of *H. sapiens* but with some archaic and ancestral features.	*Homo floresiensis*, or "Hobbit people," a dwarf human species, evolve on the Indonesian island of Flores.	The Denisovan people of Siberia exist at this time, as fragmentary fossil evidence and ancient genetic material later reveals.	Neanderthals become extinct, but not before interbreeding with European populations of *Homo sapiens*.

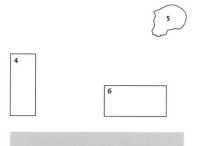

4 The remarkably complete "Nariokotome Boy" skeleton, probably *Homo ergaster*, dates to some 1¾ million years ago.

5 The "Broken Hill Skull," or Kabwe 1 cranium, was recovered from a metal mine in 1921.

6 Long before *Homo sapiens*, humans left footprints that were discovered in 2013 at Happisburgh, Norfolk, England. At least 820,000 years of age, they are the most ancient human prints outside Africa.

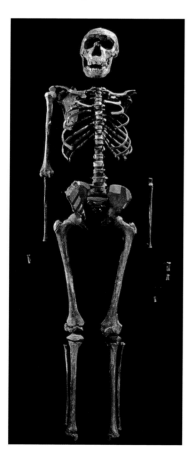

(handy man). It was the most ancient human species known at the time. Consequently, Leakey redefined the genus *Homo* by reducing the lower limit for the accepted "human" brain size from 46 to 40 cubic inches (750 to 650 mL). The discovery of *H. habilis* reestablished the possibility of an African ancestry for humans. However, questions still remained about how and when they moved out of Africa, and which species was the immediate ancestor of *H. sapiens*. The fossil remains of *H. neanderthalensis* from 350,000 to 40,000 years ago, *H. erectus* from 1¾ million to less than 100,000 years ago, and *H. heidelbergensis* from 600,000 to 200,000 years ago all appeared to be Eurasian (of mixed European and Asian origins). Therefore, the human ancestor to these species would have left Africa by 1¾ million years ago.

Until 1984, the fossil remains of *H. erectus* had been found mostly in Southeast Asia, but the discovery of a near-complete, 1¾-million-year-old skeleton of an adolescent boy in Kenya diverted attention to Africa. Known as "Turkana Boy" or "Nariokotome Boy" (see image 4), his reconstructed skeleton stands at 63 inches (160 cm), taller than any of the australopithecines. His teeth show that he grew more rapidly than modern humans and may have been as young as 8 years old when he died. At maturity, he may have reached a height of 70 inches (178 cm) and a weight of 143 pounds (65 kg): a size similar to the modern male human. However, his brain capacity was only 53 cubic inches (870 mL), which is 60 to 70 percent of the brain size of modern *H. sapiens*. He was tall, long-limbed and slender-hipped: well adapted for distance walking and running. Initially, this remarkably complete specimen was tentatively referred to as "African *Homo erectus*." Subsequently, his identity was changed to a new species, *H. ergaster* (working man), which many experts now regard as ancestral to *H. erectus* and which provided evidence of the first exodus of our human relatives out of Africa. Further discoveries, in 1999 and 2001, include remains of *H. erectus*-like *H. georgicus*, from 1¾-million-year-old material in Georgia, and Spanish fossils of *H. antecessor*, dating from 1¼ million to 800,000 years ago. They may have descended from the first human relatives to leave Africa. *H. georgicus* is seen by some experts as subspecies *H. erectus georgicus*. Similarly, *H. antecessor* may be an early form of *H. heidelbergensis*.

The African fossil record also shows that *H. sapiens* appeared around 200,000 years ago and did not move out of Africa and into Eurasia and beyond until around 120,000 years ago. Consequently, it seems that there have been two separate moves out of Africa (see p.548): the first resulted in the spread of *H. ergaster/erectus* across Eurasia, and the second saw *H. sapiens* spread as far as Australia by 40,000 years ago. Despite the knowledge that was gained from these various discoveries, it was still unclear which species was the direct ancestor of *H. sapiens*. There are very few clear anatomical links to the nearest contemporary species, *H. neanderthalensis*, nor are there any geographical links between the earliest representatives of these two species. However, they did overlap much later, around 40,000 years ago.

In recent decades, it has been discovered that the only species with the appropriate anatomical, geographical, and chronological credentials to qualify as the direct ancestor of *H. sapiens* is *H. heidelbergensis*. The first fossil remains of the species consisted of a single jawbone found in 1907 at Mauer near Heidelberg, Germany, and this dates from 610,000 to 500,000 years ago. It was named as a distinct new species, *H. heidelbergensis*, in 1908 by German paleontologist Otto Schoetensack (1850–1912). Just over a decade later, in the 1920s, the "Broken Hill Skull" was found at Kabwe in Zambia (then Northern Rhodesia) and named *H. rhodesiensis* (see image 5). Subsequent analysis has claimed that this fossil is a representative of an original African *H. heidelbergensis* population. Further remains from Boxgrove, England, have been assigned to this species. The discovery in 2013 of around 50 footprints made by five individuals in Norfolk, England (see image 6), probably predate *H. heidelbergensis* and could be those of *H. antecessor*, who may also be represented by a large number of skeletal fragments from Atapuerca, Spain.

It has been argued that *H. antecessor*, which itself may have evolved from *H. ergaster*, was the direct ancestor for *H. heidelbergensis*, while other experts consider *H. antecessor* and *H. heidelbergensis* to be the same species. *H. heidelbergensis* is currently seen as a likely direct ancestor to both *H. neanderthalensis* and *H. sapiens*. This is because anatomically, *H. heidelbergensis* had a body stature and brain size similar to those of modern humans. Facially, it looked rather different, with a prominent bony brow ridge and lack of a bony chin, similar to the Neanderthals and earlier extinct human relatives. Stone tools and animal bones found alongside fossil remains show that *H. heidelbergensis* was an adept toolmaker and butcher of large animals, which it may have either scavenged or actively hunted.

However, the most recent evolutionary relationships within the *Homo* genus have been further extended by new finds such as *H. floresiensis*, which lived between 95,000 and 12,000 years ago in Southeast Asia, and the 40,000-year-old remains of the Denisovan humans from the Altai Mountains in Siberia. The recovery of ancient DNA from Neanderthals and Denisovans has further complicated ideas about these interrelationships. The global distribution of modern human DNA was first analyzed in 1987, giving an African ancestry for *H. sapiens*. Since 2010, further fossil and genetic information indicates that modern Eurasian humans, *H. sapiens*, interbred to some extent with Neanderthals and Denisovans, who left their own genetic markers in modern humans of European and Asian—but not African—descent.

Many questions remain about these and other humans: who evolved into whom and how the different species interacted with one another over the past few million years—especially during the past 500,000 years. The discovery of a new fossil tooth or scrap of genetic material could reconfigure large segments of the human story. Among the uncertainty, it is not in doubt that, despite small genetic differences, humans around the world are a single interbreeding species—*H. sapiens*—and are the only survivors of perhaps 10 or more species from the genus *Homo*. **DP**

Upright Human
PLEISTOCENE EPOCH

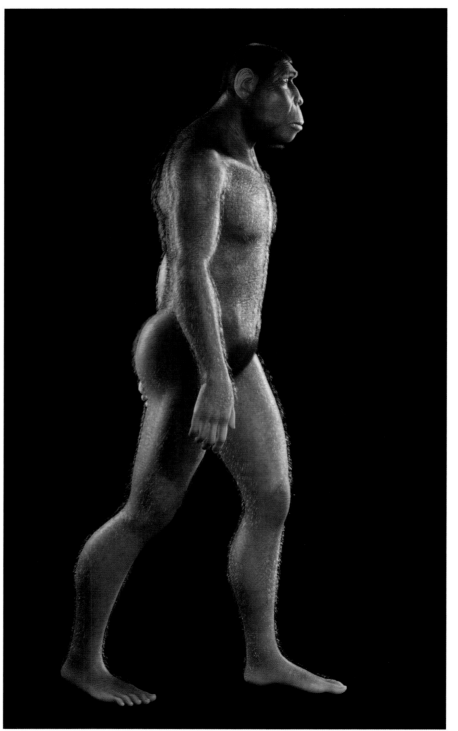

SPECIES *Homo erectus*
GROUP Hominidae
SIZE 6 feet (1.8 m)
LOCATION Eurasia, possibly Africa

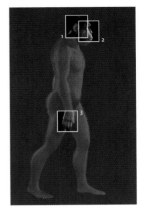
Fossil remains of the extinct human species *Homo erectus* show that it lived primarily over a wide area of Southeast Asia and East Asia, from Beijing in China to the Indonesian island of Java, from 1¾ million years ago to probably less than 100,000 years ago. They were well built, almost 6 feet (1.8 m) tall and weighed up to 165 pounds (75 kg). Apart from the skull, they were similar in appearance to modern humans. The species may have arisen in Africa or Eurasia; fossil evidence suggests that *H. ergaster* or "African *Homo erectus*," may have been in Kenya from 2 million years ago. Although widely distributed, most of the surviving Asian fossils of *H. erectus* are isolated bones and skull parts. Well-preserved remains of 40 individuals were excavated in the 1920s and 1930s and together named "Peking Man" (see panel). The discovery in 2001 of several *H. erectus*-like fossil skulls and partial skeletons from Dmanisi, Georgia, shows how the vast variation within ancient humans compounds the problem of assigning them to particular species. Dated at 1¾ million years old, the fossil skulls are undoubtedly human, yet the brain size of around 37 cubic inches (600 mL) was less than half that of modern humans, and small for *H. erectus*. Opinion is divided over whether the fossils from Georgia were a separate species, *H. georgicus*, or a subspecies of *H. erectus* named *H. erectus georgicus*. **DP**

FOCAL POINTS

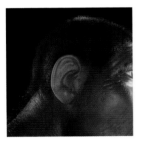

1 SKULL AND BRAIN
H. erectus brain size varies widely across its geographical spread and through its history, from as low as 46 up to 80 cubic inches (750–1,300 mL), nearing that of *H. sapiens*. However, these statistics depend on whether forms such as *H. antecessor* and *H. ergaster* are included or not.

2 FACE
The face was characterized by primitive features of a heavy bony brow ridge and, behind it, a long, low, rear-sloping cranium, lacking the more vertical forehead typical of *H. sapiens*. The skull was robustly boned with a slight forward prominence of the muzzle and a lack of a chin.

3 HANDS AND TOOLS
The hands were similar to those of *H. sapiens* and fully manipulative. Asian fossils were found with primitive choppers and scrapers, while *H. ergaster* had sophisticated hand axes of Acheulean technology. Either *H. erectus* spread to Asia before these developed or they were not needed in Asia.

PEKING MAN

Excavations in the 1920s to 1930s of the Zhoukoudian Caves, 25 miles (40 km) from Beijing, revealed remains of "Peking Man," *H. erectus* (above). Dated 750,000 to 700,000 years ago were 14 skulls, 11 jawbones, 147 teeth, limb fragments, 17,000 stone tools, and many animal bones. The stone artifacts were made of 44 different rock types, all brought to the cave from elsewhere. Most of the Acheulean-style tools were fashioned from quartz-based rocks broken into cutting flakes, hammer stones, and choppers. Animal bones had hyena chew marks with cuts from stone tools over them, suggesting that the inhabitants appropriated the hyenas' kill and cut away the flesh. Other bones were broken open for their nutritive marrow. These fossils were lost, although plaster casts still exist.

Neanderthal Human
PLEISTOCENE EPOCH

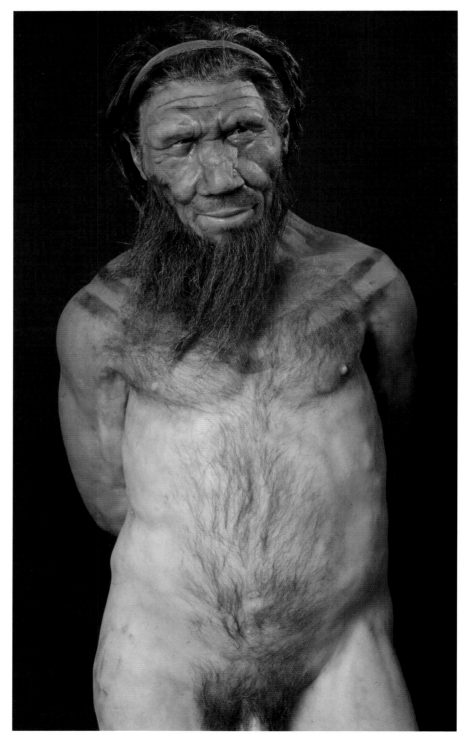

SPECIES *Homo neanderthalensis*
GROUP Hominidae
SIZE 5½ feet (1.7 m)
LOCATION Europe and southwest Asia

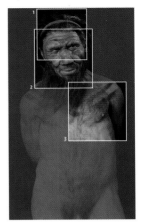

The first Neanderthal remains were found in a cave in Germany in 1856. However, they were not named as *Homo neanderthalensis* until 1864. Since then, numerous remains have been found over a wide region of Eurasia, from North Wales, Gibraltar, Israel, and the Caucasus Mountains. Despite the common misconception of the species as beastly, unintelligent, and stunted, Neanderthals were large-brained, intelligent, resourceful, big-game hunters who survived through the bleak depths of the most recent ice age. In the 19th century, Neanderthals were depicted as stooped, club-wielding cavemen due to the reconstruction of a near-complete skeleton of an adult male from La Chapelle-aux-Saints, France. The skeleton was uncovered in 1908 and the individual, around 40 years old at death, had severe arthritis, hence the stoop. Many modern reconstructions see Neanderthals as powerfully built hunters in a forbidding environment. In Europe they coexisted with incoming modern humans, *H. sapiens*, for several thousand years. Recovery of Neanderthal DNA in 2010 showed that it was a distinct species, not a subspecies of *H. sapiens* as had been claimed. It has been suggested that modern humans, being more advanced, wiped out the Neanderthals as they arrived in their territory, perhaps with the assistance of climate change. However, there is also molecular evidence that Neanderthals interbred with modern humans as recently as 40,000 years ago. **DP**

FOCAL POINTS

1 · BRAIN
The skull was large and held a brain with a capacity of more than 92 cubic inches (1,500 mL). Neanderthal brains were in fact slightly larger than the brains of modern humans, which recent evidence shows are shrinking slightly. In contrast to common preconceptions, Neanderthals were far from small-brained.

2 · FACIAL APPEARANCE
The Neanderthal face retained a number of primitive features, such as a bony brow ridge; a sloped-back receding forehead; round eye sockets and a large nasal opening. The jaw was big and robust but lacked the bony protuberance of the angular chin that characterizes *H. sapiens*.

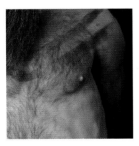

3 · ROBUST BUILD
Heavy bones and muscle attachment marks show that Neanderthals had a vigorous physique. Males weighed up to 176 pounds (80 kg) and the smaller females are estimated at 132 pounds (60 kg). The arms were long and powerful with strong hands, while the legs were short and powerfully built.

MOBILE HUNTERS

The bones and stone tools (below) found at Neanderthal sites show that they were mobile hunters who lived in small groups and tackled large animals using heavy, hand-held wooden spears tipped with stone points. Fossil evidence of healed fractures on their bones suggests a difficult lifestyle. Animal bones reveal that their diet ranged from large deer and horses to small hares and tortoises. This is supported by chemical isotope analysis of their teeth, which indicates they were predominantly meat eaters. However, starchy grains, legumes, and dates have recently been found embedded in Neanderthal tooth dental plaque, indicating their diet was more varied than previously thought. Stone tools included hand axes and spear points made in the Mousterian way. However, they did not produce such a variety of tools as the *H. sapiens* that lived alongside them.

Hobbit Human
PLEISTOCENE EPOCH

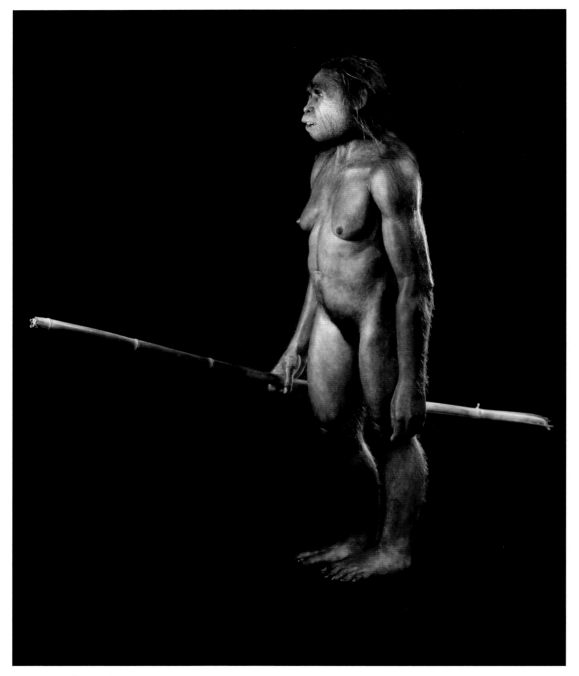

SPECIES *Homo floresiensis*
GROUP Hominidae
SIZE 39 inches (1 m)
LOCATION Indonesia

The discovery in 2003 of the skull and partial skeleton of *Homo floresiensis* within cave-floor accumulations on the Indonesian island of Flores was unexpected. The impression was a mixture of primitive and advanced features coupled with extremely small size. The partial skeleton of one female adult, "Flo," and pieces from six others indicate that *H. floresiensis* grew to no more than 39 inches (1 m) tall. Modern humans can be excluded as ancestors, because by the time *H. sapiens* reached Flores, around 50,000 years ago, *H. floresiensis* had been there for some time. The release of *The Lord of the Rings* films at the time of the discovery enabled the public to imagine these tiny humans as "hobbits." The remains were found alongside those of extinct animals, such as *Stegodon*, which dates *H. floresiensis* to between 95,000 and 12,000 years old. The discovery of stone tools and the remains of butchered bones suggest that *H. floresiensis* made tools, hunted a variety of animals, and used fire. The tools include long, narrow blades that were removed from prepared stone cores, which shows a level of intellectual development found in the genus *Homo*. The blades show microwear patterns, illustrating that they were used to work fibrous and woody plants, potentially as spear shafts. When first described as a human species, *H. floresiensis* were seen by some as modern humans with a condition that caused small stature, such as cretinism, but their primitive anatomical features contradict this. **DP**

👁 FOCAL POINTS

1 SKULL AND BRAIN
The brain, often described as "grapefruit-size," had a volume of only 24 cubic inches (400 mL), comparable to those of the australopithecines. The skull combines a mixture of archaic features, such as a bony brow ridge and low sloping forehead, with more advanced ones, such as a bony chin.

2 BODY
Like the skull, the general skeleton had a combination of archaic and advanced features. The wrist and pelvis were more apelike, while other areas resemble its probable ancestor, *H. erectus*. The body weight is estimated at only 55 pounds (25 kg), far lighter than the smallest living human groups.

3 TOOLS
The Flores tools resemble those of the Upper Paleolithic, or late Stone Age, which had begun to fade in many regions by 10,000 years ago. The shape of the fossil skull's interior indicates that, despite the miniature brain, the portions associated with thinking and intelligence were well developed.

ISLAND DWARFISM

The tiny size of *H. floresiensis* (above left) is probably due to evolutionary dwarfism, a phenomenon that occurs in isolated island populations with limited food resources. Large species either become extinct when their food supply is exhausted or natural selection alters them into smaller forms needing less food. This process produced the extinct dwarf hippopotamuses of Madagascar, the dwarf sheep-goat *Mytragus* of Majorca, dwarf woolly mammoths on Wrangel Island, and *Stegodon*, the dwarf elephant that lived on Flores in the early Pleistocene epoch. These dwarfed forms often retain juvenile features of their kind, such as short snouts and legs. This trend could account for some of the features in *H. floresiensis*, which can be seen as a small and childlike descendant of *H. erectus*.

MODERN HUMANS

1 Cave paintings of increasing sophistication occurred from 40,000 years ago in Africa, Southeast Asia, Europe, Australia, and South America.

2 The Denisova Cave site has yielded varied human and other fossils from more than 100 vertebrate species.

3 *Homo sapiens* originated about 200,000 years ago in Africa.

Anatomically modern humans—*Homo sapiens* (see image 3)—have features that distinguish them from various earlier humans and other apes. They include permanent bipedalism (walking on two legs) and the anatomical adjustments related to this vertical stance, such as feet, knees, pelvis, and a skull balanced on the spinal column rather than angled in front. The modern human skull also has a high-domed cranium and high smooth forehead, lacking bony brow ridges. The face is small and flat; the cheek teeth (molars and premolars) are also small, as is the lower jaw, which does, however, sport a distinctive bony protuberance or chin at the front. When the first ancient remains of *H. sapiens* were found in Europe in the early 19th century, theologically minded paleontologists tried to interpret them as victims of the Noachian Flood. However, it was soon evident that many of these anatomically modern humans were associated with extinct animals from the last part of the most recent ice age.

There was clear evidence from cave paintings (see image 1), other artworks, and butchered bones that animals, such as the extinct woolly mammoth (see p.554), were hunted by *H. sapiens*. By the middle of the 20th century, fossils of *H. sapiens* were found in the Middle East, Africa, and Australia that predated the remains found in Europe, most of which were no more than 40,000 years old.

KEY EVENTS

200,000 Ya	160,000 Ya	120,000 Ya	100,000 Ya	80,000–70,000 Ya	60,000 Ya
Modern humans, *Homo sapiens*, evolve in Africa, probably from an African population of *Homo heidelbergensis*.	The *Homo sapiens idaltu* subspecies live in Ethiopia, as is later shown by some of the earliest modern human fossils.	Early moves of *Homo sapiens* occur from Africa into the Middle East, but do not seem to have left permanent occupation.	*Homo erectus* begin to fade in Asia before *H. sapiens* reach the area, although fossil evidence for this is far from definite.	A second wave of *Homo sapiens* begins from Africa toward the Middle East, then towards north and east Asia.	*Homo sapiens* spread east across South Asia and into Southeast Asia.

Recent genetic studies of the global human population show that all modern humans have a common ancestry that extends back 200,000 years to an African ancestor. These molecule-based studies reinforce the archeological record of bones and stones, which has, in recent decades, supported the claim by Charles Darwin (1809–82) that humanity originated in Africa. The first tentative foray of *H. sapiens* into the Middle East occurred around 120,000 years ago when Neanderthal populations were already living there. However, it was not until 80,000 years ago that the main diffusion beyond Africa took place (see p.548), with modern humans moving through Arabia into western Asia, India, and on to Southeast Asia by 70,000 years ago, reaching Australia by 50,000 or 40,000 years ago. These modern humans also extended from western Asia into eastern and then western Europe by 45,000 years ago, when some interbreeding with Neanderthals took place. The first human colonization of the Americas did not occur until lowered sea levels and the ice age allowed *H. sapiens* to cross the Bering land bridge before 15,000 years ago. Finally, it was not until around CE 1000 that Polynesia and New Zealand were reached by modern humans.

However, this generally accepted view of the spread of modern humans was questioned by the discovery in 2008 of a 40,000-year-old fossil finger bone in Denisova Cave in the Siberian Altai Mountains (see image 2). Analysis of the bone and subsequent finds of teeth have yielded fragments of DNA that show genetic links with both modern humans and the Neanderthals. They also show that the Denisovan people shared a common ancestor with the Neanderthals, who lived around 600,000 years ago, and the divergence between the Neanderthals and Denisovans occurred around 200,000 years ago. The Denisova Cave has yielded fossils and artifacts that suggest it was, at various times, occupied by Neanderthals, Denisovans, and modern humans. Whether the Denisovans were a separate species of the *Homo* genus or a subspecies is yet to be confirmed.

Due to the lack of skeletal remains, the Denisovan people have so far been characterized by their genetics. This makes them the first extinct human species to be known predominantly from their DNA. A search for traces of Denisovan genetic material among living humans has revealed that some Melanesians and Indigenous Australians share between 3 and 5 percent of their DNA with them. It appears that, following their divergence from the Neanderthals 200,000 years ago, the Denisovan people spread over a vast range of Asia into the Southeast Asian islands. Then, when populations of modern humans spread through Southeast Asia 60,000 years ago, some of them bred with the Denisovans. This would explain the "genetic fingerprint" that persists in some living Melanesians and Indigenous Australians. It may seem extraordinary that such a widespread group of extinct humans as the Denisovans existed, yet lived and died without leaving any archeological record. This is a salutary reminder of the incomplete nature of the fossil record and the advancing power of gene analysis that shows where the human species originated and its unexpected make-up. **DP**

50,000 Ya	45,000–40,000 Ya	40,000 Ya	40,000 Ya	40,000 Ya	15,000 ya
Homo sapiens reach Australia, across many land bridges and narrow straits due to lower sea levels.	*Homo sapiens* reach western Europe and Britain as "Cro-Magnons" and interbreed with local Neanderthals.	Cave paintings begin to appear at very widely separated locations, from Europe and Africa to Australia.	Neanderthals are fading to extinction in Europe at this time, according to the latest evidence.	*Homo sapiens* encounter the Denisovan people in Asia and interbreed.	*Homo sapiens* reach the Americas via East Asian Siberia and the Bering land bridge into Alaska.

Out of Africa
PLEISTOCENE EPOCH

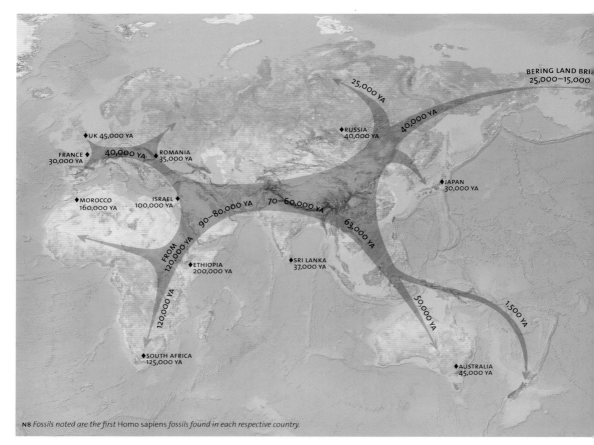

BERING LAND BRI
25,000–15,000

25,000 YA

40,000 YA

◆RUSSIA
40,000 YA

◆UK 45,000 YA

FRANCE ◆   40,000 YA   ◆ROMANIA
30,000 YA   35,000 YA

◆JAPAN
30,000 YA

◆MOROCCO   ISRAEL ◆
160,000 YA   100,000 YA

90–80,000 YA   70–60,000 YA

63,000 YA

FROM 120,000 YA

◆ETHIOPIA
200,000 YA

◆SRI LANKA
37,000 YA

120,000 YA

50,000 YA

1,500 YA

◆SOUTH AFRICA
125,000 YA

◆AUSTRALIA
45,000 YA

NB *Fossils noted are the first* Homo sapiens *fossils found in each respective country.*

SPECIES *Homo sapiens*
GROUP Hominidae
SIZE 6 feet (1.85 m)
LOCATION Africa, spreading to Asia, Australia, Europe, and the Americas

Genetic studies of living humans show that *Homo sapiens* have a common ancestor who lived in Africa 200,000 years ago and had spread through that continent by around 90,000 years ago. Fossils remains found in Ethiopia, named *H. sapiens idaltu*, are thought to represent members of this first African population. There was a brief excursion into the Middle East around 120,000 to 100,000 years ago, known from fossil remains found in Israeli caves, but the main dispersion of modern humans beyond Africa dates from less than 100,000 years ago. Analysis of DNA and other molecules reveals several identifiable groupings within the African continent, where *H. sapiens* has lived longest. Modern humans outside of Africa have a variety of low-level contributions to their DNA from human populations that were already established in Europe and Asia when *H. sapiens* first moved into these regions, including that of Neanderthals, which is most likely a result of interbreeding in the Middle East. Some Melanesians and Indigenous Australians have DNA derived from the Denisovans, for whom evidence dates to around 40,000 years ago. There does not seem to be any identifiable genetic contribution from *H. erectus* (see p.540). Presumably those early travelers lived the hunter-gatherer nomadic lifestyle, advancing at certain times by an average of ½ mile (1 km) per year. The differences today between ethnic groups of modern humans originate from, and since, those times. The total global population of *H. sapiens* 35,000 years ago is estimated to have been 3 million. It has since seen phenomenal growth, far faster than any other species, reaching more than 7 billion today. **DP**

 NAVIGATOR

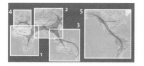

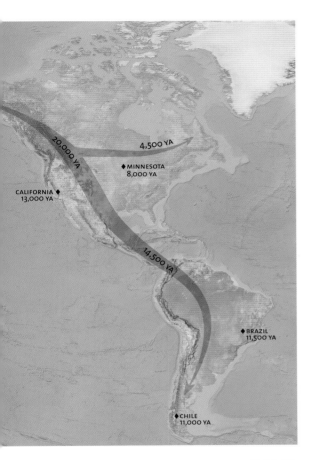

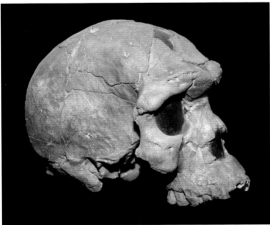

▲ The fossil remains of *H. sapiens idaltu* were found in Ethiopia in 1997 and described in 2003. Dated at around 160,000 years old, they retain some primitive features of the skull and nasal region, with more modern features such as a large brain and small face.

1 AFRICA

H. sapiens fossils from eastern Africa include *H. sapiens idaltu* from Ethiopia, at 160,000 years; Laetoli in Tanzania at 120,000 years; Border Cave in KwaZulu-Natal, northeast South Africa, at 115,000 to 90,000 years; and Klasies River Mouth, Eastern Cape, South Africa at 90,000 years.

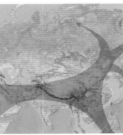

2 ASIA

From 90,000 to 80,000 years ago onward, *H. sapiens* spread east across southern Asia. Remains from the Annamite Mountains of Laos are estimated at 63,000 years old. Dispersal through these regions was linked to a short warmer period in the middle of the last ice age, 60,000 to 40,000 years ago.

3 AUSTRALIA

Most evidence points to *H. sapiens* reaching Australia by at least 40,000 years ago and possibly 50,000 years ago—before the species spread into Europe. Around this time large, native animals in Australia began to disappear in one of several megafauna extinctions (see p.550).

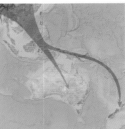

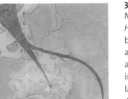

4 EUROPE

The first fossils recognized as humans were discovered in France in 1868 and excavated by Louis Lartet (1840–99). Known as "Cro-Magnons," the five large-brained skeletons had been buried in shallow graves with ornaments and stone tools 28,000 years ago. The bones preserve evidence of trauma.

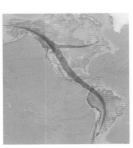

5 AMERICAS

Tools and a settlement in Chile date to 14,500 years ago, but some claim *H. sapiens* could have reached South America 20,000 years ago. Presumably, settlers crossed the Bering land bridge from Siberia, a journey of 11,000 miles (17,500 km) in a short time; parts of this route would have been ice-bound.

ICE AGE MEGAFAUNA

2

3

A huge number of large animals—mostly mammals—have become extinct within the past 50,000 years, in particular during the past 20,000 years. These creatures are known as ice age megafauna. In this context "ice age" refers to the most recent cold period (glacial) when average global climates cooled and ice sheets and glaciers spread—predominantly southward in the northern hemisphere—before warming toward the current interglacial period. The most recent glacial–interglacial series was during the Pleistocene epoch, which was within the Quaternary period (Pleistocene–Quaternary) and began around 2½ million years ago. Rocks, fossils, and drilled-out ice cores from the Arctic and Antarctic reveal glacials and interglacials that began on a 40,000-year cycle, extending to 100,000 years. These ice age cycles may well continue in the future; the present interglacial followed a glacial that ended between 14,000 and 10,000 years ago (see image 1). Possible causes of ice ages include interactions between drifting continents that rework ocean shapes and currents; trends in atmospheric make-up, including the amounts of carbon dioxide and oxygen; volcanic activity, such as the Toba eruption 70,000 years ago; and cycles in Earth's orbit and angle of tilt as it travels around the sun.

Climate cooling and an oncoming glaciation had huge effects on global landscapes. Year by year, cold-adapted vegetation such as conifer forests spread, while broad-leaved temperate and tropical habitats shrank. Some animal

1 The last ice age, 20,000 years ago, as envisioned by Swiss geologist and naturalist Oswald Heer (1809–83)

2 The age of eggshell fragments from the Australian flightless *Genyornis* shows it went extinct over a short period—too briefly to link to climate change.

3 South American *Toxodon* remains have been found with weapons, such as arrowheads, dating to around 15,000 years ago.

KEY EVENTS

110,000 Ya	76,000–70,000 Ya	50,000 Ya	47,000–45,000 Ya	25,000–15,000 Ya	20,000 Ya
The last glacial period begins as Earth enters another phase of cooling and polar freezing.	The Toba supervolcano on Sumatra, Indonesia, erupts, possibly leading to a worldwide "volcanic winter" over several decades.	Humans reach mainland Australia.	Many Australian megafauna species go extinct, but there is little evidence of drastic climatic and environmental change.	The Last Glacial Maximum occurs, with the greatest extent of ice sheets and glaciers across the northern lands of North America and Europe.	A small area on and around Mount Kosciuszko, the highest point in Australia, becomes glaciated.

species followed the climate trend and changing vegetation belts, adapting with it. Others faded under the pressures of locating food, avoiding predators, and fending off parasites and diseases—all of which were evolving, too. As more ice was contained in ice caps and glaciers, sea levels dropped. During the Last Glacial Maximum, which occurred around 25,000 to 15,000 years ago, world sea levels were 460 feet (140 m) lower than today, thus exposing many land bridges that allowed previously separated species to mix, compete, and prey.

There also came another threat. The spread of *Homo sapiens* during the late Pleistocene (see p.546), around 80,000 years ago, began from Africa into Asia and Australasia, then north into Europe, and finally from northeast Asia into North and then South America. The dates of these movements often correlate, sometimes within a few thousand years, with the last known presence of megafauna. Humans are thought to have arrived in Australia around 50,000 to 40,000 years ago, about the time the marsupial "lion," *Thylacoleo carnifex*, went extinct, alongside the giant wombat, *Diprotodon* (see p.432), the giant short-faced kangaroo, *Procoptodon*, the 6½-foot (2 m) tall thunderbird, *Genyornis* (see image 2), and the 20-foot (6 m) long lizard, *Megalania*.

In North America, the last native horses, *Equus* (see p.494), have been dated at 12,000 years ago, soon after human arrival. Also disappearing within a few thousand years were the saber-toothed cat, *Smilodon* (see p.528); dire wolf, *Canis dirus* (see p.492); Jefferson's ground sloth, *Megalonyx*; American mastodon, *Mammut americanum*; and vulturelike teratorn, *Aiolornis incredibilis*. South America lost the camel-like litoptern, *Macrauchenia*; rhinoceroslike *Toxodon* (see image 3); giant "armadillo," *Glyptodon*; and giant ground sloth, *Megatherium* (see p.480). In some of these cases (with the exception of Australia), the climate, vegetation and habitats were changing rapidly, so it is difficult to disentangle human effects from environmental ones. Another suggested contribution to the extinctions is "hyperdisease" brought by humans or their animals, but there is little evidence for this.

Humans could have caused extinction by hunting for food, for safety, or to reduce competition. The second-order predation theory argues that the killing of predators by humans meant that prey populations reproduced uncontrollably, leading to overexploited local food resources and the starvation of prey species due to ecological collapse. Yet many groups had already evolved through several cycles of environmental change, so the last cycle must have differed if humans were not the primary cause. Comparatively fewer megafauna extinctions occurred in Africa—where *Homo* had been evolving as part of the landscape for hundreds of thousands of years— although climate changes here were reduced compared to elsewhere. Plentiful megafauna survived in Africa and live there now, including the biggest land mammals: elephants, rhinoceroses, giraffes, and hippopotamuses. **SP**

20,000 Ya	20,000–15,000 Ya	19,000–15,000 Ya	13,000 Ya	12,900–11,500 Ya	12,000 Ya
Global atmospheric temperatures are an average of 7° F (4° C) lower than the present day.	Humans enter North America from north-east Asia, rapidly spreading south and east (although these dates are still contentious).	Ice sheets and glaciers in northern and northeast Asia (Siberia) reach their greatest extent.	Ice sheets are retreating rapidly from northern Europe and Scandinavia.	The Younger Dryas occurs. It is the third of the Dryas periods— brief cooler intervals, overlaid as "mini-glacials" on the general warming trend.	The end of the last glacial period and start of the current interglacial. The woolly mammoth (see p.554) numbers begin to show drastic decline.

Woolly Mammoth

PLEISTOCENE EPOCH—HOLOCENE EPOCH

SPECIES *Mammuthus primigenius*
GROUP Elephantidae
SIZE 11 feet (3.5 m)
LOCATION Asia, Northern Europe, and North America

✦ NAVIGATOR

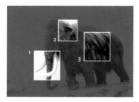

One of the most impressive and well-known recently extinct mammals is the woolly mammoth, *Mammuthus primigenius*. This giant herbivore roamed the northern forests and tundra of Asia, northern Europe, and North America from about 200,000 years ago, when it evolved from the steppe mammoth (see p.476). Many specimens have been found across northern countries, several of them extremely well preserved by the ice and permafrost of the Arctic (see panel).

This mammoth was similar in size to the African elephant, *Loxodonta africana*, but it was bulkier and the tusks were longer and more strongly curved than those of living elephants. The woolly mammoth is most closely related to the Asian elephant, *Elephas maximus*, and both of these diverged from a lineage that led separately to the African elephants. All three shared a common ancestor 8 million years ago in Africa, and the mammoths and Asian elephant lineages separated around 6 million years ago, during the late Miocene epoch. Unlike living elephant species, the body of the woolly mammoth was covered with fur, which was one of a number of adaptations to life in the cold conditions of the ice age. Further adaptations to the cold included a shorter tail, shorter ears and a layer of fat, around 4 inches (10 cm) thick, beneath the skin. As highly sociable animals, woolly mammoths lived in herds, similar to modern elephants, and fed on a diet of coarse grasses and other rough vegetation. Their numbers began to decline from the end of the last ice age, around 12,000 to 10,000 years ago, most likely due to a combination of human hunting and climate change. A small number of dwarf individuals survived until as recently as 4,000 years ago, as shown by finds from Wrangel Island, in the Arctic Ocean, off the far northeast coast of Russia. **MW**

◉ FOCAL POINTS

1 LONG TUSKS

The curved tusks reached 13 feet (4 m) long in the largest males and up to 6 feet (1.8 m) long in females. This disparity indicates that their main use among males was for display and rivalry as well as to impress potential mates. Other uses include digging up roots, knocking over small trees and sweeping snow from ground vegetation.

2 SHORT EARS AND TAIL

The ears were relatively small, as was the tail, which had 21 vertebrae as opposed to about 30 in living species. The long, trailing hairs extended the tail's apparent length. Both were adaptations to reduce heat loss and to prevent the animal from developing frostbite in its extremities.

3 HAIRY BODY

Close to the skin was a layer of dense, short, insulating wool, covered by an outer layer of long, coarse, straggly guard hairs that could reach 16 inches (40 cm) long, and would have shed snow and rain. These are known from both prehistoric cave art and frozen carcasses from Siberian permafrost, with soft tissues intact.

⊓ MAMMOTH RELATIONSHIPS

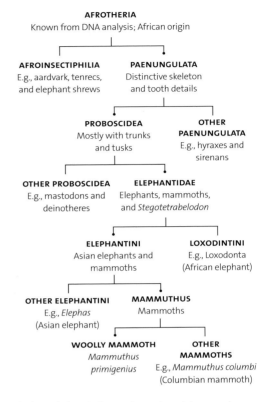

AFROTHERIA
Known from DNA analysis; African origin

AFROINSECTIPHILIA
E.g., aardvark, tenrecs, and elephant shrews

PAENUNGULATA
Distinctive skeleton and tooth details

PROBOSCIDEA
Mostly with trunks and tusks

OTHER PAENUNGULATA
E.g., hyraxes and sirenans

OTHER PROBOSCIDEA
E.g., mastodons and deinotheres

ELEPHANTIDAE
Elephants, mammoths, and *Stegotetrabelodon*

ELEPHANTINI
Asian elephants and mammoths

LOXODINTINI
E.g., Loxodonta (African elephant)

OTHER ELEPHANTINI
E.g., *Elephas* (Asian elephant)

MAMMUTHUS
Mammoths

WOOLLY MAMMOTH
Mammuthus primigenius

OTHER MAMMOTHS
E.g., *Mammuthus columbi* (Columbian mammoth)

▲ Elephants belong to the very large placental mammal group Afrotheria, whose ancient origins are in Africa. The only living members of the Proboscidea, their closest relatives today are the hyraxes, dugongs, and manatees (sirenians).

PRESENT-DAY SIGHTINGS

Occasional reports suggest that the woolly mammoth may survive to this day. Certainly, the thought that these giants may still lurk somewhere in the vast stretches of conifer forests across north Asia is tantalizing, especially to cryptozoologists. Reports have emerged of Siberian people, with no knowledge of elephants, mentioning large shaggy beasts encountered in taiga forests. However, no such sightings have been confirmed, and they could have been based on frozen mammoth corpses (right). All reliable evidence points to the woolly mammoth having gone extinct by about 4,000 years ago. Also fascinating is the possibility that the species could be recreated from DNA gathered from specimens that were frozen in permafrost. DNA could be implanted into a living elephant's egg cell, or mammoth sperm could be used to fertilize an elephant egg cell.

HUMAN ACTIVITY AND EVOLUTION

1 The global symbol of extinction is the dodo, *Raphus cucullatus*.

2 Better-camouflaged dark peppered moths thrive in industrial areas.

3 Introduced cane toads, *Rhinella marina*, wreak havoc among Australian wildlife.

S
o far, there have been five great mass extinction events, the most severe of which was the "Great Dying" at the end of the Permian period (see p.224). Many biologists and evolutionists believe that a sixth event has begun: the Holocene Extinction—the Holocene being the current geological epoch, which began 11,700 years ago. The recent decline in species numbers has been well publicized, and as further losses are documented, further species become extinct. Several projects are underway to reverse extinction caused by humans, such as the Quagga Project (see p.558). Despite the rate of extinction, new species are continually reported. These evolve due to processes such as allopatric (geographic) speciation, or polyploidy, where a species can evolve "instantly" due to the doubling of its genetic material. In general terms, the average time it takes for a new kind of plant or animal to evolve—a speciation—ranges from thousands to around 1 million years. And, as a background rate, between 10 and 100 new species evolve per year. A typical single species range is estimated to survive from 500,000 to 10 million years on average, as evidenced by fossils. Mammals are at the lower end of this range, with an estimated 500,000 to 2 million years, while invertebrates exist for 5 million to 10 million years.

KEY EVENTS

1400s	1690s	1800s	1848	1898	1930s
House mice, *Mus musculus*, arrive via sailing ships on the Atlantic island of Madeira; they will evolve into six different forms by the 1990s.	The dodo, an icon of extinction, has by now disappeared from the Indian Ocean island of Mauritius due to human hunting.	A new fly species, the honeysuckle maggot, *Rhagoletis mendax x zephyria*, evolves from two species by hybridization.	The first dark peppered moths, *Biston betularia*, are recorded in industrial Manchester, England, and become common on soot-stained trees.	The last specimens of the woolly stalked *Begonia eiromischa* are collected from Penang Island, Malaysia. The species is declared extinct in 2007.	The last wild plants of the Cry pansy or Cry violet, *Viola cryana*, are identified in the Cry region of France. The species no longer survives in cultivation.

Humans have only existed for 200,000 years, but Earth has changed enormously in the past few hundred years due to the activities of *Homo sapiens*. The best-known example is the dodo (see image 1), which was quickly hunted to extinction when humans colonized Mauritius. The human population rose from 1 billion in 1800 to more than 6 billion by 2000, and may reach 10 billion by 2100. By 2009, the number of people living in urban conditions had overtaken the number in rural environments. Twenty percent of the planet's land surface has been colonized by humans. Two-thirds of this is for industrial, urban, and cropland use, and one-third is pasture. Industrial development causes problems such as the pollution of land, soil, and water; rising greenhouse gas emissions; and vast habitat loss. Such huge environmental and ecological changes are affecting the natural world and the animals within it, such as the baiji dolphin (see p.556), and it is uncertain how evolution will fare. Extinction rates within the Holocene are estimated to be between 100 and 1,000 times greater than the average background rate of other time periods. Compared to past mass extinction events, today's rates could be between 10 and 100 times higher.

Extinction does leave room in nature for new species to evolve (see image 2). However, environments that have been cultivated by humans, such as urban sprawl, farm croplands, and fields of livestock, tend to obliterate natural habitats. This limits the available roles in nature into which new species can evolve. Worldwide, where humans spread, similar groups of animals and plants follow. Examples include domestic cats, dogs, brown rats, house mice, city pigeons, house sparrows, houseflies, and green vegetable shield bugs; plants include shepherd's purse, creeping wood sorrel, nettles, bracken fern, and shield lichen; as well as numerous microbes. These species flourish in the conditions created by human habitation and development. However, they hinder a range of local species, reduce biodiversity, and limit options for speciation.

Humans have also introduced invasive species, accidentally or deliberately, into new environments, such as rabbits, common starlings, cane toads (see image 3), Burmese pythons, common carp, zebra mussels, tiger mosquitoes, cotton (potato) whitefly, Chinese mitten crabs, tubifex sludge worms, and, among plants, kudzu vines, tamarisk trees, water hyacinths, common prickly pear cacti, and wakame edible seaweeds. Species introduced in this way tend to have rapid growth and reproduction rates, the ability to spread or disperse well, adaptable diets in animals, growing abilities in plants, and a tolerance of a broad range of physical and chemical conditions, including forms of pollution. They may thrive due to the exploitation of a previously underused food resource in their new habitat, along with a lack of predators, diseases, and competitors. So, while extinction rates rise, humans are reducing opportunities for speciation. It would be fascinating to consider our current age from a few million years in the future, to see whether the sixth extinction was the most dramatic of all and how long it took biodiversity to recover in spite of, or after, humans. **SP**

1950s	1971	1979	1982	1999	2012
Two new North American salsify (goatsbeard) plants, *Tragopogon mirus* and *T. miscellus*, are identified as hybrids of species introduced in the 1900s.	Ten Italian wall lizards, *Podarcis sicula*, arrive on Pod Mrcaru, Croatia. By 2004 they evolve larger heads and jaws and new stomach parts (caecal valves).	A new flower, York groundsel, *Senecio eboracensi*, is discovered as a hybrid of *S. squalidus* and *S. vulgaris*; it breeds with neither parent.	Large ground finches arrive on Daphne island in the Galapagos; subsequent populations of the native medium ground finch evolve smaller beaks.	The ratlike red viscacha, *Tympanoctomys barrerae*, has almost twice the amount of genetic material as its close relatives.	A new hybrid form of mimulus, *Mimulus peregrinus*, is discovered on a stream bank in Scotland.

Extinctions Today: Baiji
EARLY MIOCENE EPOCH—LATE HOLOCENE EPOCH

SPECIES *Lipotes vexillifer*
GROUP Lipotidae
SIZE 8 feet (2.5 m)
LOCATION Yangtze River, China
IUCN Extinct

NAVIGATOR

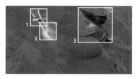

In 2005 to 2006 an intensive survey attempted to locate surviving members of the rare Chinese or Yangtze River dolphin, the baiji, *Lipotes vexillifer*. This freshwater mammal, a former resident of the Yangtze in central China, had become rarer throughout the 20th century. Its initial population was estimated at more than 5,000; then, in the middle of the 20th century only 1,000 to 2,000; in the 1980s fewer than 500 survived; reducing to 50 to 100 by the 1990s. The survey in 2006 found none, and the baiji was declared the first medium to large animal to die out in the 21st century. It was "functionally extinct," meaning that even if a few survived, they would be too scattered, too limited in genetic diversity, and facing too many environmental problems to recover in any viable way. Numerous factors contributed to the dolphin's demise: it lost its traditional revered status in China's Great Leap Forward, which promoted technology and agricultural self-sufficiency, so from 1958 it was hunted for its skin and meat as well as bones as trinkets. Fishing was a factor: prey was decimated due to local fishing; illegal but common electrofishing and accidental by-catch in fishing gear all contributed, as did increased river traffic, with the risk of collision and high noise levels that interfered with echolocation (the sonar navigation system). Dams and other barriers were constructed, such as the giant Three Gorges hydroelectric scheme, and water quality became poorer, including run-off from sewage, agricultural pesticides, and industrial chemicals. After the survey, an official report concluded that being trapped in fishing nets and similar gear was a major factor: "We are forced to conclude that the baiji is now likely to be extinct, probably due to unsustainable bycatch in local fisheries." **SP**

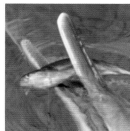

1 JAWS AND TEETH

The snout, or beak, was long and slender with about 30 to 35 small, pointed, interlocking teeth in the left and right sides of both the upper and lower jaws. Baijis fished mainly by day, poking and probing the sand and mud with their long beak in river bays, tributaries, and sandbanks.

2 SIGHT AND ECHOLOCATION

The eyes were small and functional, but in the river's cloudy waters, the main navigation sense was highly evolved echolocation. The baiji made whistles, clicks, and squeals focused by the bump in the prominent forehead, then listened for reflections or echoes from nearby objects.

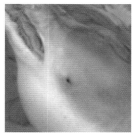

3 FLUKES AND FIN

The tail flukes were stocky and deep from front to back, adapted for quick bursts of speed rather than for cruising, as in oceanic dolphins. The baiji's fin was much reduced—it was not needed for stability, whereas the tall dorsal fin in oceanic dolphins provides steadiness while cruising.

FRESHWATER ADAPTATION

The baiji and four other still-surviving river dolphins were once classified in one group, Platanistoidea because they all inhabited fresh water. Despite this similarity, the grouping did not reflect their evolutionary origins—the five dolphins are not particularly closely related. Each species developed from a different oceanic ancestor that became adapted to river life in a series of separate but parallel evolutionary events, perhaps as a way of avoiding competition with other faster, more efficient dolphin species. For example, the tucuxi, *Sotalia fluviatilis*, of the Amazon River is included in the oceanic dolphin family Delphinidae, and its closest cousin is the costero or Guiana (estuarine) dolphin in the same genus, *S. guianensis*. The Amazon Basin's other freshwater dolphin, the boto or Amazon River dolphin, *Inia geoffrensis* (right), has its own separate family, Iniidae, more closely related to the saltwater but coastal franciscana or La Plata dolphin, *Pontoporia blainvillei*.

"De-extinction": Quagga
PLIOCENE EPOCH—HOLOCENE EPOCH

SUBSPECIES *Equus quagga quagga*
GROUP Equidae
SIZE 8 feet (2.5 m)
LOCATION Southern Africa
IUCN Extinct

In 1987 the Quagga project began in South Africa to "re-breed" a type of zebra that had gone extinct a century before. This was the quagga, a member of the horse genus *Equus*. For many years is was believed to be a separate species of zebra, commonly referred to as *E. quagga*. It looked distinctive, with typical zebra stripes only on the head, neck, and shoulders, fading along the sides, while the rear body and four legs were mostly stripeless. In the late 1870s, quaggas had been hunted to extinction by European settlers for their meat and skins, and to prevent them competing with farmed livestock for food.

From the 1980s molecular, genetic, and physical studies concluded that the quagga was a subspecies or local variant of the plains zebra, formerly *E. burchellii*. Since the name *E. quagga* was given in 1785, prior to *E. burchellii* in 1824, the former replaced the latter for the plains zebra. Various other populations of the species were also studied, showing that there are probably between five and seven subspecies, including the quagga, *E. quagga quagga*, Chapman's zebra, *E. q. chapmani*, and Grant's zebra, *E. q. boehmi*. As a subspecies, the quagga would have been able to interbreed with and share the same gene pool as other subspecies of plains zebra. This led to the Quagga Project's aim of choosing the most quagga-looking plains zebra individuals for a selective breeding program. Eventually a true-breeding population of quaggalike animals should result, at least from the perspective of phenotype: the observable features of an organism, such as its size, shape, color, pattern, development, and behavior. However, it will not be possible to confirm re-forming the true quagga genotype—the full complement of genetic information (genome)—because this is not known for original quaggas. To this extent, the quagga can be regarded as a partial "de-extinction." **SP**

⬡ NAVIGATOR

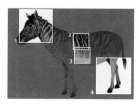

◎ FOCAL POINTS

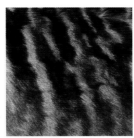

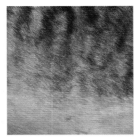

1 HEAD AND MANE
Pale or white stripes covered the head and neck, then faded along the flanks. In most quaggas, the mane was erect and striped. The quagga had good sight, hearing, and smell, front incisor teeth to crop grass, and large molars for chewing.

2 STRIPES
Preserved skins show a wide range of patterns among quaggas. Some individuals had stripes that had almost faded by the time they reached the shoulders, while others sported stripes nearly to the hind quarters.

3 BACKGROUND COLOR
Existing skins and paintings of quaggas show a background or base color of brown, which became lighter red-brown toward the rear body. However, like the base color in zebras, this would have varied among individual quaggas.

4 REAR BODY AND LEGS
In the Quagga Project, the numbers of stripes on the rear body and rear legs have steadily decreased with each generation of breeding. There are more than 80 zebras in the project, distributed across sites in South Africa and Namibia.

◄ In 1870, the only photographs of a living quagga, a mare, were taken at Regent's Park Zoo, London. At the time, nature was deemed "inexhaustible," therefore no one was truly aware that an extinction was taking place.

FROG RESURRECTION

Australia's now-extinct gastric-brooding frogs, *Rheobatrachus* (right), evolved a unique method of reproduction. The female laid her eggs, the male fertilized them, then she swallowed them into her stomach where the tadpoles developed over 6 weeks and emerged from her mouth as fully formed froglets. These frogs died out in 1983, probably due to disease. The Lazarus Project aims to bring them back from their preserved, deep-frozen tissues using the somatic (body) cell nuclear transfer (SCNT) procedure. A body cell nucleus containing the genetic material is removed from thawed frog tissue and transplanted into the egg cell of a close relation, the great barred frog, *Mixophyes fasciolatus*. In 2013, eggs developed into early embryos over a few days, but no further.

GLOSSARY

adapt
To change and alter to suit conditions of a given time, which in nature are themselves often changing.

adaptive radiation
When a group of organisms evolves by adapting to various conditions or ecological niches (roles) to produce a range of different groups, often in a relatively short time period.

advanced
A later or more derived form of an organism, but not necessarily implying long-term success. See also *archaic, derived, primitive*.

agnathan
A chordate or vertebrate animal that lacks jaws; usually reserved for various fish, including living hagfish and lampreys, and various extinct groups.

algae
An informal term for simpler plants and plant-like organisms that gain energy by photosynthesis and live in water or damp places, including seaweeds and microscopic kinds. The term is not an officially named taxon or grouping.

amniotes
Vertebrate animals, being reptiles, birds, and mammals, whose eggs are laid and survive out of water or develop in the mother, as opposed to anamniotes—most fish and amphibians—whose eggs must be laid in water.

archaic
An early, initial, or old-fashioned form, but not necessarily implying failure or maladaptation. See also *advanced, derived, primitive*.

articulated
When remains such as the fossilized bones of a skeleton are in lifelike positions, laid out as they would be in the original organism, rather than scattered, jumbled, and misshapen.

basal
Existing in an earlier or more basic, less derived (changed) form, for example, at or near the "root" of diagrams such as evolutionary trees and cladograms.

bilateral symmetry
Two-sided symmetry, with the left and right sides being mirror images of each other. Many animals have this form of basic body plan. See also *radial symmetry*.

bipedal
Moving on two legs or limbs, as in dinosaurs like the theropods (meat-eaters), birds, kangaroos, and humans.

biodiversity
The range or variety of living things in a certain place and time (rather than the numbers of each particular kind). This can be assessed in various ways, often as the number of species present.

biota
The total amount or collection of living things in a particular time and place, including both the variety (biodiversity) and numbers found.

bone-bed
A general term for a place where fossilized bones and other remains are especially dense and numerous.

bottom-dweller
An aquatic (water-dwelling) organism that tends to stay on or near the bottom, or bed, of its habitat, such as a lake or the deep sea.

carapace
The upper or dorsal part of a hard body casing, for example, in crustaceans such as crabs, insects, arachnids, and turtles and tortoises.

carbon dating
Studying the change or decay of various kinds of carbon atoms in (usually) once-living things to assess when they were alive. See also *radiometric dating*.

carnivoran
A member of the mammal group Carnivora, including cats, dogs, foxes and bears.

carnivore
An animal that eats mainly flesh or meat.

cartilaginous
Made of cartilage, a lightweight, tough, often slightly springy or bendy substance. In informal contexts it is sometimes called "gristle."

caudal
To do with the rear end of an organism, especially the tail of an animal.

chitin
A common substance in the makeup of living materials and objects, especially the outer body casings of insects and crustaceans, and other parts that provide strength and resilience.

chordate
An animal with a notochord—a stiff, rodlike structure along the upper center of the body—and certain other body parts; or, with a vertebral column or backbone, which is a derivative of the notochord. Chordate animals include fish, amphibians, reptiles, birds, mammals, and their relatives. See also *vertebrate*.

clade
A group of living things consisting of an ancestor and all of its descendants, and excluding any others. Clades are the basis of grouping organisms to reflect their evolutionary relationships. All members of a clade share unique features that are not present outside the clade. See also *synapomorphy*.

cladistics
A method of grouping organisms to reflect their evolutionary relationships. See *clade* and *phylogenetic systematics*.

class
In traditional taxonomy (the naming and classifying of organisms), a group in the ranking hierarchy: kingdom, phylum (division), class, order, family, genus, species.

coelurosaurian
A dinosaur belonging to the Coelurosauria or "hollow-tails," which were all meat-eaters.

convergent evolution
When different organisms evolve to look or work in a similar way because they are adapted to the same environment and ecological roles. For example, sharks, ichthyosaurs, and dolphins, although unrelated, all evolved similar body shapes to move swiftly through water.

Deoxyribonucleic acid, or DNA
The chemical substance from which genes are made and which is replicated to pass inherited features from parents to offspring. Living organisms share a high proportion of DNA. See also *genes*.

derived
A "new" feature or trait that evolved in an organism and is present in its descendants, but was not present in the last ancestor. See also *synapomorphy*.

division
A major rank in the grouping of plants. See also *phylum*.

domain
One of the basic, fundamental groups into which all living things are divided. In many systems the three domains are Bacteria, Archaea, and Eukarya or Eukaryota. In some systems, domains are then divided into kingdoms.

dorsal
To do with the upper side or back of an organism.

dorsoventral
To do with the upper side or back of an organism (dorsal), and the underside or belly (ventral), as opposed to, for example, the sides (lateral).

echolocation
Production of high-pitched sounds and detection of their echoes or reflections to locate surrounding objects and orientate, especially at night.

ecology
Changes in living things through time, in particular, their inherited features or traits through successive generations.

exaptation
When a feature or characteristic that has evolved for one reason, finds a new function as it adapted or co-opted for a different role.

family
In traditional taxonomy (the naming and classifying of organisms), a group in the ranking hierarchy: kingdom, phylum (division), class, order, family, genus, species. Also used as an informal term for a small group of closely related organisms.

filter-feeder
An organism that sieves or strains water to collect tiny pieces of food.

foraminiferans
Single-celled, amoeba-like, water-dwelling organisms that feed on even smaller items, such as bacteria, and make a hard shell or test, often of a very intricate form.

fossorial
Having an underground lifestyle of digging and tunneling.

genes
Lengths of DNA (or RNA) that contain heritable information, or "chemical instructions," for how a particular organism develops and functions.

genus
In traditional taxonomy (the naming and classifying of organisms), a group in the ranking hierarchy: kingdom, phylum (division), class, order, family, genus, species. The genus is a group of closely related species, with the first of their two-part official scientific names being the genus or generic name, for example for *Homo sapiens* (living humans) and *Homo neanderthalensis* (Neanderthals), the genus is *Homo*.

glabella
The upper, near-front portion of a trilobite shell, approximately equivalent to its "forehead."

Great American Interchange
An important paleozoogeographic event, peaking about 3 million years ago, in which land and freshwater fauna migrated from North America via Central America to South America, and vice versa, after the volcanic Isthmus of Panama thrust up and bridged the formerly separated continents.

habitat
A particular kind of place or environment in which organisms live—for example, a pond, mountain, desert, grassland, rocky seashore, or the bottom of the deep sea.

herbivore
An animal that eats mainly plant parts.

heterocercal
A design of fish tail or caudal fin where the vertebrae (backbones) extend into the upper of the two lobes, which is usually larger than the lower lobe.

hydrofoil
A winglike shape that produces a lifting force when moving through water.

hyperpredator
A topmost predator, usually very large, which hunts any prey in its habitat but is unlikely to be preyed upon itself.

hypocercal (reverse heterocercal)
A design of fish tail or caudal fin where the vertebrae (backbones) extend into the lower of the two lobes, which is usually larger than the upper lobe.

index fossils
Fossil remains that are common, widespread, and change relatively rapidly through time, so that their presence can be used to give dates (time periods) to their rocks and other fossils found with them. Also called "indicator fossils."

insectivore
An animal that eats mainly insects and, often, other small invertebrates such as spiders and worms.

invertebrate
An animal that lacks a backbone (vertebral column), and in general use, also one that lacks a notochord—that is, a non-chordate animal.

kingdom
In traditional taxonomy (the naming and classifying of organisms), a group in the ranking hierarchy: kingdom, phylum (division), class, order, family, genus, species. See also *domain*.

lateral
To do with the sides, usually left and right, of an organism, such as the lateral fins of a fish, rather than the upper (dorsal) or lower (ventral) surfaces.

living fossil
A vague term implying that a living organism is very similar and little changed from relatives that lived long ago and are known only from fossils.

metazoans
Animals with a body made of many cells, rather than just one single cell.

molecular clock
The notion that the molecules that make up living things change with evolution though time, and the rate of change can be estimated to find out, for example, when two living species separated from each other.

multiregional hypothesis
The proposal that human populations have evolved semi-independently around the world over a very long period of time to produce the variety seen in living humans, *Homo sapiens*.

New World
Pertaining to North and South America and their associated islands. See also *Old World*.

niche (ecological)
The role or part played by an organism in its environment and how it interacts with other organisms and its

surroundings. For example, with the lack of ground mammals in New Zealand due to its geographical isolation, several birds evolved to fill the niches usually occupied by mammals, and became flightless.

notochord
See *chordate*.

obligate
By necessity, or being unable to function without something.

Old World
Pertaining to Europe, Asia, and Africa, also perhaps Australasia and Antarctica, and their associated islands. See also *New World*.

opposable thumb
A thumb that can be placed against each of the fingertips in turn, allowing great dexterity and range of grip.

order
In traditional taxonomy (the naming and classifying of organisms), a group in the ranking hierarchy: kingdom, phylum (division), class, order, family, genus, species.

organism
Living thing.

oviparous
When a female animal lays eggs, in which the embryos develop before hatching. See also *ovoviviparous*, *viviparous*.

oviphagy
Egg-eating, as, for example, an animal specialized to eat birds' eggs.

ovoviviparous
When embryos develop in their eggs inside a female animal, and then hatch just before they are born as live young. See also *oviparous*, *viviparous*.

paleontology
The study of fossils and similar remains, and of ancient or prehistoric life in general.

pharyngeal
To do with the throat or neck area of an animal. It is usually involved in respiration or breathing, and perhaps in feeding.

phylogenetic systematics
The study and grouping or organisms to reflect their evolutionary relationships,

especially involving clades and the possession of unique features or traits. See also *clade*, *synapomorphy*.

phylum
A major grouping of living things, sometimes called a division in the case of plants. In traditional taxonomy (the naming and classifying of organisms), a group in the ranking hierarchy: kingdom, phylum (or division), class, order, family, genus, species.

placoderm
Meaning "plate-skin," a group of long-extinct fish characterized by bony plates, and one of the first fish groups to have jaws and proper fins.

primitive
The first, or an early, form of an organism, but not necessarily implying failure or maladaptation. See also *advanced*, *archaic*, *derived*.

quadrupedal
Moving on four legs or limbs, as in most mammals and reptiles.

radial symmetry
Circular or wheel-like symmetry. Animals with this basic form of body plan include anemones and corals, and starfish and other echinoderms. See also *bilateral symmetry*.

radiometric dating
Studying the change or decay of various kinds of atoms to assess, for example, when they were incorporated into rocks, thereby giving the age of the rock's formation. See also *carbon dating*.

relict
Leftover: for example, a remaining small population of a once common and widespread organism.

seeds
Reproductive items of certain plants, usually produced sexually by union of female and male cells, often tough and resistant to harsh conditions, and containing an embryo plant plus food reserves. See *spores*.

sexual dimorphism
Difference in body form and function between the female and male sexes of a species.

specialist
An organism that has narrow or restricted

needs or lifestyle, for example, a predator that can only feed on one kind of prey.

species
In traditional taxonomy (the naming and classifying of organisms), a group in the ranking hierarchy: kingdom, phylum (division), class, order, family, genus, species. The species is variously defined: for example, members of a species can breed together to produce viable offspring. For extinct groups, anatomical (structural) and other features are used. Increasingly genetic and molecular criteria are involved.

sporangia
Spore-producing parts of certain plants. See *spores*.

spores
Reproductive items of certain plants, usually produced asexually (vegetatively), often tough and resistant to harsh conditions, but rarely holding any food reserves. See *seeds*.

synapomorphy
A feature or trait that is inherited from a common ancestor and shared by all members of a clade (evolutionary grouping), and which is unique, or not present in any other clade.

taxonomy
The grouping, describing, classification, naming, and identification of organisms.

tetrapod
A vertebrate (animal with a backbone) that has four limbs.

thermoregulation
In biology, control of body temperature.

trace fossils
Fossils that are not an organism's preserved actual parts, but signs or evidence left behind, such as footprints, bite and claw marks, tunnels, and droppings (coprolites).

vertebrate
An animal with a backbone or vertebral column—a linked series of bones running from the skull in the head to the rear end or tail. Vertebrata is a subgroup of Chordata. See also *chordate*.

viviparous
When a female animal gives birth to babies that have developed inside her body, usually nourished by a placenta, rather than inside eggs.

CONTRIBUTORS

Petra Brookes (PB)
is a scuba enthusiast and underwater photographer who has visited many world-renowned sites. She studied marine invertebrates, especially molluscs, corals, and sponges, and gained qualifications in marine biology. Petra is a consultant to several marine conservation projects, publishing local developmental plans and relating these to wider issues affecting the oceans.

Leon Gray (LG)
studied zoology at University College London, specializing in vertebrate palaeontology and mammalian reproduction. In 1995 he completed his research paper at the Institute of Zoology. Leon then trained as a subeditor for a London publishing company. Since then, he has written and edited more than a hundred publications, mainly about science, technology and the natural world.

Dan Green (DG)
is a science writer of bestselling nonfiction books. He holds an MA in Natural Sciences from Cambridge University, where he specialized in palaeontology and earth sciences. Dan is a regular contributor and scientific consultant for the *Wonders of Nature* magazine series. His recent books include *The Science Book: Big Ideas Simply Explained* (Dorling Kindersley), and *The Elements* (Scholastic), which was awarded the National Science Teachers Association's Outstanding Science Book for Students 2013. Dan's books have also been shortlisted for the Royal Society Young People's Book Prize.

Tom Jackson (TJ)
is a writer and conservationist. He has helped to survey Vietnamese jungle, rescue buffaloes from African droughts, and replant Somerset woodland. He has been a nonfiction author for twenty years and has written about everything from axolotls via ecology and evolution to zoroastrianism. He studied zoology at Bristol University. Tom's work has been published by *National Geographic*, Bloomsbury, Dorling Kindersley, and BBC Magazines, and has been translated into a dozen languages.

Darren Naish (DN)
is a palaeontologist based at the University of Southampton, UK, where he works on dinosaurs, pterosaurs, and marine reptiles. In 2001, he and his colleagues named the tyrannosauroid dinosaur *Eotyrannus*, which formed the focus of his PhD. With others, he named the dinosaurs *Mirischia* and *Xenoposeidon* and the pterosaur *Vectidraco*. Together with pterosaur expert Mark Witton he proposed that giant azhdarchid pterosaurs were "terrestrial stalkers." He has argued that long-necked sauropod dinosaurs kept their necks in upright poses and that the crests of dinosaurs and pterosaurs evolved as sexual display structures.

Douglas Palmer (DP)
is a Cambridge-based science writer and lecturer for the University of Cambridge Institute of Continuing Education. Previously a lecturer and researcher in palaeontology at Trinity College, University of Dublin, he has written and contributed to more than thirty books ranging from academic palaeontology to children's books. His recent works include the *Natural History Museum: Evolution* app for iPads and *Evolution and Origins: Human Evolution Revealed* (Mitchell Beazley).

Steve Parker (SP)
holds a First Class BSc in Zoology and is a Senior Scientific Fellow of the Zoological Society of London. He worked at London's Natural History Museum on the evolution of crustaceans and interpretive exhibitions, and recent books for the Museum include *Extinction: Not the End of the World?* He has authored more than two hundred titles, mainly on life sciences, and has contributed to another hundred projects and websites. Recent awards include the British Medical Association Book Award for *Kill or Cure—An illustrated History of Medicine* (Dorling Kindersley) and the School Library Association Information Book for *Science Crazy* (QED).

Robert Snedden (RS)
has worked in science and publishing for more than thirty years, specializing in books for young people on a number of science-related topics, ranging from the environment, evolution, genetics, and cell structure, to space exploration, computers, and the Internet. He has also been involved in mounting an exhibition on robotics in conjunction with Heriot-Watt University, and authored a paper on pharmacogenetics for the Wellcome Trust.

Giles Sparrow (GS)
has a master's degree in Science Communication from London's Imperial College. He has written, edited, and contributed to many books on the history of life, including *Encyclopedia of Dinosaurs and Prehistoric Life* (Dorling Kindersley), *Evolution* (Mitchell Beazley), *Origins* (Weidenfeld and Nicolson), and *The Ancestor's Tale* by Richard Dawkins (Weidenfeld and Nicolson). He is a regular writer for magazines such as *How It Works* and *Focus*.

Mike Taylor (MT)
is a dinosaur palaeontologist specializing in sauropods. Since the turn of the millennium, he has published on the history of dinosaur research, the anatomy of sauropod dinosaurs, their taxonomy, how they evolved such long necks, the posture in which they habitually held them, how their vertebrae changed throughout their lives, and the selection pressures that influenced their evolution. He has described and named two new dinosaurs: *Xenoposeidon* and *Brontomerus*. He is an associate researcher at the University of Bristol.

Martin Walters (MW)
is a writer, editor, and naturalist based in Cambridge, UK. He studied Zoology at the University of Oxford, and his interests include birds, botany, natural history, conservation, and evolution. Martin worked for several years as Biological Sciences Editor at Cambridge University Press before becoming a freelance editor and writer. He has traveled, photographed, and published widely, exploring and documenting wildlife in Europe and many other regions around the world, from Iceland to South Africa and China.

INDEX

Page references for
illustrations are indicated
in **bold**

A

aardvarks, 470, 478
abelisaurs, 277, 281
abiogenesis, 36
Acanthisitta chloris, 391,
  394–95
acanthodians, 184–85
Acanthostega, **208**, 209
Acipenseriformes, 202
acorn worms, 110–11
Acristatherium, 425
Actinia equina, **105**
Actinistia, 196
Actinopterygii, 196, 202–5
Adapiformes, 440, 442
adaptive radiation, 92, 393
Adasaurus, 308
Aeger tipularius, **142**
Aegyptopithecus, 507, 510–11
Aepyornis maximus, **414**, 415
Afroinsectiphilia, 471
Afrotheria, 435, 470–77, 553
agamas, 347
Agassiz, Louis, 188
age of the Earth *see* Earth age
Agelenidae, 153
Aglaophyton, 62, **63**
agnathans, 178–79
Agustinia, 271
Aiolornis incredibilis, 399, 551
Aktiogavialis, 258
Alberti, Friedrich Von, 26
alethinophidians, 359
algae, 41, 42–43, 57, 58–59,
  61, 64–65
Allodesmus kelloggi, 520–21
Allosaurus, 298–99
Altiatlasius koulchii, 440
Alytidae, 221
amber, 155
Amborella, 86, **90**, 91, 94
Ambulacraria, 148
Ambulocetidae, 500–501
Ameghino, Florentino, 405
Americhelydia, 296
Amia calva, 203
ammonoids, 128, 156–59
amniotes, 230–31, 418
amoeba, 36
Amphibamus, 220
amphibians
  early development,
    214–18

limb development,
  206–13
lissamphibian
  development, 220–21
reproduction, 67, 215
Amphicyon ingens, 490–91
Amphioctopus marginatus,
  161
Amphistium, 203
Amynodontidae, 483
anamodonts, 419–420
Anaspida, 179, 180–81
Anax imperator, **114**
Anchiornis huxleyi, **372**, 373
Ancylopoda, 483
Andalgalornis, 405–6
Andreolepis, 202
Andrews, Roy Chapman, 455
Andrewsarchus, 454, 455
angiosperms, 71, 81, 86–87,
  89, 91–93, 94, 96
anguimorphs, 347
ankylosaurs, 320, 332–37
Ankylosaurus, 333
Annelida, 45, 51, 108, 109
Anning family, 246, 283–84
anoles, 347
Anomalocaris, 50, 51, 115,
  116–17, 120
Antarctopelta, 335
anteaters, 479
Anthornis melanocephala, 393
Anthozoa, 105
anthropoids, 439
antiarchs, 185
Antilocapra americana, **456**,
  457
ants, 169
anurognathids, 289
Apatosaurus, 270, 271
apes, 506–8, **512**, 513
Apis mellifera, 97
Aplodontia rufa, 462–63
Aptenodytes forsteri, **388**,
  389
Apterygota, 167
arachnids, 61, 115, 150–55
Araguainha Crater, Brazil,
  224, 225
Arandaspis, **178**, 179
Araucaria, 77
Araucariaceae, 81
Archaea, 36, **37** *see also*
  extremophiles
Archaean eon, 38–39,
  43

Archaefructaceae, 87, **88**, 89
Archaeognatha, 135
Archaeopterus, 446
Archaeopteryx, 262, 263,
  305, 310, 372, 373–75, **376**,
  377, 381
Archelon, 293, 294–95
Archicebus, 439–440
archosauromorphs, 231
archosaurs, 231, 248–49
Arctocyonidae, 453
Ardipithecus, 531, **532**, 533
Arduino, Giovanni, 26
Argentavis magnificens,
  399, **400**, 401
Argentina, Ischigualasto-
  Talampaya Formation, 253,
  264
Argentinosaurus, 271, 274–75
Argiope bruennichi, **150**
armadillos, 478–79
ARMAN organisms, 41
Arsinoitherium, **435**, **470**, 471
Artemia salina, **115**
Arthrodira, 185, 186–87
Arthropleurideans, 73, 131,
  132–33
arthropods, **48**, 49, **51**,
  114–123, 130–141
artiodactyls, 452, 454–55
Asilisaurus, **249**
Aspen, Peder, 212
Asteroidea, 163
Asteroxylon, **60**
astraspids, 179
Astrodon, 323
Atlantogenata, 435
Attercopus, 151
Audubon, John James, 457
auks, **385**
Aurornis, 384–86
Australia
  Ediacara Hills, **46**, 47
  Pilbara, **38–39**
  Riversleigh, 431
Australopithecus afarensis
  (Lucy), 531, **534**, 535, 537
Australopithecus africanus,
  399, 530, **531**
Austroraptor, 305
Aysheaia, 51
azhdarchids, 289

B

baboons, 507
backbones, 174–75, 207
Bacteria, 36, 37, 41 *see also*

cyanobacteria
badgers, 515
Baiji dolphin, 555, 556–57
Balaenoptera musculus,
  504–5
baleen whale, **501**
banded iron formations
  (BIFs), 43
Barbaturex, 347
Barychelidae, 154
Barylambda, 436
Baryonyx, 301
Basilosauridae, 501, 502–3
batfish, **199**
batoids, 191
bats, 446–451
baurusuchids, 254
beavers, 462–63
Becquerel, Henri, 31
bees, 96, 169
beetles, 169
Beipiaosaurus inexpectus,
  318–19
belemnoids, 128, 157
Belgium, Bernissart, 324
bennettitaleans, 80, 87
Bergmann, Carl, 477
Berman, David, 216
Bernissart, Belgium, 324
biarmosuchians, 420
Bilaterata, 108–9
bilaterians, 124
biofilms, **38**
biogenic graphite, 39
birds
  birds of prey, 398–402
  diversification, 384–85
  early development, 311,
    372–79
  flight, 380–81, 382
  giant, 404–15
  Passeriformes, 390–97
  song, 391
Birkenia elegans, 180–81
Bivalvia, 125, 128
Black, Davidson, 537
Blastoidea, 165
blue chaffinch, 393, 396–97
blue whale, 501, 504–5
bluebells, **93**
Boltwood, Bertram, 25, 31
Bonaparte, José, 281
Bonitasaura, 271
Boreoeutheria, 435
borophagins, 489
Bothriolepis, **184**, 185
bourgueticrinids, 165

PICTURE CREDITS

The publishers would like to thank the museums, illustrators, archives, and photographers for their kind permission to reproduce the works featured in this book. Every effort has been made to trace all copyright owners but if any have been inadvertently overlooked, the publishers would be pleased to make the necessary arrangements at the first opportunity.

(Key: top = t; bottom = b; left = l; right = r; center = c; top left = tl; top right = tr; center left = cl; center right = cr; bottom left = bl; bottom right = br)

2 © Cathy Keifer / Shutterstock **8** © Charles R. Belinky / Science Photo Library **9** © Mehau Kulyk / Science Photo Library **10** © Herve Conge, ISM / Science Photo Library **11** © Sue Daly / naturepl.com **12** © Martin Strmiska / Alamy **13** 3news.co.nz **14** © Javier Treuba / MSF / Science Photo Library **15** © John Phillips / UK Press via Getty Images **16** © Tui De Roy / naturepl.com **17** © Royal Geographical Society / Alamy **18** © Science Picture Co / Science Photo Library **19** © Jo Crebbin / Shutterstock **20** © Isabel Eeles **22–23** © Dr Karl Lounatmaa / Science Photo Library **24** © Frans Lanting, Mint Images / Science Photo Library **25 t** © Paul D Stewart / Science Photo Library **25 b** © myLoupe / Universal Images Group via Getty Images **26 t** © The Natural History Museum / Alamy **26 b** Omalius d'Halloy by unknown painter. Licensed under Public Domain via Wikimedia Commons. **27** © SSPL / Getty Images **28** © Sinclair Stammers / Science Photo Library **29 t** © Kent and Donna Dannen / Science Photo Library **29 b** © Gary Hincks / Science Photo Library **30** © Dean Conger / Corbis **31 t** © Bettmann / Corbis **31 b** © Isabel Eeles **32** © Doug Meek / Corbis **33 tr** © NPS / Alamy **33 cr** © B.A.E. Inc. / Alamy **33 b** © Tom Bean / Alamy **34** © Chris Butler / Science Photo Library **35 t** © NASA / Science Photo Library **35 b** © imageBROKER / Alamy **36 b** © E.R. Degginger / Alamy **36 t** © Dr Karl Lounatmaa / Science Photo Library **37 b** © B. Murton / Southampton Oceanography Centre / Science Photo Library **37 t** © Eye of Science / Science Photo Library **38** © AFP via AAP / David Wacey (20110821000339238573) **39 tl** Image by Nora Noffke; published in Noffke et al. (2013), Astrobiology, v. 13, p. 1 –22. **39 tc** © AAP Image / University of Western Australia, David Wacey (20110820000339334370) **39 tr** © Dr. Richard Roscoe / Visuals Unlimited, Inc. / Science Photo Library **39 bl** © El Albani / Mazurier **39 br** © Dr. Richard Roscoe / Visuals Unlimited, Inc. / Science Photo Library **40** © Bryan Mullennix / Getty Images **41** © Michael J Daly / Science Photo Library **42** © Jabruson / naturepl.com **43** © Corbin17 / Alamy **44** © Chase Studio / Science Photo Library **45 t** © age fotostock / Alamy **45 b** © The Natural History Museum / Alamy **46 t** http: // ivanov–petrov.dreamwidth.org/1669966.html **46 cl** © Sinclair Stammers / Science Photo Library **46 cr** © Ken Lucas / Visuals Unlimited / Corbis **46 br** © Natural History Museum, London / Science Photo Library **46 bl** © Museum Victoria, Melbourne **47** © Barrett & MacKay / All Canada Photos / Corbis **48** © O. Louis Mazzatenta / National Geographic / Getty Images **49 t** © www.fossilmall.com **49 b** © fossilmuseum.net **50** © T. Kitchin and V. Hurst / Getty Images **51 t** © L. Newman & A. Flowers / Science Photo Library **51 c** © The Natural History Museum / Alamy **51 cl** © Ken Lucas / Visuals Unlimited / Corbis **51 bl** © O. Louis Mazzatenta / National Geographic / Getty Images **51 br** © All Canada Photos / Alamy **52** © Dorling Kindersley / Getty Images **53 c** Courtesy of Smithsonian Institution **53 b** © Pam Baldaro **54–55** © blickwinkel / Alamy **56** © John Durham / Science Photo Library **57 t** © Michael Abbey / Science Photo Library **57 b** © Mark Conlin / Alamy **58** © Sinclair Stammers / Science Photo Library. **59** © Jan Hinsch / Science Photo Library **60** Asteroxylon fossil by Ghedoghedo – Own work. Licensed under CC BY-SA 3.0 via Wikimedia Commons. **61 t** © DeAgostini / Getty Images **61 b** © Dr Keith Wheeler / Science Photo Library **62** © Sinclair Stammers / Science Photo Library **63 c** © Richard Bizley / Science Photo Library **63 b** © Richard Bizley / Science Photo Library **64** © Bob Gibbons / ardea.com **65 l** © Pascal Goetgheluck / Science Photo Library **65 r** © M.I. Walker / Science Photo Library **66** © Wild Wonders of Europe / Lesniewski / naturepl.com **67 t** © Eye of Science / Science Photo Library **67 b** © Natural History Museum, London / Science Photo Library **68** © G.D. Carr **69** © Swedish Museum of Natural History, photo taken by Benjamin Bomfleur **70** © Kurkul / Shutterstock **71 t** © age fotostock / Alamy **71 b** © The Natural History Museum / Alamy **72** © Look and Learn / Bernard Platman Antiquarian Collection / Bridgeman Images **73** © Biophoto Associates / Science Photo Library **74** © Walter Myers / Science Photo Library **75 t** UC Museum of Paleontology Plants Specimen 150339, Photo by Dave Strauss, © 2014, License to use: CC BY 3.0 **75 b** © Adam Hutchinson **76** © Corbin17 / Alamy **77 t** © Alan and Linda Detrick / Science Photo Library **77 b** © sunnyfrog / Shutterstock **78** © The Natural History Museum / Alamy **79** © blickwinkel / Alamy **80** © Biophoto Associates / Science Photo Library **81 t** © GAP Photos / Mark Bolton **81 b** © Nature Photographers Ltd / Alamy **82** © Edward Parker / Alamy **83 tl** © imageBROKER / Alamy **83 cl** © Rowan Isaac / Alamy **83 bl** © Steffen Hauser / botanikfoto / Alamy **83 r** © Power and Syred / Science Photo Library **84** © Alberto Perer / Alamy **85** © Images of Africa Photobank / Alamy **86** © DK Limited / Corbis **87 t** © Nigel Cattlin / Visuals Unlimited / Corbis **87 b** © Sue Bishop / Flowerphotos / ardea.com **88** © David Dilcher and Ge Sun **89** © Buiten–Beeld / Alamy **90** Amborella trichopoda (3173820625) by Scott Zona from USA – Amborella trichopoda. Uploaded by bff. Licensed under CC BY 2.0 via Wikimedia. **91 b** © Custom Life Science Images / Alamy **92** © Nick Garbutt / naturepl.com **93 t** © Steve Taylor / Science Photo Library **93 b** © Dan Tucker / Alamy **94 t** © The Trustees of the Natural History Museum, London **94 b** © Jonathan Buckley / GAP **95 t** © John Kaprielian / Science Photo Library **95 b** © Valerie Giles / Science Photo Library **96** © Zoonar GmbH / Alamy **97 tr** © The Natural History Museum / Alamy **97 b** © Susumu Nishinaga / Science Photo Library **98–99** © Michael Durham / Minden Pictures / Corbis **100** © Martin Land / Science Photo Library **101 t** © Pete Oxford / naturepl.com **101 b** © Steve Gschmeissner / Science Photo Library **102** © Custom Life Science Images / Alamy **103 l** © MBARI 2012 **103 r** © PF–(usna1) / Alamy **104** © age fotostock Spain, S.L. / Alamy **105 t** © JonMilnes / Shutterstock **105 b** © Peter Scoones / Nature Picture Library **106** © Tom Uhlman / Alamy **107 tl** © Indiana Dept. of Natural Resources / Falls of the Ohio State Park **107 tr** © Indiana Dept. of Natural Resources / Falls of the Ohio State Park **107 cl** © Indiana Dept. of Natural Resources / Falls of the Ohio State Park **107 cr** © Indiana Dept. of Natural Resources / Falls of the Ohio State Park **107 b** © David Rumsey Map Collection **108** © Zoonar GmbH / Alamy **109 t** © Andreas Altenburger / Alamy **109 b** © O. Louis Mazzatenta / National Geographic / Getty Images **110** © C.B. Cameron, Université de Montréal **111** © Sue Daly **112** © Alex Hyde / naturepl.com **113** © Robert Schuster / Science Photo Library **114** © Adrian Bicker / Science Photo Library **115** © John P. Adamek / Fossilmall.com **115 b** © Kim Taylor / naturepl.com **116** © Stocktrek Images, Inc. / Alamy **117** © Alan Sirulnikoff / Science Photo Library **118 t** © Sinclair Stammers / Science Photo Library **118 b** PhacopidDevonian by Wilson44691 – Own work. Licensed under Public Domain via Wikimedia Commons. **119** © Gerald & Buff Corsi / Visuals Unlimited / Corbis **120** © Sinclair Stammers / Science Photo Library **121 t** © Brückner / Optics Express, image from Optics Express, DOI: 10.1364 / OE.18.024379 **121 b** © Walter Myers / Science Photo Library **122** Yunnanocephalus S CRF by Dwergenpaartje – Own work. Licensed under CC BY-SA 3.0 via Wikimedia Commons. **123 bl** fossilmall.com **123 br** fossilmall.com **124** © Matteis / Look at Sciences / Science Photo Library **125 t** © EyeOn / UIG via Getty Images **125 b** © Jeff Rotman / Science Photo Library **126** © Visuals Unlimited, Inc. / William Jorgensen **127 t** http: // www.mcz.harvard.edu / **127 b** © The Natural History Museum / Alamy **128** © DEA / G. Cigolinil / De Agostini / Getty Images **129 t** © Adrian Davies / naturepl.com **129 b** © blickwinkel / Alamy **130** © Nigel Cattlin / Alamy **131 t** Reproduced by permission of The Royal Society of Edinburgh from *Transactions of the Royal Society of Edinburgh: Earth Sciences* volume 90 (2000, for 1999), pp 351–375 **131 b** © Viewpoint / Alamy **132** © Walter Myers / Science Photo Library **133** © Richard Bizley / Science Photo Library **134** © The Natural History Museum / Alamy **135 c** © D. Grimaldi / AMNH / Columbia University Press **135 b** © Ian Beames / ardea.com **136** © 1999–2015, The Virtual Fossil Museum Rights and Permissions **137 t** © Biophoto Associates / Science Photo Library **137 b** © Jaime Chirinos / Science Photo Library **138** © Stocktrek Images, Inc. / Alamy **139** © Sinclair Stammers / Science Photo Library **140** © John Greim / LightRocket via Getty Images **141 c** © Sabena Jane Blackbird / Alamy **141 b** © Alexander Semenov / Science Photo Library **142** © WaterFrame / Alamy **143 t** © Jean Paul Ferrero / ardea.com **143 b** © Edmund Fellows **144** © Chase Studio / Science Photo Library **145 bl** © Bygone Collection / Alamy **145 br** © DeAgostini / Getty Images **146** © DK Limited / Corbis **147 t** © Albert Lleal / Minden Pictures / Corbis **147 b** © The Natural History Museum / Alamy **148** © Professor Billie J. Swalla **149** © David Wrobel / Visuals Unlimited, Inc. **150** © Claude Nuridsany and Marie Perennou / Science Photo Library **151 t** © Paul Selden **151 b** © Piotr Naskrecki / Getty **152 t** © SuperStock / Alamy **152 bcl** © Ger Bosma / Alamy **153 l** © Gerry Bishop / Visuals Unlimited, Inc. **153 r** © blickwinkel / Alamy **154** © James H. Robinson / Science Photo Library **155 c** © Steve Hopkin / ardea.com **155 b** © Francois Gohier / ardea.com **156** © Patrick Dumas / Look at Sciences / Science Photo Library **157 t** © Natural History Museum, London / Science Photo Library **157 b** © The Trustees of the Natural History Museum, London **158** © Jim Amos / Science Photo Library **159 t** Dactylioceras NT by NobuTamura © www.palaeocritti.com – Own work . Licensed under CC BY– SA 3.0 via Wikimedia Commons. **159 b** © Pascal Goetgheluck / Science Photo Library **160** © Image Source / Alamy **161 t** © Reinhard Dirscherl / Visuals Unlimited / Corbis **161 b** © Mike Veitch / Alamy **162** © Alan Majchrowicz / Getty **163 t** © FLPA / Alamy **163 b** © WaterFrame / Alamy **164** © DeAgostini / Getty Images **165** © The Natural History Museum / Alamy **166** © Natural History Museum, London / Science Photo Library **167 t** © Steve Gschmeissner / Science Photo Library **167 b** Reprinted by permission from Macmillan Publishers Ltd: ‹A complete insect from the Late Devonian period› Nature 488, 82–85, copyright 2012 **168 t** © blickwinkel / Alamy **168 b** © Thomas Marent / ardea.com **169 t** © Steve Hopkin / ardea.com **169 b** © Michael